Published for
The University of Texas
M. D. Anderson Hospital and Tumor Institute at Houston
Houston, Texas, by Raven Press, New York

UT M. D. Anderson Symposium on Fundamental C
Volume 37

Mediators in Cell Growth and Differentiati

UT M.D. Anderson
Symposium on Fundamental
Cancer Research
Volume 37

Mediators in Cell Growth and Differentiation

Edited by

Richard J. Ford, M.D., Ph.D.
Associate Professor of Pathology
Department of Pathology

Abby L. Maizel, M.D., Ph.D.
Associate Professor of Pathology
Chief, Section of Pathobiology
Department of Pathology

The University of Texas
M. D. Anderson Hospital and Tumor Institute at Houston
Houston, Texas

Raven Press ■ New York

Raven Press, 1140 Avenue of the Americas, New York, New York 10036

Made in the United States of America

Library of Congress Cataloging in Publication Data

Symposium on Fundamental Cancer Research (37th :
1984 : Houston, Tex.)
Mediators in cell growth and differentiation.

(UT M.D. Anderson Symposium on Fundamental
Cancer Research ; v. 37)
"A compilation of the proceedings of the University
of Texas M.D. Anderson Hospital and Tumor Institute at
Houston's 37th Annual Symposium on Fundamental Cancer
Research, held March 6–9, 1984, in Houston, Texas"—
T.p. verso.
Includes bibliographies and index.
1. Cell proliferation—Congresses. 2. Cell differentiation—Congresses. 3. Cells—Growth—Congresses.
4. Cancer cells—Congresses. 5. Cellular control
mechanisms—Congresses. 6. Growth regulators—
Congresses. I. Ford, Richard J., 1943–
II. Maizel, Abby L. III. M.D. Anderson Hospital and
Tumor Institute at Houston. IV. Title. V. Series:
UT M.D. Anderson Symposium on Fundamental Cancer
Research (Series) ; v. 37. [DNLM: 1. Cell Differentiation—congresses. 2. Growth—congresses. 3. Growth
Substances—congresses. 4. Interferons—congresses.
5. Neoplasms—physiopathology—congresses.
W3 SY5177 37th 1984m / QH 607 S9894 1984m]
QH605.S94 1984 574.87'61 84-24875
ISBN 0-88167-071-5

This volume is a compilation of the proceedings of The University of Texas M. D. Anderson Hospital and Tumor Institute at Houston's 37th Annual Symposium on Fundamental Cancer Research, held March 6–9, 1984, in Houston, Texas. The volume is indexed in the *Index Medicus* MEDLINE database and its subsets.

The material contained in this volume was submitted as previously unpublished material, except in the instances in which credit has been given to the source from which some of the illustrative material was derived.

Great care has been taken to maintain the accuracy of the information contained in the volume. However, the Editorial Staff, The University of Texas, and Raven Press cannot be held responsible for errors or for any consequences arising from the use of the information contained herein.

Preface

The subject of the 37th Annual UT M. D. Anderson Symposium on Cancer Research "Mediators in Cell Growth and Differentiation," on which this volume is based, was of particular timeliness in 1984, owing to the rapid expansion of information in these fields. This subject encompasses two of the most fundamental processes in cell biology with obvious ramifications on many pathological conditions. Since the pioneering work of Stanley Cohen, who described the first two tissue-specific growth factors (epidermal growth factor and nerve growth factor) as well as purified epidermal growth factor, its receptor, and the associated tyrosine kinase activity, there has been intense effort by scientists in a wide spectrum of disciplines to unravel the molecular basis of the control of cell division. This has been equally true in the related area of the control of cellular differentiation processes. For the first time, we are beginning to understand the biochemical nature of the control signals that stimulate cell proliferation and to be able to identify and clone the genes encoding these activities.

Concurrently, hybridoma and genetic engineering technology is becoming available for the large-scale production of the cytokines responsible for stimulating cell growth and differentiation. These allow not only for their characterization by protein sequencing, and so on, but also provide adequate amounts of the purified factors for performing biological experiments to ascertain the precise function of these molecules and where they fit into the concatenation of molecular events that results in cell proliferation or differentiation.

This volume is also timely in regard to the recent convergence of cell growth and differentiation studies and the discovery of cellular oncogenes. This discovery, coupled with the elucidation of the possible role of some oncogene products as growth factors—such as the simian sarcoma virus oncogene (v-*sis*) product, which appears to be closely related to platelet-derived growth factor—has provided a hypothetical mechanism by which an oncogene growth factor product could be involved in an autocrine stimulation model of neoplasia. Also, there is now evidence that some tumor growth factors are structurally homologous to epidermal growth factor. Knowledge of the relationship of tumor growth factors to their normal growth factor counterparts at both the genetic and functional levels could provide important information on the basic mechanisms of neoplastic transformation. It appears from information provided at this symposium and in the literature that the probes and experimental systems for answering these important questions have now been developed.

Differentiation is another fundamental cellular process on which information is now beginning to become available. Hematopoietic and lymphoid cells seem to provide the most useful experimental systems here, as at least some phenotypic markers exist for monitoring the maturation process that ensues during differentiation. These processes often result in the formation of an end-stage cell that can be identified (e.g., plasma cell, macrophage, and so on), which then produces a functional end product characteristic of the cell lineage. The B lymphocyte has recently provided a good model of such a differentiating system, in which antigen-activated, lymphokine-stimulated proliferating B cells are acted upon by another group of lymphokines (T cell–replacing factors or B cell–differentiating factors), which causes the cells to become immunoglobulin-producing plasma cells. There is also evidence that other lymphokines cause the differentiated cell to switch the isotype of the immunoglobulin produced, further diversifying the humoral antibody response to antigen.

This volume provides an exciting update of our current knowledge regarding cell growth and differentiation with some application to the pathological questions that are clearly indicated. Perhaps more important, however, it will serve as a multidisciplinary forum for scientists working in many diverse biological fields to become aware of the elegant cell biological systems developed to study the fundamental questions posed in regard to these subjects. Utilization of the sophisticated and ever-increasing molecular technology that is evident in the following chapters should provide both an intellectual challenge to the investigators involved in these studies and the prospect of a rapid growth and maturation of our understanding of these subjects in the very near future.

Richard J. Ford
Abby L. Maizel

Acknowledgments

We would like to thank many individuals who unselfishly provided advice and assistance at every stage during the planning and execution of the symposium. We thank the Symposium Organizing Committee members from The University of Texas System Cancer Center, Drs. Karel A. Dicke, Isaiah J. Fidler, Lawrence B. Lachman, Potu N. Rao, and Jordan U. Gutterman, and the external advisory committee, Drs. Samuel Baron, Renato Baserga, Stanley Cohen, Robert Gallo, Peter Nowell, and Joost Oppenheim, for their assistance in planning the meeting. We thank the session chairmen, Drs. Bill Brinkley, Garth Nicolson, Karel Dicke, Stanley Cohen, Peter Nowell, Samuel Baron, and Daniel Medina, for their guidance during the meeting.

We owe special thanks and appreciation to Mr. Jeff Rasco and the staff of UT M. D. Anderson Hospital Conference Services and Ms. Frances Goff for providing assistance throughout the symposium. Benefactors, such as Hoffman-La Roche, Inc., E. I. Du Pont De Nemours and Company, K. C. Biological, Merrell Dow Pharmaceuticals, UpJohn Company, Bristol-Meyers Company, Fisher Scientific, Cellular Products, Inc., Irvine Scientific, Syntex Research, Applied Biosciences, Inc., Merck Sharp & Dohme Research Laboratories, and Searle Research, were critical to our success. We are especially grateful to the American Cancer Society, Texas Division, Inc., for its continuing sponsorship of the symposium, which was conducted in cooperation with The University of Texas Graduate School of Biomedical Sciences at Houston. Finally, we thank the Department of Public Information and Education for its assistance to us and the professional and public media.

Contents

Cell Cycle and Control of Cell Growth

Growth Factors for Nonlymphoid Cells

Colony-Stimulating Factors, Stem Cells, and Hematopoiesis

Lymphoid Growth Factors

Growth Factors in Neoplasia

Interferon

Differentiation in Normal and Neoplastic Cells

Contributors

Stuart A. Aaronson
National Cancer Institute
Bethesda, Maryland 20205

Ramesh C. Adlakha
Department of Chemotherapy Research
The University of Texas M. D. Anderson Hospital and Tumor Institute at Houston
Houston, Texas 77030

Patricio Aller
Department of Pathology and Fels Research Institute
Temple University School of Medicine
Philadelphia, Pennsylvania 19140

Samuel Baron
Department of Microbiology
The University of Texas Medical Branch
Galveston, Texas 77550

Sam Barranco
Department of Radiation Therapy
The University of Texas Medical Branch
Galveston, Texas 77550

Renato Baserga
Department of Pathology and Fels Research Institute
Temple University School of Medicine
Philadelphia, Pennsylvania 19140

Roger J. Booth
Department of Immunobiology
School of Medicine
Auckland University
Auckland, New Zealand

Ralph A. Bradshaw
Department of Biological Chemistry
California College of Medicine
University of California
Irvine, California 92717

Miriam Brysk
Department of Dermatology
The University of Texas Medical Branch
Galveston, Texas 77550

Antony W. Burgess
Ludwig Institute for Cancer Research
Victoria, 3050 Australia

Judith Campisi
Dana-Farber Cancer Institute
Boston, Massachusetts 02115

Brent H. Cochran
Department of Microbiology and Molecular Genetics
Harvard Medical School and Dana-Farber Cancer Institute
Boston, Massachusetts 02115

Stanley Cohen
Department of Pathology
University of Connecticut Health Center
Farmington, Connecticut 06032

Paul J. Conlon
Immunex Corporation
Seattle, Washington 98101

David Danielpour
The Department of Biochemistry and Molecular Biology
The University of Texas Medical School at Houston
Houston, Texas 77030

Frances Davis
Department of Chemotherapy Research
The University of Texas M. D. Anderson Hospital and Tumor Institute at Houston
Houston, Texas 77030

Michael Dean
Frederick Cancer Research Institute
Frederick, Maryland 21701

Joan C. Dunbar
Department of Biological Chemistry
California College of Medicine
University of California
Irvine, California 92717

Nicolette Ebsworth
Imperial Cancer Research Fund
Lincoln's Inn Fields
London WC2A 3PX England

Paul B. Fisher
Institute of Cancer Research
Columbia University College of Physicians & Surgeons
New York, New York 10032

W. Robert Fleischmann, Jr.
Department of Microbiology
The University of Texas Medical Branch
Galveston, Texas 77550

Richard J. Ford
Department of Pathology
The University of Texas M. D. Anderson Hospital and Tumor Institute at Houston
Houston, Texas 77030

Janice Gabrilove
Department of Developmental Hematopoiesis
Memorial Sloan-Kettering Cancer Center
New York, New York 10021

Diethard Gemsa
Department of Molecular Pharmacology
University of Hannover
Hannover
Federal Republic of Germany

Carolyn W. Gibson
Department of Pathology and Fels Research Institute
Temple University School of Medicine
Philadelphia, Pennsylvania 19140

Steven Gillis
Immunex Corporation
Seattle, Washington 98101

Eugene Goldwasser
Department of Biochemistry
The University of Chicago
Chicago, Illinois 60637

Denis Gospodarowicz
Cancer Research Institute and the Department of Medicine and Ophthalmology
University of California Medical Center
San Francisco, California 94143

Harry E. Gray
Dana-Farber Cancer Institute
Boston, Massachusetts 02115

Vicram Gupta
Department of Internal Medicine
The University of Texas Medical Branch
Galveston, Texas 77550

Janice K. Gutowski
Department of Pathology
University of Connecticut Health Center
Farmington, Connecticut 06032

Ute Hamann
Institute for Immunology and Genetics
German Cancer Research Center
Heidelberg
Federal Republic of Germany

Anne C. Hannam-Harris
Department of Pathology and Laboratory Medicine
University of Pennsylvania School of Medicine
Philadelphia, Pennsylvania 19104

Sidney L. Hendrickson
Department of Microbiology and Molecular Genetics
Harvard Medical School and Dana-Farber Cancer Institute
Boston, Massachusetts 02115

Ronald B. Herberman
Biological Therapeutics Branch
Biological Response Modifiers Program
National Cancer Institute
Frederick Cancer Research Facility
Frederick, Maryland 21701

Ricky R. Hirschhorn
Department of Pathology and Fels Research Institute
Temple University School of Medicine
Philadelphia, Pennsylvania 19140

Tatsuhiko Ikeda
The Faculty of Nutrition
Kobe-Gakuin University
Igawadani-cho Arise
Nishi-ku, Kobe, Japan 673

Paul J. Isackson
Department of Biological Chemistry
California College of Medicine
University of California
Irvine, California 92717

Brigitte Kaltmann
Institute for Immunology and Genetics
German Cancer Research Center
Heidelberg
Federal Republic of Germany

Tsuguhiro Kaneda
DeWitt Wallace Research Laboratory and Sloan-Kettering Division
Graduate School of Medical Sciences
Memorial Sloan-Kettering Cancer Center
New York, New York 10021

Hanna C. Kelker
Biological Response Modifiers Unit
Department of Microbiology and Kaplan Cancer Center
New York University Medical Center
New York, New York 10016

Gary Klimpel
Department of Microbiology
The University of Texas Medical Branch
Galveston, Texas 77550

Gary A. Koretzky
Department of Pathology and Laboratory Medicine
University of Pennsylvania School of Medicine
Philadelphia, Pennsylvania 19104

Rozenn N. Kouchalakos
Department of Biological Chemistry
California College of Medicine
University of California
Irvine, California 92717

Peter H. Krammer
Institute for Immunology and Genetics
German Cancer Research Center
Heidelberg
Federal Republic of Germany

Sanford B. Krantz
Department of Medicine
Vanderbilt University School of Medicine and VA Medical Center
Nashville, Tennessee 37203

Claire Kubelka
Institute for Immunology and Serology
University of Heidelberg
Heidelberg
Federal Republic of Germany

Lawrence B. Lachman
Department of Cell Biology
The University of Texas M. D. Anderson Hospital and Tumor Institute at Houston
Houston, Texas 77030

Jerome A. Langer
Roche Institute of Molecular Biology
Roche Research Center
Nutley, New Jersey 07110

Junming Le
Biological Response Modifiers Unit
Department of Microbiology and Kaplan Cancer Center
New York University Medical Center
New York, New York 10016

Li Lu
Department of Developmental Hematopoiesis
Memorial Sloan-Kettering Cancer Center
New York, New York 10021

Paul A. Marks
DeWitt Wallace Research Laboratory and Sloan-Kettering Division
Graduate School of Medical Sciences
Memorial Sloan-Kettering Cancer Center
New York, New York 10021

Diane Y. Mochizuki
Immunex Corporation
Seattle, Washington 98101

Malcolm A. S. Moore
Department of Developmental Hematopoiesis
Memorial Sloan-Kettering Cancer Center
New York, New York 10021

Claudia J. Morgan
Department of Microbiology and Immunology
Albert Einstein College of Medicine
New York, New York 10461

Wolfgang Müller
Institute for Immunology and Genetics
German Cancer Research Center
Heidelberg
Federal Republic of Germany

Takashi Murate
DeWitt Wallace Research Laboratory and Sloan-Kettering Division
Graduate School of Medical Sciences
Memorial Sloan-Kettering Cancer Center
New York, New York 10021

Peter C. Nowell
Department of Pathology and Laboratory Medicine
University of Pennsylvania School of Medicine
Philadelphia, Pennsylvania 19104

Nancy E. Olashaw
Department of Pharmacology and Cancer Cell Biology Program
Lineberger Cancer Research Center
The University of North Carolina
Chapel Hill, North Carolina 27514

John Ortaldo
Biological Therapeutics Branch
Biological Response Modifiers Program
National Cancer Institute
Frederick Cancer Research Facility
Frederick, Maryland 21701

Arthur B. Pardee
Dana-Farber Cancer Institute
Boston, Massachusetts 02115

Sidney Pestka
Roche Institute of Molecular Biology
Roche Research Center
Nutley, New Jersey 07110

Robert A. Phillips
Department of Medical Biophysics
University of Toronto and The Ontario Cancer Institute
Toronto, Canada M4X 1K9

W. J. Pledger
Department of Pharmacology and Cancer Cell Biology Program
Lineberger Cancer Research Center
The University of North Carolina
Chapel Hill, North Carolina 27514

Ross L. Prestidge
Department of Immunobiology
School of Medicine
Auckland University
Auckland, New Zealand

Irma Ramirez
Department of Pediatrics
The University of Texas M. D. Anderson Hospital and Tumor Institute at Houston
Houston, Texas 77030

Potu N. Rao
Department of Chemotherapy Research
The University of Texas M. D. Anderson Hospital and Tumor Institute at Houston
Houston, Texas 77030

Jeffrey Ravetch
DeWitt Wallace Research Laboratory and Sloan-Kettering Division
Graduate School of Medical Sciences
Memorial Sloan-Kettering Cancer Center
New York, New York 10021

John C. Reed
Department of Pathology and Laboratory Medicine
University of Pennsylvania School of Medicine
Philadelphia, Pennsylvania 19104

Angela C. Reffel
Department of Microbiology and Molecular Genetics
Harvard Medical School and Dana-Farber Cancer Institute
Boston, Massachusetts 02115

Richard A. Rifkind
DeWitt Wallace Research Laboratory and Sloan-Kettering Division
Graduate School of Medical Sciences
Memorial Sloan-Kettering Cancer Center
New York, New York 10021

Keith C. Robbins
National Cancer Institute
Bethesda, Maryland 20205

Enrique Rozengurt
Imperial Cancer Research Fund
Lincoln's Inn Fields
London WC2A 3PX England

Leo Sachs
Department of Genetics
Weizmann Institute of Science
Rehovot 76100, Israel

David A. Sirbasku
The Department of Biochemistry and Molecular Biology
The University of Texas Medical School at Houston
Houston, Texas 77030

Kendall A. Smith
Department of Medicine
Dartmouth Medical School
Hanover, New Hampshire 03756

Gail Sonenshein
Boston University Medical School
Boston, Massachusetts 02115

Charles D. Stiles
Department of Microbiology and Molecular Genetics
Harvard Medical School and Dana-Farber Cancer Institute
Boston, Massachusetts 02115

Joyce Taylor-Papadimitriou
Imperial Cancer Research Fund
Lincoln's Inn Fields
London WC2A 3PX England

Steven R. Tronick
National Cancer Institute
Bethesda, Maryland 20205

Steven Tyring
Department of Microbiology
The University of Texas Medical Branch
Galveston, Texas 77550

David L. Urdal
Immunex Corporation
Seattle, Washington 98101

Jan Vilček
Biological Response Modifiers Unit
Department of Microbiology and Kaplan Cancer Center
New York University Medical Center
New York, New York 10016

Fung Fang Wang
Department of Biochemistry
The University of Chicago
Chicago, Illinois 60637

James D. Watson
Department of Immunobiology
School of Medicine
Auckland University
Auckland, New Zealand

I. Bernard Weinstein
Institute of Cancer Research
Columbia University College of Physicians & Surgeons
New York, New York 10032

Y. K. Yip
Biological Response Modifiers Unit
Department of Microbiology and Kaplan Cancer Center
New York University Medical Center
New York, New York 10016

Zhi-An Yuan
Institute of Biophysics
Academia Sinica
Beijing
People's Republic of China

Yee Pang Yung
Department of Developmental Hematopoiesis
Memorial Sloan-Kettering Cancer Center
New York, New York 10021

Cell Cycle and Control of Cell Growth

Mediators in Cell Growth and Differentiation,
edited by Richard J. Ford and Abby L. Maizel.
Raven Press, New York © 1985.

Molecular Biology of Cell Division

Ricky R. Hirschhorn, Zhi-An Yuan,* Patricio Aller,
Carolyn W. Gibson, and Renato Baserga

Department of Pathology and Fels Research Institute, Temple University School of Medicine, Philadelphia, Pennsylvania 19140

We have established that, in mammalian cells, the transition from G_0 to S phase requires unique copy gene transcription (Baserga et al. 1982). We therefore began a search for genes whose expression is induced when quiescent cells are stimulated to proliferate by serum addition. The cell line used, ts13 (derived from BHK Syrian hamster cells), is temperature-sensitive for growth, specifically arresting in G_1 at the nonpermissive temperature (Talavera and Basilico 1977). For this purpose, a library of double-stranded cDNA clones (cloned into pBR322) was constructed from poly(A)$^+$ mRNA isolated from ts13 cells 6 hr after serum stimulation. Replicas of the cDNA clones were hybridized with single-stranded cDNA probes, which were synthesized using poly(A)$^+$ mRNA from G_0 cells or from cells stimulated for 6 hr with serum. Clones that repeatedly hybridized only to G_1 cDNA were considered to be derived from G_1-specific mRNAs. Clones designated as G_1 specific were further characterized by dot blot hybridization to RNA isolated from cells at different times after serum stimulation. Five cDNA clones were verified as G_1 specific. Levels of expression of these clones were shown to be severalfold higher in G_1 than in G_0 cells. These five G_1-specific cDNA clones were further characterized by Southern and Northern blot analysis. These studies have been extended to another G_1-specific ts mutant, tsAF8 cells, and the combined results indicate that several genes, whose expression is induced by a proliferative stimulus, can be identified and can form the basis for a genetic analysis of the mammalian cell cycle.

In this chapter, we describe the isolation and characterization of cDNA clones derived from gene sequences whose expression is markedly increased in the G_1 phase of the cell cycle. The search for G_1-specific mRNAs is justified by recent findings formally demonstrating that unique copy gene transcription is necessary for the transition of cells from mitosis (or G_0) to S phase (Baserga et al. 1982). Accordingly, this chapter will be divided into three parts: (1) evidence that genes transcribed by RNA polymerase II are necessary for the transition of cells from a

*Present address: Institute of Biophysics, Academia Sinica, Beijing, People's Republic of China

resting to a growing state; (2) identification of G_1-specific cDNA clones; and (3) significance of these findings in relation to the literature.

RNA POLYMERASE II IS NECESSARY FOR ENTRY INTO S

Although it is almost self-evident that cell cycle progression is regulated by unique copy genes, until recently no formal evidence was available to support this concept. Indeed, some investigators had worked (and still work) on the assumption that cell proliferation can be under posttranscriptional controls (Cooper and Braverman 1981), whereas others believe that the entry of cells into S is simply dependent on the attainment by the cell of a critical size or a critical amount of certain cellular components (see, for instance, Darzynkiewicz et al. 1980, Stancel et al. 1981, Alberghina and Sturani 1981).

Suggestions that unique copy gene transcription was necessary for the transition of cells from a resting to a growing state, which go back to the pioneer experiments of Lieberman et al. (1963), remained just that—suggestions, because they were based on the use of drugs such as actinomycin D whose effects can, at best, be described as ambiguous. More recently, our laboratory has shown conclusively that a functional RNA polymerase II is necessary for the transition of cells from G_0 (or mitosis) to S phase. This evidence includes the observation that α-amanitin, a drug that has one site of action—one of the large subunits of RNA polymerase II (Ingles 1978)—inhibits the progression of cells through G_1 (Baserga et al. 1982). In addition, a temperature-sensitive (ts) mutant, tsAF8 cells, known to stop specifically in G_1 at the nonpermissive temperature (Burstin et al. 1974) has been shown to have a defective RNA polymerase II (Rossini et al. 1980, Waechter et al. 1984). *A functional RNA polymerase II is therefore necessary for the transition of cells from G_0 (or mitosis) to S.*

The demonstration that RNA polymerase II is required for progression through G_1 justifies the search for genes that control the transition of cells from G_0 (or mitosis) to S. As a first approach to this problem, we looked for genes whose expression is markedly and specifically increased in G_1. Genes that are preferentially expressed in G_1 do not necessarily regulate cell cycle progression, let alone control the transition from a resting to a growing state. But it is a reasonable beginning in a program aimed at elucidating the genetic basis of cell proliferation. We will equate RNA polymerase II and unique copy gene transcription in this chapter, but it should be understood that RNA polymerase II also transcribes some repetitive sequences, as well as gene families, as will be illustrated below.

cdc GENES

We propose to name the genes that are preferentially expressed in specific phases of cell cycle "cell division cycle genes" (cdc genes). This proposal is in agreement with the nomenclature proposed by Hartwell (1971) to indicate genes, defined by ts mutations, that are required for cell cycle progression in yeast. Again, we emphasize that our experiments, as well as those of Linzer and Nathans (1983) and

Cochran et al. (1983), simply identify cDNA clones derived from sequences that are preferentially expressed in a phase of the cell cycle, for instance G_1. We will probably find that some of these regulate the progression of cells through the cell cycle whereas others may have other functions not related to cell proliferation. Our hope, of course, is that one of them will eventually be identified as the "trigger" gene that releases cells from a resting state so they may proliferate, similar to the "start" gene of yeast, the cellular equivalent of the SV40 T-antigen.

IDENTIFICATION OF G_1-SPECIFIC cdc GENES

The strategy for the identification of genes preferentially expressed in G_1 is straightforward and is the same repeatedly used in a variety of experiments to identify genes preferentially expressed in different physiological conditions or different stages of development (Kramer and Anderson 1980, Mangiarotti et al. 1981, Lee et al. 1981, Fregien et al. 1983, Sargent and Dawid 1983, Kelly et al. 1983). This strategy is summarized in Figure 1.

Our choice of cells is especially important. We have selected for our studies two cell lines that are G_1-specific ts mutants of the cell cycle, tsAF8 and ts13 cells (Burstin et al. 1974, Talavera and Basilico 1977), both originally derived from BHK cells. These G_1-specific ts mutants can be advantageously used to sort out, among the preferentially expressed genes, those that may play a major role in cell cycle progression.

cdc Genes of ts13 and tsAF8 Cells

Figure 2 illustrates the general procedure used to identify cDNA clones derived from sequences preferentially expressed in G_1. Replica plates were made of a cDNA library prepared from poly(A)$^+$ mRNA of G_1 cells. The replica plates were screened with radioactive single-stranded cDNA probes prepared from two sources: RNA of G_0 cells and RNA of G_1 cells. After hybridization and autoradiography, G_1 clones were identified as those that hybridized with the single-stranded cDNA probe from G_1 cells, but not with the G_0 probe. The screening of replica plates was repeated three times, and those clones that always hybridized to a G_1 probe and never hybridized to a G_0 probe were defined as G_1-specific cDNA clones. An example of a G_1-specific cDNA clone is indicated with an arrow in Figure 2. A list of G_1-specific cDNA clones that we have identified from ts13 cells is given in Table 1. We have also identified five G_0-specific cDNA clones, but this chapter will be exclusively devoted to the G_1-specific clones.

Time Course of Expression of cdc Genes

The expression of cdc genes through the cell cycle could be quantitated using the procedure illustrated in Figure 3. Total RNA was prepared from G_0 cells and from cells at various times after stimulation. Equal amounts of RNA were spotted onto a nitrocellulose filter (Thomas 1983) and hybridized to a ^{32}P-labeled probe

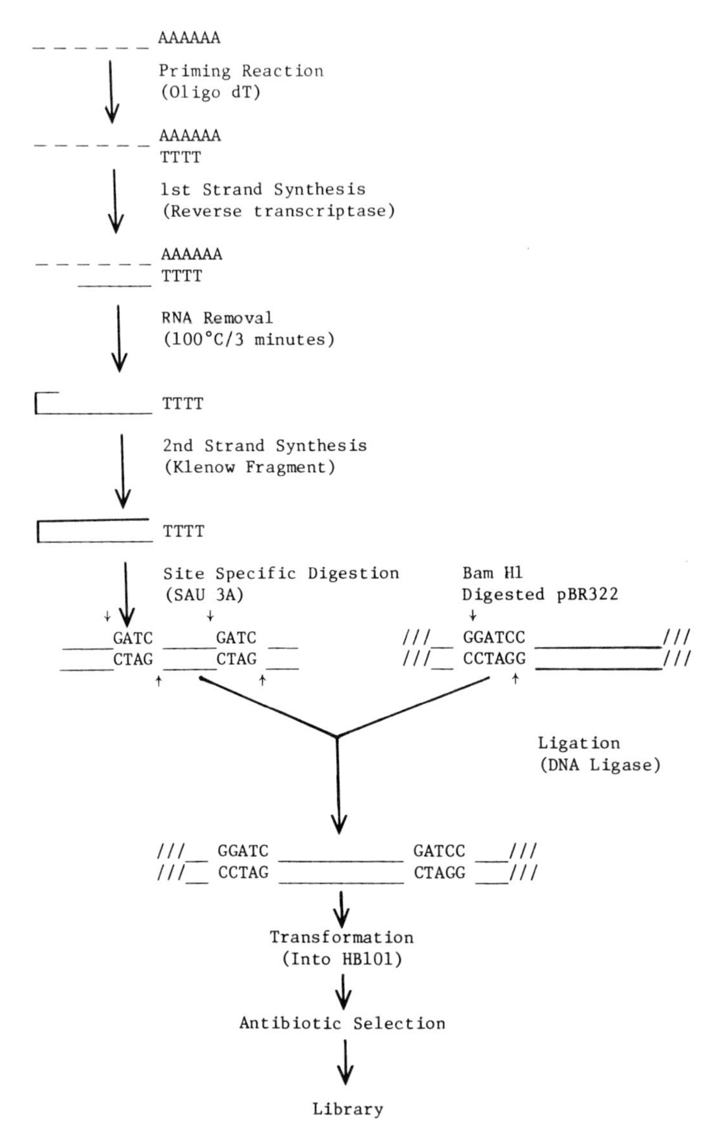

FIG. 1. Construction of the cDNA library. A library of double-stranded cDNA inserts was constructed from poly(A)$^+$ mRNA isolated from ts13 cells 6 hr after serum stimulation. After antibiotic selection, individual bacterial clones were grown in microtiter plates in L-broth containing 20 μg/ml ampicillin, then supplemented to 8% dimethylsulfoxide and stored at −70°C.

prepared from a G_1 cDNA clone. The intensities of the resulting autoradiographs were quantitated with a densitometer. In the dot blot of Figure 3, one can see that the expression of the sequence homologous to the cDNA clone, 2F1, increased after serum stimulation, but decreased again 24 hr after stimulation, when the cells were in S phase. The expression of the cDNA clone 7B5, which is not cell cycle

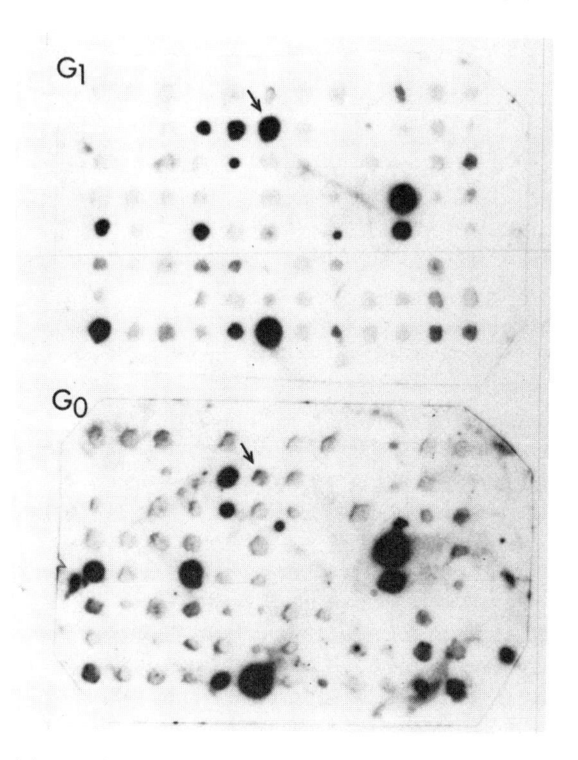

FIG. 2. Differential screening of the cDNA library. Clones from the microtiter plates were transferred, in duplicate, to nitrocellulose, using a replica plating device. After lysis of the colonies, the recombinant plasmid DNA was hybridized to ^{32}P-labeled single-stranded cDNA probes synthesized from poly(A)$^{+}$ mRNA isolated from serum-starved cells (G_0) or from cells stimulated for 6 hr with serum (G_1). Hybridization was for 3 days at 65°C in a cocktail that contained 5XSSC (standard saline citrate), 5X Denhardt's, 0.05 M phosphate buffer, pH 7.0, 0.5% sodium dodecyl sulfate (SDS), 1 mM EDTA, 100 μg/ml sheared, denatured calf thymus DNA, 10% dextran sulfate, and the radioactive probe. After hybridization the filters were washed with 2XSSC at room temperature, then 0.1XSSC containing 0.1% SDS at 50°C, and autoradiographed overnight. The arrow indicates an example of a clone (2F1) that repeatedly hybridized to G_1 single-stranded cDNA.

TABLE 1. *Cell cycle–specific cDNA clones*

Clone	Insert size (bp)*
p13-2A8	170
p13-2A9	100
p13-2A10	100
p13-2F1	420
p13-4F1	220

*Insert size was determined by migration of the *Eco*RI/*Sal* I fragment on a 4% acrylamide gel.

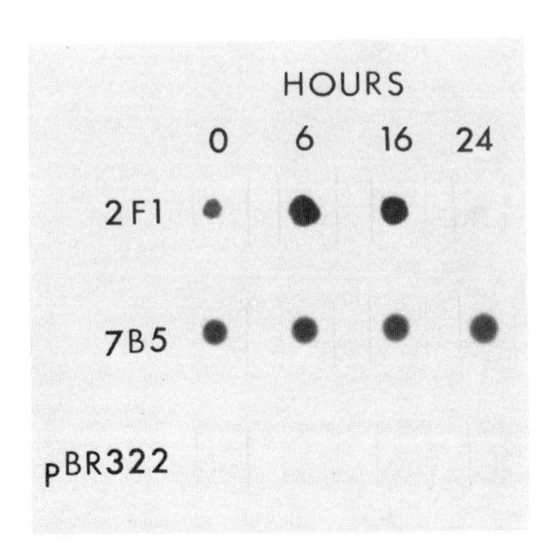

FIG. 3. Expression of a cell cycle-specific cDNA clone after serum stimulation. RNA isolated from ts13 cells at different times after serum stimulation was dotted onto nitrocellulose for hybridization. The cDNA insert from the cell cycle-specific clone was gel purified after digestion with *Eco*RI and *Sal*I and then nick-translated in the presence of [^{32}P]dCTP to 1×10^8 cpm/μg. Hybridization was at 42°C for 20 hr in a cocktail containing 50% formamide, 5XSSC, 1× Denhardt's, 0.02 M phosphate buffer, pH 6.5, 10% dextran sulfate, 100 μg/ml sheared, denatured calf thymus DNA, and ^{32}P-labeled cDNA. After hybridization, the filters were washed with 2XSSC/0.1% sodium dodecyl sulfate (SDS) at room temperature and then 0.1XSSC/0.1% SDS at 52°C. Blots were exposed to film in the presence of an intensifying screen for 24 hr. A non-cell-cycle-dependent cDNA clone and pBR322 were included as controls.

specific, remained constant from G_0 to S (Figure 3, row 2). There was no background hybridization when the RNA was probed with ^{32}P-labeled pBR322. The expression of the cDNA clones listed in Table 1 was shown to be G_1 specific in both ts13 and tsAF8 cells.

Nature of the cdc Genes

The cDNA clones listed in Table 1 have been characterized by Southern blots (Southern 1975) and Northern blots (Thomas 1983). Southern blots have revealed that these cdc genes fall into two major categories: (1) cDNA clones hybridizing to one or just a few bands, probably derived from single-copy or low-repeat frequency genes; and (2) cDNA clones hybridizing to many regions of widely varying sizes and containing either a moderately repetitive sequence or representing a member of a multigene family. These patterns are similar to those observed by other groups identifying genes preferentially expressed in different systems using the same type of experimental design (Feinberg et al. 1983, Fregien et al. 1983). cDNA clones from repetitive sequences generate, on Northern blots, a smear of mRNAs of different sizes. cDNA clones from low-frequency repeat or single-copy genes give one or two bands on Northern blots.

Relationship of the cdc Genes to the ts Block

The choice of ts13 and tsAF8 cells was deliberately made to study how the G_1 ts block affects the expression of G_1-specific cdc genes. Table 2 shows the effect of nonpermissive temperature on the expression of some of the cDNA clones in both ts13 and tsAF8 cells. For instance, p13-2A9 is a G_1-specific clone that also displays increased expression in ts13 cells stimulated at the nonpermissive temper-

TABLE 2. *Effect of the restrictive temperature on the expression of G_1-specific cdc genes*

	Expression in G_1 at the nonpermissive temperature	
cDNA clones	ts13	tsAF8
p13-2A8	=	↓
p13-2A9	↑	↓
p13-2F1	↑	↓
p13-4F1	↑	↓

Levels of specific mRNAs were determined by RNA dot blot analysis (Thomas 1983). RNA was isolated from ts13 cells or tsAF8 cells stimulated with serum at the nonpermissive temperatures. The RNA dots were hybridized to nick-translated inserts from the cDNA clones listed in the first column. The autoradiographic result was quantitated by densitometry.
↑, increased; =, same as a G_0; ↓, decreased.

ature. In contrast, p13-2A8, although elevated in G_1 cells at the permissive temperature, is not expressed above G_0 levels in ts13 cells stimulated at the restrictive temperature.

In tsAF8 cells at the nonpermissive temperature, the expression of most cdc genes was decreased, as one would expect from an RNA polymerase II mutant. Indeed, the expression of many transfected or microinjected genes is markedly reduced in tsAF8 cells at the nonpermissive temperature (Floros et al. 1981, Baserga et al. 1984). The differences in expression observed in the different cell lines may result from their different execution points.

SIGNIFICANCE OF cdc GENES

The cDNA clones we have identified represent genes whose expression is cell cycle dependent and that meet our (temporary) criteria for cdc genes. Similar genes or cDNA clones have already been described in serum-stimulated 3T3 cells (Linzer and Nathans 1983), in PDGF-stimulated 3T3 cells (Cochran et al. 1983), and in cells stimulated to proliferate or transformed by SV40 (Schutzbank et al. 1982, Scott et al. 1983). At present, nothing is known of the function of these genes. Indeed, as already mentioned, we do not even know whether or not these genes are relevant to the regulation of cell cycle progression.

Fragmentary information from the literature, however, offers some clues. The cDNA clone isolated by Linzer and Nathans (1983) has extensive homologies with the growth factor prolactin. Another growth factor, platelet-derived growth factor (PDGF), is homologous to an oncogene, v-*sis* (Waterfield et al. 1983, Doolittle et al. 1983), and part of the predicted protein sequence of the oncogene v-*erb*-B matches a portion of the sequenced avian EGF receptor (Downward et al. 1984). Another oncogene, B*lym*-1, identified in both chicken and human B-cell lymphoma

DNA, has been shown to code for a small protein with extensive homology to the amino terminus of transferrin, an absolute requirement for serum-free growth of cells in culture (Goubin et al. 1983, Diamond et al. 1983). One of the cDNA clones isolated from SV40-transformed mouse cells has homologies with a surface antigen (Brickell et al. 1983). On the other hand, some of the cdc genes from yeast (Conrad and Newlon 1983) and a cell cycle-specific ts mutation of mouse cells (Colwill and Sheinin 1983) have been identified as part of the DNA-synthesizing machinery. Other genes whose expression is cell cycle-dependent include histone genes (Plumb et al. 1983, DeLisle et al. 1983), the dihydrofolate reductase gene (Kaufman and Sharp 1983, LaBella et al. 1983), and the gene for the p53 protein (Reich and Levine 1984).

Piecing together these different bits of information, we would like to make the prediction that eventually cdc genes will fall into three major categories: genes coding for growth factors, genes coding for receptors for growth factors, and genes coding for internal proteins, such as the p53 protein or the enzymes of the DNA-synthesizing machinery. Among these, one hopes, is *the* gene (or at most a handful of genes) that truly control cell proliferation by triggering quiescent cells into the cell cycle.

ACKNOWLEDGMENTS

This work was supported by NIH Grant CA-25898 from the National Cancer Institute and by a training grant (CA-09214-06) for C.W.G.

REFERENCES

Alberghina L, Sturani E. 1981. Control of growth and of the nuclear division cycle in neurospora-crassa. Microbiol. Rev. 45:99–122.

Baserga R, Waechter DE, Soprano KJ, Galanti N. 1982. Molecular biology of cell division. Ann NY Acad Sci 397:110–120.

Baserga R, Shen YM, Yuan ZA. 1984. Cellular Transformation Systems. *In* Wolman SR, Mastromarino AJ (eds), Progress in Cancer Research and Therapy, vol 29. Raven Press, New York, pp. 377–385.

Brickell PM, Latchman DS, Murphy D, Willison K, Rigby PWJ. 1983. Activation of a Qa/tla class 1 major histocompatibility antigen gene is a general feature of oncogenesis in the mouse. Nature 306:756–760.

Burstin SJ, Meiss HK, Basilico C. 1974. A temperature-sensitive cell cycle mutant of the BHK cell line. J Cell Physiol 84:397–408.

Cochran BH, Reffel AC, Stiles CD. 1983. Molecular cloning of gene sequences regulated by platelet derived growth factor. Cell 33:939–947.

Colwill RW, Sheinin R. 1983. tsA169 locus in mouse L cells may encode a novobiocin binding protein that is required for DNA topoisomerase II activity. Proc Natl Acad Sci USA 80:4644–4648.

Conrad MN, Newlon CS. 1983. Saccharomyces-cerevisiae cdc 2 mutants fail to replicate approximately one-third of their nuclear genome. Mol Cell Biol 3:1000–1012.

Cooper HL, Braverman R. 1981. Close correlation between initiator methionyl-tRNA level and rate of protein synthesis during human lymphocyte growth cycle. J Biol Chem 256:7461–7467.

Darzynkiewicz Z, Sharpless T, Staiano-Coico L, Melamed MR. 1980. Subcompartments of the G_1 phase of the cell cycle detected by flow cytometry. Proc Natl Acad Sci USA 77:6696–6699.

DeLisle AJ, Graves RA, Marzluff WF, Johnson LF. 1983. Regulation of histone mRNA production and stability in serum stimulated mouse 3T6 fibroblasts. Mol Cell Biol 3:1920–1929.

Diamond A, Cooper GM, Ritz J, Lane MA. 1983. Identification and molecular cloning of the human B-lym transforming gene activated in Burkitt's lymphomas. Nature 302:114–119.

Doolittle RF, Hunkapiller MW, Hood LE, Devare SG, Robbins KC, Aaronson SA, Antoniades HN. 1983. Simian sarcoma virus oncogene, v-*sis*, is derived from the gene (or genes) encoding a platelet-derived growth factor. Science 221:275–277.

Downward J, Yarden Y, Mays E, Scrace G, Totty N, Stockwell P, Ullrich A, Schlessinger J, Waterfield MD. 1984. Close similarity of epidermal growth factor receptor and v-erb-B oncogene protein sequences. Nature 307:520–527.

Feinberg RF, Sun LK, Ordahl CP, Frankel FR. 1983. Identification of glucocorticoid-induced genes in rat hepatoma cells by isolation of cloned cDNA sequences. Proc Natl Acad Sci USA 80:5042–5046.

Floros J, Jonak G, Galanti N, Baserga R. 1981. Induction of cell DNA replication in G_1 specific ts mutants by microinjection of recombinant SV40 DNA. Exp Cell Res 132:215–223.

Fregien N, Dolecki GJ, Mandel M, Humphreys T. 1983. Molecular cloning of 5 individual stage and tissue specific mRNA sequences from sea urchin pluteus embryos. Mol Cell Biol 3:1021–1031.

Goubin G, Goldman DS, Luce J, Neiman PE, Cooper GM. 1983. Molecular cloning and nucleotide sequence of a transforming gene detected by transfection of chicken B-cell lymphoma DNA. Nature 302:114–119.

Hartwell LH. 1971. Genetic control of the cell division cycle in yeast. J Mol Biol 59:183–194.

Ingles CJ. 1978. Temperature-sensitive RNA polymerase II mutations in Chinese hamster ovary cells. Proc Natl Acad Sci USA 75:405–409.

Kaufman RJ, Sharp PA. 1983. Growth-dependent expression of dihydrofolate reductase mRNA from modular cDNA genes. Mol Cell Biol 3:1598–1608.

Kelly LJ, Kelly R, Ennis HL. 1983. Characterization of cDNA clones specific for sequences developmentally regulated during *Dictyostelium discoideum* spore germination. Mol Cell Biol 3:1943–1948.

Kramer RA, Anderson N. 1980. Isolation of yeast genes with mRNA levels controlled by phosphate concentration. Proc Natl Acad Sci USA 77:6541–6545.

LaBella F, Brown EH, Basilico C. 1983. Changes in the levels of viral and cellular gene transcripts in the cell cycle of SV40 transformed mouse cells. J Cell Physiol 117:62–68.

Lee AS, Delegeane A, Scharff D. 1981. Highly conserved glucose-regulated protein in hamster and chicken cells. Preliminary characterization of its cDNA clone. Proc Natl Acad Sci USA 78:4922–4925.

Lieberman I, Abrams R, Ove P. 1963. Changes in the metabolism of ribonucleic acid preceding the synthesis of deoxyribonucleic acid in mammalian cells cultured from the animal. J Biol Chem 238:2141–2149.

Linzer DIH, Nathans D. 1983. Growth regulated changes in specific mRNAs of cultured mouse cells. Proc Natl Acad Sci USA 80:4271–4275.

Mangiarotti G, Chung S, Zuker C, Lodish HF. 1981. Selection and analysis of cloned developmentally regulated *Dictyostelium discoideum* genes by hybridization competition. Nucleic Acid Res 9:947–963.

Plumb M, Stein J, Stein G. 1983. Co-ordinate regulation of multiple histone mRNAs during the cell cycle in HeLa cells. Nucleic Acid Res 11:2391–2410.

Reich NC, Levine AJ. 1984. Growth regulation of a cellular tumour antigen, p53, in nontransformed cells. Nature 308:199–201.

Rossini M, Baserga S, Huang CH, Ingles CJ, Baserga R. 1980. Changes in RNA polymerase II in a cell cycle specific temperature-sensitive mutant of hamster cells. J Cell Physiol 103:97–103.

Sargent TD, Dawid IB. 1983. Differential gene expression in the gastrula of *Xenopus laevis*. Science 222:135–139.

Scott, MRD, Westphal KH, Rigby PWJ. 1983. Activation of mouse genes in transformed cells. Cell 34:557–567.

Schutzbank T, Robinson R, Oren M, Levine AJ. 1982. SV40 large tumor antigen can regulate some cellular transcripts in a positive fashion. Cell 30:481–490.

Stancel GM, Prescott DM, Liskay RM. 1981. Most of the G_1 period in hamster cells is eliminated by lengthening the S period. Proc Natl Acad Sci USA 78:6295–6298.

Southern E. 1975. Detection of specific sequences among DNA fragments separated by gel electrophoresis. J Mol Biol 98:503–517.

Talavera A, Basilico C. 1977. Temperature-sensitive mutants of BHK cells affected in cell cycle progression. J Cell Physiol 92:425–436.

Thomas PS. 1983. Hybridization of denatured RNA and small DNA fragments transferred to nitrocellulose. Proc Natl Acad Sci USA 77:5201–5205.

Waechter DE, Avignolo C, Freund E, Riggenbach CM, Mercer WE, McGuire PM, Baserga R. 1984. Microinjection of RNA polymerase II corrects the temperature sensitive defect of tsAF8 cells. J Mol Cell Biochem 60:77–82.
Waterfield MD, Scrace GT, Whittle N, Stroobant P, Johnson A, Wasteson, A, Westermark B, Heldin CH, Huang JS, Deuel TT. 1983. Platelet derived growth factor is structurally related to the putative transforming protein $p28^{sis}$ of simian sarcoma virus. Nature 304:35–39.

Mediators in Cell Growth and Differentiation,
edited by Richard J. Ford and Abby L. Maizel.

Intracytoplasmic Signals for DNA Replication in Lymphocyte Proliferation

Stanley Cohen and Janice K. Gutowski

*Department of Pathology, University of Connecticut Health Center,
Farmington, Connecticut 06032*

The immune system is extremely complex. Induction of an immune response involves the interaction of a number of distinct helper and suppressor cells, macrophages specialized for antigen presentation or mediator production, various hormone-like lymphokines, and of course, the specific lymphocyte that will ultimately mediate the response. That response might be antibody production, lymphokine production, the generation of effector cells, and/or the creation of an expanded population of memory cells. Each of these, in turn, performs its biological task via additional cell interactions throughout the body.

Each of these aspects of the immune response is the result of two processes, proliferation and differentiation. Little is known about the basic mechanisms involved in either of these. Of the two, proliferation is the most general, since it involves events common to all cells, whereas differentiation, though it is also a general phenomenon, in this case leads to cell species and biologically active molecules unique to the immune system.

Lymphocytes can be stimulated to proliferate by a wide variety of extracellular agents, including antigens, mitogens, and growth factors. Little is known, however, about the intracellular systems that relate surface receptor binding to the induction of nuclear DNA synthesis. A number of mechanisms, including receptor-ligand internalization, ligand-induced receptor enzyme activity, ion fluxes, cytoskeletal changes, and cyclic nucleotides have been implicated as putative "second signals" for lymphocyte activation. However, the specific involvement of these processes in mechanisms of mitogenesis remains unclear. In addition, these all appear to involve the early stages of the activation pathway rather than the direct triggering of the critical nuclear events.

A number of studies demonstrate that nuclear activity is under cytoplasmic control. Indirect evidence for extranuclear regulatory events includes the observation that, in most cases, the nuclei of binucleate and multinucleate organisms exit G_1 and then enter the S phase of the cell cycle synchronously (Gonzalez-Fernandez et al. 1971). Graham et al. (1966, Graham 1966) have shown that nuclei from normally quiescent cells (such as adult frog brain or liver cells) begin synthesizing DNA when injected into the cytoplasm of frog eggs. Similar results were obtained in cell fusion studies in which nuclei from resting cells were induced to synthesize DNA when exposed to the cytoplasm from proliferating cells in the heterokaryon (Harris et al. 1966, Harris 1968, Johnson and Harris 1969). More recently, it has been shown that cytoplasmic extracts prepared from mitogenic hormone-stimulated cells (Das 1980), from early frog embryos (Benbow and Ford 1975), and from continuously proliferating cells (Jazwinski et al. 1976) can initiate DNA synthesis in isolated nuclei from adult frog liver and spleen cells. In addition, Floros et al. (1978) have demonstrated that for a cytoplasmic factor to stimulate DNA synthesis, essentially all information of G_1 cells must be present in the cells providing that factor. In the case of the study by Das (1980), epidermal growth factor (EGF) stimulation of fibroblasts was shown to lead to the appearance of a cytoplasmic protein, distinct from internalized EGF or receptor, with the capacity of initiating DNA synthesis in nuclei isolated from nondividing cells.

Our studies were undertaken to determine whether such cytoplasmic intermediates play a role in lymphocyte activation for proliferation. For this purpose we studied both spontaneously proliferating lymphoblastoid cell lines and mitogen-activated normal lymphocytes. Cytoplasmic extracts were prepared from a murine plasmacytoma cell line (X63Ag8.653, here abbreviated P3), a human B cell lymphoma line (RPMI 8392), a human T cell line (MOLT 4), and mitogen-activated normal human and murine lymphocytes. These extracts were tested for their capacity to induce DNA synthesis in quiescent nuclei isolated from *Xenopus laevis* splenocytes. Our methods for preparing the cytoplasmic extracts and isolated nuclei preparations have been described in detail (Gutowski and Cohen 1983).

The assay for DNA synthesis made use of an incorporation mixture of 2′-deoxyadenosine 5′-triphosphate (dATP), 2′-deoxyguanosine 5′-triphosphate (dGTP), 2′-deoxycytidine 5′-triphosphate (dCTP), adenosine 5′-triphosphate (ATP), phosphoenolpyruvate, pyruvate kinase, and methyl-tritiated thymidine-5′-triphosphate ([^{3}H]TTP). This is essentially the cocktail used in most of the above-referenced studies. The requirement for all four deoxyribonucleosides strongly suggests that we are measuring DNA synthesis rather than DNA repair. In addition, Jazwinski et al. (1976) demonstrated the absolute requirement for ATP by pretreating the reagents to remove the endogenous nucleotide and abolishing activity. In addition, they documented the appearance of replication forks and "eyes" by electron microscopy.

CYTOPLASMIC ACTIVITY IN SPONTANEOUSLY PROLIFERATING AND MITOGEN-ACTIVATED LYMPHOID CELLS

Isolated nuclei were prepared from adult frog spleen cells and incubated with cytoplasmic extracts prepared from spontaneously proliferating MOLT 4, P3, and 8392 cells. DNA synthesis in these cultures was measured by the incorporation of [^{3}H]TTP into trichloroacetic acid (TCA)-precipitable material. We found that cytoplasmic extracts prepared from all these cell lines induced high levels of DNA synthesis in isolated nuclei. The incorporation of [^{3}H]TTP was rapid, reaching maximal levels within 90 min. In contrast, incubation of nuclei without cytoplasmic extract, or cytoplasmic extracts without nuclei, resulted in only minimal amounts of [^{3}H]TTP incorporation. We also studied cytoplasmic extracts prepared from quiescent cells (unstimulated murine splenic T cells). Such extracts failed to induce DNA synthesis in the isolated nuclei.

To determine the pattern of dose response, we incubated 200,000 nuclei with increasing doses of cytoplasmic extract derived from proliferating MOLT 4 cells. We obtained a sigmoidal dose-response curve that plateaued at a dose at least 13-fold higher than the ED_{50}. Similar results were obtained using extracts prepared from proliferating P3 and 8392 cells. This pattern of response is of special interest, since extracellular inducers of cell proliferation such as growth hormones, antigens, or lectins have bell-shaped response curves, with depressions rather than plateaus at high doses.

We next examined the capacity of mitogen-activated normal lymphocytes to generate the cytoplasmic factor. Human peripheral blood lymphocytes were cultured with 20 μg/ml phytohemagglutinin (PHA) for 66 hr. Extracts were prepared from lymphocytes cultured in the presence or absence of PHA and were tested for their relative capacity to induce DNA synthesis in isolated nuclei. Cytoplasmic extracts prepared from PHA-stimulated lymphocytes induced high levels of DNA synthesis in the isolated nuclei. In contrast, extract prepared from unstimulated cells induced only minimal amounts of DNA-synthesizing activity; counts obtained were not substantially above the background contributed by nuclei alone. The stimulatory activity detected in the activated cell extract was not due to carryover of residual mitogen from the PHA-stimulated cultures, as PHA alone was incapable of activating nuclear DNA synthesis. Similar results were obtained using concanavalin A (ConA)-stimulated murine thymic lymphocytes. More recently, we demonstrated that stimulation of interleukin-2-responsive lymphocyte populations by that mediator is a potent means of inducing cytoplasmic stimulatory activity. These unpublished studies suggest that interleukin-2 may exert its proliferative effect by this means.

PRELIMINARY CHEMICAL PROPERTIES OF THE CYTOPLASMIC FACTOR

Some physicochemical properties of the cytoplasmic factor responsible for inducing DNA synthesis in isolated nuclei were next investigated. For these studies, we used extracts prepared from the continuously proliferating lymphoblastoid cells, because of the ease of obtaining relatively large quantities of active extract from this source. Activity was totally destroyed by trypsin treatment. This effect was not due to the soybean inhibitor used to stop the digestion nor to any residual tryptic effects on the nuclei. Other studies revealed that the factor is stable to storage at 4°C for 24 hr, to freeze/thawing, and to lyophilization. In addition, the activity was found to be nondialyzable (3500 MW cutoff dialysis tubing) and was totally abrogated by heat treatment (60°C for 20 min). Taken together, these results strongly suggest that the active component is a protein.

To determine the approximate molecular weight of the stimulatory factor, 8392 extracts were prepared and subjected to AMICON ultrafiltration. In these studies, the extract was filtered serially through XM100A and XM50 membranes and the retentate and filtrate from both filtration steps were collected. Whole extract and the four Amicon "cuts" (XM100A retentate, XM100A filtrate, XM50 retentate, and XM50 filtrate) were tested for the capacity to induce DNA synthesis in frog spleen nuclei. All of the activity present in the whole 8392 extract was retained on the XM100A membrane, suggesting that the active cytoplasmic protein has a molecular weight ≥100,000 Da. The components of the cytoplasmic extract that were <100,000 Da (those found in the XM100A and XM50 filtrates and XM50 retentate) were unable to induce significant DNA synthetic responses in quiescent nuclei. Similar results were obtained using extracts prepared from growing MOLT 4 cells.

We next subjected the XM100A retentate to ammonium sulfate fractionation. We found that the factor was precipitable at 30–50% saturation. We were able to reduce the total protein content by 73% and still retain 100% of the stimulatory activity in this fraction. We defined this cytoplasmic protein as ADR (activator of DNA replication).

BIOLOGIC PROPERTIES OF ADR

We next performed studies to determine whether ADR was secreted from the cell, as is the case for conventional lymphokines. To this end, 8392 cells were cultured at 37°C for 24 hr in serum-free medium at 5×10^6 cells/ml. The supernatants from these cultures were collected and dialyzed against the extraction buffer for 6 hr. The dialyzed supernatants were then diluted 1:2, 1:5, and 1:10 with extraction buffer and concentrated 5× and 10× over an AMICON YM5 membrane, to provide a hundredfold range of concentrations. The samples were then tested for the capacity to induce DNA synthesis in isolated nuclei. We were unable to recover stimulatory activity for isolated nuclei in the supernatants of cultured cells whose cytoplasm was known to contain high levels of such activity. Culture supernatants were not effective in initiating DNA synthesis in isolated nuclei even when concentrated 10×. These results were not due to loss of factor by dialysis, as previous experiments described above showed that the cytoplasmic stimulatory activity was nondialyzable.

In the next series of experiments, we attempted to determine whether ADR was capable of stimulating the proliferation of intact cells. ADR-containing extracts were prepared in the usual fashion from growing 8392 cells and tested for the capacity to induce the proliferation of whole frog spleen cells. Frog spleen cells were cultured for 1–4 days in medium alone, with 100 μl 8392 extract or with 100 μl ConA (10 μg/ml) as a positive control. DNA synthesis in these cultures was measured by the incorporation of [^{3}H]dTR during the last 24 hr of culture. Cytoplasmic extracts from 8392 cells failed to stimulate the proliferation of intact frog spleen cells in 1 to 4 day cultures. Stimulatory activity was not detected even when the extract was concentrated 5×. The cells were sensitive to mitogenic stimuli as seen by their positive response to ConA.

Although the cytoplasmic factor lacks species specificity with respect to its effect on isolated nuclei, it is possible that in whole cells, a step involving attachment to the cell membrane might show such specificity. Accordingly, these experiments were repeated utilizing a murine source of cytoplasmic-stimulating factor activity and intact murine lymphocytes as targets. In no instance could we detect a stimulatory effect on intact cells.

Since ADR is neither secreted by the cells that produce it nor has an effect on intact cells, it appears to be entirely an intracellular mediator.

ADR PRODUCTION BY LYMPHOCYTES FROM AGED DONORS

There is conflicting evidence regarding the relative contribution of nucleus and cytoplasm to the proliferative defects seen in aging cells. Wright and Hayflick (1975) have fused cytoplasts (enucleated cells) in various stages of their lifespans with whole cells to investigate the role of cytoplasmic components in controlling lifespan. Their results suggested that the mechanism controlling the proliferative potential of a cell is located in the nucleus. More recently, Carlin et al. (1983) have found diminished tyrosine kinase activity of the EGF receptor of senescent human fibroblasts, suggesting that prenuclear events are also involved in the aging defect. Studies of Muggleton-Harris and Hayflick (1976) involving reconstitution of karyoplasts with cytoplasts, each obtained from cells of different ages, suggested that both the nucleus and cytoplasm from the older cells could confer their lifespan characteristics on the cell.

The ADR assay provides a convenient way to address this issue in the lymphocyte. For these studies, we chose to study in vivo aging. It is known that lymphocytes from aged individuals have impaired proliferation response to PHA (Weksler and Hutteroth 1974, Hefton et al. 1980). We investigated ADR production with respect to proliferative response in peripheral blood lymphocytes from a variety of healthy elderly donors 66–72 years old (Gutowski et al. 1984). In this study, lymphocyte responsiveness to PHA, depending on the subject, ranged from 20% to 100% of that of matched young adult controls. Nevertheless, in all cases, ADR activity was found to be at normal or above-normal levels. To rule out the possibility that these results were merely due to a plateau in intracellular ADR activity at low levels of mitogenic response, we stimulated lymphocytes from young adult donors with varying suboptimal levels of PHA to generate proliferative responses similar to those seen in the aged population. Under these conditions, cytoplasmic ADR activity increased linearly with cell proliferation. Cells with strong proliferative responses contained high levels of ADR, whereas cells with suboptimal responses showed correspondingly low ADR activity. In no case did we find adult cells exhibiting relatively low proliferative activity in the face of maximal ADR activity.

In contrast to these observations, cells from aged donors did not show the same correlation between the cytoplasmic activity and cellular response. Although the PHA response of the aged cells varied from 20% to 100% of the adult controls, they consistently exhibited maximal levels of ADR.

The above results suggest that the impaired PHA response of cells from elderly individuals is not due to a failure to generate this cytoplasmic stimulatory signal. Rather, the data imply that the impairment may be due to age-related changes in the nucleus. We recently succeeded in preparing human nuclei in a manner that allows their use in the ADR assay. Preformed ADR from any source is as effective at stimulating these nuclei as frog nuclei, provided that they are obtained from young adult donors. In contrast, we have found, in unpublished observations, that the nuclei from lymphocytes of aged individuals have diminished responsiveness in standardized preparations of ADR to a degree that parallels the deficit in PHA responsiveness of the intact cell. Thus, in the case of the age-dependent loss of mitogenic response by lymphocytes, the defect appears to be at the level of the nucleus. Further studies on the mechanisms of action of ADR may also shed light on the nature of this nuclear defect.

DISCUSSION

The results described above, which have been presented elsewhere in detail (Gutowski and Cohen 1983, Gutowski et al. 1984), demonstrate the existence of a cytoplasmic factor (ADR) in proliferating human and murine lymphoid cells. This factor can induce DNA synthesis in isolated, quiescent nuclei. Frog spleen cell nuclei were used in most of these studies because they are easier to prepare and more stable in culture than mammalian nuclei. As indicated above, however, we have found it possible to demonstrate significant DNA synthetic responses in isolated quiescent mammalian nuclei as well.

The cytoplasmic state responsible for the induction of DNA synthesis appears to be characteristic of proliferating cells only. No stimulatory activity has been found in the cytoplasm of resting cells in these studies or in those of other investigators utilizing other cell types.

It is important to emphasize that we were able to detect stimulatory activity not only in continuously proliferating cells, confirming the results of Jazwinski et al. (1976), but also in normal human and murine lymphoid cells, activated by lectins in primary culture. These results are consistent with the results reported by Das (1980) by utilizing a system involving stimulation of 3T3 fibroblasts by EGF. The demonstration of the induction of this cytoplasmic state in two unrelated systems (mitogenic activation of lymphocytes and growth factor activation of fibroblasts) suggests that this may represent a general phenomenon and that the cytoplasmic protein factor involved may serve as an intracellular mitogenic signal ("second signal") in at least some forms of proliferative response mediated by activation of cell surface receptors. In this regard, it would be of interest to determine whether antigen-induced activation of mature lymphocytes is a sufficient trigger for the induction of cytoplasmic stimulatory activity. Because of the limited size of the responding subpopulation in conventional immunization, such experiments require the use of antigen-reactive clones. However, we have recently shown that another important endogenous trigger of lymphocyte proliferation, interleukin-2, can activate its target cell population for ADR production.

It was of obvious interest to investigate the relationship between the cytoplasmic inducer of DNA replication reported here and the growth factors generally produced and secreted by lymphocytes. We were unable to detect any stimulatory activity for isolated nuclei in culture supernatants from cells whose cytoplasm was known to contain high levels of such activity. This finding suggests that the factor is not released from the cell; rather, it appears to represent a strictly intracellular modulator of cell growth. In addition, we found that the cytoplasmic extracts with high activity for isolated nuclei were incapable of stimulating the proliferation of *intact* cells. These two results serve to distinguish the cytoplasmic factor from other lymphocyte-derived growth factors such as interleukin-2 and the mitogenic lymphokines.

The results from the characterization studies suggest that the active component is a protein and has a molecular weight $\geq$100,000 Da. These findings rule out the possibility that the stimulation of DNA synthesis may have been due solely to low-molecular-weight substances such as cations or cyclic nucleotides. Although the activation of quiescent nuclei has been shown by several investigators to be associated with an intranuclear influx of cytoplasmic proteins (Harris 1968, Merriam 1969, Ringertz et al. 1971, Choe and Rose 1973, Allfrey et al. 1975), it is not yet known whether the factor described in this study acts at the nuclear membrane or penetrates the nuclear membrane and acts on the nuclear material itself. Wherever the stimulatory effect is exerted, the rapidity of the reaction suggests that it represents one of the final biochemical events in the triggering of nuclear DNA synthesis.

The finding of high levels of ADR activity in lymphocytes from aged individuals when the nuclei of these cells were engaged in minimal DNA synthetic responses confirms that ADR is a true cytoplasmic factor and not merely derived from products released by active nuclei. More important, it suggests that the defect might be found at the level of the nucleus, a result confirmed by our subsequent investigations. In the study by Carlin et al. (1983), an age-related proliferative defect appears to be located at the early stages of ligand-induced activation of the cell membrane. Whether this difference reflects differences in aging in vitro and in vivo or is related in some way to the different biologic behavior of the two cell types remains to be explored.

The exact mechanism by which the cytoplasmic stimulatory factor induced DNA synthesis in isolated nuclei is unclear. It may be that this protein represents the molecular trigger for the G_0/G_1 to S transition. Alternatively, it may exert an effect by inactivating inhibitory substances that maintain the resting state and prevent a cell from proliferating. One possibility for the former situation could involve the induction of phosphorylation of nuclear proteins. For example, Nishizawa and his co-workers have shown that the stimulation of B cells with anti-immunoglobulin leads to the appearance of a cytoplasmic protein that can activate protein kinases within the nucleus to phosphosphorylate nonhistone nuclear proteins (Nishizawa et al. 1977, Kishimoto et al. 1979). Although these authors did not measure DNA synthesis, results from other studies (Pogo and Katz 1974, Johnson et al. 1974, Johnson and Hadden 1975, Kishimoto et al. 1977) have suggested a close association between phosphorylation of nuclear proteins and the differentiation and/or proliferation of lymphocytes. Thus, it will be of importance to explore not only the locus of action, but also the precise mode of action of the factor reported here. Such issues include the possibility of a nuclear receptor for the factor as well as its range of enzymatic activities. The preliminary characterization studies reported here suggest the feasibility of obtaining adequately purified preparations for such study.

SUMMARY

Cytoplasmic extracts prepared from continuously proliferating lymphoblastoid cells, as well as mitogen-activated normal lymphocytes, contain a protein or proteins (ADR) capable of inducing DNA synthesis in isolated quiescent nuclei. Lymphocytes obtained from aged donors with depressed proliferative responses have normal levels of ADR, but depressed nuclear responses to exogenous ADR, suggesting a nuclear defect in these cells. In no instance could an effect of ADR on intact cells be demonstrated. Moreover, it is not released into the extracellular milieu by cells that contain it. These observations suggest that it serves as an intracellular mitogenic signal in replicating cells. Thus, it may be an important intermediate in the pathway leading from cell membrane activation to nuclear activation.

ACKNOWLEDGMENTS

This work was supported by NIH grants AI-12477, AI-16706 and CA-35703. We wish to thank Ms. Ann West for her invaluable technical assistance.

REFERENCES

Allfrey VG, Inoue A, Karn J, Johnson E, Good R, Hadden JW. 1975. Sequence-specific binding of DNA by non-histone proteins and their migration from cytoplasm to nucleus during gene activation. *In* The Structure and Function of Chromatin, Ciba Foundation Symposium. Elsevier, Amsterdam, pp. 199–219.

Benbow RM, Ford CC. 1975. Cytoplasmic control of nuclear DNA synthesis during early development of *Xenopus laevis*: A cell-free assay. Proc Natl Acad Sci USA 72:2437–2441.

Carlin CR, Phillips PD, Knowles DB, and Cristofalo VJ. 1983. Diminished in vitro tyrosine kinase activity of the EGF receptor of senescent human fibroblasts. Nature 306:617–620.

Choe BK, Rose NR. 1973. Synthesis of DNA-binding proteins during the cell cycle of WI-38 cells. Exp Cell Res 83:271–280.

Das M. 1980. Mitogenic hormone-induced intracellular message: Assay and partial characterization of an activator of DNA replication induced by epidermal growth factor. Proc Natl Acad Sci USA 77:112–116.

Floros J, Chong H, Baserga R. 1978. Stimulated DNA synthesis in frog nuclei by cytoplasmic extracts of temperature-sensitive mammalian cells. Science 201:651–652.
Gonzalez-Fernandez A, Gimenez-Martin G, Diez JL, de la Torre C, Lopez-Saez JF. 1971. Interphase development and beginning of mitosis in the different nuclei of polynucleate homokaryotic cells. Chromosoma 36:100–111.
Graham CF. 1966. The regulation of DNA synthesis and mitosis in multinucleate frog eggs. J Cell Sci 1:363–374.
Graham CF, Arms K, Gurdon JB. 1966. The induction of DNA synthesis by frog egg cytoplasm. Dev Biol 14:349–354.
Gutowski JK, Cohen S. 1983. Induction of DNA synthesis in isolated nuclei by cytoplasmic factors from spontaneously proliferating and mitogen-activated lymphoid cells. Cell Immunol 75:300–311.
Gutowski JK, Innes J, Weksler ME, Cohen S. 1984. Induction of DNA synthesis in isolated nuclei by cytoplasmic factors. II. Normal generation of cytoplasmic stimulatory factors by lymphocytes from aged humans with depressed proliferative responses. J Immunol 132:559–562.
Harris H. 1968. Nucleus and Cytoplasm. Clarendon Press, Oxford.
Harris H, Watkins JF, Ford CE, Schoefl GI. 1966. Artificial heterokaryons of animal cells from different species. J Cell Sci 1:1–30.
Hefton JM, Darlington GJ, Casazza BA, Weksler ME. 1980. Immunologic studies of aging. V. Impaired proliferation of PHA responsive human lymphocytes in culture. J Immunol 125:1007–1010.
Jazwinski SM, Wang JL, Edelman GM. 1976. Initiation of replication in chromosomal DNA induced by extracts from proliferating cells. Proc Natl Acad Sci USA 73:2231–2235.
Johnson EM, Hadden JW. 1975. Phosphorylation of lymphocyte nuclear acidic proteins: Regulation by cyclic nucleotides. Science 187:1198–1200.
Johnson EM, Karn J, Allfrey VG. 1974. Early nuclear events in the induction of lymphocyte proliferation by mitogens. J Biol Chem 249:4990–4999.
Johnson RT, Harris H. 1969. DNA synthesis and mitosis in fused cells. II. HeLa-chick erythrocyte heterokaryons. J Cell Sci 5:625–643.
Kishimoto T, Nishizawa Y, Kikutani H, Yamamura Y. 1977. Biphasic effect of cyclic AMP on IgG production and on the changes of non-histone nuclear proteins induced with anti-immunoglobulin and enhancing soluble factor. J Immunol 118:2027–2033.
Kishimoto T, Kikutani H, Nishizawa Y, Sakaguchi N, Yamamura Y. 1979. Involvement of anti-Ig-activated serine protease in the generation of cytoplasmic factor(s) that are responsible for the transmission of Ig-receptor-mediated signals. J Immunol 123:1504–1510.
Merriam RW. 1969. Movement of cytoplasmic proteins into nuclei induced to enlarge and initiate DNA. J Cell Sci 5:333–349.
Muggleton-Harris AL, Hayflick L. 1976. Cellular aging studied by the reconstruction of replicating cells from nuclei and cytoplasms isolated from normal human diploid cells. Exp Cell Res 103:321–330.
Nishizawa Y, Kishimoto T, Kikutani H, Yamamura Y. 1977. Induction and properties of cytoplasmic factor(s) which enhance nuclear non-histone protein phosphorylation in lymphocytes stimulated by anti-Ig. J Exp Med 146:653–664.
Pogo BGT, Katz JR. 1974. Early events in lymphocyte transfer by PHA. II. Synthesis and phosphorylation of nuclear proteins. Differentiation 2:119–124.
Ringertz NR, Carlsson S, Ege T, Bolund L. 1971. Detection of human and chick nuclear antigens in nuclei of chick erythrocytes during reactivation in heterokaryons with HeLa cells. Proc Natl Acad Sci USA 68:3228–3232.
Weksler ME, Hutteroth TH. 1974. Impaired lymphocyte function in aged humans. J Clin Invest 53:99–104.
Wright WE, Hayflick L. 1975. Nuclear control of cellular aging demonstrated by hybridization of a nucleate and whole cultured normal human fibroblasts. Exp Cell Res 96:113–121.

Mediators in Cell Growth and Differentiation,
edited by Richard J. Ford and Abby L. Maizel.
Raven Press, New York © 1985.

Cellular Oncogenes, Growth Factors, and Cellular Growth Control

Arthur B. Pardee, Judith Campisi, Harry E. Gray, Michael Dean,* and Gail Sonenshein†

*Dana-Farber Cancer Institute, Boston, Massachusetts; *Frederick Cancer Research Institute, Frederick, Maryland; and †Boston University Medical School, Boston, Massachusetts*

Nearly two dozen oncogenes have been described during the past few years (Bishop 1983). These genes can be introduced into cells by viral infection or by transfection and can convert nontumorigenic animal cells into tumor-forming cells. Viral transforming genes (oncogenes) appear to be derived from normal cellular genes (proto-oncogenes) (Duesberg 1983). Proto-oncogenes are assumed to have important functions in normal differentiation and growth regulation, and after modifications they are assumed to be the genetic units responsible for improper growth regulation in cancer cells. Much has been discovered regarding the structures and cancer-inducing potentials of oncogenes, but much less is known regarding the relationship of these genes to normal and deranged regulation of growth. In this paper we describe recent experiments that relate expression of proto-oncogenes, particularly *myc* and *ras*, to the growth regulation of mouse 3T3 fibroblasts.

At this time one can relate growth, growth factors, and oncogenes at the level of cell biology in only a few instances. An oncogene may eliminate the requirement for a growth factor, or overexpression of oncogene mRNA can correlate with tumorigenicity. Subtler changes can also be observed, as will be described. Molecular relationships between oncogenes and growth largely remain to be discovered.

BACKGROUND

Cellular proliferation is controlled by external factors, among which the most important are hormones and growth factors. These are small proteins such as platelet-derived growth factor (PDGF) (Stiles 1983), the insulin-like growth factors (particularly IGF I, also known as somatomedin C) (Clemmons and Van Wyk 1981), epidermal growth factor (EGF) (Carpenter and Cohen 1979), and others (Barnes and Sato 1980). When these factors are provided, usually as components of serum, quiescent fibroblasts initiate growth; when essential factors are insufficient the cells arrest growth. Tumorigenic cells commonly have lower requirements for growth factors than do the untransformed cells from which they are derived.

Growth factors bind to specific receptors on the cell surface. Connections between certain oncogenes, growth factors, and growth factor receptors have very recently been discovered. Oncogenes can code for growth factor-like proteins (Waterfield et al. 1983), and conversely growth factors can activate expression of proto-oncogenes (Kelly et al. 1983, Campisi et al. 1984). Oncogenes can also code for growth factor receptor-like molecules (Downward et al. 1984).

Cellular nutrition is another important regulator of growth. A cell must duplicate all of its components to proliferate. Thus, nutrients such as vitamins and essential amino acids are necessary as building blocks. But beyond this obvious requirement for nutrients, highly specific requirements have been detected. They are related to the regulation of growth in a more subtle way than the above, being specifically related to events in the G_1 phase of the cell cycle (Pardee et al. 1978). These requirements are relaxed in certain transformed cells, establishing the importance of their role in physiologically normal growth control and its derangement in cancer (Campisi et al. 1982). No oncogenes are known to eliminate a nutritional requirement or to be directly responsible for decreasing such a requirement (e.g., by more effective transport).

Cells can exist in a proliferating state, in which they traverse the cell cycle and pass through the familiar events of DNA replication and mitosis, at the same time increasing their mass through production of the new proteins and other components that are needed to produce a new cell (Figure 1). Alternatively, they can be in a quiescent steady state (sometimes indicated as G_0) in which they are metabolically active but do not proliferate or show net synthesis of molecules. It is the fraction of cells in the proliferating state as contrasted to those in the quiescent state that determines the rate at which a population of cells increases. In a slowly growing population, cells are mostly in a quiescent state and in a rapidly growing population more cells are in the proliferating state. Growth is dependent on a switch from one of these states to the other that depends on external factors: nutrients, growth factors, and hormones.

We can consider growth control by starting from the quiescent state. After quiescent cells are stimulated by addition of necessary growth factors and nutrients many hours can elapse (in the case of 3T3 cells about 12 hr) before they initiate DNA synthesis. The G_0-S interval should be subdivided into two main periods. A variety of experiments involving oncogenes, growth factors, nutrients, and inhibitors show that the events in these two intervals are quite different. In the first half the cells emerge from quiescence through a competence and progression process (Pledger et al. 1978), discussed briefly below (see Hendrickson et al., pages 71–85, this volume), and synthesize new mRNAs and proteins. In this interval from G_0 to a mid-G_1 point (termed V) the cell restores its machinery for making mRNAs and proteins. Then, in the second half of the G_0-S period the cell prepares for DNA synthesis.

In contrast to G_0 cells, cycling cells already possess an active protein and RNA synthetic machinery when they emerge from mitosis. Some events that are performed as quiescent cells commence growth appear to be carried out during the

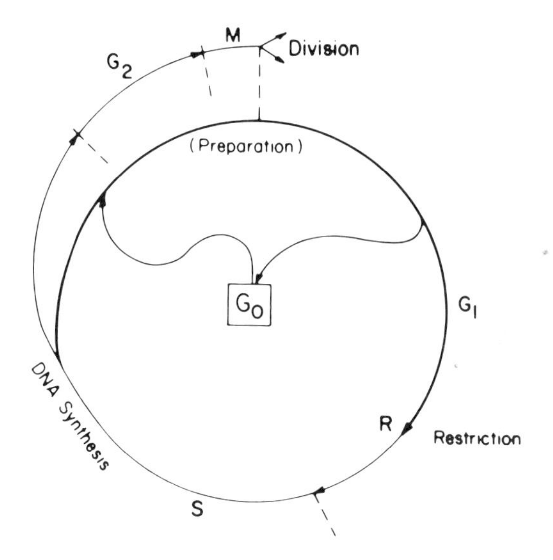

FIG. 1. A regulated cell cycle. This cycle is drawn to indicate that two separate sequences of events are taking place in the latter part of the cycle. Preparation for the next round of DNA duplication (S) occurs during the prior cycle, separately and concurrently with the terminal events leading to mitosis (M) and division. Cells placed under inadequate nutritional or growth factor conditions cannot pass the restriction point (R) of the "preparation" sequence, but can complete the mitotic and division processes. They move into the quiescent G_0 state, with an unreplicated DNA content. They emerge from G_0 if conditions are optimized.

prior cell cycle by growing cells (Figure 1) (Cooper 1979, Brooks et al. 1980). The events during the G_1, pre-DNA synthetic phase of cycling cells are quite similar to those that usually occur in the second portion of G_0-S.

Cell cycle regulation is exerted mainly in G_1, and consequently cells become growth arrested with an unduplicated DNA content. When conditions are not optimal for growth, nontransformed cells complete their current division cycle by finishing DNA synthesis and proceeding through mitosis and cell division, but are unable to initiate a new round of DNA synthesis. Thus, growth-arrested cells are unable to perform events that are preparatory for the initiation of DNA synthesis (Figure 1). It is these regulated events that are particularly susceptible to external factors, are the most important for growth regulation, and probably are affected by oncogenes. In nontransformed cells, these events depend on both growth factors and nutrients; growth-arrested nontransformed cells exhaust the supply of growth factors when they stop cycling at confluence. By contrast, some transformed cells are arrested owing only to insufficient nutrients (Moses et al. 1978). Most transformed, tumorigenic cells retain some degree of growth control; they have "leaky," defective growth control mechanisms (Pardee et al. 1978).

Progression of cells from G_0 to the midpoint of the G_0-S period requires not only PDGF but other growth factors, important among which are EGF and IGF I (Leof et al. 1982). The requirements for nutrients, i.e., essential amino acids, are not great during this interval (Stiles et al. 1979). Competent cells provided with plasma (which lacks PDGF) and in a medium lacking numerous essential amino acids can progress toward S phase for about 6 hr and then are arrested, at the V point. To progress beyond this state requires IGF I and also a complete complement of essential amino acids.

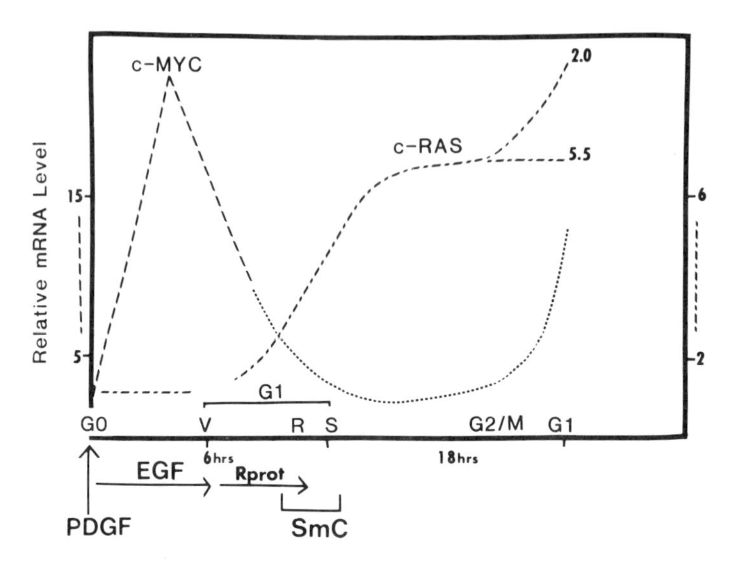

FIG. 2. Fluctuations of *myc* and *ras* proto-oncogene expression relative to cell cycle position and growth factor-dependent control points. Quiescent 3T3 cells are stimulated by the sequential action of PDGF, EGF, and somatomedin C (IGF I). In untransformed A31 cells, c-*myc* mRNA levels increase 20- to 30-fold within 3 to 4 hr of PDGF action. The level of this mRNA then falls, but it may rise again since exponentially growing cells have elevated amounts of c-*myc* mRNA. C-*ras* levels rise 5- to 7-fold much later, shortly after cells no longer require EGF for entry into S phase. These data were obtained with a Kirsten *ras* probe; Harvey *ras* mRNA levels appear to behave similarly. In the next G_1 following exit from quiescence, the level of the smaller Kirsten *ras* transcript increases another 2- to 3-fold. The results with Kirsten *ras* were obtained using transformed BPA31 cells. BPA31 seem to regulate *ras* mRNA levels similarly to A31 cells, unlike their regulation of c-*myc* transcriptional expression.

THE *myc* ONCOGENE AND GROWTH REGULATION

Expression of the cellular *myc* proto-oncogene (c-*myc*) is activated by PDGF (Kelly et al. 1983). PDGF is coded for by a cellular gene closely related to the viral *sis* oncogene (Waterfield et al. 1983, Doolittle et al. 1983). PDGF activates initial stages of emergence of quiescent 3T3 cells, producing the state known as "competence" (Pledger et al. 1978). Competent cells activated by PDGF cannot initiate DNA synthesis unless other factors (which include EGF and IGF I) are provided. PDGF does cause the cell to produce new messages and proteins (see Hendrickson et al., pages 71–85, this volume). We have found, consistent with the above reports, that *myc* mRNA appears during the first 4 hr after activation of 3T3 cells with serum (Campisi et al. 1984) (Figure 2). We have also found that the amount of *myc* mRNA diminishes in the next 4 hr, and so the activation of *myc* appears to be a transient process that occurs as cells emerge from quiescence. Furthermore, since ample *myc* mRNA is found in exponentially growing cultures, this proto-oncogene is probably expressed during each cycle. This is consistent with the finding that PDGF must be supplied during each cycle in order to permit cells subsequently to initiate another round of DNA synthesis (Scher et al. 1979). In regenerating liver, two peaks of *myc* mRNA accumulation are seen within 4 days after hepatectomy (Fausto and Shank 1983).

When we examined 3T3 cells transformed by two chemical carcinogens [benzo(a)pyrene and dimethylbenzanthracene], we found a striking difference in expression of c-*myc* compared to untransformed cells. The mRNA was present in large amounts in the quiescent transformed cells, whereas it was absent in quiescent untransformed 3T3 cells (Campisi et al. 1984). Once the cells were growing, the level of *myc* mRNA was similar in both cell types. These results suggest a novel mechanism for transformation, based on only the constitutive production of an oncogene product. This change in control at the critical time is sufficient to permit growth. This type of misregulation does not require either alterations in the structure of the *myc* gene or excessive production of its product (Little et al. 1983) (examples of both have been described in some other transformations), but rather a *genetic change in the timing mechanism.*

THE *ras* ONCOGENES

The *ras* is a family of proto-oncogenes that code for similar p21 proteins. The Harvey *ras* gene can transform NIH 3T3 mouse cells and is found to be altered in EJ human tumor cells (Land et al. 1983). We found that the normal Kirsten *ras* proto-oncogene is transcriptionally expressed in the latter part of the G_0-S interval in A31 cells, in contrast to *myc*, which is expressed early (Campisi et al. 1984) (Figure 2). Thus, *ras* expression is related to the later events involved in progression to DNA synthesis, a result also found in regenerating liver (Fausto and Shank 1983). The *ras* gene appears to be expressed similarly in BPA31 cells and the parental A31 cells (Campisi et al. 1984). This again is in contrast to the results we have obtained with *myc*, which show that *myc* is expressed constitutively in the transformed cells but not in the parental line. Thus, if *ras* is changed in these transformed cells, it is not through quantity or timing of transcription during the cell cycle. It is possible that the *ras* gene has undergone a more subtle sequence change in these cells, as has been described for other transformed cells such as EJ (Land et al. 1983).

It is in the latter part of emergence of quiescent cells into the cycle that an unstable protein of MW 68,000 is produced in A31 cells (Croy and Pardee 1983). We predicted from cycle kinetics (Campisi et al. 1982), and then observed, a change in this P68 protein. Namely, it has much greater stability in a variety of transformed cells. This could be a major factor in transformation owing to increased capacity of these cells to transit the controlled G_0-S interval. Production of this protein is dependent upon rapid protein synthesis, and hence on nutritional requirements such as essential amino acids. The requirement for essential amino acids for transit beyond the V point also indicates a protein synthesis requirement during this latter part of the G_0-S interval (Stiles et al. 1979). The relationship of the P68 protein to *myc* or to *ras* is not yet clarified.

RELATIONSHIPS OF OTHER GROWTH FACTORS TO ONCOGENES

Two growth factors important in the progression to initiation of DNA synthesis are EGF and IGF I. EGF is required by A31 cells in mid-G_0–S and it is not needed during the latter part of this period (Leof et al. 1982). The requirement for EGF, as for other growth factors, is eliminated or diminished in various transformed cells. One cause of this decreased requirement is that some transformed cells produce "tumor growth factors," which take the place of EGF (Roberts et al. 1983). In other cells this is not the case. For example, medium conditioned by a 3T3 cell line that produces transforming growth factor was able to block EGF binding by CHEF/18 cells, but medium similarly prepared from transformed CHEF/18 cells did not replace EGF or block binding (Cherington and Pardee 1982). Thus, the decreased number of measured EGF receptors could not be attributed to secretion by the cells of a competing tumor growth factor.

Receptors for growth factors are necessary for stimulation of cycling, untransformed cells, but their amount and binding activity have not been shown to limit growth of transformed cells. Chinese hamster embryo fibroblasts (CHEF/18) transformed by chemicals have specifically diminished their EGF requirements (Cherington et al. 1979). After passage in nude mice, these cells totally lose their EGF requirement. The number of EGF binding sites on the cell surface is progressively decreased in these transformed and tumorigenic cells (Cherington and Pardee 1982). This indicates that some change of the EGF receptor must be involved, one that leads to loss of affinity as well as to activation of its function.

A similar dual loss of the EGF requirement and decreased binding was seen when 3T3 cells were treated with cholera toxin (Wharton et al. 1982). Phosphorylation of growth factor receptors may be the biochemical basis for these effects. EGF activates self-phosphorylation of tyrosine residues of its receptor, a kinase (Ushiro and Cohen 1980). Receptors for other growth factors (for PDGF, insulin, and IGF I) are also known to be phosphorylated when their ligands combine with them. Cholera toxin raises the cAMP level of cells, thereby activating cAMP-dependent protein kinases that could phosphorylate the EGF receptor. The *src* oncogene causes phosphorylation of a substrate that is also phosphorylated by the EGF receptor (Ghosh-Dastidar and Fox 1983). Transformation could decrease a growth factor requirement and the available receptors simultaneously by permitting constitutive phosphorylation of a receptor.

One may speculate, based on timing of events, that EGF is triggering a sequence of events involving the activation of the *ras* gene and the MW 68,000 protein. EGF has been shown to increase mRNAs transcribed from retrovirus-related elements called VL 30 in mouse embryo cells (Foster et al. 1982). The *erb*-B oncogene may code in part for the EGF receptor (Downward et al. 1984).

IGF I is the only factor that is required during the later part of the G_0-S transition (Leof et al. 1982). Campisi and Pardee (1984) have found similarly that only IGF I (or insulin at hyperphysiological concentrations) is required for entry of all G_1 cycling cells into S phase. Could an oncogene be related to this requirement for

IGF I? Some of our recent results suggest that IGF I, unlike PDGF, may not control gene expression via a transcriptional event, but rather exerts its effects without new mRNA synthesis (Campisi and Pardee 1984). Therefore IGF I (somatomedin C), unlike PDGF, may not have a direct oncogene-activating function. On the other hand, an altered (unknown) oncogene could eliminate the requirement for IGF I; tumor cells have been shown to produce IGFs (DeLarco and Todaro 1978).

INITIATION OF DNA SYNTHESIS

The suddenness of onset of DNA synthesis in individual cells indicates a special biochemical event or set of biochemical events that are abruptly turned on. This event apparently is not the provision of DNA precursors, since permeabilized cells cannot initiate DNA synthesis unless they are already in S phase, even when they are supplied with excesses of the four deoxynucleoside triphosphates (Castellot et al. 1978). No specific enzyme is known to be involved, and indeed most of the enzymes of DNA synthesis are present in pre-S cells (G_1). A number of these enzymes, however, increase quite dramatically in parallel with incorporation of [^{3}H]thymidine into DNA. Our recent studies (Coppock and Pardee 1985) indicate that the induction of these enzymes in 3T3 cells is under control of the same labile protein system as is the onset of DNA synthesis itself. The preparation for these events appears to be building up for several hours prior to DNA synthesis.

The enzymes that function in DNA synthesis are not free, but rather they are assembled as a multienzyme complex, both in bacteria (Mathews et al. 1979) and in mammalian cells (Reddy and Pardee 1980). This multienzyme "replitase" complex assembles very near in time to the initiation of S phase. It is a large entity, about the size of a ribosome, and contains many proteins (Noguchi et al. 1983). A large number of proteins are involved in initiation of DNA synthesis even in bacteria (Kornberg 1982). A plausible possibility is that the onset of DNA synthesis depends on the assembly of this multienzyme complex, which activates key enzymes. One piece of evidence in favor of this hypothesis is that the DNA synthesis-related enzyme thymidylate synthase is active in vivo only after the complex is formed, although it is present in G_1 cells when assayed in extracts (Reddy 1982). The biochemical and genetic basis of replitase assembly remains to be discovered, and the effects of transformation and oncogenes on it are unknown.

SUMMARY

In this article we relate the functioning of oncogenes, particularly *myc* and *ras*, to current ideas regarding regulation of mammalian cell growth by growth factors. Assuming the genetic basis of transformation to be alterations of several proto-oncogenes, the mechanisms by which transformation could diminish growth control are numerous. (1) The oncogene could interact with a growth factor in several ways. Mutations could alter the quantity of an oncogene's product or its quality through primary structure or covalent modifications such as phosphorylations. (2) Oncogenes could code for a receptor for a growth factor. Various alterations parallel to the above set could then affect growth factor function via receptor changes (including abolished requirement for the factor). (3) Some oncogenes might operate during the chain of intracellular events that must follow growth stimulation. Introduction of such an oncogene (e.g., the coding region for a DNA virus T antigen) would bypass requirements for both growth factors and receptors.

Various observations regarding the oncogenes, cell cycle–timed events, and growth factors have been presented in a way we hope will suggest experiments designed to provide a basis for understanding growth regulation at the genetic, biochemical, and cellular levels in normal cells and tumorigenic cells, which have deranged growth regulation.

ACKNOWLEDGMENTS

This investigation was supported by Grants GM/CA24571 and CA36355 awarded by the National Cancer Institute, U.S. Department of Health and Human Services to A. B. P. and to G. S., respectively. We thank Marjorie Rider for preparing the manuscript.

REFERENCES

Barnes D, Sato G. 1980. Serum-free cell culture: A unifying approach. Cell 22:649–655.
Bishop JM. 1983. Cellular oncogenes and retroviruses. Annu Rev Biochem 52:301–354.
Brooks RF, Bennett DC, Smith JA. 1980. Mammalian cell cycles need two random transitions. Cell 19:493–504.
Campisi J, Pardee AB. 1984. Post-transcriptional control of the onset of DNA synthesis by insulin-like growth factor. Mol Cell Biol, 14:1807–1814.
Campisi C, Medrano EE, Morreo G, Pardee AB. 1982. Restriction point control of cell growth by a labile protein: Evidence for increased stability in transformed cells. Proc Natl Acad Sci USA 79:436–440.
Campisi J, Gray HE, Pardee AB, Dean M, Sonenshein GE. 1984. Cell cycle control of c-*myc* but not c-*ras* expression is lost following chemical transformation. Cell 36:241–247.
Carpenter G, Cohen S. 1979. Epidermal growth factor. Annu Rev Biochem 48:193–216.
Castellot JJ Jr, Miller MR, Pardee AB. 1978. Animal cells reversibly permeable to small molecules. Proc Natl Acad Sci USA 75:351–355.
Cherington PV, Pardee AB. 1982. On the basis for loss of the EGF growth requirement by transformed cells. *In* Growth of Cells in Hormonally Defined Media (Cold Spring Harbor Conferences on Cell Proliferation, Volume 9). Cold Spring Harbor Laboratory, Cold Spring Harbor, New York, pp. 221–230.
Cherington PV, Smith BL, Pardee AB. 1979. Loss of epidermal growth factor requirement and malignant transformation. Proc Natl Acad Sci USA 76:3937–3941.
Clemmons DR, VanWyk JJ. 1981. Somatomedin: Physiological control and effects on cell proliferation. Handbook of Experimental Pharmacology 57:161–208.
Cooper S. 1979. A unifying model for the G_1 period in prokaryotes and eukaryotes. Nature 280:17–19.
Croy RG, Pardee AB. 1983. Enhanced synthesis and stabilization of M_r 68,000 protein in transformed BALB/c-3T3 cells: Candidate for restriction point control of cell growth. Proc Natl Acad Sci USA 80:4699–4703.
DeLarco JE, Todaro GJ. 1980. A human fibrosarcoma cell line producing multiplication-stimulating activity (MSA) peptides. Nature 272:356–358.
Doolittle RF, Hunkapiller MW., Hood LE, Devare SG, Robbins KC, Aaronson SA, Antoniades HN. 1983. Simian sarcoma virus onc gene, v-*sis*, is derived from the gene (or genes) encoding a platelet-derived growth factor. Science 221:275–277.

Downward J, Yarden Y, Mayes E, Scrace G, Totty N, Stockwell P, Ullrich A, Schlessinger J, Waterfield MD. 1984. Close similarity of epidermal growth factor receptor and v-*erb*-B oncogene protein sequences. Nature 307:521–527.

Duesberg PH. 1983. Retroviral transforming genes in normal cells? Nature 304:219–226.

Fausto N, Shank PR. 1983. Oncogene expression in liver regeneration and hepatocarcinogenesis. Hepatology 3:1016–1023.

Foster DN, Schmidt LJ, Hodgson CP, Moses HL, Getz MJ. 1982. Polyadenylated RNA complementary to a mouse retrovirus-like multigene family is rapidly and specifically induced by epidermal growth factor stimulation of quiescent cells. Proc Natl Acad Sci USA 79:7317–7321.

Ghosh-Dastidar P, Fox CF. 1983. Epidermal growth factor and epidermal growth factor receptor-dependent phosphorylation of a Mr = 34,000 protein substrate for $pp60^{src}$. J Biol Chem 258:2041–2044.

Kelly K, Cochran B, Stiles CD, Leder P. 1983. Cell-specific regulation of the c-*myc* gene by lymphocyte mitogens and platelet-derived growth factor. Cell 35:603–610.

Kornberg A. 1980. DNA Replication. W. H. Freeman, San Francisco.

Land H, Parada LF, Weinberg RA. 1983. Cellular oncogenes and multistep carcinogenesis. Science 222:771–778.

Leof EB, Wharton W, VanWyk JJ, Pledger WJ. 1982. Epidermal growth factor (EGF) and somatomedin C regulate G_1 progression in competent Balb/c-3T3 cells. Exp Cell Res 141:107–115.

Little CD, Nau MM, Carney DN, Gazdar AF, Minna JD. 1983. Amplification and expression of the c-*myc* oncogene in human lung cancer cell lines. Nature 306:194–196.

Mathews CK, North TW, Reddy GPV. 1979. Multienzyme complexes in DNA precursor biosynthesis. *In* Weber G (ed), Advances in Enzyme Regulation, Vol 17. Pergamon Press, Oxford and New York, pp. 133–156.

Moses HL, Proper JA, Volkenant ME, Wells DJ, Getz MJ. 1978. Mechanism of growth arrest of chemically transformed cells in culture. Cancer Res 38:2807–2812.

Noguchi H, Reddy GPV, Pardee AB. 1983. Rapid incorporation of label from ribonucleoside diphosphates into DNA by a cell-free high molecular weight fraction from animal cell nuclei. Cell 32:443–451.

Pardee AB, Dubrow R, Hamlin JL, Kletzien RF. 1978. Animal cell cycle. Annu Rev Biochem 47:715–750.

Pledger WJ, Stiles CD, Antoniades HN, Scher CD. 1978. An ordered sequence of events is required before BALB/c-3T3 cells become committed to DNA synthesis. Proc Natl Acad Sci USA 75:2839–2843.

Reddy GPV, Pardee AB. 1980. Multienzyme complex for metabolic channeling in mammalian DNA replication. Proc Natl Acad Sci USA 77:3312–3316.

Reddy GPV. 1982. Catalytic function of thymidylate synthase is confined to S phase due to its association with replitase. Biochem Biophys Res Commun 109:908–915.

Roberts AB, Frolik CA, Anzano MA, Sporn MB. 1983. Transforming growth factors from neoplastic and nonneoplastic tissues. Fed Proc 42:2621–2626.

Scher CD, Stone ME, Stiles CD. 1979. Platelet-derived growth factor prevents G_0 growth arrest. Nature 281:390–392.

Stiles CD. 1983. The molecular biology of platelet-derived growth factor. Cell 33:653–655.

Stiles CD, Isberg RR, Pledger WJ, Antoniades HN, Scher CD. 1979. Control of the BALB/c-3T3 cell cycle by nutrients and serum factors: Analysis using platelet-derived growth factor and platelet-poor plasma. J Cell Physiol 99:395–405.

Ushiro H, Cohen S. 1980. Identification of phosphotyrosine as a product of epidermal growth factor-activated protein kinase in A-431 cell membranes. J Biol Chem 255:8363–8365.

Waterfield MD, Scrace GT, Whittle N, Stroobant P, Johnsson A, Wasteson A, Westermark B, Heldin C-H, Huang JS, Deuel TF. 1983. Platelet-derived growth factor is structurally related to the putative transforming protein $p28^{sis}$ of simian sarcoma virus. Nature 304:35–39.

Wharton W, Leof E, Pledger WJ, O'Keefe EJ. 1982. Modulation of the epidermal growth factor receptor by platelet-derived growth factor and choleragen: Effects on mitogenesis. Proc Natl Acad Sci USA 79:5567–5571.

Mediators in Cell Growth and Differentiation, edited by Richard J. Ford and Abby L. Maizel.
Raven Press, New York © 1985.

Mechanisms Initiating Cellular Proliferation

Nancy E. Olashaw and W. J. Pledger

Department of Pharmacology and Cancer Cell Biology Program, Lineberger Cancer Research Center, The University of North Carolina, Chapel Hill, North Carolina 27514

The capacity of serum to promote the proliferation of nontransformed cells in culture has long been recognized, and many of the serum components responsible for growth have been identified. Remaining unanswered, however, is the question of how specific growth factors signal target cells to proliferate. The decreased dependency of transformed cells on serum or serum-derived factors for growth, a generally observed phenomenon, raises the corollary question of how such cells proliferate in the apparent absence of exogenous mitogenic signals. Studies defining the growth factor requirements of the nontransformed mouse embryonic cell line BALB/c-3T3 and of its spontaneously transformed counterpart ST3T3 have established a framework amenable to biochemical investigation of the mechanisms activating cellular proliferation. These studies identified platelet-derived growth factor (PDGF) as the serum component initiating the cell cycle traverse of G_0-arrested BALB/c-3T3 cells and demonstrated the capacity of ST3T3 cells to proliferate independently of this polypeptide. More recent reports have described several unique biochemical events modulated by PDGF during the transition of BALB/c-3T3 cells from the quiescent to the cycling state. These events suggest possible answers to the questions posed above and are the subject of this discussion.

NONTRANSFORMED CELLS

Identification of the intracellular processes modulated by a particular serum component requires prior knowledge of where in the proliferative cycle the factor in question is operative. Early work established the prereplicative phase (G_0/G_1) of the cell cycle as the primary site of growth factor action (Pardee 1978). Within this phase, two discrete subphases, each controlled by different serum components, have been defined for density-arrested BALB/c-3T3 cells. According to the model of growth control proposed by Pledger and co-workers (1977), PDGF, a potent serum mitogen stored in the α granules of platelets, initiates proliferation by rendering quiescent cells "competent" to respond to "progression" factors contained in platelet-poor plasma (PPP). In the presence of PPP, competent cells traverse G_0/G_1 and, after a minimum lag of 12 hr, enter S phase. Noncompetent cells do not proliferate in response to PPP, and PDGF-treated cells in the absence of PPP do not undergo progression. Thus, competence and progression represent two functionally distinct subphases of G_0/G_1. This dichotomy is reinforced by data demonstrating that competence factors (e.g., PDGF, fibroblast growth factor (FGF), macrophage-derived growth factor, and calcium phosphate) function weakly, at best, as progression factors and vice versa (Stiles et al. 1979a, Wharton et al. 1982).

Leof et al. (1982b) elaborated on this model by demonstrating that the progression activity of PPP could be efficiently replaced by a hormonally defined medium containing nanogram levels of epidermal growth factor (EGF) and somatomedin C. Insulin at supraphysiological concentrations could substitute for somatomedin C. Tactical use of this medium further showed that these PPP-derived polypeptides regulate different aspects of progression (Leof et al. 1983). EGF, in the presence of subnanogram amounts of somatomedin C, allowed competent cells to traverse the first 6 hr of G_0/G_1, whereas somatomedin C, at nanogram levels, mediated traversal of the last 6 hr of this phase and commitment to DNA synthesis. Completion of the remainder of the cell cycle occurred in unsupplemented medium (Wharton 1983). It should be pointed out that some cell types such as human fibroblasts produce somatomedin C (Clemmons et al. 1981).

The BALB/c-3T3 system, therefore, allows the actions of three distinct growth factors to be studied at defined times during the cell cycle of nontransformed fibroblasts. Because this discussion is concerned primarily with the mechanisms initiating growth, as opposed to those mediating growth, the events occurring during progression will not be further addressed. For a more detailed account of progression, the reader is referred to the recent review by O'Keefe and Pledger (1983).

Competence is both a dose- and time-dependent phenomenon. With optimal levels of highly purified PDGF, 2–3 hr is usually sufficient to render the majority of the density-arrested population responsive to PPP. In addition, PDGF-treated BALB/c-3T3 cells remain competent for several hours following the removal of PDGF from the medium. The stability of competence cannot be attributed to residual cell-associated PDGF, as (1) antibody to PDGF does not prevent the progression of competent cells and (2) the half-life of competence (10–20 hr) is significantly greater than that of cell-bound PDGF (50 min) (Singh et al. 1982, Heldin et al. 1982). Although PDGF binds to plastic (Smith et al. 1982), the small amount adherent to culture dishes containing confluent monolayers has been calculated to be insufficient to stimulate DNA synthesis (Singh et al. 1982). These findings argue against a persistent action of PDGF and suggest instead the existence of a stable, PDGF-induced cellular mediator. The observation that competence can be transferred from PDGF-treated cells to nontreated cells via fusion techniques further supports the concept of a PDGF-generated "second signal" (Smith and Stiles 1981). The biochemical nature of this putative signal is unknown, but recent reports suggest several intriguing possibilities (see below). Knowledge of the mechanisms initiating the proliferation of BALB/c-3T3 cells may also be applicable to other nontransformed cell types, as PDGF has been shown to be mitogenic for smooth muscle cells (Ross et al. 1974), Swiss 3T3 cells (Kohler and Lipton 1974), human fibroblasts (Scher et al. 1978), and glial cells (Heldin et al. 1977).

TRANSFORMED CELLS

The common denominator of cellular transformation and of growth factors such as PDGF is the capacity of both to increase the proliferative capacity of target cells. Numerous virally transformed cell lines are highly responsive to PPP, as are PDGF-treated BALB/c-3T3 cells, and the DNA virus SV40 has been shown to possess competence activity (Scher et al. 1978). These observations suggest a functional analogy between viral gene products and PDGF or PDGF-induced cellular mediators. ST3T3 cells, spontaneously transformed clonal variants of BALB/c-3T3 cells, resemble their SV40-transformed counterparts in several respects (Scher et al. 1982). Both cell lines induce tumors in nude mice, achieve high saturation densities, and grow equally well in medium containing serum or PPP. In contrast to SV40-transformed BALB/c-3T3 cells, ST3T3 cells exhibit density-dependent inhibition of growth; unlike their nontransformed parent, these cells reinitiate G_0/G_1 traverse in response to PPP. The origin of the ST3T3 cell line precludes viral activities as the basis of its abrogated PDGF requirement. Examples of how abnormal expression of cellular genes might alter growth factor dependency are discussed below with respect to recent studies relating PDGF to retroviral oncogenes and their cellular homologs.

Gene Products Resembling Growth Factors

Retroviral oncogenes are a functionally heterogeneous group of transduced cellular sequences (proto-oncogenes) responsible for the manifestation of the transformed phenotype (reviewed by Cooper 1982, Bishop 1983). In nontransformed cells, proto-oncogene expression is extremely limited, and the role of these potentially transforming genes in normal cellular proliferation is obscure. Many of the proteins encoded by viral and cellular oncogenes have been identified and isolated. Comparison of the sequence of $p28^{sis}$, the product of the simian sarcoma virus (SSV) transforming gene, v-*sis*, to that of purified PDGF revealed near identity of the two structures (Doolittle et al. 1983, Waterfield et al. 1983). Recent immunological studies confirm this finding (Robbins et al. 1983, Niman 1984). The functional similarity of these strikingly homologous polypeptides has also been described. Lysates prepared from SSV-transformed NIH-3T3 cells were shown to stimulate the proliferation of quiescent Swiss 3T3 cells in a dose-dependent manner comparable to that of purified PDGF (Deuel et al. 1983). Antibody to PDGF abolished the mitogenic response, thus demonstrating that the growth-promoting activity of the lysates resulted from the expression of a PDGF-like molecule. The protein immunoprecipitated from SSV-transformed cell lysates by this antibody had an apparent molecular weight of 20,000, identical to that of the putative processed product of $p28^{sis}$. Whether ST3T3 cells, like SSV-transformed cells, produce a protein structurally and functionally analogous to PDGF is unknown;

activation of the cellular *sis* gene (c-*sis*), however, suggests a possible basis for the abrogation of the PDGF requirement in some cell lines.

The capacity of transformed cells and their phylogenetic equivalents, nontransformed embryonic cells, to produce growth factor-like molecules is well established. The synthesis and secretion of autostimulatory substances collectively known as "transforming growth factors" (TGFs) has been documented for a variety of transformed cell lines (reviewed by Moses et al. 1984). For example, Kaplan et al. (1982) attributed the growth factor independence of MSV-transformed cells to the production by these cells of TGF-α, a class of TGFs capable of binding to EGF receptors, and further showed that medium conditioned by these cells effectively replaced the growth factor requirements of the nontransformed parent cells. Other studies demonstrated the ability of normal human embryonic fibroblasts to produce and respond to somatomedin C (Atkinson et al. 1980). Rechler et al. (1979) reported the synthesis and release of the somatomedin-like peptide, multiplication-stimulating activity, by fetal rat liver explants grown in organ culture. The production of a PDGF-like molecule by osteosarcoma cells (Heldin et al. 1980) and by embryonic and newborn human fibroblasts (Clemmons 1983) has also been described. In the presence of PPP, medium conditioned by these cells stimulated the proliferation of nontransformed glial cells (oseteosarcoma-conditioned medium) and of fibroblasts derived from older donors (fibroblast-conditioned medium), neither of which was capable of growth in PPP-supplemented medium alone.

Gene Products Mediating the Actions of Growth Factors

As stated above, PDGF presumably initiates proliferation via a cellular mediator. Recent data by Kelly et al. (1983) suggest that the product of a second cellular oncogene, c-*myc*, may function in this respect. Density-arrested BALB/c-3T3 cells briefly treated with PDGF manifested a 40-fold enhancement of c-*myc* mRNA as compared to untreated cells. The capacity of PDGF to induce the transcription of c-*myc* was directly proportional to its mitogenic activity. As postulated by Kelly et al. (1983), the temporal expression of this gene in response to PDGF may effect the reentry of quiescent cells into the proliferative cycle. Constitutive expression of c-*myc*, therefore, represents an alternative means by which ST3T3 cells might initiate traverse independently of PDGF; this hypothesis, however, has yet to be explored. In contrast, investigations addressing the abrogation of the requirement for macrophage and granulocyte inducer (MGI) by myeloid leukemia cells have identified several proteins induced by MGI in MGI-dependent normal myeloblasts but constitutively synthesized by leukemic cells (Lieberman et al. 1980).

In addition to its induction of the c-*myc* gene, PDGF has also been shown to preferentially enhance the expression of other cellular, presumably nononcogenic genes in BALB/c-3T3 cells. Thus, the products of these genes also represent potential cellular mediators of PDGF in this cell line, and in the ST3T3 variant, potential constitutively synthesized proliferative signals.

CELLULAR MEDIATORS

The search for potential protein mediators of PDGF was approached from a number of directions. Using molecular cloning techniques, Cochran et al. (1983) identified several unique PDGF-inducible gene sequences in BALB/c-3T3 cells. The messages for two of these sequences, termed JE and KC, were induced 10- to 20-fold within 1 hr of addition of PDGF to quiescent cells; the addition of inhibitors of protein synthesis did not preclude this response. The in vitro translation products of JE and KC mRNAs had apparent molecular weights of 10,000 and 19,000, respectively. Although the functions of these genes have yet to be determined, their rapid expression suggests an early activity in PDGF-mediated growth initiation.

Preferential transcription in response to PDGF has also been demonstrated by in vitro translation of cytoplasmic mRNA isolated from competent BALB/c-3T3 cells (Hendrickson and Scher 1983). Unique messages translating to proteins with molecular weights of 29,000, 55,000, and 35,000 were demonstrable within 30, 90, and 240 min, respectively, of exposure to PDGF. Concomitant treatment of cultures with PDGF and cycloheximide did not prevent the accumulation of the mRNA for the 29,000 MW protein but did inhibit the appearance of the messages encoding the 55,000 MW and 35,000 MW proteins. Thus, translation of the mRNA for the 29,000 MW protein may be required for the transcription or posttranscriptional processing of the mRNAs for the 55,000 and 35,000 MW proteins.

To detect proteins selectively synthesized in response to PDGF, we electrophoretically analyzed cellular extracts prepared from competent, [^{35}S]methionine-labeled BALB/c-3T3 cells. Five unique proteins, designated pI through pV, were detected by this method; the induction of these proteins by PDGF required de novo RNA synthesis (Figure 1). Although pV (70,000 MW), pIV (60,000 MW), and pIII (45,000 MW) have not as yet been characterized, pII (35,000 MW) has been identified as a secreted glycoprotein immunologically similar to the "major excreted protein" of virally transformed cells (Gottesman 1978, Scher et al. 1982). The production of pII may be necessary but is insufficient for competence formation; as reported by Scher et al. (1982), maximal pII synthesis is manifested at concentrations of PDGF suboptimal for the stimulation of DNA synthesis.

Of the proteins induced by PDGF, we have most extensively studied pI (29,000 MW); the growth-related properties of this protein were consistent with its putative role as the cellular mediator of PDGF. pI was related to competence in three key respects. First, the PDGF dose-response curves for pI induction and DNA synthesis were comparable (Pledger et al. 1981). Second, competence factors (PDGF, FGF, and calcium phosphate) stimulated pI production, whereas progression activities (PPP, EGF, and insulin) did not (Figure 2 and our unpublished data). Third, pI synthesis and competence formation were temporally correlated. Maximal pI production occurred between 2 and 4 hr after addition of PDGF to quiescent cells (Pledger et al. 1982, Olashaw and Pledger 1983a); as mentioned above, this length of time is sufficient to induce competence in most cells. pI synthesis then declined

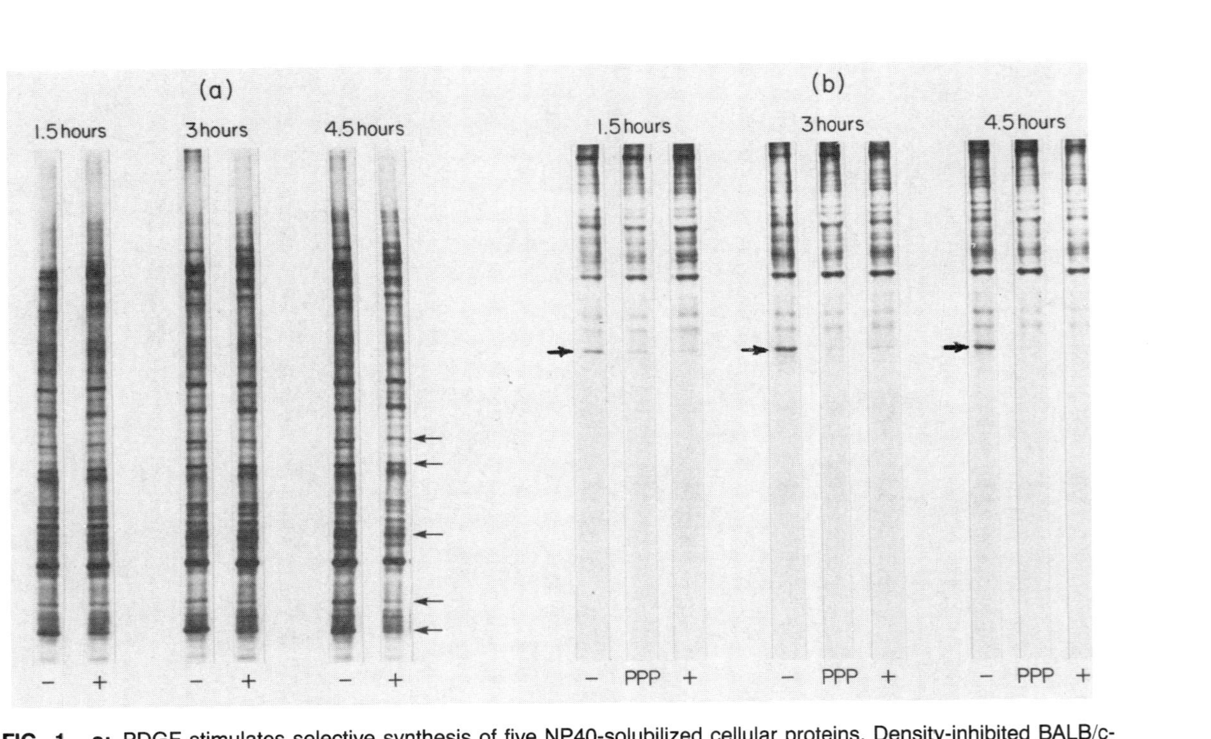

FIG. 1. a: PDGF stimulates selective synthesis of five NP40-solubilized cellular proteins. Density-inhibited BALB/c-3T3 were transferred to medium supplemented with either 5% PPP (+) or partially purified PDGF at 25 μg/ml (−). At the indicated times, [^{35}S]methionine was added and 20 min later the cultures were harvested for gel electrophoresis on a 6–18% exponential gradient. The arrows indicate proteins that were preferentially synthesized in response to PDGF. **b:** Actinomycin D inhibits PDGF-stimulated pI synthesis. Density-inhibited cells were treated with 25 μg/ml of partially purified PDGF (+ and −) or 5% PPP for the indicated times prior to addition of [^{35}S]methionine. The cultures were harvested 20 min later. Some PDGF-treated cultures (+) were incubated with actinomycin D at 5 μg/ml for 30 min prior to growth factor addition, whereas other PDGF-treated cultures (−) received no actinomycin D. Plasma-treated cultures received no actinomycin D. The arrow indicates pI. A 15% acrylamide gel was used.

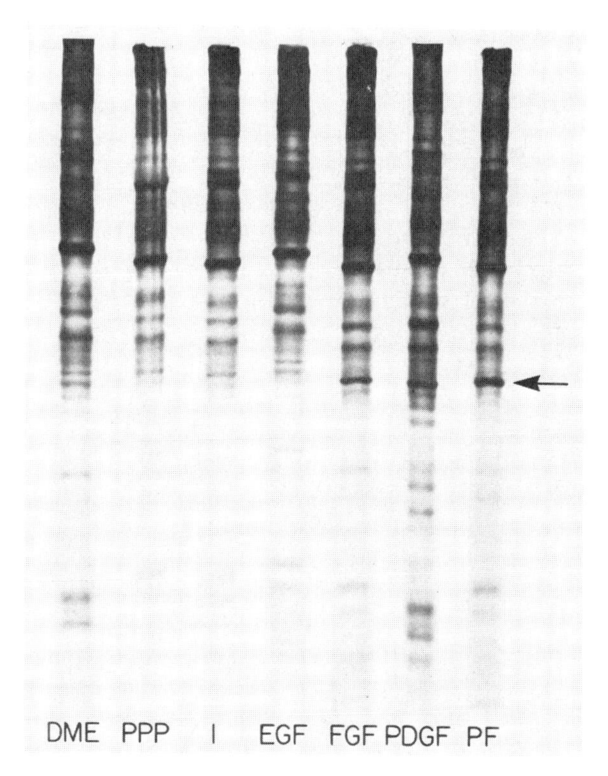

FIG. 2. Induction of pI specific for growth factors inducing competence. Density-arrested cells were refed with medium (Dulbecco's modified Eagle's (DME)) with no additions, 5% PPP, 5×10^{-6}M insulin (I), 50 ng/ml EGF, 50 ng/ml FGF, highly purified PDGF (PF), or pure PDGF (PDGF). After a 1-hr incubation, [^{35}S]methionine was added and cells were harvested 20 min later. NP40-solubilized cell preparations were analyzed electrophoretically on a 15% acrylamide gel. Arrow indicates pI.

rapidly and at 6 hr was no longer demonstrable. The cessation of pI synthesis did not result from an inadequate supply of PDGF, as cells receiving more PDGF at 6 hr did not synthesize pI.

Density-arrested BALB/c3T3 cells stimulated to proliferate in medium lacking essential amino acids or containing agents elevating cellular cAMP levels arrest at a distinct point in G_0/G_1 located 6 hr prior to S phase (Pledger et al. 1978, Stiles et al. 1979b, Leof et al. 1982b). Consistent with the time course described above, pI was not detected in cells released from these blocks into PDGF-supplemented medium (Pledger et al. 1982). pI synthesis during the remainder of the cell cycle has also been examined. BALB/c-3T3 cells synchronized at the G_1/S border by methotrexate were refed with medium containing PPP; PDGF was added at this time or at various times thereafter. Using a similar protocol, Scher et al. (1979) had previously demonstrated that treatment of cells in S or G_2 with PDGF allowed daughter cells to reinitiate DNA synthesis approximately 6 hr after mitosis. Cells receiving PPP alone completed the cell cycle but did not undergo a second round of replication. As the lag between mitosis and S phase in the PDGF-treated cells was similar to that of actively proliferating cells (6 hr), but less than that of stimulated G_0-arrested cells (12 hr), Scher et al. (1979) concluded that PDGF, in addition to promoting the exit of quiescent cells from G_0, prevented the entry of cycling cells into G_0. Administration of PDGF to cells in S or G_2, however, did not induce pI synthesis, thus indicating that the effects of this polypeptide on replicating cells are not mediated by pI (Pledger et al. 1982). Instead, pI may function exclusively to facilitate the reentry of G_0 cells into the proliferative cycle.

It is possible that proteins synthesized subsequent to pI prevent its induction by PDGF at later times in the cell cycle. This concept is supported by the aforementioned studies of Hendrickson and Scher (1983) in which the accumulation of a 29,000 MW translation product, perhaps identical to pI, was examined. These investigators demonstrated that although cycloheximide did not affect the appearance of the mRNA encoding this protein, it did inhibit the decline in the production of this mRNA normally occurring within 4 hr of addition of PDGF to cells. Implicit in these observations is the existence of a translation product, perhaps that of the cycloheximide-sensitive mRNA for the 55,000 MW or 35,000 MW protein, that represses the transcription of the mRNA for the 29,000 MW protein in a cell cycle-dependent manner.

In the experiments described above, the nonionic detergent Nonidet P-40 (NP40) was used to extract pI from whole cells. As this detergent primarily solubilizes nonnuclear components, pI was presumed to be a cytosolic or plasma membrane-associated protein. However, because some nuclear proteins are soluble in NP40, we reinvestigated the subcellular location of pI using a more rigorous means of fractionation. This protocol involved rupturing cells by mechanical homogenization, removal of intact nuclei by low-speed centrifugation, and purification of the crude nuclear pellet by centrifugation through sucrose (see legend to Figure 3 for details). Nuclei prepared in this manner were subsequently solubilized in NP40. Polyacrylamide gel electrophoresis of the nuclear and nonnuclear (the supernatant from the low-speed spin) samples revealed that pI was predominantly associated with nuclei; pI was present in the NP40-soluble but not insoluble nuclear fraction (Figure 3).

Experiments monitoring the nuclear accumulation of pI demonstrated that this protein was detectable in nuclei as early as 40 min after addition of PDGF to cells and continued to increase for up to 3 hr (Olashaw and Pledger 1983a). This time frame is compatible with that of competence formation and suggests that a nuclear function is an early event in growth initiation. Further evidence linking pI to competence was obtained from pulse/chase experiments assessing the stability of this protein. Following its induction by PDGF, pI was especially prominent in nuclei for up to 6 hr, declined between 9 and 12 hr, and was not observable at 24 hr. Thus, the stability of pI is greater than that of cell-bound PDGF ($t_{1/2} = 50$ min) and approximates that of the competence state ($t_{1/2} = 10$–20 hr). It should also be noted that pI localized in nuclear material did not subsequently accumulate in the nonnuclear fraction.

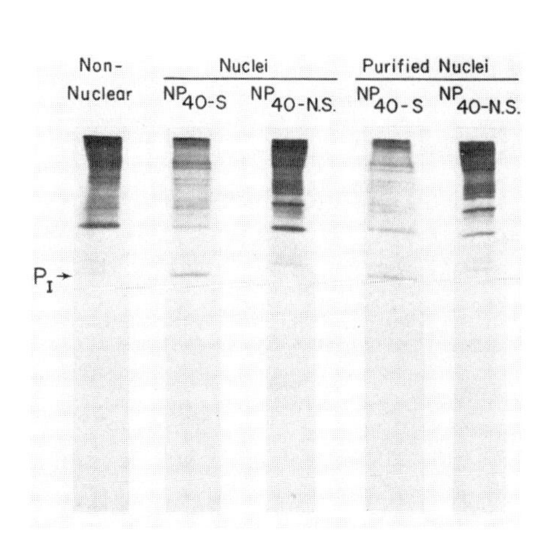

FIG. 3. Association of PDGF-induced protein (pI) with nuclear material. Density-arrested BALB/c-3T3 cells were refed with medium containing [^{35}S]methionine and 1 μg/ml PDGF. Following a 2.5-hr incubation, cells were collected, resuspended in reticulocyte saline buffer (RSB), and homogenized. Nuclei were separated from nonnuclear material by low-speed centrifugation. The pelleted nuclei were rinsed three times with RSB and an aliquot solubilizied in 1% NP40. The remainder of this crude nuclear material was resuspended in RSB containing 2 M sucrose, overlaid on 1.8 M sucrose, and centrifuged at 35,000 rpm for 30 min in an SW50.1 rotor. The pellet of purified nuclei was solubilized in 1% NP40 and samples were electrophoretically analyzed on 15% polyacrylamide gels. NP40-S, NP40-soluble material; NP40-NS, NP40-insoluble material.

pI synthesis has also been examined in ST3T3 cells. In these studies, density-arrested cells were refed with medium alone or medium containing PPP or PDGF. Regardless of treatment, the amount of pI synthesized by these cells was equal to or greater than that synthesized by PDGF-treated BALB/c-3T3 cells (Pledger et al. 1981, 1982). pI, therefore, appears to be constitutively produced in this spontaneously transformed cell line. A similar result was also observed for pII. The mechanisms allowing the constitutive synthesis of pI in ST3T3 cells are unknown; the inability of these cells to synthesize the putative "repressor" (see above), the production of a nonfunctional repressor, and the constitutive expression of the c-*sis* gene are among the possibilities.

The data presented above strongly support the causal involvement of pI in the growth initiation of both density-arrested BALB/c-3T3 and ST3T3 cells. Specific nuclear proteins capable of binding to DNA have been previously implicated in growth control (Mercer et al. 1982, Azizkhan and Klagsburn 1980, Sen and Todaro 1978). Preliminary data suggest that pI may also be associated with chromatin (Olashaw and Pledger 1983b); firm resolution of the subnuclear location of pI will provide a focus for future studies aimed at elucidating the means by which this putative proliferative signal increases cellular sensitivity to PPP.

INVOLVEMENT OF PROTEIN KINASE

Although the binding of PDGF to specific receptors on the cell surface is well documented, the events bridging this initial interaction of PDGF with target cells and the induction of unique gene products remain obscure (Heldin et al. 1981, 1982, Huang et al. 1981, Bowen-Pope and Ross 1982, Singh et al. 1982, Williams et al. 1982, Nilsson et al. 1983). Possible processes effecting the transfer of mitogenic information from the plasma membrane to the nucleus include changes in intracellular calcium concentration (Tucker et al. 1983), redistribution of vinculin from adhesion plaques to the perinuclear region (B. Herman, personal communication), and phosphorylation of cytoplasmic and membrane-associated proteins (Nishimura and Dueul 1983, Chambard et al. 1983, Ek et al. 1982, Ek and Heldin 1982, Nishimura et al. 1982, Pike et al. 1982, Harrington et al. 1983). These reactions are all extremely rapid, occurring within minutes of addition of PDGF to cells, and thus would allow the prompt induction of proteins such as pI.

Of the rapid responses elicited by PDGF in target cells, the phosphorylation of a 170,000–185,000 MW membrane protein is of particular interest for several reasons. First, the phosphorylation of this protein by PDGF-activated protein kinases occurred primarily on tyrosine residues (Ek and Heldin 1982, Nishimura et al. 1982, Pike et al. 1982, Harrington et al. 1983). In contrast, the 27,000 MW cytoplasmic protein and ribosomal protein S6, both of which are also phosphorylated by PDGF, contained only phosphoserine or phosphothreonine, or both (Nishimura and Dueul 1981, Chambard et al. 1982). Tyrosine-specific phosphorylating activity is characteristic of most membrane-associated protein kinases encoded by many retroviral oncogenes and has been correlated with retroviral induction of the transformed phenotype (Lau et al. 1981). Second, although definitive evidence is lacking, the 170,000–185,000 MW substrate of the PDGF-induced phosphorylation reaction is, most likely, the PDGF receptor (Pike et al. 1982, Harrington et al. 1983). The capacity of EGF, insulin, and somatomedin C to phosphorylate their receptors, also on tyrosine residues, has been previously described and attributed to the intrinsic kinase activity of the receptors themselves (Cohen et al. 1980, Shia et al. 1983, Jacobs et al. 1983). The spatial relationship of the PDGF receptor and the PDGF-induced kinase, however, has not been clearly resolved as yet. Third, in BALB/c-3T3 cells, the amount of putative PDGF receptor protein phosphorylated in response to PDGF was directly proportional to the mitogenic activity of this polypeptide (M. Harrington and W. J. Pledger, unpublished data). These observations suggest that the activation of a tyrosine-specific protein kinase and the consequent phosphorylation of the PDGF receptors or other as yet unidentified substrates, or both, may be fundamental to the subsequent changes in gene expression observed in PDGF-treated cells. Tyrosine phosphorylation may, in addition, represent a common means by which different growth factors and retroviruses initiate the chain of events leading to the proliferative response.

The linkage between tyrosine-specific phosphorylation and mitogenesis is supported by two recent studies addressing the structure and function of EGF receptors. Carlin et al. (1983) demonstrated that EGF receptors isolated from senescent EGF-insensitive WI-38 cells, in contrast to those obtained from young EGF-responsive cells, did not exhibit tyrosine-specific autophosphorylating activity. As Phillips et al. (1983) had previously reported the lack of effect of in vitro age on the binding and processing of EGF by WI-38 cells, the findings of Carlin et al. (1983) represent, to date, the only EGF-specific functional difference between young and old WI-38 cells. The loss of EGF receptor-kinase activity in senescent cells may, therefore, be causally related to the inability of these cells to respond proliferatively to EGF.

The second investigation, performed by Downward et al. (1984), revealed the close sequence homology of the v-*erb*-B oncogene product and the transmembrane, kinase-associated region of the EGF receptor. Peptides derived from the external, EGF-binding domain, in contrast, did not match sequences of the v-*erb*-B-encoded protein. As postulated by Downward et al. (1984), the truncated EGF receptor-like protein encoded by v-*erb*-B may function analogously to the ligand-bound, and thus activated, EGF receptor-kinase. Unlike the intact receptor, the v-*erb*-B-encoded protein would not be subject to control by EGF and thus could allow the generation of a continuous mitogenic signal. Although kinase activity has yet to be detected for the v-*erb*-B protein (Bishop 1983), this study suggests a third means by which transformation might abrogate growth factor requirements—by promoting the synthesis of proteins corresponding to the growth-stimulating regions of polypeptide receptors.

In summary, PDGF appears to initiate the proliferation of density-arrested BALB/c-3T3 cells by inducing the production of a stable cellular mediator capable of triggering G_0/G_1 traversal in response to PPP. Although several lines of evidence suggest that the nuclear protein pI may function in this respect, the biological activity and mechanism of action of this putative proliferative signal have yet to be determined. As competence formation presumably does not result from a single event, but instead involves a series of events, future studies must also define the functional interrelationships of pI and the other PDGF-inducible proteins described above. Furthermore, the attainment of a cohesive, overall picture of growth control requires the elucidation of the processes (e.g., tyrosine-specific phosphorylations) occurring prior to and culminating in the selective transcription of these proteins.

Although three mechanisms by which transformation might abrogate growth factor requirements were proposed, only one—the constitutive synthesis of PDGF-induced proteins—is, at this time, supported by experimental data in ST3T3 cells. As previously mentioned, this spontaneously transformed cell line produces both pI and pII in the absence of PDGF. However, it is likely that more than one genetic alteration is responsible for the increased responsiveness of this and other transformed cell lines to PPP. In conclusion, although the studies described above suggest a number of possible answers to the questions posed at the start of this discussion, a unifying hypothesis of growth regulation has yet to be completed.

ACKNOWLEDGMENT

This investigation was supported by grant CA24193, awarded by the National Cancer Institute, U.S. Department of Health and Human Services.

REFERENCES

Atkinson PR, Weidman ER, Bhaumick B, Bala RM. 1980. Release of somatomedin-like activity by cultured WI-38 human fibroblasts. Endocrinology 106:2002–2112.

Azizkhan JC, Klagsburn M. 1980. Chondrocytes contain a growth factor that is localized in the nucleus and is associated with chromatin. Proc Natl Acad Sci USA 77:2762–2766.

Bishop JM. 1983. Cellular oncogenes and retroviruses. Annu Rev Biochem 52:301–354.

Bowen-Pope DF, Ross R. 1982. Platelet-derived growth factor. II. Specific binding to cultured cells. J Biol Chem 257:5161–5171.
Carlin CR, Phillips PD, Knowles BB, Cristofalo VJ. 1983. Diminished in vitro tyrosine kinase activity of the EGF receptor of senescent human fibroblasts. Nature 306:617–619.
Chambard JC, Franchi A, Le Cam A, Pouysségur J. 1983. Growth factor-stimulated protein phosphorylation in G_0/G_1-arrested fibroblasts. Two distinct classes of growth factors with potentiating effects. J Biol Chem 258:1706–1713.
Clemmons DR. 1983. Age dependent production of a competence factor by human fibroblasts. J Cell Physiol 114:61–67.
Clemmons DR, Underwood LE, Van Wyk JJ. 1981. Hormonal control of immunoreactive somatomedin production by cultured human fibroblasts. J Clin Invest 67:10–19.
Cochran BH, Reffel AC, Stiles CD. 1983. Molecular cloning of gene sequences by platelet-derived growth factor. Cell 33:939–947.
Cohen S, Carpenter G, King L. 1980. Epidermal growth factor-receptor-protein kinase interactions. Co-purification of receptor and epidermal growth factor-enhanced phosphorylating activity. J Biol Chem 255:4834–4862.
Cooper GM. 1982. Cellular transforming genes. Science 218:801–806.
Doolittle RF, Hunkapiller MW, Hood LE, Devare SG, Robbins KC, Aaronson SA, Antoniades HN. 1983. Simian sarcoma virus onc gene, v-*sis*, is derived from the gene (or genes) encoding a platelet-derived growth factor. Science 221:275–277.
Downward J, Yarden Y, Mayes E, Scrace G, Totty N, Stockwell P, Ullrich A, Schlessinger J, Waterfield MD. 1984. Close similarity of epidermal growth factor receptor and v-*erb*-B oncogene protein sequences. Nature 307:521–527.
Dueul TF, Huang JS, Huang SS, Stroobant P, Waterfield MD. 1983. Expression of a platelet-derived growth factor-like protein in simian sarcoma virus transformed cells. Science 221:1348–1350.
Ek B, Heldin C-H. 1982. Characterization of a tyrosine-specific kinase activity in human fibroblast membranes stimulated by platelet-derived growth factor. J Biol Chem 257:10486–10492.
Ek B, Westermark B, Wasteson A, Heldin C-H. 1982. Stimulation of a tyrosine-specific phosphorylation by platelet-derived growth factor. Nature 295:419–420.
Gottesman MM. 1978. Transformation-dependent secretion of a low molecular weight protein by murine fibroblasts. Proc Natl Acad Sci USA 75:2767–2771.
Harrington MA, Leof EB, Pledger WJ. 1983. PDGF stimulation of a specific phosphorylation in BALB/3T3 cells. (Abstract) J Cell Biol 97:342a.
Heldin C-H, Wasteson A, Westermark B. 1977. Partial purification and characterization of platelet factors stimulating the proliferation of normal human glial cells. Exp Cell Res 109:429–437.
Heldin C-H, Westermark B, Wasteson A. 1980. Chemical and biological properties of a growth factor from human-cultured osteosarcoma cells: Resemblance with platelet-derived growth factor. J Cell Physiol 105:235–246.
Heldin C-H, Westermark B, Wasteson A. 1981. Specific receptors for platelet-derived growth factor on cells derived from connective tissue and glia. Proc Natl Acad Sci USA 78:3664–3668.
Heldin C-H, Wasteson A, Westermark B. 1982. Interaction of platelet-derived growth factor with its fibroblast receptor. Demonstration of ligand degradation and receptor modulation. J Biol Chem 257:4216–4221.
Hendrickson SL, Scher CD. 1983. Platelet-derived growth factor-modulated translatable mRNAs. Mol Cell Biol 3:1478–1487.
Huang JS, Huang SS, Kennedy B, Dueul TF. 1981. Platelet-derived growth factor. Specific binding to target cells. J Biol Chem 257:8130–8136.
Jacobs S, Kull F, Svoboda ME, Cuatrecasas P. 1983. Somatomedin-C stimulates the phosphorylation of the beta-subunit of its receptor. Endocrine Society Abstracts, 65th Annual Meeting, p. 312.
Kaplan PL, Anderson M, Ozanne B. 1982. Transforming growth factor(s) production enables cells to grow in the absence of serum: An autocrine system. Proc Natl Acad Sci USA 79:485–489.
Kelly K, Cochran BH, Stiles CD, Leder P. 1983. Cell-specific regulation of the c-*myc* gene by lymphocyte mitogens and platelet-derived growth factors. Cell 35:603–610.
Kohler N, Lipton A. 1974. Platelets as a source of fibroblast growth-promoting activity. Exp Cell Res 87:297–301.
Lau AF, Krzyzek RA, Faras AJ. 1981. Loss of tumorogenicity correlates with a reduction in $pp60^{src}$ kinase activity in a revertant subclone of avian sarcoma virus-infected field vole cells. Cell 23:815–823.

Leof EB, Wharton W, O'Keefe E, Pledger WJ. 1982a. Elevated intracellular concentrations of cyclic AMP inhibited serum-stimulated density-arrested BALB/c-3T3 cells in mid G_1. J Cell Biochem 19:93–103.

Leof EB, Wharton W, Van Wyk JJ, Pledger WJ. 1982b. Epidermal growth factor (EGF) and somatomedin C regulate G_1 progression in competent BALB/c-3T3 cells. Exp Cell Res 141:107–115.

Leof EB, Van Wyk JJ, O'Keefe EJ, Pledger WJ. 1983. Epidermal growth factor is required only during the traverse of early G_1 in PDGF-stimulated density-arrested BALB/c-3T3 fibroblasts. Exp Cell Res 147:202–208.

Liebermann D, Hoffman-Lieberman B, Sachs L. 1980. Molecular dissection of differentiation in normal and leukemic myeloblasts: separately programmed pathways of gene expression. Dev Biol 79:46–63.

Mercer WE, Nelson D, Deleo AB, Old LJ, Baserga R. 1982. Microinjection of monoclonal antibody to protein p53 inhibits serum-induced DNA synthesis in 3T3 cells. Proc Natl Acad Sci USA 79:6309–6312.

Moses HL, Childs CB, Halper J, Shipley GD, Tucker RF. 1984. Role of transforming growth factors in neoplastic transformation. *In* Veneziale C (ed), Control of Cell Growth and Proliferation, Van Nostrand Reinhold, New York, in press.

Nilsson J, Thyberg H, Heldin C-H, Westermark B, Wasteson A. 1983. Surface binding and internalization of platelet-derived growth factor in human fibroblasts. Proc Natl Acad Sci USA 80:5592–5596.

Niman HL. 1984. Antisera to a synthetic peptide of the sis viral oncogene product recognize human platelet-derived growth factor. Nature 307:180–183.

Nishimura J, Dueul TF. 1983. Platelet-derived growth factor stimulates the phosphorylation of ribosomal protein S. FEBS Lett 156:130–134.

Nishimura J, Huang JS, Dueul TF. 1982. Platelet-derived growth factor stimulates tyrosine-specific protein kinase activity in Swiss mouse 3T3 cell membranes. Proc Natl Acad Sci USA 79:4303–4307.

O'Keefe EJ, Pledger WJ. 1983. A model of cell cycle control: Sequential events regulated by growth factors. Mol Cell Endocrinol 31:167–186.

Olashaw NE, Pledger WJ. 1983a. Association of platelet-derived growth factor-induced protein with nuclear material. Nature 306:272–274.

Olashaw NE, Pledger WJ. 1983b. Platelet-derived growth factor induced protein is associated with nuclei. (Abstract) J Cell Biol 97:342a.

Pardee AB, Dubrow RD, Hamlin JL, Kletzien RE. 1978. Animal cell cycle. Annu Rev Biochem 47:715–750.

Pike LJ, Bowen-Pope DF, Ross R, Krebs EG. 1983. Characterization of a platelet-derived growth factor-stimulated phosphorylation in cell membranes. J Biol Chem 258:9383–9390.

Phillips PD, Kuhnle E, Cristofalo VJ. 1983. [^{125}I]EGF binding is stable throughout the replicative lifespan of WI-38 cells. J Cell Physiol 114:311–316.

Pledger WJ, Stiles CD, Antoniades HN, Scher CD. 1977. Induction of DNA synthesis in BALB/c-3T3 cells by serum components: Reevaluation of the commitment process. Proc Natl Acad Sci USA 74:4481–4485.

Pledger WJ, Stiles CD, Antoniades HN, Scher CD. 1978. An ordered sequence of events is required before BALB/c-3T3 cells become committed to DNA synthesis. Proc Natl Acad Sci USA 75:2839–2843.

Pledger WJ, Hart CA, Locatell KL, Scher CD. 1981. Platelet-derived growth factor-modulated proteins: Constitutive synthesis of a transformed cell line. Proc Natl Acad Sci USA 78:4358–4362.

Pledger WJ, Howe PH, Leof EB. 1982. The regulation of cell proliferation by serum growth factors. Ann NY Acad Sci 397:1–10.

Rechler MM, Eisen HJ, Higa OZ, Nissley SP, Moses AC, Fenoy I, Bruni CB, Phillips LS, Baird KL. 1979. Characterization of a somatomedin (insulin-like growth factor) synthesized by fetal rat liver organ culture. J Biol Chem 254:7942–7950.

Robbins KC, Antoniades HN, Devare SG, Hunkapiller MW, Aaronson SA. 1983. Structural and immunological similarities between simian sarcoma virus gene product(s) and human platelet-derived growth factor. Nature 305:605–608.

Ross R, Glomset J, Kariya B, Harker L. 1974. A platelet-dependent serum factor that stimulates the proliferation of arterial smooth muscle cells in vivo. Proc Natl Acad Sci USA 71:1207–1210.

Scher CD, Pledger WJ, Martin P, Antoniades HN, Stiles CD. 1978. Transforming viruses directly reduce the cellular growth requirement for a platelet-derived growth factor. J Cell Physiol 97:371–380.

Scher CD, Stone ME, Stiles CD. 1979. Platelet-derived growth factor prevents G_0 arrest. Nature 281:390–392.

Scher CD, Hendrickson SL, Whipple AP, Gottesman MM, Pledger WJ. 1982. Constitutive synthesis of platelet-derived growth factor modulated proteins by a tumorigenic cell line. *In* Cold Spring Harbor Conference on Cell Proliferation, Vol 9. Cold Spring Harbor Press, Cold Spring Harbor, New York, pp. 289–304.

Scher CD, Dick RI, Whipple AP, Locatell KL. 1983. Identification of a BALB/c-3T3 cell protein modulated by platelet-derived growth factor. Mol Cell Biol 3:70–81.

Sen A, Todaro GJ. 1978. Species specific cellular DNA-binding proteins expressed in mouse cells transformed by chemical carcinogens. Proc Natl Acad Sci USA 75:1647–1651.

Shia MA, Rubin JB, Pilch PF. 1983. The insulin receptor protein kinase. Physiochemical requirements for activity. J Biol Chem 258:14450–14455.

Singh JP, Chaiken MA, Pledger WJ, Scher CD, Stiles CD. 1982. Persistence of the mitogenic response to platelet-derived growth factor does not reflect a long-term interaction between the growth factor and the target cell. J Cell Biol 96:1457–1502.

Smith JC, Stiles CD. 1981. Cytoplasmic transfer of the mitogenic response to platelet-derived growth factor. Proc Natl Acad Sci USA 78:4363–4367.

Smith JC, Singh JP, Lillquist JS, Goon DS, Stiles CD. 1982. Growth factors adherent to cell substrata are mitogenically active in situ. Nature 296:154–156.

Stiles CD, Capone GT, Scher CD, Antoniades HN, Van Wyk JJ, Pledger WJ. 1979a. Dual control of cell growth by somatomedins and platelet-derived growth factor. Proc Natl Acad Sci USA 76:1279–1283.

Stiles CD, Isberg RR, Pledger WJ, Antoniades HN, Scher CD. 1979b. Control of the BALB/c-3T3 cell cycle by nutrients and serum factors: Analysis using platelet-derived growth factor and platelet-poor plasma. J Cell Physiol 99:395–406.

Tucker RW, Snowdowne KW, Borle AB. 1983. Platelet derived growth factor produces transient increases in the intracellular concentration of free calcium in BALB/c-3T3 cells. (Abstract) J Cell Biol 97:343a.

Waterfield MD, Scrace T, Whittle N, Stroobant P, Johnsson A, Wasteson A, Westermark B, Heldin C-H, Huang JS, Dueul TF. 1983. Platelet-derived growth factor is structurally related to the putative transforming protein $p28^{sis}$ of simian sarcoma virus. Nature 304:35–39.

Wharton W. 1983. Hormonal regulation of discrete portions of the cell cycle: Commitment to DNA synthesis is commitment to cellular division. J Cell Physiol 117:423–429.

Wharton W, Gillespie GY, Russell SW, Pledger WJ. 1982. Mitogenic activity elaborated by macrophage-like cell lines acts as competence factor(s) for BALB/c-3T3 cells. J Cell Physiol 110:93–100.

Williams LT, Tremble P, Antoniades HN. 1982. Platelet-derived growth factor binds specifically to receptors on vascular smooth muscle cells and the binding becomes nondissociable. Proc Natl Acad Sci USA 79:5867–5870.

Mediators in Cell Growth and Differentiation,
edited by Richard J. Ford and Abby L. Maizel.
Raven Press, New York © 1985.

Chromosome Condensation and Decondensation Factors in the Life Cycle of Eukaryotic Cells

Potu N. Rao and Ramesh C. Adlakha

Department of Chemotherapy Research, The University of Texas M. D. Anderson Hospital and Tumor Institute at Houston, Houston, Texas 77030

Over two decades ago, while trying to explain the mechanisms involved in the control of DNA synthesis in eukaryotic cells, Mazia (1963) suggested that the condensation state of chromatin might have a role in the replication of DNA. He further speculated that the progress of chromosome decondensation and the earliest stages of condensation that cannot be visualized under the microscope might extend far into interphase. According to this proposal, chromosome decondensation extends throughout G_1 phase to a critical point when DNA synthesis is initiated, and condensation begins again as soon as the replication is completed. His speculation regarding the chromosome condensation cycle within the cell cycle became prophetic. With the advent of new techniques, the chromosome condensation cycle has been firmly established on the basis of biochemical, biophysical, and cytological evidence (Pederson 1972, Pederson and Robbins 1972, Sperling and Rao 1974, Schor et al. 1975, Hildebrand and Tobey 1975, Nicolini et al. 1975, Moser et al. 1975, 1981, Rao et al. 1977, Hittelman and Rao 1978, Rao and Hanks 1980, Adlakha et al. 1983).

Studies on the openness of chromatin and hence its accessibility to intercalating drugs such as actinomycin D or to pancreatic DNase I revealed that the amount of the drug bound per unit of DNA or its sensitivity to the enzyme gradually increased throughout G_1, reached a maximum level by early to mid-S phase, and declined during G_2, reaching the basal value at mitosis when the chromatin was highly condensed in the form of chromosomes (Pederson and Robbins 1972, Pederson 1972). Thus, these probes reveal the conformational changes in the structure of chromatin that are closely related to the progression of cells through the division cycle. The accessibility of DNA to intercalating drugs and its sensitivity to DNase are indicative of the degree of chromatin condensation.

These results suggest chromatin gradually decondenses during G_1 and S, and condenses during G_2 and mitosis. Nicolini et al. (1975), who studied the circular dichroism spectra and ethidium bromide-binding capacity of the isolated chromatin of HeLa cells as a function of time after mitosis, observed that the positive ellipticity at 272 nm increased from 3380 degrees cm^2/dmol in G_1 to 5350 degrees cm^2/dmol in S, and decreased to a minimum of 1900 degrees cm^2/dmol in mitosis. Similar cyclical changes in the ethidium bromide-binding capacity of chromatin were observed during the HeLa cell cycle (Nicolini et al. 1975).

Nuclear magnetic reasonance (NMR) relaxation times (T_1) of the intracellular water protons and water content were measured in synchronized HeLa cells by Beall et al. (1976). Mitotic cells had a mean maximum T_1 value of 1020 msec $\pm$ 84 msec. The T_1 value continued to decrease throughout G_1 and reached its mean minimum value of 534 $\pm$ 43 msec in S phase. As cells progressed through G_2, T_1 started to increase from 621 $\pm$ 25 to 690 $\pm$ 4 msec and ultimately returned to the maximum as cells reentered mitosis. The cyclical pattern of T_1 appears to be inversely related to the degree of chromatin condensation, as revealed by the amount of actinomycin D bound to DNA during the cell cycle. Since the NMR data seem to be related to the surface area of chromatin that is available to affect the motional freedom of water, and since chromatin is a sizeable fraction (30%) of the macromolecular component of the cell, it is quite likely that not only the amount of macromolecules but also their conformational state play an important role in the variations of T_1.

Direct visualization of the chromosome condensation cycle has been made possible by the discovery of the phenomenon of premature chromosome condensation (Johnson and Rao 1970). The technique is based on the fact that fusion by UV-inactivated Sendai virus or polyethylene glycol (PEG) of a mitotic cell with an interphase cell results in premature chromosome condensation of the interphase nucleus. The morphology of the prematurely condensed chromosomes (PCC) indicates the position of a cell in the cell cycle at the time of fusion (Johnson and Rao 1970). G_1 PCC consist of single chromatids that vary considerably in length and thickness, depending on the degree of advancement of the cell toward S phase before fusion. The closer a G_1 cell gets to S phase, the more extended are its PCC (Hittelman and Rao 1976, 1978, Rao et al. 1977). Immediately before entry into S phase, individual chromosomes are no longer discernible, as the PCC take on a highly diffuse appearance.

The increasing amount of decondensation in PCC as cells progress through G_1 reflects intranuclear changes in the chromatin organization and thus provides direct visual evidence for the chromosome condensation cycle. PCC of S-phase cells exhibit a characteristic "pulverized" appearance attributable to the presence of both condensed and diffuse regions. Active sites of replication in S-phase PCC appear as gaps under the light microscope (Sperling and Rao 1974). PCC from G_2 cells resemble extended prophase chromosomes. The distinctly large heterochromatic X chromosomes of *Microtus agrestis* proved to be the ideal material to demonstrate the progressive condensation of chromatin during G_2. Sperling and Rao (1974) measured the lengths of the long arms of X chromosomes in mitotic chromosomes and the PCC of cells at various points in G_2. The average lengths were 48 μm, 39.6 μm, and 16.4 μm in early, mid, and late G_2-phase PCC, respectively, as compared to 11.3 μm in mitotic chromosomes. The progressive decrease in the accessibility of DNA to actinomycin D during G_2 reported by Pederson (1972) correlates well with progressive condensation and shortening of the G_2-phase PCC (Sperling and Rao 1974).

Even though subtle changes take place in the condensation of chromatin throughout the cell cycle, the most precipitous events of condensation and decondensation of chromatin occur at the G_2–M and M–G_1 transitions, respectively. What is the molecular mechanism for the chromosome condensation cycle? In this chapter we will examine the role of mitotic factors and the inhibitors of mitotic factors (IMF), which we have discovered recently, in the regulation of the chromosome condensation cycle and try to elucidate the molecular mechanism that drives this cycle.

THE CHROMOSOME CONDENSATION FACTORS

The fact that an interphase nucleus is transformed into chromosomes immediately following fusion between a mitotic and an interphase cell suggests that there are chromosome condensation factors in mitotic cells. Extracts of mitotic cells, when injected into *Xenopus laevis* oocytes, can also induce meiotic maturation characterized by germinal vesicle breakdown (GVBD) and condensation of chromosomes (Sunkara et al. 1979a,b, Nelkin et al. 1980, Kishimoto et al. 1982, Weintraub et al. 1983). Whether the maturation-promoting activity (MPA) and the chromosome condensing capacity of the mitotic factors are due to one and the same molecular species remains to be elucidated. However, the kinetics of accumulation of MPA and chromosome condensation factors during the cell cycle go hand in hand. MPA is absent in the extracts of HeLa cells synchronized in G_1 or S phase (Sunkara et al. 1979a). MPA accumulates slowly in the beginning of G_2 but at a progressively more rapid rate during late G_2 and reaches a threshold at the G_2-mitotic transition, when the nuclear membrane breaks down and chromatin condenses into chromosomes (Figure 1). Extracts from HeLa cells irreversibly blocked in G_2 by exposure of cells to an alkylating agent (Al-Bader et al. 1978) do not exhibit MPA (Sunkara et al. 1979a).

Using the amphibian oocyte system as a bioassay, we tried to characterize the mitotic factors from the extracts of HeLa cells synchronized in mitosis. These studies revealed that the mitotic factors were heat- and Ca^{2+}-sensitive, Mg^{2+}-dependent, high-molecular-weight proteins (Sunkara et al. 1979b). Differential extraction of proteins from cytoplasm and chromosomes of mitotic cells indicated that the MPA was present both in the cytoplasm and on the chromosomes, but the specific activity of the chromosomal fraction was at least three times greater than that of the cytoplasmic fraction (Table 1), suggesting that a major portion of the mitotic factors is localized on the metaphase chromosomes (Adlakha et al. 1982a). Differential extraction of proteins from the cytoplasm and the nuclei of early- and mid-G_2-phase HeLa cells and their injection into oocytes indicated that MPA was present in the nuclear fractions but not in the cytoplasmic fractions. However, in late G_2-phase cells, both the cytoplasmic and nuclear extracts exhibited MPA (Table 2). These data tend to support the conclusion that the mitotic factors preferentially

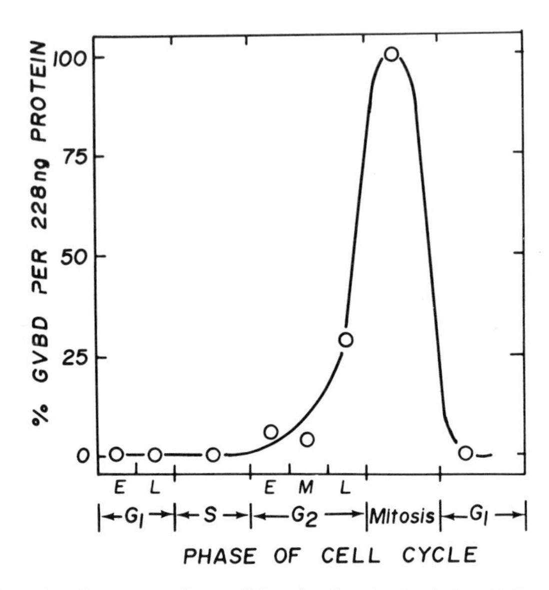

FIG. 1. Meiotic maturation-promoting activity of cell extracts during HeLa cell cycle. The data for this graph are derived from Table 1. Because 228 ng of mitotic protein induced GVBD in 100% of the cases, the percentage activity for other phases of the cell cycle was normalized to that amount of protein. E, early; M, mid; and L, late. (Reproduced from Sunkara et al. 1979a.)

TABLE 1. *Maturation-promoting activities of cytoplasmic and chromosomal fractions of mitotic HeLa cells*

Oocytes injected with:	Extract dilution	Protein injected* (ng)	No. oocytes injected	No. oocytes showing GVBD	GVBD induction (%)
Cytoplasmic fraction I	0	585	30	30	100
	1/2	292	25	25	100
	1/3	195	25	25	100
	1/4	146	25	7	28
	1/8	73	25	0	0
Cytoplasmic fraction II	0	69	25	0	0
Chromosomal fraction	0	179	30	30	100
	1/2	90	25	25	100
	1/3	60	25	23	92
	1/4	45	20	5	25
	1/8	22	25	0	0

*A total volume of 65 nl of the extracts was injected into each oocyte, and oocytes were scored for GVBD 2–3 hr after injection. A concentration of 40 × 10^6 cells/ml was used for preparation of these extracts. (Reproduced from Adlakha et al. 1982a, with permission of the Company of Biologists Ltd.)

TABLE 2. *Maturation-promoting activity of the cytoplasmic and nuclear extracts from early-, mid-, and late- G_2 HeLa cells*

Oocytes injected with:*	Protein injected (ng)	No. oocytes injected	No. oocytes showing GVBD	GVBD induction (%)
Early G_2				
Cytoplasmic extract	686	15	0	0
Nuclear extract	201	15	10	67
Mid G_2				
Cytoplasmic extract	721	25	0	0
Nuclear extract	218	25	23	92
Late G_2				
Cytoplasmic extract	545	15	12	80
Nuclear extract	176	15	13	87
Buffer alone	—	10	0	0

*A volume of 65 nl of the extracts was injected into each oocyte and GVBD was scored 2–3 hr after injections. In some cases GVBD was scored after overnight incubation. 40×10^6 to 60×10^6 cells/ml were used for preparation of these extracts. (Reproduced from Adlakha et al. 1982a, with permission of the Company of Biologists Ltd.)

bind to chromatin as soon as they are synthesized, and since the cell synthesizes more of these factors in preparation for mitosis, increasing amounts of them are retained in the cytoplasm (Adlakha et al. 1982a). The chromosome-bound mitotic factors can be released by digestion of metaphase chromosomes with micrococcal nuclease or DNase II (Adlakha et al. 1982b). The fact that even after extensive digestion of the chromosomes with these nucleases mitotic factors remain bound to the chromosomes implies that the mitotic factors are probably bound to the DNA in the core particle. It is not clear, however, at what specific sites and by what forces these factors are bound to chromatin.

Recognition of Mitosis-Specific Proteins by Monoclonal Antibodies

Using the extracts from HeLa cells synchronized in mitosis as antigens, we obtained two hybridoma clones by following the standard procedures originally described by Kohler and Milstein (1975). These two clones, viz, MPM-1 and MPM-2, secrete antibodies that specifically reacted with the mitotic cells of all the species tested, which included cells of human, rodent, bird, insect, nematode, and plant origin (Davis et al. 1983). Chromosomes and cytoplasm in mitotic cells reacted with the antibodies as detected by indirect immunofluorescence. Measurement of the fluorescence intensity from individual cells in populations synchronized in various phases of the cell cycle revealed that the immunoreactivity was tightly coupled to mitosis (Figure 2). The peak intensity of fluorescence was found in metaphase cells, and it was at least 10 times greater than that from G_1 or S-phase cells. The antigenic reactivity detected by indirect immunofluorescence was first observed in the nuclei of G_2 cells and rapidly increased in mitosis. This pattern is

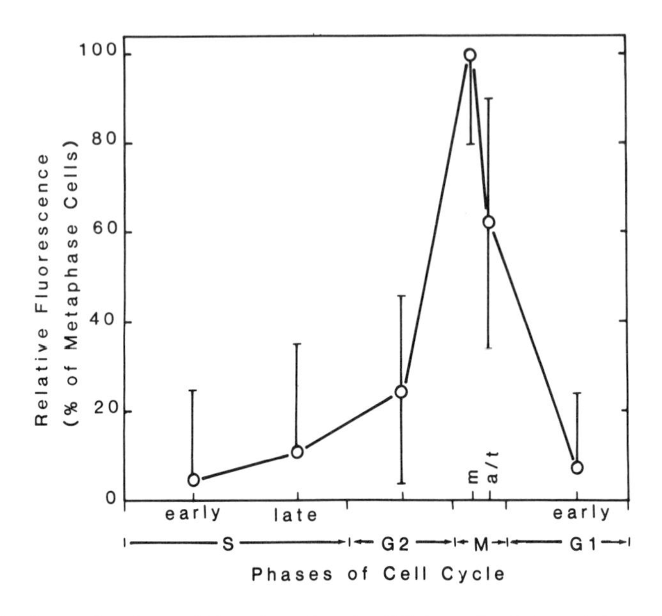

FIG. 2. Relative antigenic reactivity of synchronized HeLa cells. Cytocentrifuge preparations of synchronized HeLa cells were stained by indirect immunofluorescence with antibody MPM-1. The fluorescence intensity from 25 cells from each population was measured with a Leitz MPV microscope photometer. The fluorescence from blank areas on the slide adjacent to the cells was subtracted from each measurement. The fluorescence intensity from the nonmetaphase populations was normalized to that from 25 metaphase cells included on the same slide. The mean fluorescence intensities were plotted, and the bars represent the standard deviation. Metaphase cells (m) fluoresced more intensely than anaphase-telophase cells (a/t), and mitotic cells fluoresced more intensely than interphase cells. (Reproduced from Davis et al. 1983.)

very similar to the accumulation of MPA shown in Figure 1. There was a significant decrease in the intensity of fluorescence as cells progressed into anaphase/telophase, and it practically disappeared as the cells entered G_1. These two sets of data clearly suggest that mitosis-specific proteins that exhibit MPA and react with the monoclonal antibodies begin to accumulate from early G_2, first in the nucleus and then in the cytoplasm, and reach their maximum level at mitosis. At the end of mitosis these factors are either degraded or modified in such a way that they lose MPA and are no longer recognized by the antibodies.

When the polypeptides from mitotic HeLa cell extracts were separated by electrophoresis on sodium dodecyl sulfate–polyacrylamide gels, electrophoretically transferred to nitrocellulose paper, incubated with antibodies to mitotic cells, and stained by the immunoperoxidase method, we observed a family of at least 16 polypeptide bands with an apparent molecular mass range of 40 to 200 kDa (Davis et al. 1983). Three major bands of 182 kDa, 118 kDa, and 70 kDa were recognized by both the antibodies. The antigens recognized by these antibodies were all phosphoproteins, and their antigenic reactivity was lost when the phosphate groups were removed by digestion with alkaline phosphatase. Therefore, it is likely that these antibodies recognize a common or similar phosphorylated site that is shared by this subset of mitosis-specific proteins. These data suggest that the condensation of chromosomes and the initiation of mitosis may be regulated through the phosphorylation of a specific subset of nonhistone proteins (Davis et al. 1983; for a review, see Adlakha et al. 1984a).

Association of Mitotic Factors with Interphase Chromatin during Premature Chromosome Condensation

Does premature chromosome condensation result from the direct binding or association of mitotic proteins with interphase chromatin or from the modification, i.e., phosphorylation, acetylation etc., of the proteins that are constituents of the interphase chromatin? To answer this question, we labeled HeLa cells synchronized in G_2 phase with a mixture of ^3H-labeled amino acids for 3 hr, after which the radioactive medium was replaced with regular medium containing Colcemid to block cells in mitosis. These prelabled mitotic cells were fused with unlabeled interphase cells, chromosome spreads were made, and the slides were processed for autoradiography. We observed that label was associated not only with metaphase chromosomes but also with the PCC (Rao and Johnson 1974). Blockage of protein synthesis by the addition of cycloheximide (25 μg/ml) to the fusion mixture had no effect on either the induction of PCC or the association of labeled mitotic protein with the PCC. These data suggest that premature chromosome condensation is probably caused by the direct binding of the mitotic proteins to interphase chromatin.

This conclusion has received further support from our recent studies. Antibodies raised in rabbits to extracts of mitotic HeLa cells reacted specifically with mitotic chromosomes from cells of human origin but not from those of other species (Davis and Rao 1982). HeLa or Chinese hamster ovary (CHO) nuclei undergoing premature chromosome condensation after fusion of interphase cells with mitotic HeLa cells became antigenically reactive. In contrast, the PCC resulting from the fusion of interphase HeLa with mitotic CHO cells did not react with these antibodies. Antigenic reactivity apparently resides in the protein component of the DNA-protein complex, since HeLa mitotic cells can induce antigenic reactivity to the PCC of interphase CHO cells to which they are fused, whereas the converse is not true for mitotic CHO cells.

These data are consistent with the hypothesis that the antibodies recognize a species-specific determinant on a protein associated with chromosome condensation and that this determinant is available for antibody binding only when the protein is chromatin bound. Furthermore, this protein becomes associated with chromatin undergoing condensation in such a manner that the antigenic determinant becomes available for antibody binding. These studies seem to suggest that specific binding of mitotic factors to chromatin is a prerequisite for the condensation of chromosomes during mitosis. The association between the mitotic factors and the chromatin begins after DNA replication, when chromosome condensation is initiated, but reaches a supersaturation level at the G_2-mitosis transition, when it triggers dramatic changes in the degree of chromosome condensation.

THE CHROMOSOME DECONDENSATION FACTORS

If the mitotic factors bind to chromatin to bring about chromosome condensation, do they dissociate from chromosomes or become inactivated at the mitosis-G_1 transition?

The Fate of Mitotic Factors during Mitosis-G_1 Transition

If the chromosome decondensation that occurs in G_1 phase is due to the presence of some factors, do they have any effect on the chromosome condensation factors present in mitotic cells? Mitotic cell extracts were mixed with extracts from G_1-phase HeLa cells in different proportions, and the mixtures were injected into *Xenopus* oocytes to determine the effect of G_1 cell extracts on the MPA of the mitotic cell extracts (Adlakha et al. 1983). Mitotic cell extracts diluted with the extraction buffer in corresponding proportions served as control. No significant loss of MPA was seen until the concentration of the mitotic cell extract was lowered to 20% by dilution with the extraction buffer (Figure 3). In contrast, the mitotic cell extract lost its MPA at a concentration of 66.6% when diluted with G_1 cell extracts with equal protein content, and complete inactivation of the mitotic cell extract occurred as its concentration was reduced to 50%. The results obtained after the dilution of mitotic cell extracts with the extraction buffer containing bovine serum albumin (BSA) as a carrier protein were identical with those obtained by using buffer alone for dilution.

These data suggest that G_1-phase cell extracts have specific factors that inactivate MPA in mitotic cell extracts; these are called inhibitors of mitotic factors (IMF). Extracts from HeLa G_1 cells collected at any time from 2 to 7 hr after reversal of the N_2O blockade exhibited strong IMF activity. However, cells collected at later times were less active (Figure 3). Furthermore, since IMF are detectable as early as 1.5 hr after reversal of the N_2O block, when 10–15% of the cells have yet to complete mitosis, it is likely that IMF are either activated or newly synthesized at the end of mitosis, i.e., at telophase.

Similar studies were performed to determine whether the IMF are present during other phases of the cell cycle. These studies revealed that extracts from early S-phase cells were more effective in neutralizing the activity of the mitotic factors than those from cells either in mid or late S phase (Figure 4). No significant inhibitory activity was observed in late S-phase cell extracts even when the S-phase cell protein–mitotic cell protein ratio was increased to 5:1. Extracts from early-, mid-, and late-G_2-phase HeLa cells also had no inhibitory effect on the mitotic factors.

Activation of IMF in Quiescent (G_0) Human Diploid Fibroblasts (WI-38)

To answer the question of whether noncycling cells arrested in G_1 or G_0 contain IMF, extracts were made from quiescent WI-38 cells at 7 days after they had

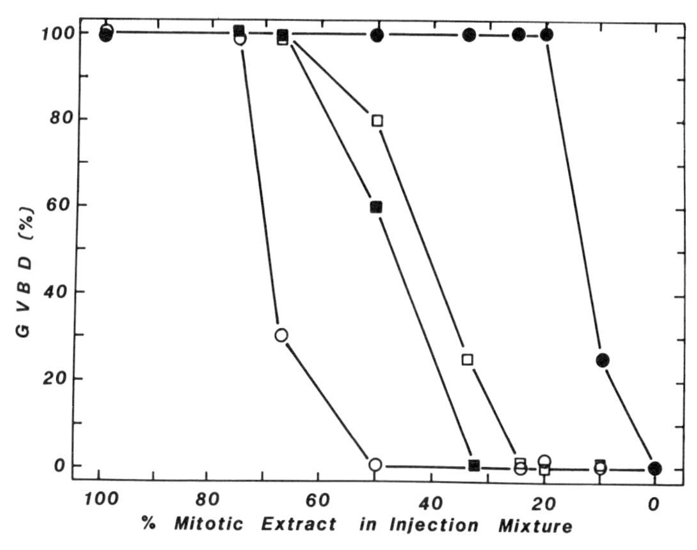

FIG. 3. Effect of G_1 cell extracts on the MPA of the mitotic cell extracts. HeLa cells in G_1 phase were collected at different times (i.e., at 1.5, 2, 3, 4, 5, 6, 7, and 7.5 hr) after reversal of the N_2O block. G_1 cell extracts were made using $6–8 \times 10^7$ cells/ml with a protein content ranging from 8 to 11 mg/ml. Similarly, mitotic cell extracts were made using 4×10^7 cells/ml containing 8 mg/ml of protein. Whenever necessary, extracts were concentrated to give a protein content of 8 mg/ml by ultrafiltration using Amicon YM-10 filters. Extracts of G_1 cells at different points during G_1 were separately mixed with mitotic extracts in various proportions so as to obtain a mitotic extract concentration of 100, 75, 66.6, 50, 33.3, 25, 20, 10, and 0% in the injection mixture. These mixtures were incubated for 1 hr at 4°C prior to injection into *Xenopus* oocytes. For each dilution a minimum of 10 oocytes was injected. A volume of 65 nl (containing ~50–550 ng of proteins) was usually injected. The percentage of GVBD in the injected oocytes was determined at 2–3 hr after injection by scoring the oocytes for the appearance of a white spot in the animal hemisphere. In doubtful cases, oocytes were fixed in 7.5% TCA and dissected to check for the breakdown of the germinal vesicle. Mitotic cell extracts diluted with the extraction buffer (in the presence or absence of 10 mg/ml of bovine serum albumin) to give corresponding concentrations served as controls. These data represent an average of five experiments and the difference in % GVBD between the experiments did not exceed 10%. Typically, oocytes from the same female were used for a given experiment. Dilution of mitotic extract with extracts from G_1 cells at 1.5 hr (□), 2 hr (○), and 7.5 hr (■) after reversal of the N_2O block. (●), Dilution with buffer. The data obtained with extracts from G_1 cells collected at 3, 4, 5, 6, or 7 hr after reversal were identical to those of G_1 cells at 2 hr, and hence these data are not presented. (Reproduced from Adlakha et al. 1983, J Cell Biol 97:1707–1713, by copyright permission of the Rockefeller University Press.)

reached confluency. When these extracts were mixed with mitotic cell extracts and tested for MPA as described earlier, they exhibited minimal inhibitory effect on mitotic factors. However, UV irradiation of G_0 cells and subsequent incubation for 2–4 hr increased the inhibitory activity of extracts from these cells (Figure 5). Irradiation of cells with UV has been shown to induce chromosome decondensation and unscheduled DNA synthesis (for a review, see Johnson et al. 1982). The presence of cycloheximide during incubation after UV irradiation had no effect on IMF activity. IMF activity was further enhanced if hydroxyurea and arabinosylcytosine (ara-C) were added during incubation after UV treatment (Figure 5).

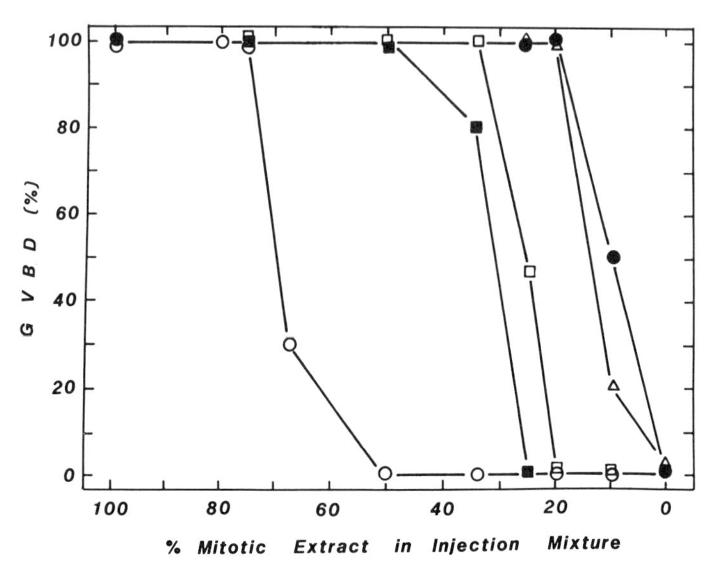

FIG. 4. Effect of S-phase cell extracts on the MPA of mitotic extracts. HeLa cells were synchronized in S phase by double thymidine block method. Early-, mid-, and late-S-phase cells were collected at 2, 4, and 6 hr, respectively, after reversal of the second thymidine block. Extracts from these S-phase cells were prepared using 12–15 × 10^7 cells/ml with a protein concentration of 22–27 mg/ml, whereas the mitotic extract was prepared by using 4 × 10^7 mitotic cells/ml, with 8 mg/ml of protein, thus giving an S-phase/mitotic protein ratio of 3:1. Aliquots from each of these S-phase extracts were separately mixed with mitotic extracts in various proportions, and mixtures were injected into oocytes to test for MPA. (●), Dilution with buffer (negative control); (○), dilution with mid-G_1 cell extract (positive control); dilution with extracts from early (■), mid (□), and late (△) S-phase cells. (Reproduced from Adlakha et al. 1983, J Cell Biol 97:1707–1713, by copyright permission of the Rockefeller University Press.)

HeLa cells synchronized in mitosis by selective detachment after N_2O blockade are able to divide normally, whether in the presence or absence of cycloheximide. At 3 hr after the reversal of the mitotic block, when 95% of the cells were in G_1, extracts were made from the control and cycloheximide-treated cultures, separately mixed with mitotic extracts in various proportions, and tested for MPA. The extracts from cells that were allowed to divide in the presence of cycloheximide were as inhibitory as those of the control G_1 cells. These results indicate that the IMF are not newly synthesized but are rather activated at the end of mitosis (Adlakha et al. 1983).

Presence of IMF in a Cell Line (V79-8) that Lacks Both G_1 and G_2 Periods in its Cell Cycle

In HeLa cells, which exhibit a G_1 period of about 10 hr, the levels of IMF are maximum during G_1, minimum during S, and absent during G_2. Are the IMF expressed in S phase if the G_1 period is absent from the cell cycle? Since the Chinese hamster cell line V79-8 has no measurable G_1, we decided to test it for the presence of IMF during S phase. The results presented in Figure 6 clearly show

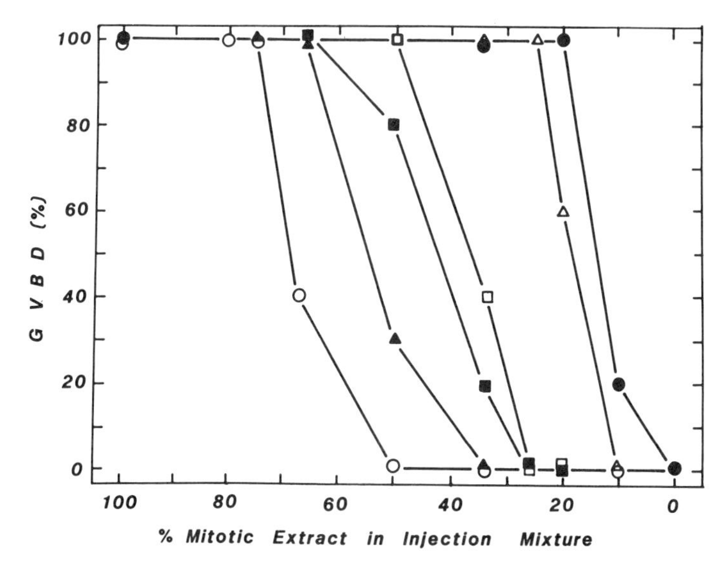

FIG. 5. Activation of IMF in quiescent (G_0) WI-38 human diploid fibroblasts by UV irradiation. WI-38 cells in G_0 phase were collected at 7–10 days after they had reached confluency. G_0 cells were UV irradiated for 10 sec at 90 ergs/cm sec and incubated for various times in the presence or absence of cycloheximide (25 μg/ml), arabinosylcytosine (10^{-4} M), or hydroxyurea (10^{-2} M). Extracts from the control and treated G_0 cells were prepared so as to contain a protein concentration equal to that of mitotic extracts (8 mg/ml). Extracts of G_0 cells from the different treatments were separately mixed with mitotic extracts and tested for MPA. (●), Dilution with buffer (negative control); (○), dilution with mid-G_1 cell extract (positive control); (△), extracts from untreated G_0 cells pretreated with Ara-C and hydroxyurea; extracts from G_0 cells, UV-irradiated and incubated in the presence or absence of cycloheximide for 2 hr (□), and 4 hr (■); extracts from G_0 cells, UV-irradiated and incubated for 2 hr in the presence of Ara-C and hydroxyurea (▲). The data presented here are an average from two experiments. (Reproduced from Adlakha et al. 1983, J Cell Biol 97–1707–1713 by copyright permission of the Rockefeller University Press.)

that maximum IMF activity is present in early S phase, and this activity decreases rapidly as the cells traverse S phase. Little or no activity is present in the extracts of late S-phase cells. When V79-8 cells were grown to confluency they became arrested in G_1 phase; the extracts of these cells possessed as much inhibitory activity as early S-phase cells (Figure 6). These results suggest that the IMF, which are usually present during G_1 in other cell types, are manifest during early S phase in the V79-8 cell line that lacks the G_1 period.

Cellular Localization of IMF

To determine whether IMF in G_1 cells are localized in the cytoplasm or the nucleus, we prepared cytoplasmic extracts by lysing G_1 cells in a low-salt hypotonic buffer and separating nuclei by centrifugation. The nuclear pellet was washed twice with the same buffer before nuclear proteins were extracted with the high-salt (0.2 M NaCl) buffer. The cytoplasmic and nuclear extracts thus obtained were used in mixing experiments with the mitotic extract and tested for MPA. The results of

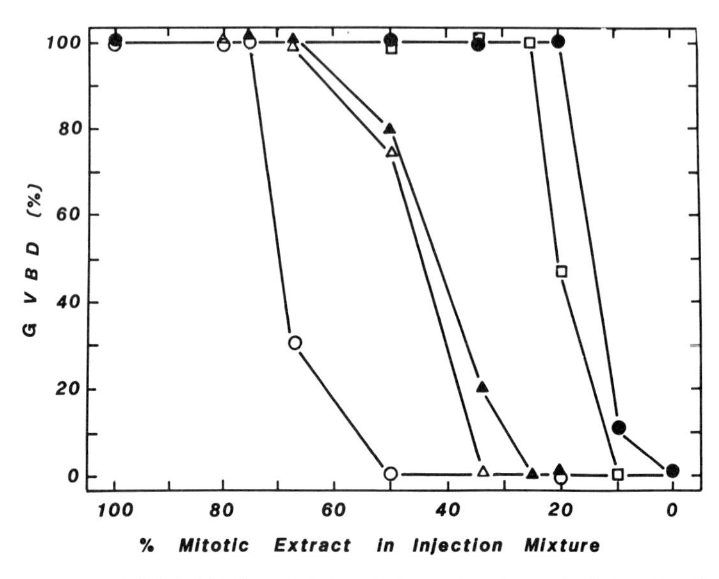

FIG. 6. Presence of IMF in V79-8 cells that lack G_1 and G_2 phases in their cell cycles. V79-8 cells were synchronized in mitosis by selective detachment after a 4-hr Colcemid block. Early-, mid-, and late- S-phase cells were obtained by collecting cells at 2, 5, and 8 hr, respectively, after the reversal of the Colcemid block. Extracts from confluent cultures of V79-8 cells and from those in early, mid, and late S phase were prepared so as to obtain a protein concentration equal to that of the mitotic extract. These extracts were separately mixed with extracts from mitotic HeLa cells in various proportions and tested for MPA. (●), Buffer (negative control); (○), extracts of HeLa G_1 cells (positive control); extracts from V79-8 cells in early S (△), mid S (□), and confluent cultures (▲). Data from experiments involving extracts from late-S-phase V79-8 cells were identical to those of the buffer control (Rao and Adlakha, unpublished data).

these experiments indicate that the cytoplasmic extracts are as effective as the whole-cell extracts (in high-salt buffer) in neutralizing the mitotic factors. Nuclear extracts did not contain any significant inhibitory activity, suggesting that the IMF may be predominantly cytoplasmic.

Preliminary Characterization of IMF

The studies by Adlakha et al. (1983) revealed that IMF are nondialyzable, nonhistone proteins sensitive to inactivation by proteases, but not by RNase, EGTA, or EDTA (Table 3; Figure 7). The IMF activity is less likely to be due to a protease since three different protease inhibitors, phenylmethylsulfonyl fluoride, antipain, and trypsin inhibitor, were present in the extraction buffer. Unlike the mitotic factors, IMF are heat stable (at 65°C for 15 min). They are stable over a broad pH range (6.0–10.0) but are extremely sensitive to low pH (5.0). The apparent molecular weight is 12,000 Da. Some of these characteristics distinguish them from the mitotic factors (Table 4). The pH dependency of these factors is also in agreement with the results of the early cell fusion experiments of Obara and co-workers (1973), which suggested that high pH favors "telophasing" and low pH "prophasing"

TABLE 3. *Preliminary characterization of IMF*

Treatment of G_1 cell extract	Relative IMF activity (%)*
None	100
Papain†	0
RNase‡	100
Temperature§	
0°C for 2 days	80
−70°C for 2 months	100
25–65°C for 15 min	100
75°C for 15 min	75
100°C for 15 min	0
pH#	
4.0	20
5.0–5.5	75
6.0–8.0	100
8.5–10.0	75–80

*A freshly prepared mid-G_1 cell extract from 8×10^7 cells/ml was mixed with mitotic extract from 4×10^7 cells/ml to give a M:G_1 protein ratio of 1:1, and the mixture was injected into *Xenopus laevis* oocytes for MPA determination. A mixture of mitotic extract with extraction buffer served as control. If the G_1 cell extract completely inactivated the MPA of the mitotic extract at a 50% dilution, i.e., a protein ratio of 1:1, then the activity of the G_1 cell extract was considered to be 100%. An average of 10 oocytes was injected for each sample.

†G_1 cell extract was treated with the protease papain (500 μg/ml) in extraction buffer containing cystein (5 mM) and β-mercaptoethanol (0.05 mM) for 1 hr at 25°C. Antipain (80 μg/ml) was then added for 15 min at 25°C to neutralize the excess papain, before being mixed with mitotic extract in various proportions and eventually injected into oocytes for MPA determination. In these experiments, we made certain that papain activity was completely neutralized by antipain so that it could not inactivate the mitotic factors, which are proteins.

‡G_1 cell extracts were incubated with RNase (1.5 units/ml) at 25°C for 1 hr. At the end of incubation, G_1 extracts were mixed with mitotic extracts and tested for MPA as above.

§G_1 cell extracts were incubated at different temperatures for 15 min, and at the end of the incubation extracts were centrifuged at 10,000 *g* for 15 min at 4°C to remove any precipitate before being mixed with mitotic extract for injection into oocytes.

#G_1 cell extracts were dialyzed overnight against buffers of different pH with three changes each and redialyzed for 12 hr against the extraction buffer (pH 6.5) before the mixing experiments were carried out to test the activity as explained above.

(Reproduced from Adlakha et al. 1983, J Cell Biol 97:1707–1713, by copyright permission of the Rockefeller University Press.)

or premature chromosome condensation. Our own studies (W. N. Hittelman and P. N. Rao, unpublished data) showed that the frequency of induction of premature chromosome condensation was much higher if a low pH was maintained during the collection of mitotic cells and subsequent fusion procedures. Thus, it appears that at low pH, the mitotic factors are active, whereas the G_1 factors are either inactive or less active.

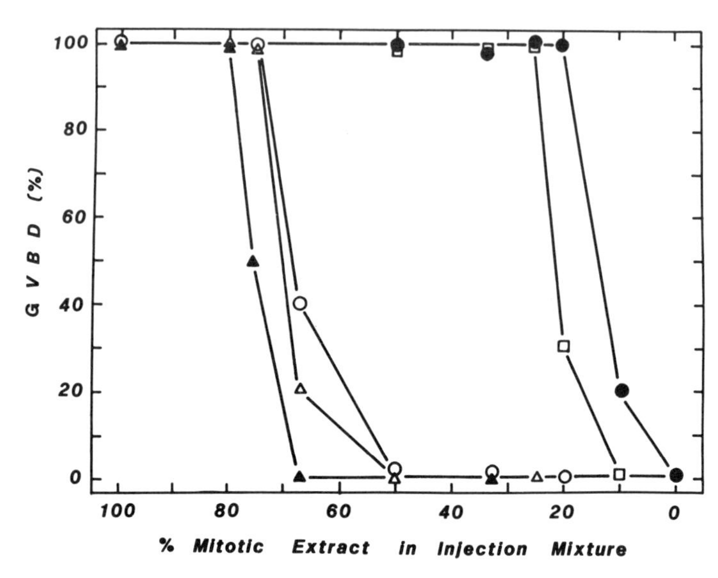

FIG. 7. Effect of chelating agents on IMF activity. Extracts from early G_1 HeLa cells were prepared and incubated for 1 hr at 4°C with various concentrations of EGTA or EDTA (1 mM–5 mM). These EGTA- or EDTA-treated G_1 cell extracts were separately mixed with extracts from mitotic HeLa cells and tested for MPA. (●), Buffer (negative control); (○), HeLa G_1 cell extracts (positive control); (□), buffer containing 5 mM EGTA or EDTA; (△), HeLa G_1 cell extracts incubated with 1, 2, 3, or 4 mM EGTA and EDTA prior to being mixed with mitotic extracts; (▲), HeLa G_1 cell extracts incubated with 5 mM EGTA or EDTA. These data represent an average from two experiments. (Reproduced from Adlakha et al. 1983, J Cell Biol 97:1707–1713, by copyright permission of the Rockefeller University Press.)

Inactivation of Mitotic Factors by UV Irradiation of Mitotic Cells

UV irradiation of Colcemid-arrested mitotic HeLa cells and subsequent incubation in the presence of inhibitors of DNA synthesis, i.e., ara-C (10^{-4} M) and hydroxyurea (10^{-2} M), resulted in extensive attenuation or decondensation of metaphase chromosomes (Schor et al. 1975, Adlakha et al. 1984b). The degree of decondensation was dependent on both the dose of UV irradiation and the duration of incubation (Figure 8). A 2-hr posttreatment incubation caused a maximum degree of chromosome decondensation. If UV-irradiated mitotic cells were not incubated, the morphology of the metaphase chromosomes remained unchanged, being indistinguishable from those of unirradiated cells.

Adlakha et al. (1984b) have shown that extracts made from mitotic HeLa cells exposed to 2500 ergs of UV and incubated for 2 hr in the presence of ara-C and hydroxyurea had very little MPA when injected into *Xenopus* oocytes (Figure 9). Inhibition of protein synthesis by addition of cycloheximide to the incubation medium had no effect on the UV-induced inactivation of the mitotic factors. Extracts from UV-irradiated mitotic cells that were not incubated with ara-C and hydroxyurea had almost as much MPA as the control mitotic cell extracts (Figure 9). The

TABLE 4. *Comparison of the mitotic factors (MF) with the inhibitors of mitotic factors (IMF)*

Property	MF	IMF
Induces GVBD in amphibian oocytes	Yes	N.T.
Induces amphibian meiotic chromosome condensation	Yes	No
Induces premature condensation in interphase nuclei	Yes	Induces decondensation ("telophasing")
Affinity to chromatin	Very high	No
Dependent on new protein synthesis	Yes	No
Inactivated by proteases	Yes	Yes
Inactivated by RNase	No	No
Inactivated by EGTA or EDTA	Yes, at concentrations higher than 5 mM	No
Activity sensitive to Ca^{2+}	Yes	No
Activity dependent on Mg^{2+}	Probably yes	No
Heat sensitivity	Over 37°C	Over 75°C
Effect of pH	Inactivated at pH 7.5 and above	Inactivated at pH 5.0 and below but active up to pH 10.0
Activity stabilized by the presence of phosphatase inhibitors (ATP, NaF, and Sodium β-glycerolphosphate)	Greatly	No effect
Molecular size	Approx. 100,000 Da	>12,000 Da

N.T., Not tested.
Data drawn from Sunkara et al. (1979 a,b, 1982) and Adlakha et al. (1982 a,b, 1983).

inactivation of the mitotic factors was less pronounced if ara-C and hydroxyurea were not added to the medium during posttreatment incubation.

As was the case with chromosome attenuation (Figure 8), the UV-induced inactivation of mitotic factors was also dependent on the dose of radiation and the duration of incubation (Figure 10). Exposure of mitotic cell extracts (not whole cells) to even a very high dose (5000 ergs) of UV followed by incubation for 2 hr in the presence of ara-C and hydroxyurea failed to inactivate the mitotic factors. These data indicate that inactivation of mitotic factors is not directly due to UV irradiation but rather is due to the mitotic cell's response to UV exposure. This response is not dependent on new protein synthesis. In contrast, X-irradiation of mitotic cells at high doses, which caused extensive chromosome fragmentation rather than attenuation, resulted only in a slight inactivation of MPA (Figure 11). These results suggest that UV-induced decondensation of metaphase chromosomes is causally related to the loss of MPA of mitotic cell extracts. The UV-induced inactivation of mitotic factors could be partially neutralized in dose-dependent

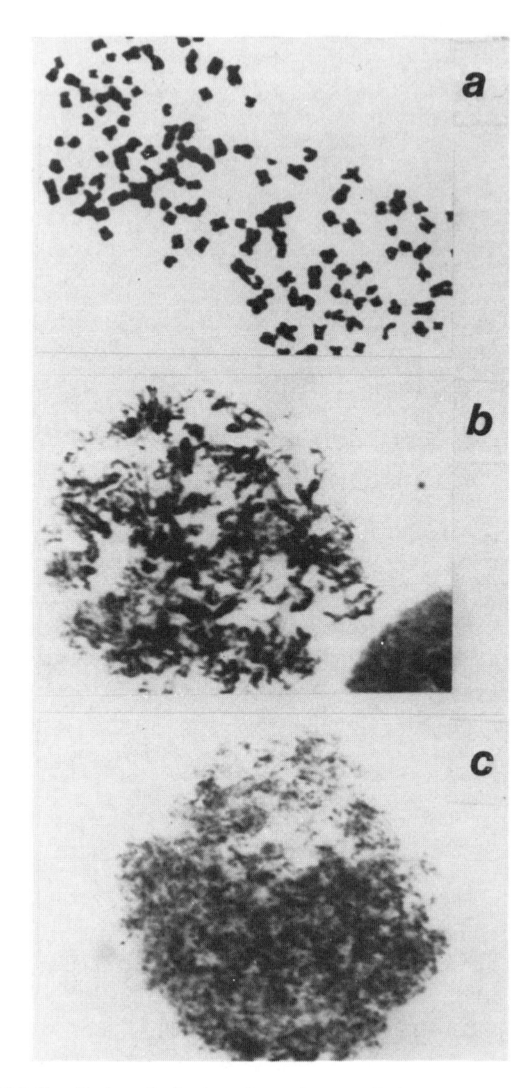

FIG. 8. Effect of UV irradiation of mitotic cells on the morphology of metaphase chromosomes. Nitrous oxide-blocked mitotic HeLa cells were irradiated with different doses of UV (500–3000 ergs). After irradiation, cells were incubated for 2 hr at 37°C in fresh medium containing Colcemid (to hold the cells in mitosis) and hydroxyurea (10^{-2} M) and ara-C (10^{-4} M) (to block DNA repair synthesis). Following incubation, cells were thoroughly washed with Hanks' balanced salt solution, and chromosome spreads prepared and stained with Giemsa. Structural morphology of metaphase chromosomes from unirradiated mitotic cells (A) and mitotic cells irradiated with 500 ergs (B) and 3000 ergs (C) of UV. Note the different degrees of attenuation in the chromosomes of UV-treated mitotic cells. (Reproduced from Adlakha et al. 1984b, with permission of the Company of Biologists Ltd.)

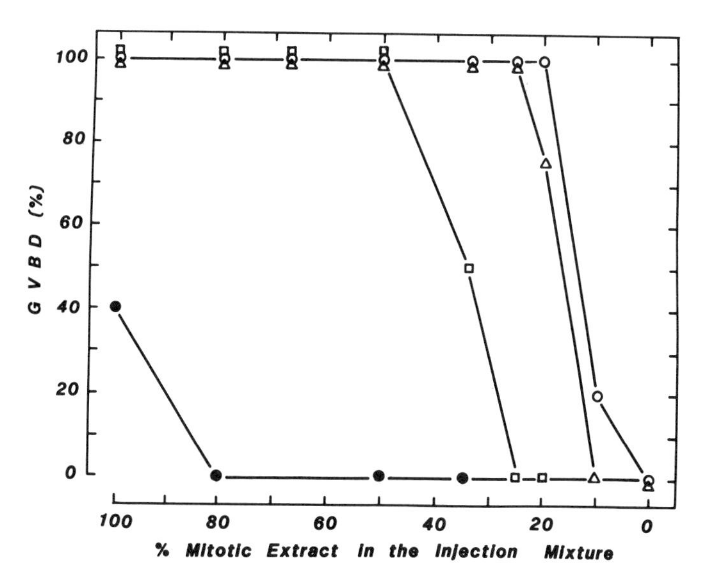

FIG. 9. Effect of UV irradiation of mitotic HeLa cells on the MPA of the mitotic cell extracts. Extracts from the untreated and UV-irradiated (2500 ergs) mitotic cells were diluted with various proportions of buffer, injected into *Xenopus laevis* oocytes, and scored for % GVBD. Symbols: (○), extracts from control mitotic cells; extracts from mitotic cells UV irradiated: (△), with no post-treatment incubation; (□), with 2-hr incubation in regular medium; (●), with 2-hr incubation in the presence of ara-C and hydroxyurea. (Reproduced from Adlakha et al. 1984b, with permission of the Company of Biologists Ltd.)

manner by the addition of Mg^{2+} or polyamines (i.e., putrescine, spermidine, or spermine) to the incubation medium (Figure 12). Among the polyamines, which are known to promote chromatin condensation, spermine was most effective in reversing UV effects. However, these positively charged compounds did not completely reverse the UV-induced inactivation of the mitotic factors even at very high concentrations.

The fusion of UV-treated mitotic cells (2500 ergs followed by 2 hr incubation in the presence of ara-C and hydroxyurea) with untreated mitotic cells resulted in the recondensation of chromosomes that had been attenuated by UV irradiation (Figure 13). The recondensation effect was much more pronounced when one UV-treated mitotic cell was fused with two untreated mitotic cells. Furthermore, the UV-treated mitotic cells failed to induce premature chromosome condensation when fused with interphase cells. These results lend further support to our earlier conclusions that UV irradiation of mitotic cells resulted in a complete inactivation of the mitotic factors.

How are the mitotic factors inactivated by UV-irradiation? Earlier in this section we showed that IMF, which are activated at the end of mitosis, continue to be active throughout G_1. Furthermore, IMF are activated in quiescent (G_0) human diploid fibroblasts by UV irradiation (Figure 5). It is therefore likely that IMF are also activated in mitotic cells by UV irradiation. They interact or form a complex

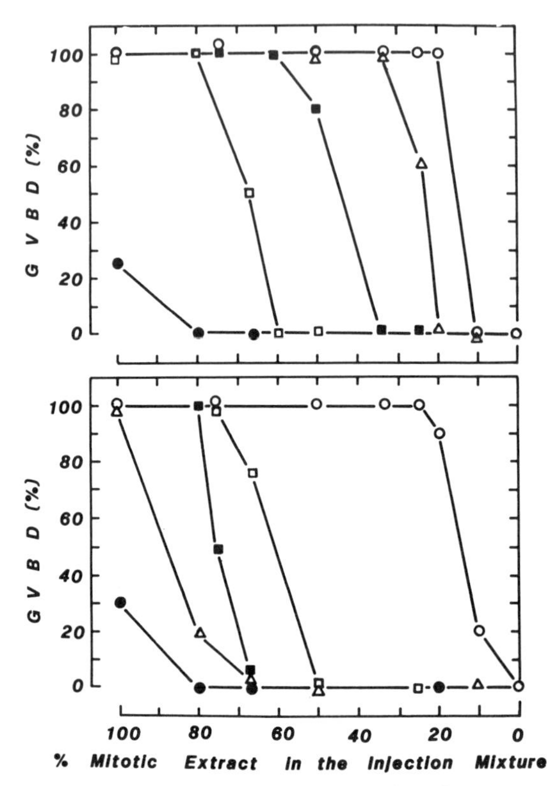

FIG. 10. Effect of the dose of UV irradiation and the duration of post-treatment incubation of mitotic HeLa cells on the MPA of the mitotic cell extracts. **A:** Dosage effects of UV. Mitotic HeLa cells were exposed to various doses of UV and incubated for 2 hr in medium containing Colcemid, hydroxyurea, and ara-C. Extracts prepared from these cells were diluted with various proportions of the extraction buffer and tested for GVBD. Symbols: (○), extracts from the untreated mitotic cells; extracts from cells exposed to: (△), 500; (■), 1000; (□), 2000; and (●), 3000 ergs of UV irradiation. **B:** Effect of duration of posttreatment incubation. Extracts were made from mitotic HeLa cells that were exposed to 2500 ergs of UV and incubated in the presence of Colcemid, hydroxyurea, and ara-C for: (□), 30 min; (■), 60 min; (△), 90 min; and (●), 120 min and tested for GVBD. (○), control, same as in (A). (Reproduced from Adlakha et al. 1984b, with permission of the Company of Biologists Ltd.)

with mitotic factors and thus lose their individual activities. To support this thesis, Adlakha et al. (1984b) mixed G_1 cell extracts, known to contain IMF, with extracts from untreated (control) mitotic cells and heated them for 15 min at 60°C. This treatment, which removed nearly 60% of the protein as precipitate from the mixture of G_1 and mitotic cell extracts, destroyed the IMF and MPA (Table 5).

When extracts from G_1 cells that were prelabeled with ^{3}H-labeled amino acids during the G_2 period of the preceding cycle were mixed with partially purified mitotic factors and heated for 15 min at 60°C, about 60% of the radioactivity was found in the precipitate (Table 6). No significant amount of radioactivity was found in the precipitate when the extract from prelabeled G_1 cells was mixed with

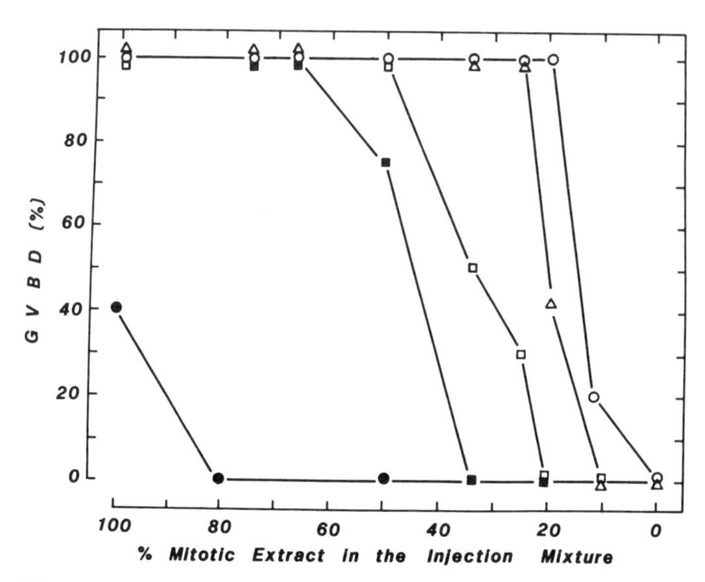

FIG. 11. Effect of X irradiation of mitotic HeLa cells on the MPA of the mitotic cell extracts. Extracts were made from mitotic cells irradiated with different doses of X rays and incubated for 2 hr in medium containing Colcemid, hydroxyurea, and ara-C and tested for GVBD as described above. Symbols: (○), extracts from untreated mitotic cells. Extracts from mitotic cells irradiated with: (△), 500 or 1000; (□), 2000; (■), 3000 rad of X rays or (●), 2500 ergs of UV (positive control). (Reproduced from Adlakha et al. 1984b, with permission of the Company of Biologists Ltd.)

unlabeled S phase cell extract or BSA and heated for 15 min at 60°C. Furthermore, if the mitotic factors were first allowed to bind to DNA-cellulose, and then this complex was incubated with G_1 cell extract for 45 min at 4°C and thereafter eluted with high salt (0.2 M NaCl), neither IMF nor mitotic factor activity was recovered (Adlakha et al. 1984b). These results suggest that mitotic factors are inactivated when they form a complex with IMF, and the IMF are activated at telophase or by exposure of cells to UV irradiation.

ROLE OF IMF IN THE REGULATION OF CHROMOSOME DECONDENSATION

The data presented in the preceding section indicate that the IMF are activated at telophase and are present throughout the G_1 period, thus coinciding with the process of chromosome decondensation, which is known to begin at telophase and continue throughout G_1 phase. It is tempting to speculate that the IMF, which are antagonistic to mitotic factors, may serve the reverse function of the mitotic factors, i.e., the regulation of chromosome decondensation. Our results suggest that activation of IMF at telophase may lead to a rapid inactivation of the mitotic factors and consequently may result in the decondensation of chromosomes.

This proposition is further strengthened by our observations on the activation of IMF in quiescent (G_0 phase) and mitotic cells by UV irradiation and also by some

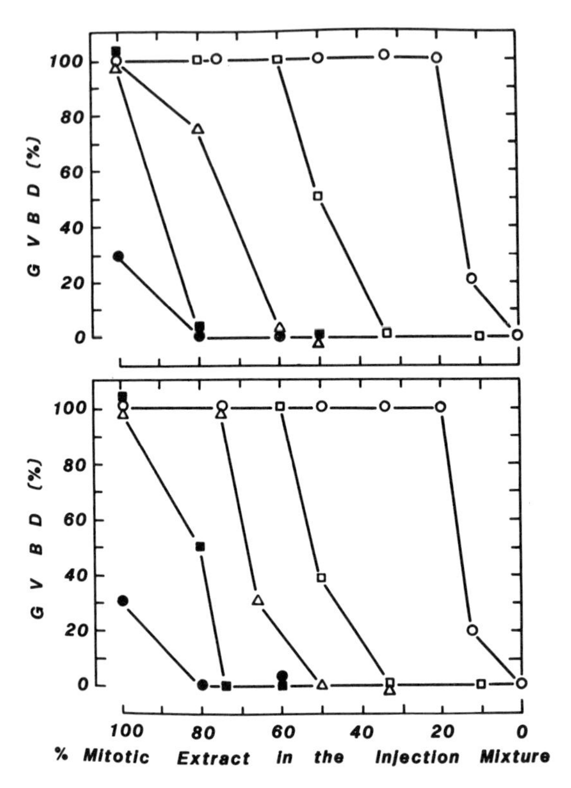

FIG. 12. Effect of Mg^{2+} and polyamines on the UV-induced inactivation of mitotic factors. Conditions for irradiation and incubation are the same as in the legend to Figure 10, except Mg^{2+} or polyamines were present in the medium during irradiation and posttreatment incubation. **A:** Reversal of the UV-induced inactivation of mitotic factors by various doses of Mg^{2+}. Symbols: (○), control, extracts of untreated mitotic cells. Extracts of UV-treated cells incubated in the presence of $MgCl_2$ at: (□), 5 mM; (△), 0.5 mM; (■), 0.05 mM; (●), zero concentrations. **B:** Reversal of the UV-induced inactivation of mitotic factors by polyamines. Symbols: (○), control, extracts from untreated mitotic cells. Extracts of UV-treated mitotic cells incubated in the presence of: (□), spermine; (△), spermidine; and (■), putrescine, all at a concentration of 5 mM; (●), no polyamines. (Reproduced from Adlakha et al. 1984b, with permission of the Company of Biologists Ltd.)

of our earlier observations. (1) In cell fusion experiments, Rao and Johnson (1970) and Rao et al. (1975) observed that the entry of G_2 cells into mitosis was delayed after fusion with G_1- or S-phase cells. The G_2 nucleus in the heterophasic binucleate cell would wait until the G_1- or S-phase nucleus had completed DNA synthesis; both nuclei then entered mitosis synchronously. It was speculated at that time that G_1- and S-phase components were causing decondensation of chromatin in G_2 nuclei, thus blocking them from entering mitosis. Our present results seem to provide the experimental evidence to support that assumption. (2) Rao and Smith (1981) noted that in binucleate cells formed by the fusion of HeLa cells in G_2 phase with G_0 human diploid fibroblasts (HDF), the G_0 component had no effect

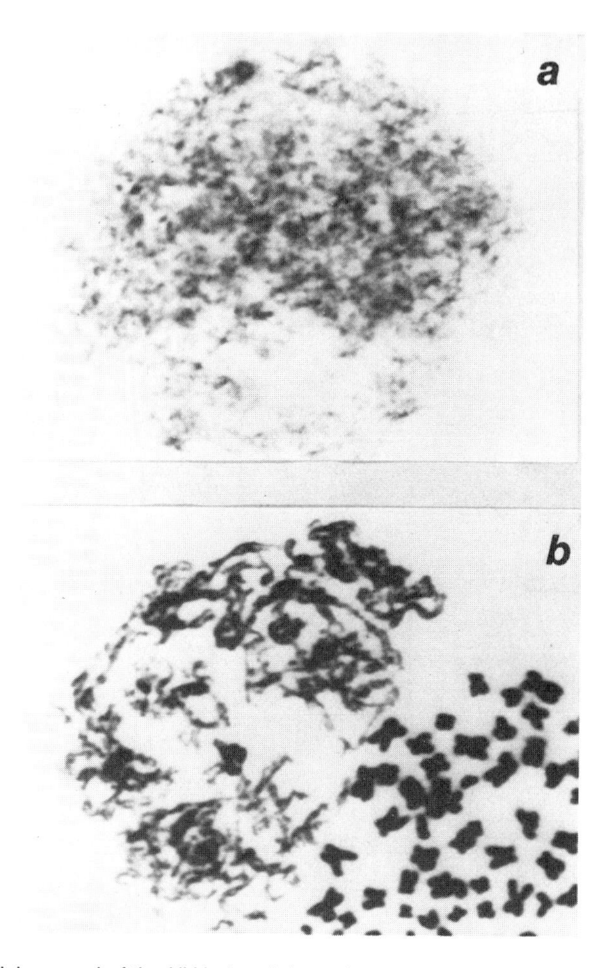

FIG. 13. Partial reversal of the UV-induced decondensation of metaphase chromosomes by fusion with unirradiated mitotic cells. HeLa mitotic cells treated with 2500 ergs of UV and incubated for 2 hr in the presence of hydroxyurea and ara-C were washed and fused with untreated mitotic HeLa cells using UV-inactivated Sendai virus. No inhibitors were present in the fusion mixture during the 45-min incubation at 37°C. After fusion, cells were processed for chromosome spreads. Chromosomes of UV-treated mitotic cells before fusion **(A)**, and after fusion **(B)** with untreated mitotic cells. The darkly stained chromosomes in (B) are of the untreated mitotic cells. (Reproduced from Adlakha et al. 1984b, with permission of the Company of Biologists Ltd.)

on the progression of the G_2 nucleus into mitosis. In contrast, fusion of G_2 cells with a UV-treated G_0 HDF delayed the entry of the G_2 nucleus into mitosis. Knowing now that UV irradiation causes chromatin decondensation and also activates IMF, we can readily explain the G_2 delay in the G_0-UV/G_2 binucleate cells by assuming the presence of IMF in UV-irradiated G_0 cells, which would neutralize the mitotic factors progressively accumulating in the G_2 cells. Hence the G_2 component needs extra time to compensate for the loss and build up the factors to a

TABLE 5. *Effects of heating a mixture of mitotic and G_1 cell extracts at 60°C for 15 min on the activities of IMF and mitotic factors*

Extracts	Amounts of proteins present in the supernatant (mg/ml)*		Relative amount of protein lost as precipitate after heating (%)	IMF activity† (%)		MF activity† (%)	
	Before heating	After heating		Before heating	After heating	Before heating	After heating
G_1 cell extract (no treatment)‡	9.2	8.95	2.7	100	100	—	—
G_1 cell extract + BSA§	9.6	9.30	3.1	100	100	—	—
Mitotic cell extract (no treatment)	9.0	6.40	28.9	—	—	100	0
Mitotic cell extract + BSA	9.5	8.00	16.7	—	—	100	0
G_1 cell extract (0.5 ml) + mitotic cell extract (0.5 ml)	9.1	3.70	59.3	—	0	0	0

*After each treatment, extracts were cleared by centrifugation and proteins determined in the supernatants.

†IMF and mitotic factor (MF) activities were determined as explained in text. An average of 10 oocytes were injected for each extract, and 65 nl of the extracts was injected into each oocyte.

‡Early G_1 cells were obtained by collecting cells at 3 hr after the reversal of N_2O-blocked mitotic cells and extracts were prepared by using 7×10^7 cells/ml as described in Materials and Methods.

§G_1 cell extracts were mixed with equal volumes (0.5 ml each) of mitotic cell extracts or bovine serum albumin (10 mg/ml) in extraction buffer.

(Reproduced from Adlakha et al. 1984b, with permission of the Company of Biologists Ltd.)

TABLE 6. *Effect of heating a mixture of partially purified unlabeled mitotic extract and G_1 cell extract prelabeled with ^{3}H-labeled amino acids at 60°C for 15 min on the activities of IMF and mitotic factors*

Treatment	TCA precipitable cpm/100 μl*	Relative loss of cpm due to heating (%)	IMF activity† (%)	MF activity† (%)
G_1 cell extract (no treatment)‡	147,584		100	0
G_1 cell extract + 60°C for 15 min§	133,105	9.8	100	0
G_1cell extract + extraction buffer + 60°C for 15 min	73,687		100	0
G_1 cell extract + partially purified mitotic factors# + 60°C for 15 min	27,714	62.4	0	0
G_1 cell extract + crude mitotic extract + 60°C for 15 min	29,429	60.0	0	0
G_1 cell extract + BSA (10 mg/ml) + 60°C for 15 min	75,331	0	100	0
G_1 cell extract + S-phase cell extract (11.0 mg/ml)‖ + 60°C for 15 min	74,830	0	100	0

*After each treatment, proteins in the supernatant were precipitated by addition of sufficient 100% trichloroacetic acid (TCA) to give a final concentration of 10%. After two washings with 10% TCA and one with methanol, the proteins were dissolved in Soluene and a sample was counted in a liquid scintillation counter.

†IMF and mitotic factor (MF) activities were determined as explained in footnote to Table 5.

‡HeLa cells were synchronized into S phase by the double thymidine block method. At 7 hr after reversal of the second thymidine block, when the cells were in either late S or early G_2 phase, ^{3}H-labeled L-amino acid mixture (1 μCi/ml) was added. G_1 cells were collected at 3 hr after the reversal of N_2O block. G_1 cell extract was prepared as described in footnote to Table 5.

§This heat-treated G_1 cell extract was used in subsequent dilutions (listed in this table) with an equal volume of the extraction buffer or other cell extracts.

#Mitotic cell extracts were prepared by using 4×10^7 cells/ml as described in Materials and Methods. Extracts were dialyzed overnight (with 3 changes of buffer) against 10 mM Na_2HPO_4/NaH_2PO_4 buffer containing 1 mM PMSF, 1 mM ATP, 5 mM NaF, and 5 mM sodium glycerolphosphate, pH 6.5. Mitotic factors were partially purified from the dialyzed extracts by affinity chromatography on a 2-ml column of DNA-cellulose.

‖Mid-S-phase cells were collected at 4 hr after the reversal of second thymidine block and extracts were prepared by using 5.5×10^7 cells/ml as described under Methods.

critical concentration required for entry into mitosis. In light of all these data, it appears that IMF play an important role in the regulation of chromosome decondensation. However, whether the activation of IMF is the cause or the result of chromosome decondensation remains to be elucidated.

SUMMARY

In this chapter, we have attempted to review our recent work pertaining to the regulation of the chromosome condensation–decondensation cycle within the life cycle of mammalian cells. The results summarized here strongly suggest that this sequence of events may be regulated by different protein factors. Mitotic factors injected into fully grown *X. laevis* oocytes induce meiotic maturation, i.e., GVBD and chromosome condensation. These factors, which accumulate slowly in the beginning of G_2 and reach a threshold at the G_2–mitosis transition, have a great affinity for chromatin and are localized on metaphase chromosomes, as well as in the cytoplasm. They are nondialyzable, heat- and Ca^{2+}-sensitive, Mg^{2+}-dependent nonhistone proteins with an approximate molecular mass of 100,000 Da. At the telophase of mitosis, the mitotic factors are rapidly inactivated by another set of factors, IMF. IMF are also nondialyzable nonhistone proteins, but unlike mitotic factors, are heat-stable. They are also stable over a broad pH range, but are extremely sensitive to low pH. IMF are activated at telophase and remain active throughout the G_1 period, thus coinciding with the process of chromosome decondensation. Although evidence implicating IMF in the regulation of chromosome decondensation is still largely circumstantial, data summarized here nevertheless suggest a strong correlation between these two phenomena.

The way in which mitotic factors and IMF might bring about the condensation–decondensation of chromosomes has not been established. Our studies on the role of protein phosphorylation and the use of monoclonal antibodies specific for mitotic cells have provided some evidence implicating nonhistone protein phosphorylation–dephosphorylation in the regulation of mitosis. A causal link between these events is suggested, but remains to be established. Characterization of these factors will help us learn about their functions, as well as lead to a better understanding of the events regulating the chromosome condensation–decondensation cycle in eukaryotic cells.

REFERENCES

Adlakha RC, Sahasrabuddhe CG, Wright DA, Lindsey WF, Rao PN. 1982a. Localization of mitotic factors on metaphase chromosomes. J Cell Sci 54:193–206.

Adlakha RC, Sahasrabuddhe CG, Wright DA, Lindsey WF, Smith ML, Rao PN. 1982b. Chromosome-bound mitotic factors: Release by endonucleases. Nucl Acids Res 10:4107–4117.

Adlakha RC, Sahasrabuddhe CG, Wright DA, Rao PN. 1983. Evidence for the presence of inhibitors of mitotic factors during G_1 period in mammalian cells. J Cell Biol 97:1707–1713.

Adlakha RC, Davis FM, Rao PN. 1984a. Role of phosphorylation of nonhistone proteins in the regulation of mitosis. *In* Boyton AL, Leffert HL (eds), Cell Proliferation: Recent Advances Vol. 1. Academic Press, New York, in press.

Adlakha RC, Wang YC, Wright DA, Sahasrabuddhe CG, Bigo H, Rao PN. 1984b. Inactivation of mitotic factors by ultraviolet irradiation of HeLa cells in mitosis. J Cell Sci 65:279–295.

Al-Bader AA, Orengo A, Rao PN. 1978. G_2-phase specific proteins of HeLa cells. Proc Natl Acad Sci USA 75:6064–6068.

Beall PT, Hazelwood CF, Rao PN. 1976. Nuclear magnetic resonance patterns of intracellular water as a function of HeLa cell cycle. Science 192:904–907.

Davis FM, Rao PN. 1982. Antibodies specific for mitotic human chromosomes. Exp Cell Res 137:381–386.

Davis FM, Tsao TY, Fowler SK, Rao PN. 1983. Monoclonal antibodies to mitotic cells. Proc Natl Acad Sci USA 80:2926–2930.

Hildebrand CE, Tobey RA. 1975. Cell cycle-specific changes in chromatin organization. Biochem Biophys Res Commun 63:134–139.

Hittelman WN, Rao PN. 1976. Premature chromosome condensation: Conformational changes of chromatin associated with phytohemagglutin-stimulation of peripheral lymphocytes. Exp Cell Res 100:219–222.

Hittelman WN, Rao PN. 1978. Mapping G_1 phase by the structural morphology of the prematurely condensed chromosomes. J Cell Physiol 95:333–341.

Johnson RT, Rao PN. 1970. Mammalian cell fusion: Induction of premature chromosome condensation in interphase nuclei. Nature 226:717–722.

Johnson RT, Collins ARS, Waldren CA. 1982. Prematurely condensed chromosomes and the analysis of DNA and chromosome lesions. *In* Rao PN, Johnson RT, Sperling K (eds), Premature Chromosome Condensation: Application in Basic, Clinical, and Mutation Research. Academic Press, New York, pp. 253–308.

Kishimoto T, Kuriyama R, Kondo H, Kanatani K. 1982. Generality of the action of the various maturation-promoting factors. Exp Cell Res 137:121–126.

Kohler G, Milstein C. 1975. Continuous culture of fused cells secreting antibodies of predefined specificity. Nature 246:495–497.

Mazia D. 1963. Synthetic activities leading to mitosis. J Cell Comp Physiol 62 (Suppl I):123–140.

Moser GC, Muller H, Robbins E. 1975. Differential nuclear fluorescence during the cell cycle. Exp Cell Res 91:73–78.

Moser GC, Fallon RJ, Meiss HK. 1981. Fluorometric measurements and chromatin condensation patterns of nuclei from 3T3 cells throughtout G_1. J Cell Physiol 106:293–301.

Nelkin B, Nichols C, Vogelstein B. 1980. Protein factor(s) from mitotic CHO cells induce meiotic maturation in *Xenopus laevis* oocytes. FEBS Lett 109:233–238.

Nicolini C, Ajiro K, Borun TW, Baserga R. 1975. Chromatin changes during the cell cycle of HeLa cells. J Biol Chem 250:3381–3385.

Obara Y, Yoshida H, Chai LS, Weinfeld H, Sandberg AA. 1973. Contrast between the environmental pH dependencies of prophasing and nuclear membrane formation in interphase-metaphase cells. J Cell Biol 62:104–113.

Pederson T. 1972. Chromatin structure and the cell cycle. Proc Natl Acad Sci USA 69:2224–2228.

Pederson T, Robbins E. 1972. Chromatin structure and the cell division cycle. Actinomycin binding in synchronized HeLa cells. J Cell Biol 55:322–327.

Rao PN, Hanks SK. 1980. Chromatin structure during the prereplicative phases in the life cycle of mammalian cells. Cell Biophys 2:327–337.

Rao PN, Johnson RT. 1970. Mammalian cell fusion: I. Studies on the regulation of DNA synthesis and mitosis. Nature 225:159–164.

Rao PN, Johnson RT. 1974. Regulation of cell cycle in hybrid cells. *In* Clarkson B, Baserga R (eds), Cold Spring Harbor Conference on Cell Proliferation, Vol. 1, Control of Proliferation in Animal Cells. Cold Spring Harbor Laboratory, Cold Spring Harbor, New York, pp. 785–800.

Rao PN, Smith ML. 1981. Differential response of cycling and noncycling cells to inducers of DNA synthesis and mitosis. J Cell Biol 88:649–653.

Rao PN, Hittelman WN, Wilson BA. 1975. Mammalian cell fusion. VI. Regulation of mitosis in binucleate HeLa cells. Exp Cell Res 90:40–46.

Rao PN, Wilson BA, Puck TT. 1977. Premature chromosome condensation and cell cycle analysis. J Cell Physiol 91:131–142.

Schor SL, Johnson RT, Waldren C. 1975. Changes in the organization of chromosomes during the cell cycle: Response to ultraviolet light. J Cell Sci 17:539–565.

Sperling K, Rao PN. 1974. Mammalian cell fusion. V. Replication behaviour of heterochromatin as observed by premature chromosome condensation. Chromosoma 45:121–131.

Sunkara PS, Wright DA, Rao PN. 1979a. Mitotic factors from mammalian cells induce germinal vesicle breakdown and chromosome condensation in amphibian oocytes. Proc Natl Acad Sci USA 76:2799–2802.

Sunkara PS, Wright DA, Rao PN. 1979b. Mitotic factors from mammalian cells: A preliminary characterization. J Supramol Struct 11:189–195.

Sunkara PS, Wright DA, Adlakha RC, Sahasrabuddhe CG, Rao PN. 1982. Characterization of chromosome condensation factors of mammalian cells. *In* Rao PN, Johnson RT, Sperling K (eds), Premature Chromosome Condensation: Application in Basic, Clinical, and Mutation Research. Academic Press, New York, pp. 233–251.

Weintraub H, Buscaglia M, Ferrez M, Weiller S, Boulet A, Fabre F, Baulieu EE. 1983. "MPF" activity in *Sacchromyces cerevisiae*. CR Hebd Seanc Acad Sci, Paris, Serie III 295:787–790.

Growth Factors for Nonlymphoid Cells

Mediators in Cell Growth and Differentiation,
edited by Richard J. Ford and Abby L. Maizel.
Raven Press, New York © 1985.

Regulation of Gene Expression by Serum and Serum Growth Factors

Sidney L. Hendrickson, Brent H. Cochran, Angela C. Reffel, and Charles D. Stiles

Department of Microbiology and Molecular Genetics, Harvard Medical School and Dana-Farber Cancer Institute, Boston, Massachusetts 02115

In recent years, there have been suggestions that alterations in the regulation of gene functions normally controlled by growth factors might lead to oncogenic transformation. Indeed, there are growing indications that the studies of growth factors and those of oncogenes are converging on a common set of metabolic pathways. At least three levels of the mitogenic cascade initiated by growth factors contain candidate oncogene functions (Stiles 1983).

The first candidate function lies at the outset of the mitogenic cascade, wherein oncogenes may cause the production of a growth factor within the very cells that are responsive to it. Simian sarcoma virus (SSV) appears to have acquired its transforming gene (v-*sis*) (Devare et al. 1983) from the structural gene for platelet-derived growth factor (PDGF) (Doolittle et al. 1983, Waterfield et al. 1983). The cellular homologue of v-*sis* (c-*sis*) is expressed in many connective tissue tumors (Eva et al. 1982), and these same tumors secrete a PDGF-like mitogen into cell culture medium (Heldin et al. 1980, 1981).

A second candidate function is at the receptor level. A tyrosine-specific protein kinase is intimately associated with the receptor for PDGF (Ek and Heldin 1982, Nishimura et al. 1982, Cooper et al. 1982, Heldin et al. 1983). This same activity is encoded by several viral oncogenes and their cellular homologues (Bishop 1983). One of these oncogenes, v-*erb* B, has been shown to have a high degree of homology with the EGF receptor (Downward et al. 1984), itself a tyrosine kinase (Cohen et al. 1982a,b).

The third candidate function lies within the control of cell cycle genes. It has been known for many years that de novo mRNA synthesis is required in order for quiescent cells to leave the G_0 phase of the cell cycle, traverse G_1, and enter S phase; paradoxically, there is little difference in the overall rate of mRNA synthesis during this transition. Moreover, the kinds of mRNAs found in growth-stimulated and quiescent cells are virtually the same (Williams and Penman 1975). It appears that normal cells contain a discrete subset of genes, transcription of which must be enhanced for growth to occur. These genes must be rare and their cognate mRNAs must be low in abundance in order to account for the quantitative and qualitative constancy observed during the G_0-to-S transition.

Recent developments in molecular biology and gene cloning technology have allowed detection and isolation of representative cell cycle genes. These genes may be induced by serum or by the individual growth factors contained within serum. At least one of the growth factor–inducible cell cycle genes has been recently identified as an oncogene (c-*myc*). In this review, we discuss the recent progress in isolation and characterization of mammalian cell cycle genes.

Genes Induced by Serum

Serum, as a complex mixture of growth factors, has been used for many years as a source of these factors for stimulation of quiescent cells. Johnson and co-workers first studied induction of dihydrofolate reductase (DHFR) as a function of cell cycle traverse upon serum stimulation of serum-deprived quiescent 3T6 mouse fibroblasts (Johnson et al. 1978, Wiedemann and Johnson 1979). Both level of functional enzyme and amount of DHFR capable of binding [^{3}H]methotrexate (Mtx, a potent inhibitor of DHFR) increased, beginning at about 8–10 hr after stimulation. DNA synthesis began at about 12 hr after stimulation. Thymidylate synthetase (TS) and thymidine kinase (TK) have been shown to respond similarly (Navalqund et al. 1980, Johnson et al. 1982) to serum stimulation.

Serum induction of mRNA for DHFR was also investigated as a function of serum stimulation in 3T6 cells (Hendrickson et al. 1980). The level of DHFR mRNA relative to total mRNA increased by 8 hr after serum stimulation. It peaked at about 16 hr. Furthermore, the stability of DHFR mRNA was unchanged, implying an increased synthetic rate. Later studies showed that, indeed, an increase in the DHFR pre-mRNA transcription rate could account for the increase in DHFR mRNA content (Wu and Johnson 1982).

DHFR cDNA was cloned using cDNA synthesized from a partially purified mRNA for DHFR (Chang et al. 1978). The mRNA was obtained from AT-3000 mouse cells, a cell line that is resistant to Mtx (Alt et al. 1978) and overproduces the mRNA and protein. A cDNA probe was similarly made from highly purified DHFR mRNA. Using this probe, 40% of the original bacterial transformants were found to be positive.

TS mRNA was recently cloned using a similar methodology. TS overproducer cells were first obtained using sequential selection in increasing levels of 5-fluoro-deoxyuridine (Rossana et al. 1982). These cells were found to overproduce TS as well as its mRNA. mRNA from these cells was used to synthesize the cDNA used for cloning, as well as the probe used for differential colony hybridization (Geyer and Johnson 1984).

Several other genes have been found to be responsive to serum stimulation in BALB/c-3T3 cells (Linzer and Nathans 1983) and have been isolated by molecular cloning. The time course of induction of these genes ranges widely. The amount of cognate mRNA for one gene increased by 6 hr and peaked at 12 hr after stimulation.

A second gene produced no increase in mRNA until between 6 and 12 hr; thereafter, a slow deinduction was noted. A third gene was induced between 12 and 18 hr and remained relatively constant thereafter.

Epidermal Growth Factor-Regulated Sequences

Epidermal growth factor (EGF) is synthesized at high levels in the submaxillary gland in adult mice. This high production level is dependent on stimulation of the gland by androgens. Although release of EGF from the submaxillary into the blood can be stimulated by alpha-adrenergic agents, the levels of EGF in plasma do not reflect submaxillary levels, indicating alternative sites of EGF synthesis in vivo that regulate plasma levels (for review, see Carpenter and Cohen 1979).

The biological activities of EGF are many, including acceleration of differentiation of epithelia/epidermis, induction of enzymatic activity, transport, and protein and RNA synthesis. EGF is also a powerful mitogen. Its mitogenic "action spectrum" is broad and includes fibroblasts, glia, and epithelial cells, as well as endothelial cells, chondrocytes, hepatocytes, and granulosa cells. The mitogenic response of these cells requires prolonged exposure to EGF. If EGF is removed from the culture medium or if anti-EGF antibodies are added to the medium prior to the onset of S phase, replicative DNA synthesis is much reduced. It has been suggested that EGF is one of the "progression" factors required after PDGF treatment in order for mouse fibroblasts to enter S phase (Leof et al. 1982).

Genomic sequences have been cloned that correspond to mRNAs induced by EGF in AKR-2B cells (Foster et al. 1982). The technique of differential plaque filter hybridization was used. Partially digested genomic *Eco*RI fragments were linked to charon 4A lambda phage. Replica filters containing phage DNA were probed with cDNA made to mRNA from either nongrowing cultures of AKR-2B cells or similar cultures treated with EGF at 10 ng/ml for 6 hr. From 10^5 plaques, three proved to be consistently positive for differential expression.

Quiescent cells express low levels of these sequences. One hour after EGF stimulation, their levels relative to other sequences increase only marginally. By 4 hr, however, the levels increase markedly, and by 6 hr the levels are 5- to 12-fold higher relative to the general population of mRNAs. Relative to quiescent levels, this corresponds to a 20- to 50-fold increase, since the content of total mRNA increases 4- to 5-fold by 6 hr of EGF treatment (Getz et al. 1976). Because the three cloned cDNAs used in these studies bore homology to a class of retrovirus- or transposition-like sequence elements, termed VL30, a VL30-specific probe was used to examine possible induction of VL30-related mRNAs by EGF. These mRNA sequences were also found to be induced moderately by 1 hr, but much more strongly by 4 hr of EGF treatment.

PDGF-Regulated Gene Sequences

PDGF is a polypeptide contained in the α-granules of blood platelets (Kaplan et al. 1979a,b, Gerrard et al. 1980). It is released when blood clots (Ross et al. 1974,

Kohler et al. 1974). Its biological activities are several: it acts as a chemoattractant for fibroblasts (Seppa et al. 1982), for lymphocytes (Deuel et al. 1982), and for smooth muscle cells (Grotendorst et al. 1981); causes cytoskeletal changes in cultured fibroblasts (Westermark et al. 1983); and is a powerful mitogen for smooth muscle and fibroblastic cells in vitro (Ross et al. 1974, Scher et al. 1978, Heldin et al. 1979). As a mitogen, PDGF induces in fibroblasts a stable state, termed "competence." For cells to enter into DNA synthesis (S phase), it is necessary for the growth factors in plasma to be present continuously, but the continued presence of PDGF is not required (Pledger et al. 1977). PDGF induces the competent state by causing the activation or synthesis of a cytoplasmic factor. Cytoplasts prepared from PDGF-treated cells can induce DNA synthesis when fused to quiescent cells (Smith and Stiles 1981). The induction of this transferrable competent state can be blocked by pretreatment of the donor cells with actinomycin D (AMD), an inhibitor of RNA synthesis.

To investigate the possibility that PDGF regulated specific gene expression, Pledger and co-workers (1981) examined the synthesis of cellular polypeptides after PDGF treatment of quiescent BALB/c-3T3 mouse fibroblasts. They found that several specific polypeptides were preferentially synthesized in response to PDGF but not in response to growth factors found in plasma. This induction could be blocked with AMD. Furthermore, a transformed cell line that did not require PDGF for growth also synthesized these polypeptides constitutively. More recently, in vitro translation has been used to show that several specific mRNAs are induced by PDGF (Hendrickson and Scher 1983), although the relation of the translation products of these mRNAs to the polypeptides synthesized in vivo is not fully known. The induction of these mRNAs is also blocked by AMD. Pretreatment with cycloheximide, a protein synthesis inhibitor, leads to differential induction of different classes of mRNA. The induction of one class is blocked, one is relatively unaffected, and a third is strongly enhanced. This superinduction suggests a feedback mechanism of regulation of one class of mRNA. Cloning of low-abundance mRNAs responsive to PDGF, as discussed below, has further amplified the understanding of mRNA induction by PDGF.

PDGF by itself does not promote cell cycling. Quiescent cells treated only with PDGF remain at least 12 hr away from DNA synthesis. The growth factors present in platelet-poor plasma are required for cells to "progress" into S phase. Therefore, biochemical changes in cells treated with PDGF are independent of those "events" cells undergo in the process of cell cycle traverse. This dissociation of PDGF-induced events from cell cycle events presents an opportunity to isolate gene sequences that are expressed as a direct consequence of PDGF action.

The cloning strategy is illustrated in Figure 1: Differential colony hybridization was used to screen cloned gene sequences from a cDNA library constructed from the mRNA of quiescent cells treated with PDGF. Sequences were selected that reacted weakly with a cDNA probe prepared from mRNA from quiescent cells, but reacted strongly with a cDNA probe made using PDGF-treated cell mRNA. Figure 2 shows a pair of filters screened with cDNA prepared from mRNA from

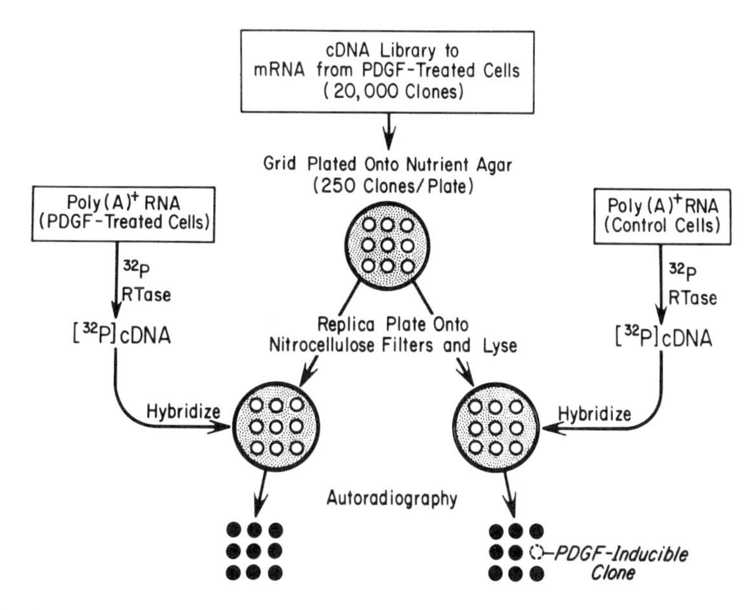

FIG. 1. Strategy for isolation of PDGF-inducible genes. (Reproduced from Cochran et al. 1984, with copyright permission of Cold Spring Harbor Laboratories.)

either quiescent or PDGF-stimulated cells. The detection limits of this screening protocol were determined by comparison with equivalent filters on which were grown bacteria carrying a plasmid-containing β-globin. These filters were screened with a dilution series of ^{32}P cDNA synthesized from globin mRNA (Figure 2, inset). The data suggest that cloned sequences present in the induced state at a level greater than 0.01% can be easily detected with cDNA containing induced sequences. Similarly, sequences would need to be present in the uninduced state at a level of $<.01\%$ to appear negative upon probing with quiescent cDNA. The sequences obtained are therefore probably low-abundance sequences (less than 0.01%, or 50 copies per cell) in the quiescent state and middle-abundance sequences in the induced state. These clones were subjected to a second and third round of screening.

Of 8000 cDNA clones screened by the protocol illustrated in Figure 1, 55 clones were found to be positive through three screenings (Table 1). Of these, 46 were found to represent only five independent sequences. For technical reasons the nine other clones could not be analyzed. Thus a minimum of five and maximum of 14 clones were obtained from the library of 8000 clones. One might expect that sequences present at a level of 0.01% would be present about 0.8 times in a library of 8000. It is possible that other sequences potentially above the threshold of detectability were not present in the library.

If one assumes that about half of the 8000-clone library is composed of sequences present only once in the library, then the independent clones isolated are 0.1% to 0.3% of the total. Thus, we estimated that PDGF regulates about 0.1 to 0.3% of

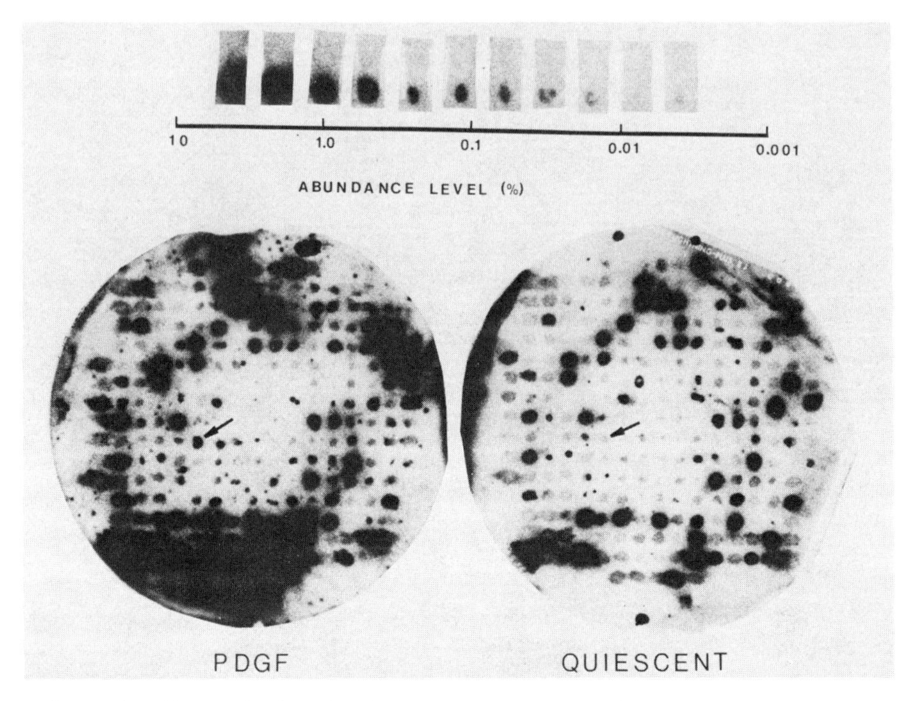

FIG. 2. Sensitivity and selectivity of the screening procedure. The two large circles are autoradiographs of replica-plated colony grids that have been lysed and hybridized to 2.5×10^6 cpm of [^{32}P]cDNA from quiescent or PDGF-treated 3T3 cells as described in Cochran et al. (1983). (→) The location of a clone containing a PDGF-inducible gene sequence. Top, Calibration strip showing autoradiographs of filters containing one colony each of pBR322 (upper) and pSV-β-globin (lower). Each filter was probed with twofold dilutions of β-globin [^{32}P]cDNA (100-200,000 cpm) under the same conditions as the large filters. The abundance level was calculated in percent as (cpm [^{32}P]globin/2.5×10^6) $\times$ 100. The pBR322 colonies do not hybridize β-globin to any significant degree and therefore are not visible in the autoradiographs. (Reproduced from Cochran et al. 1983, with copyright permission of MIT Press.)

the gene sequences expressed in 3T3 cells. This estimate is in fair agreement with that obtained from inspection of two-dimensional gels of proteins synthesized by cells treated with PDGF (Pledger et al. 1981).

Regulation of mRNA Content and Synthesis

Two of the gene sequences induced by PDGF, pBC-JE and pBC-KC (JE and KC), were used for further investigation. mRNA content was quantitated by using the dot blot technique of White and Bancroft (1982), which measures the extent of hybridization of cloned cDNA fragments to cytoplasmic mRNA. The results are shown in Table 2. Within quiescent, density-arrested cultures, KC and JE sequences are represented by low-abundance mRNAs. The levels found (70–100 copies/cell) correspond to the steady-state mRNA levels for phenobarbital-regulated genes within rat liver, which were measured by similar methods (Gonzalez and Kasper

TABLE 1. *Characterization of PDGF-inducible clones*

Plasmid	Insert size (bp)*	Times isolated from 8000-clone library†
pBC-JE	750	27
pBC-CIB	250	7
pBC-JB	1050	5
pBC-KC	820	4
pBC-CC	310	3
Others	—	9

*The size of the insert was determined by *Pst*1 digestion of the plasmid and electrophoresis on 5% polyacrylamide gels.

†Inserts were eluted from gels, nick-translated, and hybridized to all of the PDGF-inducible clones identified by differential colony hybridization.

Reproduced from Cochran et al. 1983, with copyright permission from MIT Press.

1982). PDGF treatment raises the content of mRNA for these clones to a moderate-abundance level (3000 copies/cell for JE and 700 copies/cell for KC). The relative levels of these two mRNAs are consistent with their relative frequencies in the cDNA library (27 times for JE and 4 times for KC; see Table 1), as well as the amount of induction on northern blots.

PDGF is relatively specific as an inducer of these sequences. Quiescent cultures were stimulated with a variety of growth factors. After 4 hr, total cytoplasmic RNA was extracted, and the levels of pBC-CB (CB), pBC-MF (MF), and KC mRNAs quantitated by cytoplasmic dot blot analysis. The induction of all three is highly dependent on the growth factor used (Figure 3). PDGF gives by far the strongest induction. Insulin evokes no response even at pharmacological concentrations. EGF gives a weak induction, the significance of which is not understood at present.

Of great interest is the mechanism of action of PDGF in increasing the level of these mRNA sequences. One may ask the following: Are these mRNAs induced directly in response to PDGF treatment, or are they a secondary response of cells that must first synthesize a regulatory protein? Clearly the former situation would be more interesting, for then the mode of action of PDGF is likely to be a direct coupling to the mRNA synthetic machinery. Since the cDNA library from which JE and KC were isolated was directed to mRNA induced by PDGF alone, one might expect to avoid selecting for most sequences induced during the mitogenic response, which is presumably a multistep process (Leof et al. 1982). Therefore, one may expect to shed light on whether or not competence formation, which is presumably due to mRNA induction (Smith and Stiles 1981), is also a multistep process.

To determine whether or not the regulatory system responsible for JE and KC mRNA induction required de novo protein synthesis, we treated quiescent cells

TABLE 2. *Abundance of KC and JE genes*

Gene	Input (cpm)*	Total hybridization (cpm)†	pBR322 hybridization†	Percentage of total poly (A) + RNA‡	mRNA copies/ PDGF-treated cell§	mRNA copies/ quiescent cell‖
JE	2.5×10^6	5225 ± 254	735 ± 101	0.9	3000	100
KC	2.5×10^6	1811 ± 107	735 ± 101	0.2	700	70

*Input is [^{32}P]cDNA made to poly (A) + RNA from cells treated with PDGF for 4 hr.

†Hybridization was performed as described. Each value is an average of three determinations.

‡This value was calculated as described in the text. The hybridization efficiency was 20% as measured by determining the proportion of β-globin cDNA that hybridized to a lysed colony harboring a β-globin plasmid at several different levels of input cDNA. The efficiency was constant in the range of 1000-6000 cpm bound, indicating both that the hybridization had gone to completion and that the plasmid DNA was bound to the filter in excess of the input cDNA.

§This number was calculated by assuming approximately 350,000 mRNAs per cell (Lewin 1980).

‖Copies of mRNA per cell in the induced state were determined by dividing the copy number per PDGF-treated cell by the relative induction from the uninduced to the induced state.

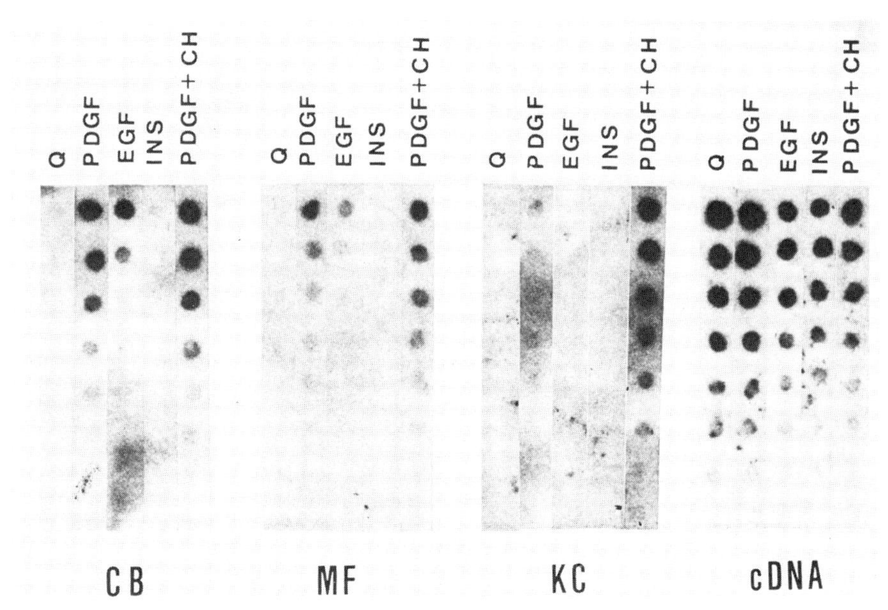

FIG. 3. Quiescent (Q) BALB/c-3T3 cells were incubated at 37°C for 4 hr with PDGF (500 units/ml), EGF (50 ng/ml), insulin (INS) (5 μg/ml), or PDGF (500 units/ml) and cycloheximide (CH) (10 μg/ml). RNA was then extracted from the cells, taken through serial twofold dilution, and blotted onto nitrocellulose filter paper. The filters were probed with [^{32}P]cDNA corresponding to the plasmid inserts from three representative PDGF-inducible gene sequences, CB, MF, and KC. As a control, to demonstrate that roughly the same amount of RNA was present in each dilution series, one filter was probed with ^{32}P-labeled total cellular cDNA. The filters were then processed by autoradiography. Details for these procedures are in Cochran et al. (1983). (Reproduced from Cochran et al. 1984, with copyright permission of Cold Spring Harbor Laboratories.)

with PDGF in the presence of cycloheximide, a protein synthesis inhibitor. Protein synthesis is required for progression of quiescent 3T3 cells from G_0/G_1 into S phase (Brooks 1977, Rossow et al. 1979). As shown in Figure 3, however, cycloheximide does not block the induction of these genes by PDGF. Quite surprisingly, one of these sequences, KC, was superinduced 20- to 60-fold over the levels found during PDGF treatment alone. Other protein synthesis inhibitors, such as puromycin, gave a similar result. These observations suggest the possibility, furthermore, that induction of KC mRNA may be feedback inhibited by a translation product normally synthesized in response to PDGF treatment.

The time course of specific mRNA induction is shown in Figure 4. The JE and KC sequences are both induced over 15-fold within the first hour of treatment by PDGF. Furthermore, KC sequences are rapidly turned over after 1 hr, leveling by about 2 hr of treatment. These results suggest the following two events as consequences of PDGF treatment. First, PDGF induces a dramatic increase in either the synthesis rate or stability of KC-related sequences. Second, after approximately 1 hr of treatment, there is a rapid change in these parameters, leading to the decrease in content seen by 2 hr of treatment. The first of these events is consistent with

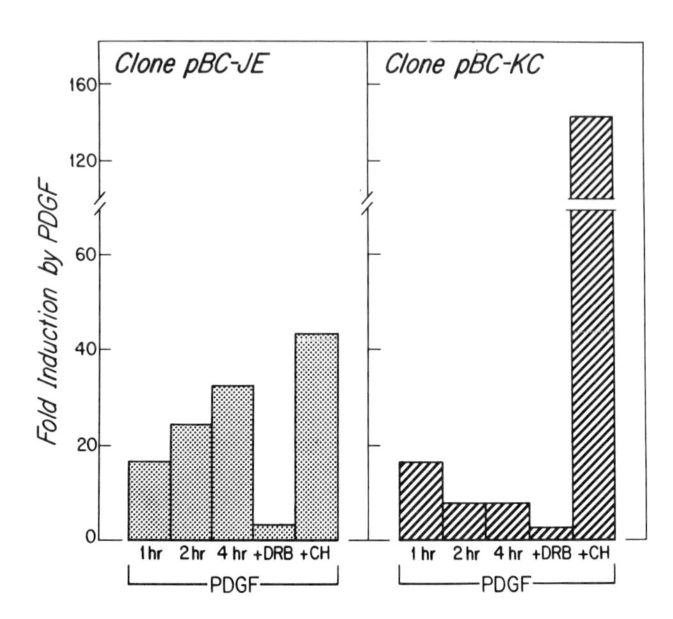

FIG. 4. Time course of KC and JE induction and the effect of metabolic inhibitors. Quiescent 3T3 cells were treated with 500 units/ml of crude PDGF, and at the indicated time the cells were lysed and prepared for dot blots as described in Figure 3. Each data point is an average of two experiments. DRB shows the effect of a 4-hr treatment of PDGF plus 15 μg/ml 5,6-dichloro-β-D-ribofuranosyl benzimidazole. CH is a 4-hr treatment with PDGF + 10 μg/ml cycloheximide. DRB alone had no effect on gene induction. Cycloheximide by itself produced a onefold to fourfold increase in gene transcripts. (Reproduced from Cochran et al. 1983, with copyright permission of MIT Press.)

the hypothesis that PDGF acts directly to increase KC mRNA content without the intermediary synthesis of a regulatory protein. The second event is consistent with the concept that a de novo synthesized regulatory protein is used to regulate the level of KC mRNA. It might require some time to increase the levels of this protein, and hence the delay in destabilization of KC mRNA. A composite of these hypotheses would predict the superinduction of KC mRNA in the presence of cycloheximide. Such an effect is illustrated again in Figure 4. Figure 4 also shows that 5,6-dichloro-β-D-ribofuranosylbenzimidazole (DRB), an inhibitor of mRNA synthesis, prevents the induction of both JE and KC. Hence de novo RNA synthesis is required for the induction response.

THE c-*myc* GENE IS INDUCED BY PDGF

Alteration of oncogene expression has been shown to contribute to tumorigenesis in a variety of neoplasms. In particular, an increased level of c-*myc* mRNA has been implicated in the growth transformation of murine plasmacytomas and human Burkitt's lymphomas (for a review, see Leder et al. 1983). The level of c-*myc* mRNA has been shown to respond to PDGF in BALB/c-3T3 cells (Kelly et al. 1983). Quiescent cultures were treated with a variety of growth factors and the isolated RNA probed using a highly sensitive Sl-nuclease assay (Kelly et al. 1983). Untreated controls showed very low levels of c-*myc* mRNA in comparison with resting cultured spleen cells. After 3 hr of PDGF treatment, with either crude or pure PDGF, the level of c-*myc* mRNA increased about 40-fold. Factors that mimic the competence formation ability of PDGF, namely fibroblast growth factor (FGF) and tetradecanoylphorbol acetate (TPA) were also able to induce c-*myc* mRNA. Other factors that are required later in the proliferative response—EGF, insulin, or 5% platelet poor plasma—were not able to induce c-*myc*. PDGF treatment in the presence of cycloheximide led to a slight superinduction.

DISCUSSION

Regulation of Gene Expression by PDGF

It is estimated that 0.1–0.3% of the genes expressed in 3T3 cells are induced by PDGF. This would represent 10–30 independent sequences. Of these, 5–14 have been isolated. These cloned sequences represent low-abundance mRNAs in quiescent 3T3 cells. The abundance increases up to 20-fold within 1 hr of PDGF treatment. A repressor protein also induced by PDGF may be involved in feedback inhibition of one of the PDGF-inducible sequences. In the absence of protein synthesis, up to 100-fold increases over basal level have been seen for sequences related to clone KC. Further experiments may define whether or not this repressor protein is synthesized from preexisting mRNA or requires the induction of one of the PDGF-regulated mRNAs.

The mechanism by which PDGF induces specific mRNA synthesis is of great interest. PDGF stimulates a number of immediate events, detectable within minutes of PDGF addition, including intracellular alkalinization via H^+/Na^+ exchange (Cassel et al. 1983), changes in phospholipase activity (Habenicht et al. 1981, Shier 1980), alteration of cytoskeletal architecture (Westermark et al. 1983), and protein phosphorylation specifically on tyrosine residues (Ek and Heldin 1982, Nishimura et al. 1982, Cooper et al. 1982, Heldin et al. 1983). Some of these actions, for example tyrosine-specific protein phosphorylation, are mimicked by other growth factors that do not regulate expression of the PDGF-induced gene sequences (Kasuga et al. 1983, Cohen et al. 1982a, b). In the case of tyrosine phosphorylation, differences in the target proteins that are phosphorylated by the growth factor receptors might account for the specificity of gene induction.

The induction of specific gene sequences as a consequence of EGF or serum stimulation is problematic in that the exact relation of induction to growth factor treatment is ill defined. The induction itself may even be unrelated to growth factor treatment per se; it might be a consequence of cell growth or metabolic changes. Pledger and co-workers have defined two "restriction" points during G_1 at which specific growth factors are required for progression of cells into S phase (Leof et al. 1982). In BALB/c-3T3 cells these are at 6 and 10 hr after PDGF treatment. Although these times correspond roughly with the induction of either EGF-inducible

(Foster et al. 1982) or serum-inducible (Linzer and Nathans 1983) mRNAs, much better definition of the timing of the induction events and requirements for growth factors will be necessary. The possibility nonetheless exists that some of these mRNAs may be induced as a result of synergistic action between growth factors. PDGF and EGF have been shown to act synergistically in the induction of DNA synthesis (Leof et al. 1982). One might speculate that the synergism is due to a synergistic induction of mRNA by EGF in cooperation with one of the gene products of PDGF-inducible genes.

Regulation of the Proliferative Response by Growth Factor-Induced mRNAs

Competence

It must be kept in mind that there are cellular responses to growth factors other than mitogenesis. Besides those activities described above, PDGF induces phospholipid synthesis (Habenicht et al. 1981) and chemotaxis (Seppa et al. 1982). Gene sequences induced by PDGF may also play a role in the expression of these activities. Nonetheless, there is evidence suggesting that PDGF-inducible mRNAs, or their translation products, play a functional role in the chain of events leading to DNA synthesis and replication. Somatic cell fusion shows that the mitogenic response to PDGF can be transferred to another via a cytoplasmic intermediate. Furthermore, de novo RNA synthesis is required for the donor cell to generate this signal (Smith and Stiles 1981). Interpretation of metabolic inhibitor studies is always risky, but a simple view of the cell fusion data suggests that PDGF may stimulate the synthesis of new mRNAs and that the translation products are responsible for initiation of the mitogenic response. At least one PDGF-induced sequence has been directly implicated in the growth response to PDGF—the cellular oncogene homologue c-*myc*. Increased levels of c-*myc* mRNA have been associated with the loss of growth regulation in murine plasmacytomas and Burkitt's lymphomas (Leder et al. 1983). Furthermore, the level of c-*myc* mRNA can be raised in cultured 3T3 fibroblasts by treatment with PDGF (Kelly et al. 1983). The avian c-*myc* gene product is a 58,000-MW polypeptide associated with the nucleus (Alitalo et al. 1983). It seems possible that c-*myc* would assert its effects by a second round of induction of specific gene sequences.

Cell Cycle

A commitment point in the cell cycle occurs shortly before the onset of replicative DNA synthesis. Variously called the restriction point (Rossow et al. 1979) or "W" point (Leof et al. 1982), cells do not need growth factors past this point to continue through the entire S phase of the cell cycle. One may also speculate that here, too, specific gene induction is responsible. Candidate mRNAs in the form of cloned cDNA sequences respond to serum induction between 6 and 12 hr after stimulation.

Of interest is the finding that PDGF-inducible genes are possibly feedback inhibited near the W point. By 12 hr after addition of PDGF to quiescent monolayers (when the first cells have entered S), the levels of PDGF-inducible mRNAs drop, even with continued PDGF treatment. This drop in mRNA abundance may be related to the commitment events that occur immediately before DNA synthesis.

The level of these mRNAs in sparse, exponentially growing 3T3 cell cultures is only moderately higher than in quiescent cultures (B. H. Cochran et al., personal communication). These observations reflect the possibility that PDGF-inducible genes—as with the other serum-induced genes—are expressed only intermittently during the cell cycle, and that during each cycle they "fire" as required to transmit the growth regulatory signal of the corresponding growth factors.

ACKNOWLEDGMENTS

Work from the authors' laboratory summarized here was supported by grants from the National Institutes of Health. BHC is supported by a fellowship from the Noble Foundation. CDS is supported by a Faculty Research Award from the American Cancer Society.

REFERENCES

Alitalo K, Ramsay G, Bishop JM, Pfeifer SO, Colby WW, Levinson AD. 1983. Identification of nuclear proteins encoded by viral and cellular *myc* oncogenes. Nature 306:274–277.

Alt FW, Kellems RE, Bertino JR, Schimke RT. 1978. Selective multiplication of dihydrofolate reductase genes in methotrexate-resistant variants of cultured murine cells. J Biol Chem 253:1357–1370.

Bishop JM. 1983. Cancer genes come of age. Cell 32:1018–1020.

Brooks RF. 1977. Continuous protein synthesis is required to maintain the probability of entry into S phase. Cell 12:311–317.

Brooks RF, Bennett DC, Smith JA. 1980. Mammalian cell cycles need two random transitions. Cell 19:493–504.

Carpenter G, Cohen S. 1979. Epidermal growth factor. Annu Rev Biochem 48:193–216.

Cassel D, Rothenberg P, Zhuang Y, Deuel TF, Glaser L. 1983. Platelet-derived growth factor stimulates Na^+/H^+ exchange and induces cytoplasmic alkalinization in NR6 cells. Proc Natl Acad Sci USA 80:6224–6228.

Chang ACY, Nunberg JH, Kaufman RJ, Erlich HA, Schimke RT, Cohen SN. 1978. Phenotypic expression in *E. coli* of a DNA sequence coding for mouse dihydrofolate reductase. Nature 275:617–624.

Cohen S, Ushiro H, Stoscheck C, Chinkers M. 1982a. A native 170,000 epidermal growth factor receptor-kinase complex from shed plasma membrane vesicles. J Biol Chem 257:1523–1531.

Cohen S, Fava RA, Sawyer ST. 1982b. Purification and characterization of epidermal growth factor receptor/protein kinase from normal mouse liver. Proc Natl Acad Sci USA 79:6237–6241.

Cochran BH, Reffel AC, Stiles CD. 1983. Molecular cloning of gene sequences regulated by platelet-derived growth factor. Cell 33:939–947

Cochran BH, Reffel AC, Callahan MA, Zullo JN, Stiles CD. 1984. Rapidly inducible cell cycle genes regulated by platelet-derived growth factor. *In* Cancer Cells, Vol. I. Cold Spring Harbor Laboratory Press, Cold Spring Harbor, New York, pp. 51–56.

Cooper JA, Bowen-Pope DF, Raines E, Ross R, Hunter T. 1982. Similar effects of platelet-derived growth factor and epidermal growth factor on the phosphorylation of tyrosine in cellular proteins. Cell 31:263–273.

Deuel TF, Senior RM, Huang JS, Griffin GL. 1982. Chemotaxis of monocytes and neutrophils to platelet-derived growth factor. J Clin Invest 69:1046–1049.

Devare SG, Reddy EP, Law JD, Robbins KC, Aaronson SA. 1983. Nucleotide sequences of the simian sarcoma virus genome: demonstration that its acquired cellular sequences encode the transforming gene product p28sis. Proc Natl Acad Sci USA 80:731–735.

Doolittle RF, Hunkapiller MW, Hood LE, Devare SG, Robbins KC, Aaronson SA, Antoniades HN. 1983. Simian sarcoma virus onc gene, v-sis, is derived from the gene (or genes) encoding a platelet-derived growth factor. Science 221:275–276.

Downward J, Yarden Y, Mayes E, Scrace G, Totty N, Stockwell P, Ullrich A, Schlessinger J, Waterfield MD. 1984. Close similarity of epidermal growth factor receptor and v-erb-B oncogene protein sequences. Nature 307:521–527.

Ek B, Heldin CH. 1982. Characterization of a tyrosine-specific kinase activity in human fibroblast membranes stimulated by platelet-derived growth factor. J Biol Chem 257:10486–10492.

Eva A, Robbins KC, Andersen PR, Srinivasan A, Tronick SR, Reddy EP, Ellmore NW, Galen AT, Lautenberger JA, Papas TS, Westin EH, Wong-Staal F, Gallo RC, Aaronson SA. 1982. Cellular genes analogous to retroviral onc genes are transcribed in human tumour cells. Nature 295:116–119.

Foster DN, Schmidt LJ, Hodgson CP, Moses HL, Getz MJ. 1982. Polyadenylated RNA complementary to a mouse retrovirus-like multigene family is rapidly and specifically induced by epidermal growth factor stimulation of quiescent cells. Proc Natl Acad Sci USA 79:7317–7321.

Gerrard JM, Phillips DR, Rao GHR, Plow EG, Waltz DA, Ross R, Harker LA, White JG. 1980. Biochemical studies of two patients with Grey platelet syndrome. Selective deficiency of platelet alpha-granules. J Clin Invest 66:102–109.

Getz MJ, Elder PK, Benz EW Jr, Stephens RE, Moses HL. 1976. Effect of cell proliferation on levels and diversity of poly(A)-containing mRNA. Cell 7:255–265.

Geyer PK, Johnson LF. 1984. Molecular cloning of DNA sequences complementary to mouse thymidylate synthase messenger RNA. J. Biol Chem 259:7206–7211.

Gonzalez FJ, Kasper CB. 1982. Cloning of cDNA complementary to rat liver NADPH-cytochrome c(P-450) oxidoreductase and cytochrome P-4506 mRNAs. J Biol Chem 257:5962–5968.

Grotendorst GR, Seppa HEJ, Kleinman HK, Martin GR. 1981. Attachment of smooth muscle cells to collagen and their migration toward platelet-derived growth factor. Proc Natl Acad Sci USA 78:3669–3672.

Habenicht AJR, Glomset JA, King WC, Nist C, Mithcell CD, Ross R. 1981. Early changes in phosphatidylinositol and arachidonic acid metabolism in quiescent Swiss 3T3 cells stimulated to divide by platelet-derived growth factor. J Biol Chem 256:12329–12335.

Heldin CH, Westermark B, Wasteson A. 1979. Platelet-derived growth factor: purification and partial characterization. Proc Natl Acad Sci USA 76:3722–3726.

Heldin CH, Westermark B, Wasteson A. 1980. Chemical and biological properties of a growth factor from human cultured osteosarcoma cells: resemblance with PDGF. J Cell Physiol 105:235–246.

Heldin CH, Westermark B, Wasteson A. 1981. Specific receptors for platelet-derived growth factor on cells derived from connective tissue and glia. Proc Natl Acad Sci USA 78:3664–3668.

Heldin CH, Ek B, Ronnstrand L. 1983. Characterization of the receptor for platelet-derived growth factor on human fibroblasts. Demonstration of an intimate relationship with a 185,000 dalton substrate for the PDGF-stimulated kinase. J Biol Chem 258:10054–10061.

Hendrickson SL, Scher CD. 1983. Platelet-derived growth factor-modulated translatable mRNAs. Mol Cell Biol 3:1478–1487.

Hendrickson SL, Wu JR, Johnson LF. 1980. Cell cycle regulation of dihydrofolate reductase mRNA metabolism in mouse fibroblasts. Proc Natl Acad Sci USA 77:5140–5144.

Johnson LF, Fuhrman CL, Wiedemann LM. 1978. Regulation of dihydrofolate reductase gene expression in mouse fibroblasts during the transition from the resting to the growing state. J Cell Physiol 97:397–406.

Johnson LF, Rao LG, Muench AJ. 1982. Regulation of thymidine kinase enzyme level in serum-stimulated mouse 3T6 fibroblasts. Exp Cell Res 138:79–85.

Kaplan DR, Chao FC, Stiles CD, Antoniades HN, Scher CD. 1979a. Platelet alpha-granules contain a growth factor for fibroblasts. Blood 53:1043–1052.

Kaplan KL, Broekman JM, Chernoff A, Lesznick GR, Drillings M. 1979b. Platelet alpha-granule proteins: studies on release and subcellular localization. Blood 53:604–615.

Kasuga M, Fujita-Yamaguchi Y, Blithe DL, Kahn CR. 1983. Tyrosine-specific protein kinase activity is associated with the purified insulin receptor. Proc Natl Acad Sci USA 80:2137–2141.

Kelly K, Cochran BH, Stiles CD, Leder P. 1983. Cell-specific regulation of the c-*myc* gene by lymphocyte mitogens and platelet-derived growth factor. Cell 35:603–610.

Kohler N, Lipton A. 1974. Platelets as a source of fibroblast growth promoting activity. Exp Cell Res 87:297–301.

Leder P, Battey J, Lenoir G, Moulding C, Murphy W, Potter H, Stewart T, Taub R. 1983. Translocations among antibody genes in human cancer. Science 222:765–771.
Leof EB, Wharton W, Van Wyk JJ, Pledger WJ. 1982. Epidermal growth factor and somatomedin C regulate G_1 progression in competent BALB/c-3T3 cells. Exp Cell Res 141:107–115.
Lewin B. 1980. Gene Expression 2, 2nd edition. John Wiley & Sons, New York.
Linzer DIH, Nathans D. 1983. Growth-related changes in specific mRNAs of cultured mouse cells. Proc Natl Acad Sci USA 80:4271–4275.
Navalqund LG, Rossana C, Muench AJ, Johnson LF. 1980. Cell cycle regulation of thymidylate synthetase gene expression in cultured mouse fibroblasts. J Biol Chem 255:7386–7390.
Nishimura J, Huang JS, Deuel TF. 1982. Platelet-derived growth factor stimulates tyrosine-specific protein kinase activity in Swiss mouse 3T3 membranes. Proc Natl Acad Sci USA 79:4303–4307.
Pledger WJ, Stiles CD, Antoniades HN, Scher CD. 1977. Induction of DNA synthesis in BALB/c-3T3 cells by serum components: reevaluation of the commitment process. Proc Natl Acad Sci USA 74:4481–4485.
Pledger WJ, Hart CA, Locatell KL, Scher CD. 1981. Platelet-derived growth factor-modulated proteins: Constitutive synthesis by a transformed cell line. Proc Natl Acad Sci USA 78:4385–4362.
Ross R, Glomset J, Kariya B, Harker L. 1974. A platelet-dependent serum factor that stimulates the proliferation of arterial smooth muscle cells in culture. Proc Natl Acad Sci USA 71:1207–1210.
Rossana C, Rao LG, Johnson LF. 1982. Thymidylate synthetase overproduction in 5-fluorodeoxyuridine-resistant mouse fibroblasts. Mol Cell Biol 2:1118–1125.
Rossow PW, Riddle VGH, Pardee AB. 1979. Synthesis of labile, serum-dependent protein in early G_1 controls animal cell growth. Proc Natl Acad Sci USA 76:4446–4450.
Scher CD, Pledger WJ, Martin P, Antoniades HN, Stiles CD. 1978. Transforming viruses directly reduce the cellular growth requirement for a platelet-derived growth factor. J Cell Physiol 97:371–380.
Seppa H, Grotendorst G, Seppa S, Schiffmann E, Martin GR. 1982. Platelet-derived growth factor is chemotactic for fibroblasts. J Cell Biol 92:584–588.
Shier WT. 1980. Serum stimulation of phospholipase A2 and prostaglandin release in 3T3 cells is associated with platelet-derived growth promoting activity. Proc Natl Acad Sci USA 77:137–141.
Smith JC, Stiles CD. 1981. Cytoplasmic transfer of the mitogenic response to platelet-derived growth factor. Proc Natl Acad Sci USA 78:4363–4367.
Stiles CD. 1983. The molecular biology of platelet-derived growth factor. Cell 33:653–655.
Waterfield MD, Scrace GT, Whittle N, Stroobant P, Johnsson A, Wasteson A, Westermark B, Heldin CH, Huang JS, Deuel TF. 1983. Platelet-derived growth factor is structurally related to the putative transforming protein p28sis of simian sarcoma virus. Nature 304:35–39.
Westermark B, Heldin CH, Ek B, Johnsson A, Mellstrom K, Nister M, Wasteson A. 1983. Biochemistry and biology of platelet-derived growth factor. *In* G. Guroff (ed), Growth and Maturation Factors. J. Wiley and Sons, Inc., New York, pp. 75–114.
White BA, Bancroft FC. 1982. Cytoplasmic dot hybridization—simple analysis of relative mRNA levels in multiple small cell or tissue samples. J Biol Chem 257:8569–8572.
Wiedemann LM, Johnson LF. 1979. Regulation of dihydrofolate reductase synthesis in an overproducing cell line during transition from resting to growing state. Proc Natl Acad Sci USA 76:2818–2822.
Williams JG, Penman S. 1975. The messenger RNA sequences in growing and resting fibroblasts. Cell 6:197–206.
Wu JR, Johnson LF. 1982. Regulation of dihydrofolate reductase gene transcription in methotrexate-resistant mouse fibroblasts. J Cell Physiol 110:183–189.

Mediators in Cell Growth and Differentiation,
edited by Richard J. Ford and Abby L. Maizel.
Raven Press, New York © 1985.

Nerve Growth Factor: Mechanism of Action

Ralph A. Bradshaw, Joan C. Dunbar, Paul J. Isackson,
Rozenn N. Kouchalakos, and Claudia J. Morgan*

*Department of Biological Chemistry, California College of Medicine,
University of California, Irvine, California 92717*

Among the rapidly expanding group of growth regulating substances, collectively referred to as polypeptide growth factors, nerve growth factor (NGF) has many distinguishing features. Historically, NGF was the first substance to bear the designation "growth factor," following its discovery in the early 1950s (Levi-Montalcini and Hamburger 1951, Levi-Montalcini 1952), albeit several substances now appreciated to act at least in part as growth factors were discovered prior to this, e.g. insulin. Research on NGF also led to the discovery of epidermal growth factor (EGF) by Cohen (1962), a molecule that has been equally important in advancing understanding of growth factor structure and function. NGF also presaged the general discovery of receptor-mediated endocytosis of substances of this type through studies on retrograde axonal transport (Hendry et al. 1974a) and was instrumental in elucidating the hormonelike character of the polypeptide growth factor subset (Frazier et al. 1972). More recently, the discovery of NGF has also, by example, provided inspiration to search for and identify other neurotrophic substances with neuronal targets in both the peripheral and central nervous systems (Barde et al. 1983).

As depicted schematically in Figure 1, four major types of neurotrophic factors can be envisioned based on the localization of their sites of synthesis and responsive cells. This diagram does not distinguish between neurons of the peripheral and central nervous systems nor does it draw a distinction between nonneuronal cells located inside or outside neural tissue. The introduction of these qualifying features would considerably expand the number of categories.

The basis of this matrix is the fundamental principle of endocrinology that hormonal messengers are elaborated in one cell type and travel to a second cell type to exert their effect. Coupled with the observations of Hendry, Thoenen, and their colleagues (Hendry et al. 1974a,b, Stöckel et al. 1974, 1975a,b, Paravicini et al. 1975) that neurons that are responsive to NGF (sympathetic and sensory neurons derived from the neural crest) are capable of specifically binding, inter-

*Present address: Department of Microbiology and Immunology, Albert Einstein College of Medicine, New York, New York 10461

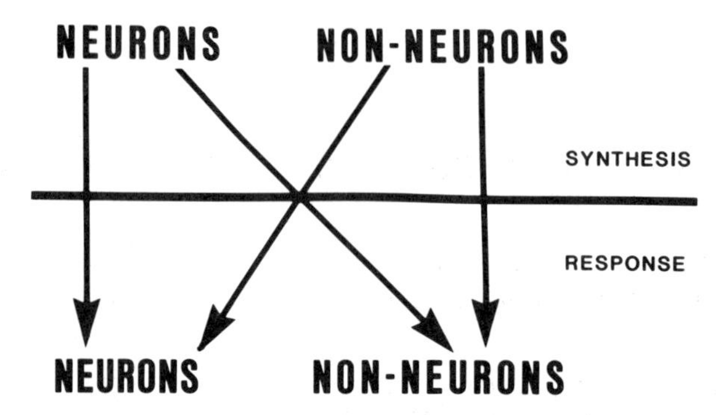

FIG. 1. Schematic representation of predicted neurotrophic growth factor families.

nalizing, and transporting it (and its receptor) to the cell bodies in a retrograde fashion, a widely accepted model for the action of NGF that is a specialized case of this tenet has evolved (Bradshaw 1978, Andres and Bradshaw 1980, Greene and Shooter 1980).

In the main, this model proposes that NGF (1) is produced by cells that are enervated by responsive neurons, (2) diffuses to the presynaptic membrane and binds to specific cell surface receptors, (3) is internalized by receptor-mediated endocytosis, and (4) is translocated to the perikaryon by retrograde axonal transport, where the ligand, receptor, or third molecule (as yet unidentified) generated as a second messenger at the cell surface membrane is responsible for both the short- and long-term effects of the hormone. Although substantial evidence in support of this model has accumulated, rigorous proof in any physiologically relevant system remains to be determined. Principal among its unconfirmed aspects are the exact location and nature of the NGF synthesized for these endocrine functions (in contradistinction to the well-characterized exocrine sources (Bradshaw 1978)) and the importance of internalization (and therefore retrograde axonal transport) in the expression of activity.

BIOSYNTHESIS OF NGF

NGF is elaborated in a wide variety of tissues that can be conveniently grouped as exocrine and endocrine. The exocrine sources, including the adult male mouse submandibular gland, the prostate glands of several higher vertebrates, and a wide variety of snake venoms, have provided the vast majority of NGF for chemical (and functional) characterization (Server and Shooter 1977, Bradshaw 1978, Hogue-Angeletti and Bradshaw 1979, Thomas and Bradshaw 1980, Rubin and Bradshaw 1981). The role of the hormone in these tissues and their secretions is not understood. The endocrine sources of NGF, which represent the putative physiologically relevant ones, remain poorly characterized primarily because of their much lower levels of synthesis (Bradshaw and Young 1976).

The best-characterized form of NGF is that expressed and released by the adult male mouse submandibular gland. This high-molecular-weight complex (designated

7S NGF) (Varon et al. 1967) is composed of three types of polypeptide chains. The complete covalent structure of each is now known (Angeletti and Bradshaw 1971, Thomas et al. 1981b, Isackson et al. 1984). They are assembled into a symmetrical aggregate in which two α- and two γ-subunits are noncovalently attached to the β dimer (Varon et al. 1968). The β-subunit (in its dimeric form) can be isolated directly from tissue homogenates and is referred to in this form as 2.5S NGF (Bocchini and Angeletti 1969). It differs from the β-subunit only by minor proteolytic modifications at the N- and C- termini; these modifications do not affect biological activity (Server and Shooter 1977, Thomas and Bradshaw 1980). The amino acid sequence of the mature, active form of the hormone contains 118 amino acids commencing with an N-terminal serine and terminating in a carboxyl terminal arginine (Angeletti and Bradshaw 1971). There are three internal disulfide bonds. The prepro sequence, predicted from cDNA structures (Scott et al. 1983, Ullrich et al. 1983), suggests a molecule with molecular mass of 33,000 Da and with the active hormone sequence positioned at the carboxyl terminus. An alternative initiator methionine site downstream would produce a preproβNGF some 5,000 Da smaller (Ullrich et al. 1983). There is a two-residue (Arg-Gly) extension at the carboxyl terminus that is also removed at some stage in the processing of the precursor molecule. The sequence for human preproβNGF has been determined from genomic sequences (Ullrich et al. 1983) and the gene localized to the short arm of chromosome 1 (Francke et al. 1983).

The mature βNGF shows a statistically significant relatedness to the pancreatic hormone insulin (proinsulin) (Frazier et al. 1972) and is thus similar to three other insulin-related molecules (insulinlike growth factor I and II and relaxin) (Bradshaw and Niall 1978). This relationship is manifested in some similarities in function and may well be reflected in homology at the receptor level too (see below).

The γ-subunit, which has been proposed to be a processing enzyme for a preproβNGF (Angeletti and Bradshaw 1971, Greene and Shooter 1980), is a protease with high specificity for arginine residues that is homologous to the trypsinlike enzymes (serine proteases) (Greene et al. 1969, Thomas et al. 1981b). Structural studies revealed that the molecule occurs in two or three chain structures that arise from limited proteolysis. These and subsequent exoproteolytic modifications result in the variety of electrophoretic variants that have been observed (Server and Shooter 1976, Thomas et al. 1981a).

The mature protein contains a carbohydrate side chain attached to asparagine 78, which is also variably processed to a heterogeneous group of glycosyl moieties that do not, however, contribute to the aforementioned electrophoretic variations (Edelman M, Baenziger JU, Thomas KA, Baglan NC, and Bradshaw RA, unpublished observations). The predicted amino acid sequence of the prepro γ-subunit has been determined from the corresponding cDNA (Ullrich A, personal communication). This sequence suggests an amino terminal extension of 24 residues that is presumably divided into a pre- and pro-segment, the pre-segment being required for insertion of the pro-γ-subunit into vesicles and the removal of the pro-segment, by analogy to other serine proteases, being required for the formation of active enzyme. The cDNA sequence also predicts that the mature protein is initially formed as a single polypeptide chain and that a four-residue segment is removed to form the two-chain structure, which can in turn be further processed to the three-chain structure that characterizes the subunit as isolated (Thomas et al. 1981b).

Partial amino acid sequence analysis of the α-subunit revealed the surprising result that it was highly similar to the γ-subunit (Isackson and Bradshaw 1984). As shown in Figure 2, segments corresponding to 200 residues of the α-subunit were partially sequenced (146 residues), revealing only 27 differences between the α- and γ-subunits. Catalytic measurements failed to reveal any hydrolysis activity for the α-subunit and prolonged reaction with diisopropylfluorophosphate showed very modest reactivity relative to the γ-subunit or bovine trypsin. The amount of incorporation was consistent with that observed for inactive zymogens of other serine proteases (Raben D, Isackson PJ, and Bradshaw RA, unpublished observations).

Synthetic oligonucleotide probes were prepared for several regions of the α molecule that appeared unique by comparison either to the γ-subunit or to a glandular kallikrein, another serine protease of the mouse submandibular gland with a similar structure (Figure 2) (Mason et al. 1983). The synthetic probes corresponding to residues 44 to 48 and 161 to 165 allowed the identification of three cDNA clones in a mouse submandibular library. The complete sequence of α has been predicted from the nucleic acid sequence of these clones and has revealed additional interesting structural features of the molecule (Isackson et al. 1984). The most obvious difference between α- and γ-subunits (and other serine proteases) is in the amino terminal region. The α-subunit contains an identical pre- and pro-leader sequence to that found in the γ-subunit (Ullrich A, personal communication) except that the arginine residue, which immediately precedes the N-terminus of the mature protein, is replaced with a glutamine residue. In addition, residues 2 through 5 of the γ-subunit are deleted from the α-subunit. Amino terminal analysis of whole α-subunit, as well as of the two chains that compose the mature form, have failed to reveal any open N-termini (Isackson and Bradshaw 1984), and the exact position of the bond separating the pre- and pro-segments is not obvious from the cDNA analysis results. Thus, the nature of the blocking group is unclear at present. These variations in structure, as well as other modifications in the sequence (such as a substitution of the histidine for glycine at position 187, a change that occurs two residues to the amino side of the active-site serine), probably account for the lack of catalytic activity of the molecule.

The similarity of α- and γ-subunits in terms of amino acid sequence is in keeping with the observations of Mason et al. (1983) that the adult male mouse submandibular gland expresses a large family of highly related, serine proteases or serine proteaselike molecules. However, these observations do not provide further insight into the role of α (or γ) in the synthesis or action of NGF. The availability of appropriate nucleotide probes should allow the detailed examination of the endocrine sources of NGF to ascertain which of the three subunits are in fact synthesized in these tissues. It should be noted that the synthetic oligonucleotide probes required

```
        1                     10                        20
mGK-1   I-V-G-G-F-K-C-E-K-N-S-Q-P-W-H-V-A-V-Y-R-Y-K-E-Y-I-
γ-NGF   I-V-G-G-F-K-C-E-K-N-S-Q-P-W-H-V-A-V-Y-R-Y-T-Q-Y-L-
α-NGF  X-[

                30                        40                50
mGK-1   C-G-G-V-L-L-D-A-N-W-V-L-T-A-A-H-C-Y-Y-E-K-N-N-V-W-
γ-NGF   C-G-G-V-L-L-D-P-N-W-V-L-T-A-A-H-C-Y-D-D-N-Y-K-V-W-
α-NGF                  ]N-W-V-L-T-A-A-H-C-Y-N-D-K-Y-Q-V-W-

                            60                    70
mGK-1   L-G-K-N-N-L-Y-Q-D-E-P-S-A-Q-H-R-L-V-S-K-S-F-L-H-P-
γ-NGF   L-G-K-N-N-L-F-K-D-E-P-S-A-Q-H-R-F-V-S-K-A-I-P-H-P-
α-NGF   L-G-K-N-N-F-L-E(B,Z,P,S,B,Z,H,R)L-V-S-K-A-I-P-H-P-

                80                     90                 100
mGK-1   C-Y-N-M-S-L-H-R-N-R-I-Q-N-P-Q-D-D-Y-S-Y-D-L-M-L-L-
γ-NGF   G-F-N*-M-S-L-M-R          F-L-E-Y-D-Y-S-N-D-L-M-L-L-
α-NGF   D-F-N*-M-S-L-L-N-E-H-T-P-Q-P-E-D-D-Y-S-N-D-L-M(L,L,

                            110                   120
mGK-1   R-L-S-K-P-A-D-I-T-D-V-V-K-P-I-A-L-P-T-E-E-P-K-L-G-
γ-NGF   R-L-S-K-P-A-D-I-T-D-T-V-K-P-I-T-L-P-T-E-E-P-K-L-G-
α-NGF   R)L-S-K-P-A-D-I-T-D-V-V-K-P-I-T-L-P-T-E-E-P-K-L-G-

                130                  140                 150
mGK-1   S-T-C-L-A-S-G-W-G-S-I-I-P-V-K-F-Q-Y-A-K-D-L-Q-C-V-
γ-NGF   S-T-C-L-A-S-G-W-G-S-I-T-P-T-K-F-Q-F-T-D-D-L-Y-C-V-
α-NGF   S-T-C-L-A-S-G-W-G(S,I,T,P,T)  X-[  ]Y-P-D-D-L-Q-C-V-

                            160                   170
mGK-1   N-L-K-L-L-P-N-E-D-C-D-K-A-Y-V-Q-K-V-T-D-V-M-L-C-A-
γ-NGF   N-L-K-L-L-P-N-E-D-C-A-K-A-H-I-E-K-V-T-D-A-M-L-C-A-
α-NGF   N-L-K-L-L-P-N-E-D-C-D-K-A-H-E-M-K-V-T-D-A-M-L-C-A-

                180                  190                  200
mGK-1   G-V-K-G-G-G-K-D-T-C-K-G-D-S-G-G-P-L-I-C-D-G-V-L-Q-
γ-NGF   G-E-M-D-G-G-K-D-T-C-K-G-D-S-G-G-P-L-I-C-D-G-V-L-Q-
α-NGF   G-E-M-D-G-G-S-Y-T-C-E-H-D-S-G-G-P-L-I-C-D(G,V,L,Z,

                            210                   220
mGK-1   G-L-T-S-W-G-Y-N-P-C-G-E-P-K-K-P-G-V-Y-T-K-L-I-K-F-
γ-NGF   G-I-T-S-W-G-H-T-P-C-G-E-P-D-M-P-G-V-Y-T-K-L-N-K-F-
α-NGF   G,I,T,S,W,G,H,T,P,C,G,Z,P,B,M,P,G,P,Y,T,Z,L,B,Z,G,

                230             237
mGK-1   T-S-W-I-K-D-T-L-A-Q-N-P
γ-NGF   T-S-W-I-K-D-T-M-A-K-N-P
α-NGF   S,S,W,I,Z,Z,S,M,A,G,Z,P)
```

FIG. 2. Comparison of the partial amino acid sequence of the α-subunit of 7S NGF with the complete amino acid sequence of the γ-subunit of the same complex (Thomas et al. 1981b) and glandular kallikrein (mGK-1) (Mason et al. 1983), all from the adult male mouse submandibular gland. Parentheses indicate compositional data only and brackets unidentified regions. Boxes indicate *differences* between the α- and γ-subunits only in the regions sequenced. The asterisk represents the site of a carbohydrate side chain in the γ-subunit and a similar probable site in the α-subunit (at no. 78). Numbering is for mGK-1, as it represents the only continuous sequence of the three proteins depicted. Symbols: A, Ala; B, Asx; C, Cys; D, Asp; E, Glu; F, Phe; G, Gly; H, His; I, Ile; K, Lys; L, Leu; M, Met; N, Asn; P, Pro; Q, Gln; R, Arg; S, Ser; T, Thr; V, Val; W, Trp; Y, Tyr; Z, Glx; X, unidentified derivative blocking α-amino group. (Taken from Isackson and Bradshaw 1984 with permission from American Society of Biological Chemists.)

to locate the α and γ cDNAs in the library will be far more useful as probes in northern analyses than the cDNA molecules themselves because of the high degree of similarity in this family.

MECHANISM OF ACTION

Functional Responses of NGF

Several of the more important functional responses of NGF are summarized in Table 1. Several of these represent classical responses to NGF and have become commonly applied criteria for other neurotrophic factors, e.g. neurite proliferation and the maintenance of cellular viability (Nieto-Sampedro et al. 1983). Other responses, such as specific changes in enzymes or proteins through modulation of gene expression, are much more general, as are such responses as metabolite uptake and alterations in ion fluxes at the plasma membrane (Greene and Shooter 1980). This implies but certainly does not prove that there are certain similarities in mechanism of action among polypeptide growth factors even if target tissues are diverse (Bradshaw and Rubin 1980). These common features may be considerably more pronounced among the neurotrophic factors when mechanistic features of these agents are elucidated. These similarities may also be expected to manifest themselves in the structural properties of their cell surface receptors. It is commonly agreed that the initiation of response for all hormones and growth factors of this class (where the ligand is proteinaceous) is initiated through formation of specific high-affinity complexes with receptor molecules on the surface of the target cell. Although the events that follow the formation of these complexes have not been elucidated in molecular terms, some structural and functional properties of cell surface receptors have been determined, primarily utilizing photoaffinity cross-linking reagents and monoclonal antibodies. Ultimately, of course, more detailed information regarding the structure will come from cloning experiments.

TABLE 1. *Metabolic/morphologic responses to NGF by peripheral neurons/ PC12 cells*

Elongation of neurites*
Maintenance of viability
Changes in specific proteins and enzymes*
Increased synthesis of RNA and protein*
Regulation of Na^+ gradient
Uptake of metabolites*
Alteration of substrate adhesiveness†
Changes in cell surface morphology†

*Demonstrated in both neurons and PC12 cells.
†Demonstrated only in PC12 cells.

Properties of NGF Receptors

Photoaffinity Cross-Linking

The NGF receptor from adult rabbit superior cervical neurons was initially characterized by hydrodynamic measurements (Costrini and Bradshaw 1979, Costrini et al. 1979). The experiments suggested that the solubilized receptor had a molecular weight of 135,000 ± 15,000 and was composed of a single, substantially asymmetric polypeptide chain. Interactions with lectins suggested, not unexpectedly, that it was a glycoprotein (Costrini and Kogan 1981). Using N-hydroxysuccinimidyl-p-azidobenzoate (HSAB), ^{125}I-labeled NGF was covalently attached to similar receptor preparations and the resultant complexes examined by sodium dodecyl sulfate gel electrophoresis and autoradiography (Massague et al. 1981a). These experiments revealed two receptor forms of 143 and 112 kDa, suggesting, after correction for the contribution of the ligand, receptor molecular masses of 130 and 100 kDa, respectively. Similar results were obtained from chicken dorsal root ganglia (Morgan 1982) and a rat pheochromocytoma cell line (PC12) (Massague et al. 1982). Utilizing an immunoblotting procedure, Fernandez-Pol et al. (1982) presented evidence for the same receptor doublet in mouse melanoma cells. The receptor in the human melanoma cell line (A875) appears somewhat different. Based on isolation experiments (Puma et al. 1983) and covalent cross-linking (Grob et al. 1983), the predominant species had molecular masses (following ligand subtraction) of approximately 85 kDa, and there was some evidence for much smaller amounts of a 200-kDa species. Using the same cross-linking procedure, Grob et al. (1983) also found a similar structure for the PC12 receptor, in contrast to earlier reports of Massague et al. (1982). Table 2 summarizes the properties of the NGF receptors from the peripheral neurons and neoplastic cell lines examined so far. Because the tumor cells have variable and sometimes distinctive responses compared with peripheral neurons, the apparent differences in receptor structures may be of some significance.

The structure for the sympathetic receptor determined from hydrodynamic and cross-linking experiments is compared to that of some other growth factors in Figure 3 (Bradshaw 1982). The upper portion schematically represents the sodium dodecyl sulfate gel electrophoresis-autoradiography experiments for each receptor, performed both in the presence and absence of the reducing agent, dithiothreitol (DTT). The NGF receptor shows two bands (as described above) that are only marginally affected by the reducing agent, which is consistent with the presence of intra- but not interdisulfide bonds (Massague et al. 1981a). Results for the insulinlike-growth factor receptor II, which is also a single polypeptide chain entity, were similar (Massague et al. 1981b). In contrast, the insulin receptor, shown in the left side of the figure, is markedly affected by the reducing agent, which is consistent with a tetrameric structure composed of two types of polypeptide chains, all associated by disulfide bonds (Jacobs and Cuatrecasas 1981).

Schematic representation of models for these receptors, based on the covalent cross-linking data, are presented in the lower portion of Figure 3. As shown, the

TABLE 2. *NGF receptor properties from various tissues*

Species/tissues	Molecular mass, kDa*	Remarks†	Ref‡
Rabbit sympathetic	>200, 135	Hydrodynamic	1
neurons	>200, 130§, 100§	PAXL	2
Chick sensory neurons	127, 97	PAXL	3
Rat pheochromocytoma	127, 97	PAXL	3
cells (PC12)	145–135	PAXL	4
	190, 138, 98‖	PAXL	5
	225, 87	PAXL; XL	6
Human melanoma	200, 85	Isolation; PAXL	5
cells (A875)	225, 87	PAXL; XL	6
Mouse melanoma cells	130, 100	Immunoblotting	7

*Ligand molecular mass (13 kDa) subtracted unless noted.
†Abbreviations: XL, covalent cross-linking; PAXL, photoaffinity cross-linking.
‡References: (1) Costrini et al. 1979, (2) Massague et al. 1981a, (3) Morgan 1982, (4) Massague et al. 1982, (5) Puma et al. 1983, (6) Grob et al. 1983, (7) Fernandez-Pol et al. 1982.
§The 130 and 100 kDa forms contain the same sequence.
‖Inclusion of ligand weight not specified.

NGF receptor can be viewed as a 130-kDa entity that is further processed by limited proteolysis to a 100-kDa species. It should be noted, however, that this conclusion is based on the presence of common sequences, as revealed by Cleveland gel experiments (Cleveland et al. 1977). Such experiments cannot exclude the possibility of a two-polypeptide model in which a 100-kDa species becomes covalently attached to a 30-kDa species to form the higher-molecular-weight form. This association would presumably be noncovalent in situ and could represent a regulating interaction. If such a model proved correct, then the lower-molecular-mass (100 kDa) species would be the principal form of the receptor; the receptors of peripheral neurons and the neoplastic cell lines would thus be more alike (see Table 2).

For the insulin receptor, proteolytic modification of the smaller subunit apparently produces a polypeptide half as large. Since this subunit of the insulin receptor bears the tyrosine-specific kinase activity (Kasuga et al. 1983, Roth et al. 1983), which is inactivated by the cleavage, a two-polypeptide chain model seems less likely here. Proteolysis has not been observed for the IGF-II receptor, but it clearly occurs with the epidermal growth factor receptor in a variety of tissues (Linsley and Fox 1980).

The potential subunit stuctures for the NGF and insulin receptors do not reveal any obvious similarities that might reflect their parallel evolution. However, these results may be misleading. The large (α-subunit) of the insulin receptor, which appears to contain most of the insulin binding site, is similar in size to the NGF

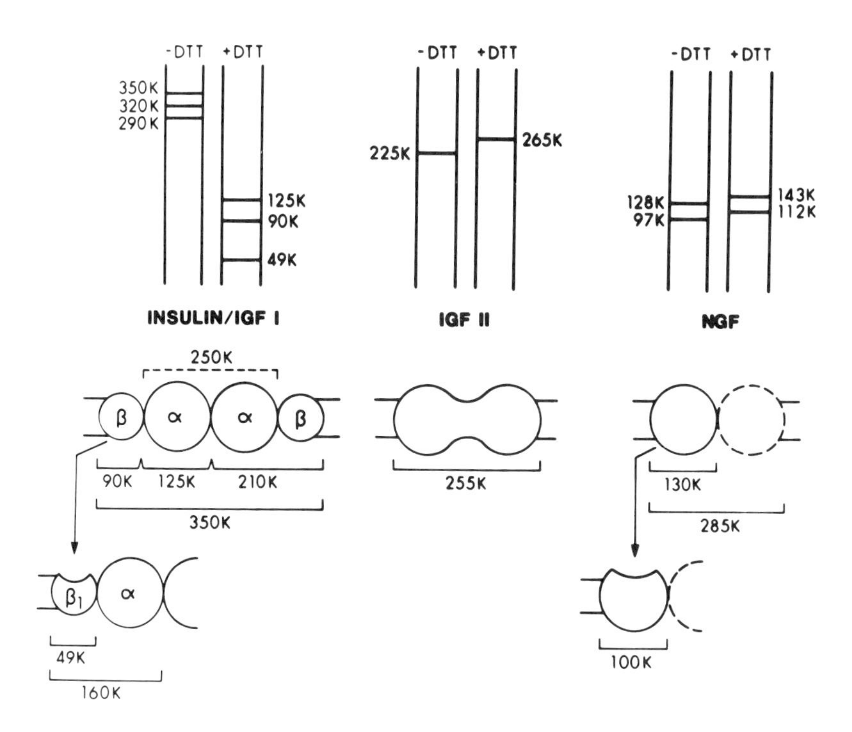

FIG. 3. Schematic representations of the receptors for insulin, the insulinlike growth factors (IGFs), and NGF. The upper diagrams represent the sodium dodecyl sulfate-gel electrophoretograms (after autoradiography) of appropriately cross-linked hormone-receptor complexes in the presence and absence of dithiothreitol (DTT). The numerical designations represent the estimated molecular weights based on comparisons to standard proteins (K = 1000). The lower diagrams are pictorial presentations of the receptor species in an in situ format. The models for NGF and IGFs are from Massague et al. (1981a,b). The insulin model is based on the results from several laboratories as summarized by Jacobs and Cuatrecasas (1981). (Taken from Bradshaw (1982) with permission from Academic Press, Inc.)

subunit receptor. If both of these recognition elements are synthesized as larger precursors, then the potential similarity between the two becomes more obvious. Thus, if the α- and β-subunits of the insulin receptor are synthesized as a single-chain precursor and if the transient, high-molecular-weight form of the NGF receptor observed in sympathetic neurons and, in greater quantities, in neoplastic cells also represents a precursor, then the two structures might indeed be nearly alike, even to the extent of sequence homology. Such a model might also predict that the portion of NGF receptor removed to generate the mature form would contain tyrosine-specific protein kinase activity. The principal difference would be the lack of a disulfide bond between the kinase and the recognition entity. This raises the interesting possibility that tyrosine-specific kinase moieties could freely associate with the binding portions of receptors in the membrane, not unlike adenylcyclase-linked systems. Alternatively, the loss of tyrosine-specific kinase activity might be related to the fact that neurons responsive to NGF are postmitotic and unable to undergo cell division. Cleavage of the kinase from the NGF receptor may be a major cause for the loss of cell division activity of the neuron.

Monoclonal Antibodies Directed Against the NGF Receptor

It has been appreciated for some time that several autoimmune diseases are characterized in part by circulating antibodies directed against cell surface receptors, e.g. myasthenia gravis, Graves' disease, and some forms of insulin-resistant diabetes. These naturally occurring antireceptor antibodies have been useful in determining functional properties of their receptors. With the advent of monoclonal antibody technology, it has been possible to make more-specific antibody reagents, which have extended the immunological approach to understanding receptor structure and function. Such an antibody has been prepared against the NGF receptor from rabbit sympathetic neurons (Morgan 1982, Morgan and Bradshaw 1984). Spleen cells from BALB/c mice that were immunized with the plasma membrane fraction from this tissue were fused with the mouse myeloma NS1. Four hybrid clones were obtained and one was further subcloned, based on the inhibition of a soluble receptor assay. The measurements were originally confused by the presence of a protein in both the medium of the hybridoma cells and the ascites fluid that specifically bound ^{125}I-NGF with high affinity. This molecule was eventually separated from the monoclonal antibody, which was shown to be an IgG type by DEAE cellulose chromatography. The monoclonal antibody bound the receptor with high affinity; half-maximal binding occurred at a concentration of approximately 1 nM. The interaction of the antibody with the receptor did not interfere with ligand binding; thus, ternary complexes composed of receptor, iodinated hormone, and antibody can be formed. Antibodies of both types (blocking and nonblocking) have been reported for other growth factor systems (James and Bradshaw 1984).

The most interesting aspect of this molecule is its ability to mimic the induction of neurite outgrowth by NGF in chick dorsal root ganglia. This observation suggests that the NGF receptor is immunologically similar across species and tissue barriers. The relevance of this observation to the internalization problem is discussed below.

Second Messengers

Polypeptide hormones that exert their effect by binding to cell surface receptors can be conveniently subdivided into two broad classes (Bradshaw and Niall 1978). The first of these, referred to as "messengers" by Sutherland and his colleagues (Robison et al. 1971), induce their response by raising the levels of intracellular cAMP. This occurs through the stimulation of membrane-bound adenylcyclase, which interacts with the hormone recognition unit via a third protein that binds guanyl nucleotides. In keeping with the criteria developed by these workers, the action of these hormones in their responsive target cells can be mimicked by cAMP derivatives and by reagents that block cAMP destruction intracellularly and can stimulate adenylcyclase in broken cell preparations.

For those hormones, including the polypeptide growth factor subset, that do not utilize cAMP, no comparable second messenger has been developed. Referred to as a class as "secondary" hormones (Bradshaw and Niall 1978) to emphasize the long-term effects that in the main distinguish them from the adenylcyclase-dependent (or "primary") hormones, these molecules fail to meet one or more of the criteria for cAMP systems. Nonetheless, considerable controversy continues to surround this area, and the role of cAMP in polypeptide growth factor action remains obscure. While it can be safely dismissed as a second messenger of the same sort as found with messenger or primary hormones, it cannot be ruled out that secondary or even tertiary effects of growth factor interactions with their receptors result in the partial induction of cAMP changes (Bradshaw 1978, Greene and Shooter 1980, Yankner and Shooter 1982).

The possibility that other molecules are generated at the internal face of the plasma membrane as a result of the formation of hormone (or polypeptide growth factor) receptor complexes has remained an avid arena of research. Suggestions of changes in membrane permeability, changes in ion fluxes, or the production of small molecules (including peptides) have characterized one or more of the polypeptide growth factor systems. The broad spectrum of activities that are induced rapidly are certainly consistent with the production of a second messenger, which may well be common to a large number of the secondary hormones. However, considerable evidence suggests that the existence of this messenger cannot by itself explain the entire scope of responses, particularly those involved in gene expression, giving rise to much of the speculation about other mechanisms that might explain the observations. Chief among these is the possibility that receptor-mediated endocytosis, a process that appears associated with most, if not all, of the polypeptide growth factors, may play some role.

Internalization of Polypeptide Growth Factors and Their Receptors

NGF, like other polypeptide growth factors, is readily internalized in responsive cells at all levels of development (Bradshaw 1978, Greene and Shooter 1980, Yankner and Shooter 1982). The most dramatic demonstrations of this phenomenon involve uptake at the presynaptic membrane and retrograde axonal transport (sometimes over considerable distance) of the internalized material to reach the cell body. These experiments have been generally performed with iodinated hormone, and the fate of the isotopes detected in the cell body are those from the ligand. Thus, it is at present largely an article of faith that the NGF receptors with which the ligand initially interacted in the plasma membrane also are internalized and travel in the same vesicles over the same pathway. By analogy to other polypeptide growth factor systems, however, this seems a reasonable conclusion.

The final destination of the internalized vesicles (with NGF or any other system) appears mainly to be lysosomal fusion and proteolytic degradation. Nonetheless, the possibility remains that the ligand, the receptor, or a limited proteolytic product derived from either might serve in a message-bearing capacity following internalization (Bradshaw 1978, Bradshaw and Rubin 1980, James and Bradshaw 1984). Information derived from naturally occurring polyclonal antibodies or specifically directed monoclonal antibodies to receptor entities, which show variable degrees of biological activity, strongly suggests that if internalization is important in the information transfer then the ligand itself is not involved. They do not, however, eliminate a role for the receptor or fragment thereof. In this regard, intracellular receptors, which have been identified through specific binding assays, could either act as sites for further activity or represent translocated molecules from the plasma membrane (Goldfine 1981, Goldfine and Shooter 1983, James and Bradshaw 1984). Receptors associated with the nucleus for NGF have been reported for sensory neurons (Andres et al. 1977) and PC12 cells (Yankner and Shooter 1979), the latter being more consistent with a translocation model than an independently generated site of activity. However, their relationship to the plasma membrane remains to be established.

These nuclear binding sites, as translocated molecules, could play a variety of roles in the gene modulation process. These molecules might be associated with the transfer of enzymatic activity inherent in the plasma membrane receptor that could now act in the nuclear compartment. The discovery of tyrosine-specific protein kinase activity in a number of polypeptide growth factor receptors is consistent with this notion. However, as noted above, the NGF receptor, as it occurs in sympathetic neurons, appears to be devoid of this activity. However, other activities have not been excluded and may be more important in this regard. These receptors may also play a role in the modulation of mRNA production, which would be consistent with their putative location in the nuclear envelope. A particularly attractive model for such an interaction would be to affect nuclear pores and the transport of processed message from the nuclear compartment to the cytoplasm, where translation ultimately occurs. Such a model has been suggested for the insulin system and could have broad applicability, with modifications, to other polypeptide growth factors (Purrello et al. 1983).

SUMMARY

Although the detailed mechanism of action of NGF as it occurs in vivo still requires considerable clarification, a model in which NGF is generated by end organs of responsive neurons and interacts with these neurons to provide trophic stimulation and maintenance of viability seems highly plausible. The demonstration of NGF synthesis by these cells, including the details of the process, is required, and information regarding the events following the formation of the hormone-receptor complex at the plasma membrane must be obtained. However, the contribution of internalization and the possible role of the internalized receptor as a second messenger for generating the long-term effects of the hormone are particularly appealing. Any findings with NGF will almost certainly provide additional important insights into other polypeptide growth factors, particularly those substances found in the neurotrophic subset.

ACKNOWLEDGMENTS

The research emanating from the authors' laboratory cited herein was supported by research grant NS 19964—formerly NS 10229. P.J.I. and R.K. are NIH postdoctoral fellows, NS 07068 and NS 07347, respectively.

REFERENCES

Andres Y, Bradshaw RA. 1980. Nerve growth factor. *In* Kumar S. (ed), Biochemistry of Brain. Pergamon Press, Ltd., Oxford, pp. 545–562.

Andres RY, Jeng I, Bradshaw RA. 1977. Nerve growth factor receptors: identification of distinct classes in plasma membranes and nuclei of embryonic dorsal root neurons. Proc Natl Acad Sci USA 76:1269–1273.

Angeletti RH, Bradshaw RA. 1971. The amino acid sequence of 2.5S mouse submaxillary gland nerve growth factor. Proc Natl Acad Sci USA 68:2417–2420.

Barde Y-A, Edgar D, Thoenen H. 1983. New neurotrophic factors. Ann Rev Physiol 45:601–612.

Bocchini V, Angeletti PU. 1969. The nerve growth factor: purification as a 30,000 molecular weight protein. Proc Natl Acad Sci USA 64:787–794.

Bradshaw RA. 1978. Nerve growth factor. Ann Rev Biochem 47:191–216.

Bradshaw RA. 1982. Structure, function and evolution of insulin-related polypeptides. *In* Bradshaw RA, Hill RL, Tang J, Liang C-c, Tsao T-c, Tsou C-l (eds), Proteins in Biology and Medicine. Academic Press, New York, pp. 1–13.

Bradshaw RA, Niall HD. 1978. Insulin-related growth factors. Trends in Biochemical Sciences 3:274–278.

Bradshaw RA, Rubin JS. 1980. Polypeptide growth factors: some structural and mechanistic considerations. Journal of Supramolecular Structure 14:183–199.

Bradshaw RA, Young M. 1976. Nerve growth factor—recent developments and perspectives. Biochem Pharmacol 25:1445–1449.

Cleveland DW, Fisher SG, Kirschner ME, Laemmli UK. 1977. Peptide mapping by limited proteolysis in sodium dodecyl sulfate and analysis by gel electrophoresis. J Biol Chem 252:1102–1106.

Cohen S. 1962. Isolation of a mouse submaxillary gland protein accelerating incisor eruption and eye lid opening in the newborn animal. J Biol Chem 237:1555–1562.

Costrini NV, Bradshaw RA. 1979. Binding characteristics and apparent molecular size of the detergent solubilized nerve growth factor receptor of sympathetic ganglia. Proc Natl Acad Sci USA 76:3242–3245.

Costrini NV, Kogan M. 1981. Lectin induced inhibition of nerve growth factor binding by receptors of sympathetic ganglia. J Neurochem 36:1175–1180.

Costrini NV, Kogan M, Kukreja K, Bradshaw RA. 1979. Physical properties of the detergent extracted nerve growth factor receptor of sympathetic ganglia. J Biol Chem 254:11242–11246.

Fernandez-Pol JA, Klos DJ, Hamilton PD. 1982. Identification of nerve growth factor receptor proteins in mouse melanoma cell membranes after electrophoretic transfer from gels to diazo-paper. Biochemistry International 5:213–217.

Francke U, de Martinville B, Coussens L, Ullrich A. 1983. The human gene for the β-subunit of nerve growth factor is located on the proximal short arm of chromosome I. Science 222:1248–1251.

Frazier WA, Angeletti RH, Bradshaw RA. 1972. Nerve growth factor and insulin. Science 176:482–486.

Goldfine ID. 1981. Interaction of insulin, polypeptide hormones and growth factors with intracellular membranes. Biochim Biophys Acta 650:53–67.

Goldfine ID, Shooter EM. 1983. Intracellular receptors: workshop report. *In* Bradshaw RA, Gill GN (eds), Evolution of Hormone Receptor Systems. Alan R. Liss, New York, pp. 383–387.

Greene L, Shooter EM. 1980. The nerve growth factor: biochemistry, synthesis, and mechanism of action. Ann Rev Neurosci 3:353–402.

Greene LA, Shooter EM, Varon S. 1969. Subunit interaction and enzymatic activity of mouse 7S nerve growth factor. Biochemistry 8:3735–3741.

Grob PM, Berlot CH, Bothwell MA. 1983. Affinity labeling and partial purification of nerve growth factor receptors from rat pheochromocytoma and human melanoma cells. Proc Natl Acad Sci USA 80:6819–6823.

Hendry IA, Stöckel K, Thoenen H, Iverson LL. 1974a. The retrograde axonal transport of nerve growth factor. Brain Res 68:103–121.
Hendry IA, Stach R, Herrup K. 1974b. Characteristics of the retrograde axonal transport system for nerve growth factor in sympathetic system. Brain Res 82:117–128.
Hogue-Angeletti RA, Bradshaw RA. 1979. Nerve growth factor in snake venoms. *In* Lee CY (ed), Handbook of Experimental Pharmacology, Vol. 52, Snake Venoms. Springer-Verlag, Berlin, pp. 276–294.
Isackson PJ, Bradshaw RA. 1984. The α-subunit of mouse 7S growth factor is an inactive serine protease. J Biol Chem 259:5380–5383.
Isackson PJ, Ullrich A, Bradshaw RA. 1984. Mouse 7S nerve growth factor: Complete sequence of a cDNA coding for the α-subunit precursor and its relationship to serine proteases. Biochemistry (in press).
Jacobs S, Cuatrecasas P. 1981. Insulin receptor: structure and function. Endocr Rev 2:251–263.
James R, Bradshaw RA. 1984. Polypeptide growth factors. Ann Rev Biochem 53:259–292.
Kasuga M, Fugita-Yamaguchi Y, Blithe DL, Kahn CR. 1983. Tyrosine specific protein kinase activity is associated with the purified insulin receptor. Proc Natl Acad Sci USA 80:2137–2141.
Levi-Montalcini R. 1952. Effects of mouse tumor transplantation on the nervous system. Ann NY Acad Sci 55:330–344.
Levi-Montalcini R, Hamburger V. 1951. Selective growth stimulating effects of mouse sarcoma on the sensory and sympathetic nervous system of the chick embryo. J Exp Zool 116:321–362.
Linsley PS, Fox CF. 1980. Controlled proteolysis of EGF receptors: evidence for transmembrane distribution of the EGF binding and phosphate acceptor sites. Journal of Supramolecular Structure 14:461–471.
Mason A, Evans BA, Cox DR, Shine J, Richards RI. 1983. Structure of mouse kallikrein gene family suggests a role in specific processing of biologically active peptides. Nature 303:300–307.
Massague J, Guillette BJ, Czech MP, Morgan CJ, Bradshaw RA. 1981a. Identification of a nerve growth factor receptor protein in sympathetic ganglia membranes by affinity labeling. J Biol Chem 256:9419–9424.
Massague J, Guillette BJ, Czech MP. 1981b. Affinity labeling of multiplication stimulating activity receptors in membranes from rat in human tissues. J Biol Chem 256:2122–2125.
Massague J, Buxser S, Johnson G, Czech MP. 1982. Affinity labeling of a nerve growth factor receptor component on rat pheochromocytoma (PC12) cells. Biochim Biophys Acta 693:205–212.
Morgan CJ. 1982. Probes of the nerve growth factor receptor structure. Doctoral Dissertation, Washington University, St. Louis, Missouri.
Morgan CJ, Bradshaw RA. 1984. Production of a monoclonal antibody directed against the nerve growth factor receptor from sympathetic membranes. J Cell Biochem (in press).
Nieto-Sampedro M, Manthrope M, Barbin G, Varon S, Cotman CW. 1983. Injury induced neural trophic activity in adult rat brain: correlation with survival of delayed implants in the wound cavity. J Neurosci 3:2219–2229.
Paravicini U, Stöckel K, Thoenen H. 1975. Biological importance of retrograde axonal transport of nerve growth factor in adrenergic neurons. Brain Res 84:279–291.
Puma P, Buxser SE, Watson L, Kelleher DJ, Johnson GL. 1983. Purification of the receptor for nerve growth factor from A875 melanoma cells by affinity chromatography. J Biol Chem 258:3370–3375.
Purrello F, Burnham DB, Goldfine ID. 1983. Insulin regulation of protein phosphorylation in isolated rat liver nuclear envelopes: potential relationship to mRNA metabolism. Proc Natl Acad Sci USA 80:1189–1193.
Robison GA, Butcher RW, Sutherland EW. 1971. Cyclic AMP. Academic Press, New York.
Roth RA, Mesirow ML, Cassell DJ. 1983. Preferential degradation of the β-subunit of purified insulin receptor: effect on insulin binding and protein kinase activities of the receptor. J Biol Chem 258:14456–14460.
Rubin JS, Bradshaw RA. 1981. Isolation and partial amino acid sequence analysis of nerve growth factor from guinea pig prostate. J Neurosci Res 6:451–464.
Server AC, Shooter EM. 1976. A comparison of the arginine esteropeptidases associated with the nerve and epidermal growth factors. J Biol Chem 25:165–173.
Server AC, Shooter EM. 1977. Nerve growth factor. Adv Prot Chem 31:339–409.
Scott J, Selby M, Urdea M, Quiroga M, Bell GI, Rutter WJ. 1983. Isolation and nucleotide sequence of a cDNA encoding the precursor of mouse nerve growth factor. Nature 302:538–540.

Stöckel K, Paravicini U, Thoenen H. 1974. Specificity of the retrograde axonal transport of nerve growth factor. Brain Res 76:413–421.

Stöckel K, Schwab M, Thoenen H. 1975a. Retrograde transport of nerve growth factor (NGF) in sensory neurons: a biochemical and morphological study. Brain Res 89:1–14.

Stöckel K, Schwab M, Thoenen H. 1975b. Comparison between the retrograde axonal transport of nerve growth factor and tetanus toxin in motor, sensory and adrenergic neurons. Brain Res 99:1–16.

Thomas KA, Bradshaw RA. 1980. Nerve growth factor. *In* Bradshaw RA, Schneider DM (eds), Proteins of the Nervous System (2nd ed). Raven Press, New York, pp. 213–230.

Thomas KA, Silverman RE, Jeng I, Baglan NC, Bradshaw RA. 1981a. Electrophoretic heterogeneity and polypeptide chain structure of the γ-subunit of mouse submaxillary 7S nerve growth factor. J Biol Chem 256:9147–9155.

Thomas KA, Baglan NC, Bradshaw RA. 1981b. The amino acid sequence of the γ-subunit of mouse submaxillary gland 7S nerve growth factor. J Biol Chem 256:9156–9166.

Ullrich A, Gray C, Berman T, Dull TJ. 1983. Human β-nerve growth factor gene sequence is highly homologous to that of mouse. Nature 303:821–825.

Varon S, Nomura J, Shooter EM. 1967. The isolation of the mouse nerve growth factor protein in a high molecular weight form. Biochemistry 6:2202–2209.

Varon S, Nomura J, Shooter EM. 1968. Reversible dissociation of the mouse nerve growth factor protein into different subunits. Biochemistry 7:1296–1303.

Yankner BA, Shooter EM. 1979. Nerve growth factor in the nucleus: interaction with receptors on the nuclear membrane. Proc Natl Acad Sci USA 76:1269–1273.

Yankner BA, Shooter EM. 1982. The biology and mechanism of action of nerve growth factor. Ann Rev Biochem 51:845–868.

Mediators in Cell Growth and Differentiation,
edited by Richard J. Ford and Abby L. Maizel.
Raven Press, New York © 1985.

Erythropoietin and Erythroid Differentiation

Eugene Goldwasser, Sanford B. Krantz,* and Fung Fang Wang

Department of Biochemistry, The University of Chicago, Chicago, Illinois 60637

Erythropoietin is one of the few characterized substances that are unequivocally known to have a specific effect on a single pathway of differentiation. With only rare exceptions, normal red cell formation in mammals is absolutely dependent on erythropoietin. It is, therefore, worthwhile to study the cellular and molecular modes of its action, as well as the structural basis for its biological effect. This last subject is made difficult by the very limited amounts of pure erythropoietin available for study. Now that its gene has been cloned and expressed (Lin et al. 1984, Lee-Huang 1984), this difficulty should be materially minimized in a short time. The data discussed here, however, were obtained using erythropoietin purified, by the method of Miyake et al. (1977), from the urine of patients with aplastic anemia, a source that is in short supply.

CHEMISTRY OF ERYTHROPOIETIN

Human urinary erythropoietin is a glycoprotein with a molecular weight of about 34,000. There are two forms (α and β), separated at the last stage of purification, that have the same biological activity, potency, and amino acid composition. They differ in behavior on hydroxyapatite chromatography, in electrophoretic mobility (Miyake et al. 1977), and in carbohydrate composition (Dordal 1982). The β form has fewer sialic acid and N-acetyl glucosamine residues than the α. It is also possible that there is a difference in amino acid amidation. The β form may well be a breakdown product of the α form; whether it is formed in vivo or during the purification procedures is still to be determined.

The carbohydrate portion (31% in the case of α erythropoietin) is in the form of N-linked, complex oligosaccharides and is required for biological activity when such activity is measured in vivo. As is the case for native and asialo erythropoietin, interaction with target cells of the marrow is not dependent on the presence of the oligosaccharide chains. In the former case, desialation results in an increase in activity when measured with marrow cell cultures (Goldwasser et al. 1974). When deglycosylation is effected by a mixture of exoglycosidases, about 60% of the

*Permanent address: Department of Medicine, Vanderbilt University School of Medicine and VA Medical Center, Nashville, Tennessee 37203

biological activity in vitro (and 40% of the original reactivity with an antibody) is retained. Glycosidase treatment, however, causes rapid aggregation of erythropoietin and we find that the aggregate is inactive in the culture assay. After a relatively short period, only about 10% of the original erythropoietin is in the monomeric state. This 10% accounts for the 60% of the activity found, indicating that removal of most of the carbohydrate results in an increased biological activity (Dordal 1982).

We found that erythropoietin has four cysteines in internal disulfide bridges. These disulfides cannot be reduced unless erythropoietin is first denatured. When it is treated with 6 M guanidine and reduced with dithiothreitol, it can be reoxidized and renatured, with recovery of at least 85% of its original activity as assayed in vivo. If the sulhydryls are blocked with iodoacetic acid or N-ethylmaleimide, however, the activity is lost, indicating that either the disulfides are involved in the active site or that the proper conformation to permit the active site of erythropoietin to interact with receptors requires the S-S bridging (Wang and Goldwasser, unpublished data).

Limited proteolysis of native erythropoietin causes loss of biological and immunological activities. This loss follows the cleavage to two compact domains containing the N- and C-terminal regions and to a small, protease-accessible region connecting the two. The two regions of the molecule that are relatively resistant to trypsin, pepsin, V-8 protease, and chymotrypsin have molecular weights of about 16,000. The N-terminal domain contains about 75% of the carbohydrate, labeled with ^{3}H in the sialic acids, whereas the C-terminal part contains about 25%. There is no detectable label in the smaller peptides derived from the bridging region, which does contain all of the easily iodinatable tyrosine and a lysine residue. We think the connecting region contains the active site, with respect to interaction both with target cells and with a neutralizing antibody. Formation of a complex by erythropoietin and this antibody prevents the tryptic hydrolysis, whereas use of an unrelated antibody in a similar experiment results in the expected cleavage and loss of activity.

Binding of Erythropoietin

Study of erythropoietin binding to its target cells is made difficult by the fact that substitution of as little as one atom of ^{125}I per molecule is sufficient to inactivate the molecule, as is the reaction of α and ϵ amino groups with the Bolton-Hunter reagent (Goldwasser 1981). This inactivated erythropoietin is not detectably bound to marrow cells. We have used two other expedients in order to study binding. The reagent N-7-dimethylaminomethyl coumarinyl maleimide (DACM) reacts with sulfhydryl groups to form a fluorescent adduct (Yamamoto et al. 1977). Although erythropoietin appears to have no free -SH groups, DACM does react with some functional group to form a fluorescent, biologically active erythropoietin. We used this derivative to study the frequency of erythropoietin-responsive cells (ERC) by fluorescence microscopy and found that about 1.5% of nucleated cells of normal marrow can bind sufficient erythropoietin to be detected. This frequency is about three times greater than that of erythroid colony-forming units (CFU-E) in the same sample of cells, suggesting that cells in the differentiation pathway between CFU-E and mature red cells (which do not bind erythropoietin) retain receptors for erythropoietin. This finding was confirmed by indirect immunofluorescence with monoclonal anti-erythropoietin (Weiss TL, Kung CKH, and Goldwasser E, unpublished data).

The sialic acids of erythropoietin can be labeled with ^3H by oxidation with periodate followed by reduction with NaB^3H_4 (Van Lenten and Ashwell 1972). Tritium-labeled erythropoietin retains its biological activity and has marginally enough radioactivity to permit the demonstration of binding to ERC (Weiss et al. 1983).

Large numbers of homogeneous ERC can be obtained after infection of mice with the anemia-producing variant of the Friend virus (Bondurant et al. 1983). Spleen cells from such infected mice bind erythropoietin to an easily detectable extent. We have measured binding of erythropoietin by these ERC and found that about two thirds of the binding can be competed for with the unlabeled molecule. Other proteins and glycoproteins had no effect on either specific or nonspecific binding by these ERC. Apparent equilibrium was attained at about 2 hr at 37°C and between 3.5 and 4 hr at 10°C. At 10°C, the apparent dissociation constant is about 5 nM; the Scatchard plot is linear and yields a mean value of about 660 receptors per cell. Since the biological activity of erythropoietin is expressed at about 0.06 nM, these data suggest that a very small number of erythropoietin receptor complexes may be sufficient to trigger these ERC.

Erythropoietin-Responsive Cells

The relationship between ERC and pluripotent hematopoietic stem cells is still not completely understood. One view is that ERC are committed only to erythroid differentiation and that stem cells are not affected by inducers of hematopoietic differentiation such as erythropoietin or colony-stimulating factor (CSF).

Data from this laboratory (Van Zant and Goldwasser 1977, 1979) tend to weaken the conventionally held view that erythropoietin acts solely on irreversibly committed cells. We have found what appears to be competition for the same, or very closely related, cells, when both erythropoietin and CSF are presented to marrow cells in vitro. In addition, Van Zant and Chen (1983) have shown that erythropoietin can cause the down-regulation of CSF receptors. One of the perplexing aspects of these "competition" experiments is the marked dependency of the phenomenon on cell concentration. At the cell concentration usually used in semi-solid cultures to measure erythroid bursts (10^5/ml), we found essentially no suppressive effect of CSF on erythropoietin-induced burst formation. At 2×10^5/ml or greater, however, suppression by CSF was easily demonstrated. Since the cell concentration in the bone marrow cavity is about 2×10^9/ml, we reasoned that the responses to effectors at high cell concentration may be more closely related to the normal control mechanisms than the responses at low cell concentration, but still had to unravel the question of dependence on cell concentration. We have used two approaches: in one we postulated that there may be cellular interactions that are favored at high concentration and that may be important for regulation of cellular function. To test this idea, we added 5 U/ml of erythropoietin or 5 U/ml and 280 U/ml of CSF to 10 ml of a cell suspension containing 10^8 mouse marrow cells at 37°C. The cells were then pelleted at $220 \times g$ and incubated under the original medium for 1 hr at 37°C. After that time the medium was removed, the cells washed with 10 ml of fresh cold medium without erythropoietin or CSF, resuspended, and plated in semisolid medium at a concentration of 10^5/ml with 5 U/ml of erythropoietin. Bursts were scored 8 days later with the help of benzidine staining. The controls (erythropoietin only in the pellet) had 14 ± 3 bursts per 10^5 cells, whereas the addition of CSF at the pellet step caused a 57% suppression (6 ± 2 bursts/10^5 cells). This finding suggests that study of regulatory functions in hematopoietic cells may require techniques that permit us to investigate their short-range interactions.

In the second approach we took advantage of our previous finding that pure interleukin-3 (IL3) (Ihle et al. 1982) has all the properties of burst-promoting activity, including the ability to maintain erythroid burst-forming units in vitro in the absence of erythropoietin-induced differentiation (Goldwasser et al. 1983). Since we originally found that colony formation from CFU-E is not suppressed by CSF (Van Zant and Goldwasser 1979) and is decreased by IL-3 (Goldwasser et al. 1983), we grew marrow cells in the presence of IL-3 for 7 days, then plated them at 10^5/ml in methyl cellulose with either erythropoietin or erythropoietin plus CSF and scored erythroid bursts 7 days later. The results very clearly showed an 80% suppression by CSF of erythropoietin-induced bursts. Thus, early cells of the hematopoietic system can be acted on by either CSF or erythropoietin.

Taken together, all of these findings very strongly suggest that either bipotent or tripotent hematopoietic cells exist that can be affected by both CSF and erythropoietin. Early, pluripotent progenitor cells appear to have receptors to interact with inducers of differentiation. We suggest that commitment to a specific hematopoietic pathway is the result of such interaction, as proposed some years ago (Goldwasser 1975).

ACKNOWLEDGMENTS

We are indebted to Mrs. Nancy Pech for excellent technical assistance, to Dr. James Ihle for generously providing interleukin-3, and to Yvonne Price for her expert typing of the manuscript.

This work was supported in part by Grant number CA 18375 awarded by the National Cancer Institute, by Grant number HL 21676 awarded by the National Heart, Lung and Blood Institute, by Grant number AM-15555 awarded by the National Institute of Arthritis, Metabolism and Digestive Diseases, U.S. Department of Health and Human Services, and by The Blossom and Irving Levin Fund for Cancer Research.

REFERENCES

Bondurant M, Koury M, Krantz SB, Blevins T, and Duncan DT. 1983. Isolation of erythropoietin sensitive cells from Friend virus-infected marrow cultures: characteristics of the erythropoietin response. Blood 65:751–758.

Dordal MS. 1982. The function and composition of the carbohydrate portion of human erythropoietin. Ph.D dissertation, The University of Chicago, Chicago, Illinois.

Goldwasser E. 1975. Erythropoietin and the differentiation of red blood cells. Fed Proc 34:2285–2292.

Goldwasser E. 1981. Erythropoietin and red cell differentiation. *In* Cunningham D, Goldwasser E, Watson J, Fox CF (eds), Control of Cell Division and Development. A. R. Liss, New York, pp. 487–494.

Goldwasser E, Kung CKH, Eliason JF. 1974. On the mechanism of erythropoietin-induced differentiation. XIII. The role of sialic acid in erythropoietin action. J Biol Chem 249:4202–4206.

Goldwasser E, Ihle JN, Prystowsky MB, Rich I, Van Zant G. 1983. The effect of interleukin 3 on hematopoietic precursor cells. *In* Golde D, Marks P (eds), Normal and Neoplastic Hemopoiesis. A. R. Liss, New York, pp. 301–309.

Ihle JN, Keller J, Henderson L, Klein F, Palaszynski EW. 1982. Procedures for the purification of interleukin 3 to homogeneity. J Immunol 129:2336–2431.

Lee-Huang S. 1984. Cloning of human erythropoietin gene. Biophys J 45:30a.

Lin K, Lin CH, Lai PH, Egrie J, Goldwasser E, Wang FF, Castro M. 1984. Cloning of the monkey erythropoietin gene. J Cell Biochem (Suppl 8b):45a.

Miyake T, Kung CKH, Goldwasser E. 1977. Purification of human erythropoietin. J Biol Chem 252:5554–5564.

Van Lenten L, Ashwell G. 1972. Tritium-labeling of glycoproteins that contain sialic acid. Methods Enzymol 28:209–211.

Van Zant G, Chen BDM. 1983. Erythropoietin causes down regulation of colony-stimulating factor (CSF-1) receptors on peritoneal exudate macrophages of the mouse. J Cell Biol 97:1945–1949.

Van Zant G, Goldwasser E. 1977. The effects of erythropoietin in vitro on spleen colony forming cells. J Cell Physiol 90:241–252.

Van Zant G, Goldwasser E. 1979. Competition between erythropoietin and colony-stimulating factor for target cells in mouse marrow. Blood 53:935–946.

Weiss TL, Kung CKH, Goldwasser E. 1983. Erythropoietin binding to bone marrow and spleen cells. *In* Golde D, Marks P (eds), Normal and Neoplastic Hemopoiesis. A. R. Liss, New York, pp. 455–463.

Yamamoto K, Sekine T, Kanaoka Y. 1977. Fluorescent thiol reagents. Anal Biochem 79:83–94.

Mediators in Cell Growth and Differentiation,
edited by Richard J. Ford and Abby L. Maizel.
Raven Press, New York © 1985.

Biological Activity In Vivo and In Vitro of Pituitary and Brain Fibroblast Growth Factor

Denis Gospodarowicz

Cancer Research Institute and the Departments of Medicine and Ophthalmology, University of California, Medical Center, San Francisco, California 94143

Among newly identified growth factors are those present in pituitary and brain of vertebrates. In the late 1960s to early 1970s, various laboratories reported the presence in partially purified pituitary hormone preparations of potent mitogens for cultured fibroblasts and chondrocytes (Holley and Kiernan 1968, Armelin 1973, Clark et al. 1972, Corvol et al. 1972). A comparison of the mitogenic activity of various crude extracts from different organs confirmed that the pituitary could be a new source of growth factors (Gospodarowicz 1974). Brain crude extract proved to be equally potent (Gospodarowicz 1974). The purification from both tissues of mitogens, respectively named pituitary and brain fibroblast growth factor (FGF), followed (Gospodarowicz 1975, Gospodarowicz et al. 1978a). Since the function of both neural and pituitary tissue is to release various trophic factors that can control homeostasis as well as to maintain the harmonious and coordinated development and maturation of various organs, the presence in both tissues of mitogens from mesoderm and neuroectoderm cells (Gospodarowicz et al. 1978c, d, e) could have wide implications for various developmental processes.

PURIFICATION AND CHARACTERIZATION OF PITUITARY AND BRAIN FGF

Pituitary FGF

FGF has been purified from whole bovine pituitaries by a four-step procedure that takes advantage of the highly cationic nature (pI 9.6) of the mitogen (Gospodarowicz 1974, Gospodarowicz et al. 1978a, 1984a, b). These steps include: (1) salt extraction of bovine pituitaries at acidic pH (4.5), (2) carboxymethyl Sephadex chromatography, (3) exclusion gel chromatography on Sephadex G75, followed by Bio Gel P-10, and (4) ion exchange chromatography on a Pharmacia Mono S cation exchanger column (Gospodarowicz et al. 1984b). This procedure results in a 200,000-fold purification over the initial crude extract and yields 60 to 100 μg of FGF per kg of pituitary gland. Pituitary FGF is a single-chain polypeptide, as indicated by its similar migratory behavior when submitted to slab gel electrophoresis under reducing and nonreducing conditions (Figure 1A,B). It is composed of 137 amino acids with two intradisulfide bridges and has a pI of 9.6 (Figure 1C) (Gospodarowicz et al. 1984b). A partial primary sequence has recently been

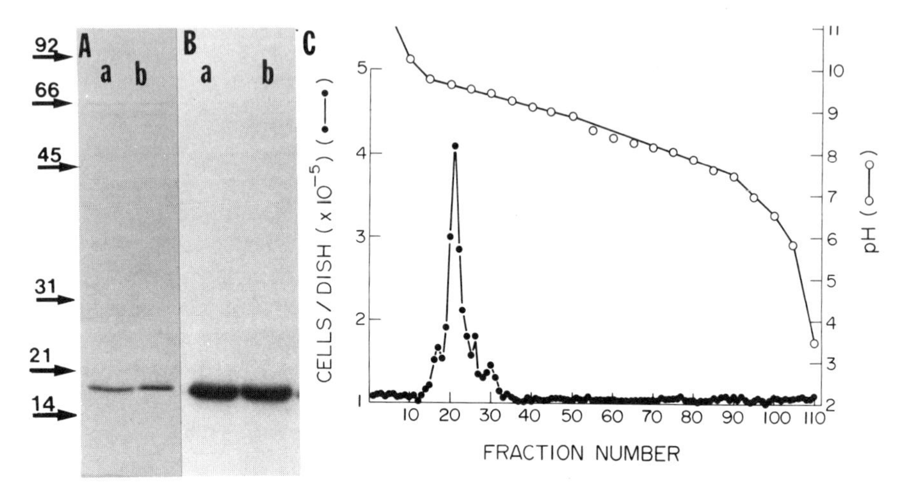

FIG. 1. Dodecyl sulfate polyacrylamide gel electrophoresis and isoelectric focusing of purified pituitary FGF. **A:** Samples containing 0.5 μg of fractions 50 to 52 (lane a) or 0.5 μg of fractions 53 and 54 (lane b) of the final purification step were added to a sample buffer composed of 15% glycerol, 0.1 M dithiothreitol, 2% sodium dodecyl sulfate, 75 mM Tris-HCl (pH 6.8), 2 mM phenylmethylsulfonylfluoride, 2 mM EDTA, 1 mM N-ethylmaleimide, and 1 mM iodoacetic acid. Samples were boiled for 3 min and then applied to an exponential gradient (10-18%) polyacrylamide slab gel with a 3% stacking gel (Gospodarowicz et al. 1984a, b). Electrophoresis was performed for 4 hr at 20 mA. After electrophoresis, the gels were fixed and stained with silver nitrate as described. The migration of the samples was compared to that of protein standards, which included phosphorylase (M_r = 92,500), bovine serum albumin (M_r = 66,000), carbonic anhydrase (M_r = 31,000), soybean trypsin inhibitor (M_r = 21,500), and lysozyme (M_r = 14,400). A similar migration pattern was observed regardless of whether or not the samples were run under reducing conditions. **B:** ^{125}I-labeled FGF samples containing 1.2×10^5 cpm for fractions 50 to 52 (lane a) and 1.6×10^5 cpm for fractions 53 and 54 (lane b) of the final purification step were submitted to sodium dodecyl sulfate polyacrylamide gel electrophoresis as described above. After electrophoresis, the gels were stained with Coomassie blue, destained overnight, dried, and subjected to autoradiography using NS-2T film. After a 6-hr exposure, the film was then developed, fixed, and photographed. Similar results, but with more intensely labeled bands, were observed after a 24-hr exposure. **C:** Isoelectric focusing of purified bovine pituitary FGF. Aliquots (100 μl) of the pool fractions of the final purification step (gradient elution on Mono S cation exchange chromatography, fractions 49 to 54) and containing 10 μg of purified bovine FGF were added to a 2 ml gradient solution (containing 1 mg of inactive fraction Bio Gel P-10 fraction as described; Gospodarowicz et al. 1984a, b) and applied at the mid-position of a pH-gradient density gradient that was formed by 3 ml of Pharmacia ampholytes (pH 8.5 to 10), and stabilized by a 5 to 50% sorbitol gradient. The separation was carried out at 4°C for 48 hr at 500 V. The gradient was then collected in 1.1-ml fractions. The pH of every five tubes was determined. Ten microliters of each fraction were further diluted at a ratio of 1 to 30 with Dulbecco's minimum essential medium (DMEM) 0.5% bovine serum albumin. Ten microliters of the final solution was added to low density vascular endothelial cell cultures (2×10^4 cells/35 mm dish) exposed to 2 ml of DMEM supplemented with 10% calf serum and antibiotics as described in Gospodarowicz et al. (1984a,b). After 4 days in culture, triplicate dishes were trypsinized and cells were counted with a Coulter counter. The final cell density of vascular endothelial cell cultures exposed to 10% calf serum was 1.02×10^5 cells/35 mm dish. Standard deviation was less than 10% of the mean.

reported (Bohlen et al. 1984a). In its native form, pituitary FGF is an extremely potent mitogen. When tested on vascular endothelial cells, it triggers cell proliferation at concentrations as low as 5 pg/ml, with half-maximal effect occurring at 22.5–39 pg/ml (1.5–2.6 pM) and saturation at 140–280 pg/ml (9.3–18.6 pM) (Gospodarowicz et al. 1984) (Figure 2). Those concentrations are 20- to 60-fold lower than those required to see comparable effects on cell growth or initiation of DNA synthesis with α-transforming growth factor (α TGF), epidermal growth factor (EGF), or platelet-derived growth factor (PDGF). Similar activity has been reported when native pituitary FGF is tested on other cell types ranging from established cell lines such as BALB/c-3T3 to normal diploid cell strains such as bovine granulosa and adrenal cortex cells, corneal endothelial cells, vascular smooth muscle cells, and human umbilical endothelial cells (Gospodarowicz et al. 1984b) (Figure 3). From those studies it can be predicted that the K_D for the FGF interaction with its receptor will be less than 10^{-12} and that the number of FGF binding sites per cell will be less than 10^3. Unlike platelet-derived growth factor (PDGF), its biological activity is strongly reduced by acid treatment (below pH 3.0) and is destroyed by heat treatment (70°C for 5 min).

The inactivation by acid treatment of pituitary FGF as well as its inability to stand exposure to organic solvents has prevented the use of reverse-phase high

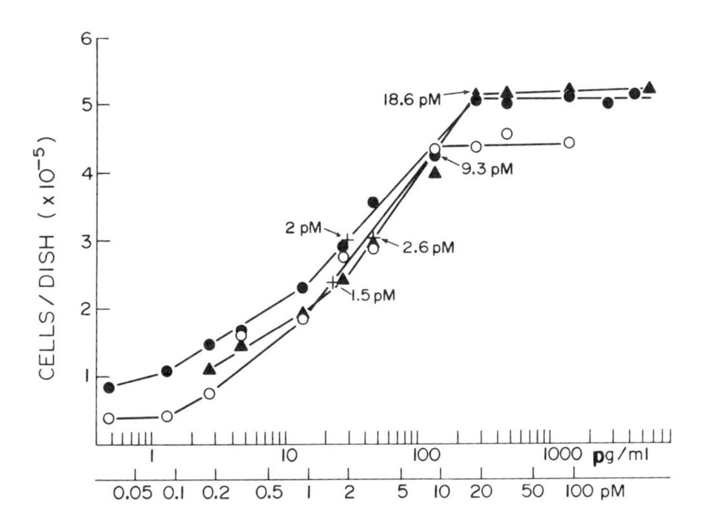

FIG. 2. Effect of increasing concentrations of pituitary FGF on the proliferation of low-density bovine vascular endothelial (ABAE) cell cultures. Low-density ABAE cells (2×10^4 cells per 35 mm dish) were seeded on plastic dishes and exposed to Dulbecco's minimum essential medium supplemented with 10% calf serum and increasing concentrations of pituitary FGF ranging from 2.7 pg to 5.6 ng/ml (0.12 pM to 333 pM) added every other day. After 4 days in culture, triplicate dishes were trypsinized and cells were counted in a Coulter counter. The crosses indicate the concentrations at which a half-maximal response was observed. The final cell density of cultures exposed to 10% calf serum alone was 1.04×10^5 cells. The results of three independent tests are shown. Saturation was observed at 18.6, 18.6, and 9.3 pM with ED_{50} at 2.0, 2.6, and 1.5 pM, respectively.

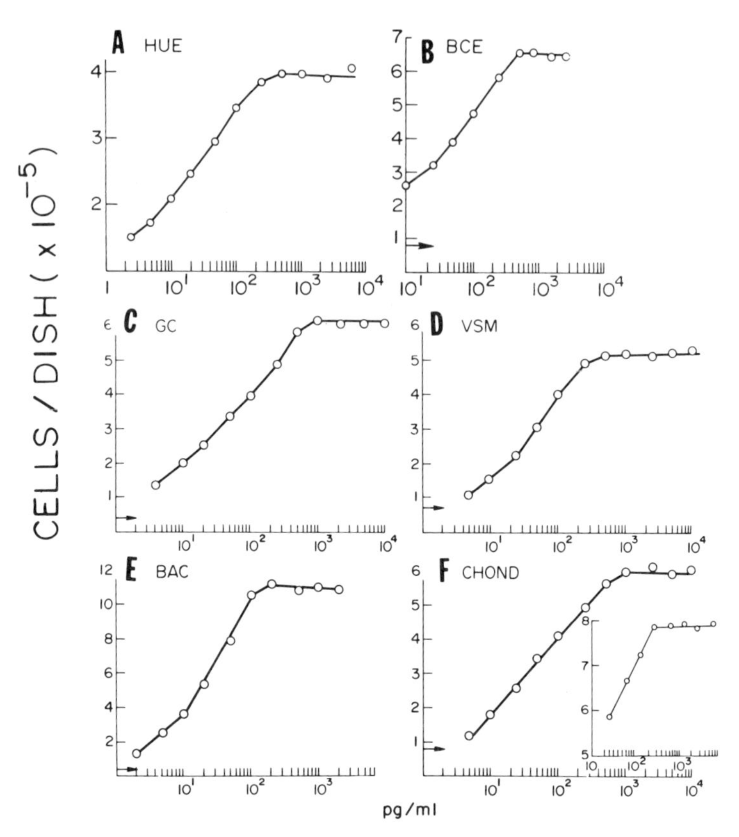

FIG. 3. Comparison of the proliferation of human umbilical endothelial cells, bovine corneal endothelial cells, granulosa cells, adrenal cortex cells, vascular smooth muscle cells, and rabbit costal chondrocytes exposed to increasing concentration of pituitary FGF purified by Mono S cation exchange chromatography. **A:** Low-density human umbilical endothelial (HUE) cells (4×10^4 cells per 35 mm dish) in their second passage were seeded on fibronectin-coated dishes and exposed to HEPES (24 mM) buffered medium 199 supplemented with 10^{-8}M selenium, 20% fetal calf serum, and increasing concentrations of pituitary FGF added every other day in 10-μl aliquots of medium. After 6 days in culture, triplicate dishes were trypsinized and counted in a Coulter counter. The final cell density of cultures exposed to 20% fetal calf serum was 3.4×10^4 cells/35 mm dish. **B–F:** Cells were seeded at an initial density of 2×10^4 cells/35 mm dish. Bovine corneal endothelial cells (BCE) were maintained in Dulbecco's minimum essential medium (DMEM) supplemented with 10% calf serum (B), while bovine granulosa cells (GC) were maintained in DMEM supplemented with 1% calf serum (C). Bovine adrenal cortex (BAC) cells were maintained in F-12 medium supplemented with 10% calf serum (D). Bovine vascular smooth muscle (VSM) cells were maintained in DMEM supplemented with 5% defibrinated plasma (E). Rabbit costal cartilage (CHOND) were maintained either in the presence of DMEM/F-12 medium (v/v) supplemented with 10% calf serum (E), or as shown in the insert in DMEM/F-12 medium (v/v) supplemented with transferrin (10 μl/ml), HDL (300 μg protein/ml), hydrocortisone (10^{-6} M), and EGF (30 ng/ml). In that latter case, chondrocytes were seeded on extracellular matrix coated dishes instead of plastic. Pituitary FGF was added every other day to the dishes at concentrations ranging from 2 pg to 5 or 10 ng/ml. Six days after seeding, triplicate cultures were trypsinized and counted with a Coulter counter. The final cell density of corneal endothelial cells maintained in 10% calf serum and in the absence of FGF was 9.3×10^4 cells per 35 mm dish (arrow), that of granulosa cells exposed to 1% calf serum was 4.2×10^4 cells per dish (arrow), that of adrenal cortex cells was 3.6×10^4 cells/35 mm dish (arrow), that of vascular smooth muscle exposed to 5% bovine plasma was 4×10^4 cells/35 mm dish, while that of chondrocytes exposed to 10% calf serum was 9.6×10^4 cells/35 mm (arrow) and when exposed to transferrin, HDL, hydrocortisone, and EGF it was 5.4×10^5 cells/35 mm dish.

performance liquid chromatography (HPLC) for the purification of FGF in its native form. For structural studies, however, where biological activity is a lesser consideration, reverse-phase HPLC has been used to purify FGF from partially purified preparations of FGF. The final product, although it had lost 95% of its activity (Bohlen et al. 1984a,b), was shown to be identical by a number of criteria to the material prepared by Mono S cation exchange chromatography (Gospodarowicz et al., 1984b). This includes similar behavior on slab gel electrophoresis, identical amino acid composition, identical NH_2 terminal amino acid sequence, and inactivation in both cases by antibodies raised against a synthetic peptide representing the first nine amino acids of the NH_2 terminal of the material purified by HPLC (Bohlen et al. 1984a, b).

Brain FGF

During the course of the purification of pituitary FGF, it was realized that brain tissue contains a similar activity (Gospodarowicz 1974; Gospodarowicz et al. 1978a). Although it was first believed that brain FGF represented degradation products of myelin basic protein (Westall et al. 1978), this was later shown to be erroneous (Gospodarowicz et al. 1984a). When brain FGF was purified to homogeneity by heparin-Sepharose affinity chromatography (Gospodarowicz et al. 1984a), it was found to be identical to pituitary FGF on the basis of similar molecular mass, amino acid composition, biological activity, and cross-reactivity with either a synthetic peptide representing the first nine amino acid sequences of the N-terminal of pituitary FGF or with murine monoclonal IgM antibodies raised against pituitary FGF (Gospodarowicz et al. 1984).

IN VIVO BIOLOGICAL EFFECT OF FGF

FGF and Blastema Cell Recruitment and Proliferation

Certain amphibian species have a well-developed capacity to regenerate lost appendages. The dependence of limb regeneration on nerves was reported as early as 1823 by Todd, and studies by Singer have clarified and extended the concept of neurotrophic control in this process (Singer and Craven 1948, Singer 1952, 1954). Denervation of an adult newt limb at the time of amputation or prior to the formation of the regeneration blastema effectively blocks the regeneration process. If denervation is delayed until after a blastema has formed, however, differentiation of the blastema cells and morphogenesis of new limb parts occur despite the lack of nerves (Singer 1974, Powell 1969). This finding implies that the neurotrophic effect is needed for the cell proliferation involved in blastema formation but not for the later events of regeneration, which include differentiation of blastema cells into muscle, cartilage, and other tissues of the new limb. The molecular basis of neurotrophic phenomena is poorly understood, but the available evidence suggests that such control is mediated by means of neurosecretion products unrelated to nerve impulses of neurotransmitters (Dresden 1969, Gutmann 1976).

Blastema cells are thought to originate from either dedifferentiating myoblasts or chondrocytes (Hay 1977). These are two cell types for which brain FGF has been shown to be a potent mitogen (Gospodarowicz et al. 1978c, e, Gospodarowicz et al. 1976b, Gospodarowicz and Mescher 1977). Infused FGF promotes the resumption of mitotic activity in denervated newt forelimb blastemas in vivo (Gospodarowicz et al. 1978e, Gospodarowicz and Mescher 1981, Mescher and Gospodarowicz 1979), and when added to cultured blastemas, brain FGF at a concentration of 10 ng/ml is as effective as an optimal dose of brain extract in promoting [^{3}H]thymidine incorporation (Mescher and Loh 1981), ^{14}C-labeled amino acid incorporation, and mitotic activity (Carlone and Foret 1979). Similar concentrations of brain FGF have also been shown to maintain total acetylcholinesterase activity in cultured newt triceps muscle above that in untreated contralateral controls after 1 week (Carlone et al. 1981). It has also been reported (Gospodarowicz et al. 1975) that administration of FGF promotes heterotrophic growth of amputated adult frog *(Rana pipiens)* forelimbs comparable to that observed by others (Singer 1954) after surgical augmentation of the nerve supply. These observations lend support to the possibility that brain FGF is involved in one of the earlier steps of regeneration, namely recruitment of primitive cells leading to blastema formation.

Two other groups of investigators have identified "trophic" factors with chemical properties similar to those reported for FGF. Jennings et al. (1979) reported purification of a mitogenic polypeptide, with a pI of 9.5–9.8 and a molecular weight of 11,000, from neonatal calf spinal cord. A similar basic factor (MW 13,500), which stimulates protein synthesis in cultured regeneration blastemas, has been partially purified from embryonic chick brain by Choo et al. (1981). The exact relationship of these factors to FGF remains to be determined. The ability of FGF to stimulate in vivo blastema formation is in agreement with its ability to induce in vivo cartilage regeneration (Jentzsch et al. 1980), since blastema cells can redifferentiate into chondrocytes.

Angiogenic Activity of FGF In Vivo and Its Effect on Granulation Tissue

Both pituitary and brain FGF have been shown to be potent angiogenic factors in vivo. As little as 0.3 ng/day induces massive capillary formation in both the rabbit cornea (Gospodarowicz et al. 1979) and the hamster cheek pouch (Schreibert and Gospodarowicz, manuscript in preparation). In agreement with those observations, brain FGF also increases formation of granulation tissue in vivo (Buntrock et al. 1982a, b). This increase is associated with stimulation of the synthetic function of fibroblast and myofibroblast (Buntrock et al. 1982a, b).

IN VITRO BIOLOGICAL EFFECT OF FGF

Effect on Cell Transformation, Migration, Proliferation, Differentiation, and Senescence

FGF has both acute and long-term effects on the morphology and growth pattern of responsive cells. Human skin fibroblasts maintained with FGF in serum-supplemented medium become extremely elongated with long slender projections—these are retraction fibrils resulting from an increase in locomotory activity (Gospodarowicz and Moran 1975). The same is true for vascular endothelial and smooth muscle cells, which in the presence of FGF become bipolar (Gospodarowicz 1984). Even more striking is the effect of FGF on cell morphology when it is added to a confluent and resting monolayer of BALB/c 3T3 cells. Such cells, which show a regular cobblestone pattern in confluent culture exposed to serum-supplemented medium, become spindle shaped and grow in an irregular crisscross pattern when exposed to FGF for 6–12 hr (Gospodarowicz and Moran 1974). Large membrane ruffles, often associated with macropinocytotic vesicles, can be seen on most of the cells. Morphologically, the FGF-treated cells look "transformed" since a reduced cell-substratum adhesion, growth in crisscross pattern, and increased membrane ruffling are features typical of transformation. The similarity in phenotype of transformed and FGF-treated cells is consistent with the hypothesis that transformation and FGF might at least partially share a common metabolic pathway. Comparison of the locomotor activity of arterial smooth muscle cells and endothelial cells in media supplemented with serum and FGF, respectively, has given circumstantial evidence that FGF stimulates cell migration (Gospodarowicz 1984).

Both pituitary and brain FGF have been shown to be potent mitogens for mesoderm-derived cells (reviewed in Gospodarowicz 1983a, b, 1984) (Table 1). Until now, pituitary and brain FGF have mostly been used in vitro to develop new cell strains (Gospodarowicz 1979, Gospodarowicz and Zetter 1977), particularly from the vascular and corneal endothelia (Gospodarowicz et al. 1976a, 1977, 1978b, c, 1979a, 1980b, 1981a, b, c). FGF is mitogenic both for cells seeded at clonal density and for low-density cultures (Gospodarowicz and Zetter 1977). Although the response of the cells can vary widely, depending on the types of sera and media used (Gospodarowicz and Lui 1981), addition of FGF to the cultures of most mesoderm-derived cells results in a greatly reduced average doubling time; in the case of vascular endothelial cells, for example, it can decrease from 72 to 18 hr (Gospodarowicz et al. 1981a, Duthu and Smith 1980). This results primarily from a shortening of the G_1 phase of the cell cycle (Gospodarowicz et al. 1980a, b, Gospodarowicz et al. 1981c, d).

FGF stabilizes the phenotypic expression of cultured cells. This is a particularly interesting characteristic of FGF, since it has made possible the long-term culturing of various cell types that otherwise would lose their normal phenotypes in culture when passaged repeatedly at low cell density. This biological effect of FGF has been best studied using vascular or corneal endothelial cell cultures cloned and maintained in the presence of FGF and then deprived of it for various time periods (Gospodarowicz et al. 1981b, Vlodavsky and Gospodarowicz 1979, Vlodavsky et al. 1979, Greenburg et al. 1980, Tseng et al. 1982, Gospodarowicz 1983a, b). It has been related to the ability of FGF to control the synthesis of various extracellular matrix components that would in turn affect cell polarity and gene expression (Gospodarowicz et al. 1979b, Gospodarowicz 1983b). In the case of sheep preadipocyte fibroblasts, it has been reported that FGF promotes their differentiation

TABLE 1. *Cell types sensitive to the mitogenic effect of FGF when maintained in vitro on plastic dishes and exposed to serum-supplemented medium*

Cell type	Animal from which cells are derived
Swiss 3T3	Mouse
BALB/c 3T3	
3T3 Thermosensitive mutant	
Foreskin and skin fibroblasts	Human, cow
Glial cells, astrocytes	Human, mouse
Kidney fibroblasts	Cow
Amniotic cells (fibroblasts)	Human, cow
Articular, ear, and costal chondrocytes	Rabbit, human
Myoblasts	Cow (fetus), rat
Vascular smooth muscle (aortic arch)	Monkey, human, cow
Vascular endothelial cells (umbilical vein)	Cow, human
Vascular endothelial cells (fetal and adult aortic arch)	Cow
Endocardium	Cow fetus
Corneal endothelial cells	Cow, rabbit, cats, human (fetus)
Lens epithelial cells	Cow, cats
Trabecular meshwork cells	Human
Y1 Adrenal cortex cells	Mouse
Adrenal cortex cells	Cow, pig
Granulosa cells	Cow, pig, rabbit
Luteal cells	Cow
Blastemal cells	Frog
Blastemal cells	*Triturus viridescens*

into adipocytes (Broad and Ham 1983). A similar observation has been made by Serrero and Khoo (1982) using the 1246 cell line. In the case of a rat pheochromocytoma (PC12 cell line) still able to respond to nerve growth factor (NGF) by initiating neurite outgrowth, both brain and pituitary FGF act as differentiating agents, inducing both neurite outgrowth and ornithine decarboxylase activity (Togari et al. 1983). This effect seemed to be additive with that of NGF (Togari et al. 1983).

FGF can also significantly delay the ultimate senescence of cultured cells. In the case of granulosa cells, addition of FGF to either clonal or mass cultures of granulosa cells can extend their lifespan in culture from 10 generations to 60 generations. Adrenal cortex cell lines cloned in the presence of FGF show a similar dependence on this growth factor during their limited in vitro lifespan. Its removal from the culture medium results not only in a greatly extended doubling time but also in rapid cell senescence. In the case of vascular and corneal endothelial cells, FGF has been shown to extend the lifespan of the cultures. Corneal endothelial cells maintained in the absence of FGF have a lifespan of 20 to 30 generations, whereas in its presence they can proliferate for 200 generations (Gospodarowicz et al. 1982a). In summary, the following effects of FGF have been observed with various cell types: it stimulates cell migration, it increases the rate of cellular proliferation, it stabilizes their phenotypic expression, and it extends the culture lifespan.

Mechanism of Action

Using cultures of vascular endothelial cells, the early effect of FGF on membrane motility has been studied (Gospodarowicz et al. 1979a). The interaction between various lectins and glycoproteins that are capable of lateral diffusion in the plane of the membrane can lead to a selective redistribution of lectin-receptor complexes in a process that results in "patch" and subsequent "cap formation" (segregation of the cross-linked patches to one pole of the cell by an active, microfilament-dependent process). Long-range receptor movement involved in patching and capping can be directly observed by looking at the distribution of fluorescein-labeled lectins such as concanavalin A. Short-range lateral movements like those involved in lectin-mediated receptor clustering and cell agglutination can be indirectly quantified by measuring the ability of cells to bind to nylon fibers coated with different densities of lectin molecules. This binding requires a short-range lateral mobility of the appropriate receptors to allow their alignment with the lectin molecules on the fiber. By using both of these techniques it was observed that both the short- and the long-range types of concanavalin A receptor mobility were increased in bovine vascular endothelial cells preincubated with FGF. A 6- to 9-hr preincubation with FGF gave a nearly maximal increase (two- to fourfold) in the binding of cells to concanavalin A molecules on the fibers, and already had a significant effect at 3 hr. A longer preincubation period (12–24 hr) with FGF is required to induce, in a large proportion (60–80%) of the cells, the ability to form caps upon addition of concanavalin A. Such changes were not obtained when cells were preincubated with EGF, which does not bind and has no mitogenic effect on bovine vascular endothelial cells (Gospodarowicz et al. 1978b, 1979a).

The early effects of FGF on increased membrane fluidity and increased ruffling activity correlate, as has been shown in the case of EGF (Schlessinger and Geiger 1981) and PDGF (Westermark et al. 1982), with rapid changes in the dynamic structure of the actin cytoskeleton (Gospodarowicz 1984).

The pleiotypic response of the cells to mitogenic factors includes sequential changes, such as stimulation of cellular transport systems, polyribosome formation, protein synthesis, ribosomal and tRNA synthesis and, eventually, DNA synthesis followed by cell division. The induction of the "pleiotypic" and mitogenic response by FGF has been analyzed using confluent cultures of BALB/c 3T3 cells.

The addition of FGF to confluent and resting cultures of BALB/c 3T3 cells leads to an increased uptake of uridine, total amino acids, and thymidine (Rudland et al. 1974). The time course of the increase in net incorporation into the cell is almost identical for both FGF and serum additions. Amino acid uptake increased immediately, followed by a 15-min lag time for uridine and a 3- to 5-hr lag time for thymidine. The rate of protein synthesis increased (by threefold) within 5 hr, that of mRNA (about 15–20%) within 5 hr, and DNA (60- to 100-fold) within 24 hr. These increases are almost identical for cultures activated with FGF or serum, both in the kinetics and extent of the increases. Finally, cell division occurs in both cases 25 to 30 hr later. The dose-response curves for the effect of varying concentrations of FGF on the stimulation of protein synthesis at 6 hr, ribosomal RNA synthesis at 14 hr, and induction of DNA synthesis are roughly parallel. The abilities of FGF and serum to induce polysome re-formation measured 6 hr after additions are also virtually identical. Approximately 28% of the mRNA in resting BALB/c 3T3 cells is located in polysomes and 72% in the 30S–80S messenger ribonucleoprotein complex. Addition of either FGF or fresh serum results in greater than 85% of the mRNA being located in the polysomal region of the sucrose gradient. This increase in polysome formation correlates with an increase in protein synthesis.

Based on data indicating that proteins need to be made for the rate of DNA synthesis to increase in response to growth stimulation, Nilsen-Hamilton et al. (1980, 1981, 1982) have identified proteins that are specifically synthesized in growth-arrested BALB/c 3T3 in response to FGF and to serum. The most dramatic, specific increases in labeling with [^{35}S]methionine were of secreted rather than intracellular proteins. The secreted levels of these proteins begin to rise within 2 hr of the addition of FGF to the growth-arrested 3T3 cells (Nilsen-Hamilton et al. 1982). The secreted levels of one of these proteins with a molecular weight of 39,000, which is also called a "major excreted protein" (MEP) (Gottesman 1978), increases about eightfold in response to FGF (Nilsen-Hamilton et al. 1980). FGF also increases the secreted levels of a set of five proteins that share the property of being "superinduced" by cycloheximide; that is, their secreted levels are increased two- to fivefold by concentrations of cycloheximide that inhibit total protein synthesis by about 85% when present during induction, but not present during labeling with [^{35}S]methionine (Nilsen-Hamilton et al. 1982). Cycloheximide acts in synergism with FGF. These cellular proteins have been named "superinducible proteins" or SIPs. They have molecular weights ranging from 12,000 (SIP 12) to 62,000 (SIP 62), and they may be the same proteins that increase intracellularly in these cells in response to stimulation by PDGF (Pledger et al. 1981). By contrast, cycloheximide when present during the induction period prevents FGF from raising the secreted levels of MEP.

The secreted levels of MEP and the SIPs are also regulated differently in that NH_4Cl, which inhibits lysosomal proteolysis, increases the secreted level of MEP but not the SIPs. Finally, the ability of FGF to raise the secreted levels of MEP and the SIPs depends on whether the BALB/c 3T3 cells are sparse and growing or confluent and quiescent. FGF is much more effective at raising the secreted level of MEP in growing BALB/c 3T3 cells than in quiescent 3T3 cells; the opposite is true for the SIPs (Nilsen-Hamilton et al. 1982).

As with PDGF, FGF has been shown by Jimenez de Asua and Rudland to be a competence factor for BALB/c, as well as for Swiss 3T3 cell lines (Jimenez de Asua et al. 1977a, b, Rudland and Jimenez de Asua 1979, Jimenez de Asua 1980). This was later confirmed by Stiles et al. (1979). Among the plasma components that could be held responsible for the progression of cells committed to divide by brief exposure to FGF or PDGF are the somatomedins, or insulin growth factor (IGF) (Stiles et al. 1979), and epidermal growth factor (Leof et al. 1982). However, these factors are not the only plasma factors required. First, when competent cells are transferred to medium containing no plasma, only somatomedin C cells do not progress through their G_0/G_1 phase. Second, although confluent cultures of BALB/c 3T3 cells exposed to synthetic medium will enter into a progression phase after being exposed to PDGF when they are exposed to somatomedins and EGF, sparse cultures exposed to similar conditions do not initiate DNA synthesis. These observations clearly demonstrate a requirement for progression factors other than somatomedins and EGF when cells are sparse (Leof et al. 1982).

Recent studies using sparse cultures of normal diploid cells maintained in serum-free medium have demonstrated that high-density lipoproteins (HDL) together with transferrin could act as a major progression factor (Gospodarowicz et al. 1982a, Cohen et al. 1982a, Longenecker et al. 1984). Insulinlike growth factors, except in the case of granulosa and adrenal cortex cells, were not required and had only minimal effect on other cell types (Table 2). These include vascular aortic smooth muscle and endothelial cells, corneal endothelial, lens epithelial, and kidney epithelial cells, and a number of tumor or established cell lines (Gospodarowicz 1983b, Gospodarowicz et al. 1982a, 1983b) (Table 2). The ability of HDL to support progressing cells through the cycle does not lie only in its ability to donate cholesterol (Longenecker et al. 1984). HDL more likely acts through the reverse process (cholesterol egress). This would activate the cellular hydroxy-3-methylglutaryl coenzyme A (HMG CoA) reductase (Cohen et al. 1982a, b), resulting in an increased synthesis of mevalonic acid, which is later utilized for the synthesis of dolichol, ubiquinone, isopentenyl adenyl adenine, and cholesterol (reviewed in Brown and Goldstein 1980).

FGF, as well as PDGF, has been shown recently to regulate cellular oncogenes. There is ample evidence that perturbations of cellular oncogenes contribute to specific tumorigenesis (Cooper 1982, Hayward et al. 1981). The *c-myc* gene, identified originally as the cellular homolog of the transforming determinant carried by avian myelocytomatosis virus, is altered in association with a broad spectrum of neoplasms. Activation of *c-myc* by avian leukosis virus has been implicated in the genesis of bursal lymphomas (Eva et al. 1982), and elevations in *c-myc* expression occur in at least some isolates of various tumor types (Eva et al. 1982). It has been suggested that an altered *c-myc* gene operationally encodes an immortalization function, expression of which is contingent upon coordinate expression of an altered *c-ras* gene (Land et al. 1983). Immortalization may result from the ability of cells to traverse specific restriction points in the cell cycle (Pledger et al. 1981, Pardee 1974) that allow growth in vitro. Therefore, it is reasonable to suggest that *c-myc* is involved in the progression of many or all cells through the cell cycle.

In agreement with those proposals, it has recently been demonstrated that *c-myc* is an inducible gene that is regulated by specific growth signals in a cell-cycle-dependent manner (Kelly et al. 1983). Specifically, agents that initiate the first

TABLE 2. *Growth factors required for the proliferation of normal diploid and transformed cells maintained on ECM-coated dishes and exposed to defined medium*

	HDL (μg protein/ml)	Insulin or Somato C (ng/ml)		FGF or EGF (ng/ml)		Transferrin (μg/ml)
Normal diploid cells						
Vascular endothelial cells	500	—	—	—	—	10
Corneal endothelial cells	250	2500	100	1	50	10
Vascular smooth muscle cells	250	2500	100	1	50	10
Granulosa cells	30	1000	100	1	50	10
Adrenal cortex cells	30	50	10	1	—	5
Lens epithelial cells	250	2500	100	1	—	10
Kidney tubule cells	750	—	—	—	—	50
Embryo fibroblasts (Rat-1)	500	5000	—	—	25	25
Transformed cells (tumor)						
A-431 carcinoma	500	—	—	—	toxic	10
Colon carcinoma cells	500	—	—	—	—	10
Ewing sarcoma cells	500	—	—	—	—	10
Rhabdomyosarcoma	500	—	—	—	—	10
MDCK (kidney-derived)	500	—	—	—	—	10
B-31 cell line	1000	—	—	—	—	25

phase of a proliferative response in lymphocytes (lipopolysaccharide or concanavalin A) and fibroblasts (PDGF and FGF) induce *c-myc* mRNA (Kelly et al 1983). Within 3 hr after the addition of these mitogens to the appropriate cells, *c-myc* mRNA concentration is increased between 10- and 40-fold. This induction of a *c-myc* mRNA concentration increase occurs in the presence of cycloheximide and therefore does not require the synthesis of new protein species (Kelly et al. 1983). Other possible oncogenes could exhibit cell-cycle-associated roles and be induced by either FGF or PDGF. The results relating oncogenes and growth factors such as FGF or PDGF suggest a rational mechanism by which these genes may contribute to the growth of normal or cancer cells. It could also provide a rational explanation for the increased lifespan of cultured cells exposed to FGF, since oncogene activation or insertion in normal cells can lead to immortalization. However, one should recognize that little is presently known about the precise function of oncogenes (Bishop 1983).

FGF AND THE CONTROL OF PRODUCTION OF EXTRACELLULAR MATRIX COMPONENT

Among the cell types for which FGF is mitogenic are endothelial cells derived either from various vascular territories or from the cornea. It is with these cells that the effect of FGF on extracellular matrix (ECM) production and the effect of ECM on cell proliferation and phenotypic expression were first investigated (Gospodarowicz et al. 1980b, 1981c). Vascular endothelial cells grown in the presence of FGF adopt, upon reaching confluence, the morphology of a cell monolayer composed of tightly packed and flattened cells. This cell layer, as it does in vivo, has asymmetrical cell surfaces. While the apical cell surface is a nonthrombogenic surface to which platelets do not bind, the basal cell surface is involved in the synthesis of a highly thrombogenic ECM.

In contrast to vascular endothelial cell cultures grown in the presence of FGF, similar cultures maintained in its absence and passaged at low cell density lose within three passages their ability to form at confluence a monolayer of closely apposed and flattened cells. Instead, the cultures adopt a multilayer configuration consisting of large and overlapping cells that are no longer contact inhibited (Vlodavsky and Gospodarowicz 1979, Vlodavsky et al. 1979). Parallel to these changes in cell morphology, cell surface polarity is lost, both the apical and basal cell surfaces becoming covered with an ECM.

These observations suggest that vascular endothelial cells maintained in the absence of FGF exhibit, in addition to a much slower growth rate, morphological and structural alterations that mostly involve changes in the composition and distribution of the ECM (Greenburg et al. 1980). These changes result in a loss of phenotypic expression, since cells are no longer contact inhibited and have a thrombogenic apical cell surface. Vascular endothelial cell cultures will regain their characteristic phenotypic expression if sparse or subconfluent cultures previously maintained without FGF are reexposed to it. It could therefore be concluded that the addition of FGF to growing cultures of vascular endothelial cells stabilizes the phenotypic expression of that cell type. This capability correlates with its ability to modulate the synthesis and distribution of various ECM components such as collagen, fibronectin, and proteoglycans (Tseng et al. 1982, Vlodavsky et al. 1979, Kato and Gospodarowicz 1984c). It raises the possibility that one of the ways FGF could promote cell proliferation and stabilization of phenotypic expression is through its effect on ECM production. A similar effect of FGF on ECM production by cultured rabbit chondrocytes has also recently been reported (Kato and Gospodarowicz, 1984a, b, c). In that case it is likely that FGF helps costal chondrocytes to express their normal phenotypic expression when maintained in tissue culture (Kato and Gospodarowicz, 1984a, b, c) by preventing, through increased matrix production (Kato and Gospodarowicz 1984c), the close contact of chondrocytes with plastic a substrate that induces increased chondrocyte motility and dedifferentiation (Benya et al. 1978).

THE EXTRACELLULAR MATRIX AND THE CONTROL OF CELL PROLIFERATION

In vivo an intact ECM scaffold is required for the maintenance of orderly tissue structure and regeneration. By its presence, it defines the spatial relationships among similar and dissimilar types of cells. Such a substrate plays an important role not only in cell attachment and migration (Hay 1980, 1981a, b, Kleinman et al. 1981, Yamada 1983), but also in cell response to various growth-promoting factors in natural fluids such as plasma, lymph, or interstitial fluid (Gospodarowicz and Tauber 1980). It also enables multicellular organisms to reconstitute histological structures of most tissues and organs after cell loss (Vrako 1974).

In the early stage of embryonic development, when different tissues composing a given organ are formed as a result of strictly timed and spatially interrelated proliferative and differentiative events, interaction of cells with newly formed ECM has been shown to promote cell proliferation and to stabilize new phenotypic expression (Hay 1980, 1981b, Wessels 1977, Saxen et al. 1982). This has been studied best in the formation of the epithelial part of the kidney nephron induced by the invasion of ureter bud into kidney mesenchyme. The metanephric mesenchymal cells are embedded in an ECM composed mostly of collagen types I and III and fibronectin. Under the direction of the invading ureter bud, mesenchymal cells condense and start to synthesize an ECM composed mainly of collagen type IV, heparan sulfate proteoglycan, and laminin. This shift in ECM component production correlates closely with the synchronized development of the epithelial part of the glomeruli and associated nephron tubule (Ekblom 1981, Ekblom et al. 1980, 1981a, b). Induction therefore stimulates the production of laminin, type IV collagen, and basement membrane proteoglycan, resulting in the formation of a proper substratum for epithelial cell attachment and further differentiation.

ECM components have also been implicated in inductive tissue interaction in somite chondrogenesis (Kratochwil 1972), in the differentiation of corneal epithelium (Hay 1981a, Meier and Hay 1974, 1975), and in salivary gland morphogenesis (Bernfield 1978, Bernfield et al. 1972, Cohn et al. 1977), as well as in tooth germ development (Thesleff and Hurmerinta 1981, Thesleff et al. 1981). It thus appears that during the whole of embryogenesis, as well as in neonatal and adult life, an intact ECM scaffold is required for maintenance of orderly tissue structure. The substrate upon which cells rest in vitro could therefore be a decisive element in their proliferative response to various factors. This possibility has been explored using dishes coated with an ECM produced by cultured corneal endothelial cells (Gospodarowicz et al. 1982a, b, Gospodarowicz and Tauber 1980, Gospodarowicz and Giguere 1982).

When the ability of the ECM to support cell proliferation was analyzed with various cell types, it proved capable of replacing the competence factors, in particular FGF (Gospodarowicz et al. 1980a, Gospodarowicz and Tauber 1980, Gospodarowicz and Greenburg 1981). This was best seen with vascular smooth muscle cells, which, when maintained on plastic, proliferate actively only when exposed to serum, but not when exposed to plasma unless PDGF or FGF is added to the medium. In contrast, when cells are maintained on an ECM and exposed to plasma, they proliferate actively and no longer require PDGF or FGF (Gospodarowicz and Ill 1980). In fact, plasma is even more mitogenic for cells maintained on an ECM than is serum for cells maintained on plastic. The growth rate and the final cell density of cultures maintained on an ECM and exposed to either plasma or serum are the same. The final cell density of cultures maintained on an ECM and exposed to plasma was found to be a direct function of the plasma concentration to which cultures were exposed (Gospodarowicz and Ill 1980). It is therefore likely that the proliferation of vascular smooth muscle cells maintained on an ECM is controlled by factors present in plasma and that the ECM has only a permissive role (Gospodarowicz and Ill 1980, Gospodarowicz et al. 1981d).

The permissive effect of the ECM produced by corneal endothelial cells on cell proliferation has also been observed in the case of fibroblasts, granulosa cells, adrenal cortex cells, corneal endothelial cells, corneal and lens epithelial cells, vascular endothelial cells (Gospodarowicz 1983a, b, Gospodarowicz et al. 1982a, b), and chondrocytes (Kato and Gospodarowicz 1984 a, b). Although these cell types require serum or FGF, or both, in order to proliferate when maintained on plastic, they no longer require FGF when maintained on an ECM: the addition of plasma alone is sufficient to make them proliferate at an optimal rate. The ECM produced by bovine corneal endothelial cells has been shown to support the proliferation and differentiation of all cell types known to respond to FGF. It does not support the growth of cell types that are not responsive to FGF. For example, in order to induce kidney epithelial cells to divide and to express their morphogenesis properly when confluent (tubule formation), one has to maintain the cells on an ECM with a composition similar to that of kidney tubules in vivo (Gospodarowicz et al. 1984c). Such matrices are produced by the mouse-derived teratocarcinoma PF-HR-9 cell line and are composed exclusively of collagen type IV, heparan sulfate proteoglycans, laminin, and entactin, as well as a 185 kDa glycoprotein whose function is at present unknown (Lewo et al. 1982, Strickland et al. 1980, Hogan et al. 1982).

These observations raise the possibility that growth factors such as FGF can replace the cell requirement for an adequate substratum such as the ECM and make the cells fully responsive to progression factors present in serum and plasma when cells are maintained on plastic. This effect could reflect the ability of FGF to induce the synthesis and secretion of the cell's ECM. Such induction could in turn make the cells sensitive to progression factors present in serum or plasma. This possibility is supported by previous observations that FGF can control the production by vascular endothelial cells of extracellular and cell surface components such as fibronectin, various collagen types, and proteoglycans. Since sparse cultures of endothelial cells proliferate poorly when maintained on plastic but not when maintained on an ECM, it may be that, at very low cell density, cultures maintained on plastic are unable to produce enough extracellular material to support further growth. The mitogenic effect of FGF on these cells could be the result not only of an increased synthesis of the ECM by vascular endothelial cells but also of the right components.

EXTRACELLULAR MATRIX COMPONENTS INVOLVED IN CELL ATTACHMENT VERSUS CELL PROLIFERATION AND DIFFERENTIATION

Little is now known about the ECM factors that have a permissive effect on cell proliferation and could restore the cell response to plasma factors. These factors could be intrinsic components of the ECM or, since cells producing it are grown on medium supplemented with serum or plasma and FGF, they could consist of as-yet-unidentified serum or plasma factors, as well as such growth factors as FGF or PDGF, that have become part of the structure during its formation of the ECM. Also, cellular growth-promoting factors released during cell lysis could be adsorbed onto it and be responsible for its growth-promoting properties.

The possibility that adsorbed growth factor such as FGF or PDGF are the active agents has been eliminated on the basis of inactivating treatment (Greenburg and Gospodarowicz 1982, Gospodarowicz et al. 1983b). Likewise, cytoplasmic factors that would have adsorbed to the ECM during the lytic process are not likely to be involved, since denudation by 2 M urea, which results in little or no cell death, yields ECMs that are as potent in supporting cell growth as those prepared by detergent or weak alkali treatment (Gospodarowicz et al. 1983b). The possibility that the factors involved are derived from serum or plasma can also be eliminated by the demonstration that ECM produced by corneal endothelial cells growing in defined medium supplemented with HDL, transferrin, and insulin are as potent in supporting cell growth as ECM produced by cells growing in serum- or plasma-supplemented medium (Gospodarowicz et al. 1983b). The sole remaining possibility is that the factors involved in permitting cell growth are intrinsic components of the ECM. Among the various components that have been analyzed for their ability to support cell growth are interstitial (types I, II, and III), basement membrane (type IV) collagen, and fibronectin. None of these components alone promoted an increased rate of cell proliferation, and in all cases they induced an aberrant morphological appearance, the cultures being composed of large cells of which a high proportion were binucleated (Gospodarowicz et al. 1980a).

The possibility exists that, although no single ECM component is active in supporting cell growth, a cooperative effect between components could issue in such an effect. The intricate nature of the ECM is exemplified by the complex interaction of its known components (collagen, glycoproteins, proteoglycans, and glycosaminoglycans), which form a highly stable scaffolding upon which cells rest in vivo. This complexity makes its reconstitution from its known constituents a formidable task. The role that the various ECM components play in cell proliferation has therefore been investigated by indirect methods that rely on selective inactivation by chemical, enzymatic, or heat treatment (Greenburg and Gospodarowicz 1982, Gospodarowicz et al. 1983b). Of all the treatments used, only three were effective in inactivating the ability of ECM to support cell proliferation. Exposure of ECM-coated dishes to 4 M NH_4OH (pH 13.8), which cleaves proteoglycoproteins or glycopeptides at the O-glycosidic bond between the protein and carbohydrate moieties, inhibited the rate of proliferation of vascular endothelial cells by 97%. Likewise, treatment of ECM-coated dishes with 4 M Guanidine-HCl, which extracts up to 80% of the proteoglycan, caused a 92% reduction in cell growth. Treatment of ECM-coated dishes with nitrous acid (HNO_2), which degrades heparan or heparan sulfate into sulfated disaccharides and nonsulfated oligosaccharides, lowered cell proliferation by 90% and led to the release of $^{35}SO_4$-

labeled macromolecules, 50% of which were disaccharides. Although the effect of HNO_2 on intact ECM is not known, its specific degradation of isolated heparan and heparan sulfate glycosaminoglycans could indicate that it has a similar effect on ECM, one that may correlate with its adverse effect on the ability of that structure to support cell growth.

The importance of cell shape in proliferation (Gospodarowicz et al. 1979b, Folkman and Moscona 1978) suggests that attachment factors within the ECM, such as fibronectin and laminin, could modify cell shape because of their direct contact with the cell membrane and could therefore play a dual role in controlling cell proliferation. To investigate the relationship between growth and attachment, the ability of vascular endothelial cells to attach to ECM-coated dishes following treatment at alkaline pH and by heat has been compared. In the absence of serum, only 10% of the cells attach to plastic after 1 hr, and maximal cell attachment (40%) is not reached before 18 hr. In contrast, cells seeded on ECM-coated dishes rapidly attach and spread, and most (90%) have attached by 1 hr. No significant difference is seen in the rate of cell attachment after treatment of ECM-coated dishes at alkaline pH, although it did destroy 97% of their ability to support cell proliferation. In fact, initial cell attachment (1 hr) was slightly enhanced by alkaline treatment. It is therefore unlikely that the component of the ECM involved in cell proliferation and removed or destroyed by high pH treatment is involved in cell attachment (Greenburg and Gospodarowicz 1982).

These studies provide evidence regarding the separate nature of the components of the ECM, those that are responsible for cell attachment versus those involved in conveying its permissive effect on cell proliferation. It is likely that the active components involved in cell attachment are represented by both laminin and fibronectin. The factor involved in conveying the signal for cell proliferation is likely to be a sulphated glycoprotein or proteoglycan that is susceptible to extraction by 4 M Guanidine-HCl, is degraded by nitrous acid, and contains an O-glycosidic bond and either glucosamine or galactosamine. It is also possible that the factors involved are tightly bound to heparan sulfate proteoglycans and are released into the medium when this proteoglycan is degraded by nitrous acid (Greenburg and Gospodarowicz 1982).

Recent studies by Shing et al. (1984) suggest that cationic polypeptides tightly bound to proteoglycans could be responsible for the growth-promoting effect of ECM. These authors have recently described the purification from a chondrosarcoma ECM extract of a capillary endothelial cell growth factor that bound tightly to a column of heparin Sepharose. Since heparan sulfate is structurally (Kraemer 1971) and at high concentrations biologically similar to heparin (Wasteson et al. 1977), it is conceivable that the capillary growth factor would have been closely associated in the ECM with heparan sulfate proteoglycans. It will also be of interest to compare the component of the ECM that is active in cell proliferation to the alkaline-labile component of demineralized bone matrix described by Reddi (1976) as inducing differentiation of fibroblasts.

Among factors that could play a role in the permissive effect of ECM on cell proliferation is calcium. ECM are very rich in calcium, which is tightly bound to the sulfated group of proteoglycans. This is reflected by the ECM's ability to stain intensely with Alizarin red. It has been shown that calcium, in the form of microcrystalline precipitate, is as good a commitment factor as FGF or PDGF (Stiles et al. 1979, Dulbecco and Elkington 1975). Recent studies by Betsholtz and Westermark (1984) have shown that PDGF's mitogenic effect depends on the extracellular calcium concentration. In sparse fibroblast cultures, a mitogenic effect of PDGF could only be seen in suboptimal Ca^{++} concentrations. When cells were exposed to physiological calcium concentrations, PDGF was no longer required in order for cells to divide (Betsholtz and Westermark 1984). Similar observations have also been reported by Balk (1971) and Ohno and Kaneko (1981). The ability of basement membrane to accumulate calcium may therefore represent an evolutionary mechanism through which the commitment factors are automatically incorporated into the substrate upon which cells will migrate, proliferate, and differentiate (Gospodarowicz and Greenburg 1981).

The ways in which the ECM could influence cells to proliferate is open to speculation. One possibility could operate through their ability to induce and support the appearance of cell surface transferrin receptor sites. Transferrin in vivo is the only source of iron for the cells. Modulation of transferrin receptor sites correlates closely with the cell's ability to proliferate in vivo (Ekblom et al. 1983). Transferrin has been shown to be an absolute requirement in vitro if cells are to divide in serum-free conditions (Ozawa et al. 1983). Recently, Ekblom et al. (1981c, 1983) have shown that metanephric mesenchymal cells do not proliferate in organ culture when exposed to synthetic medium supplemented with transferrin. In contrast, when mesenchymal rudiments are induced by the spinal cord to differentiate into nephrons, transferrin will support the proliferation of the cells. This ability to proliferate in response to transferrin corresponds to changes in basement membrane composition during the induction process (Ekblom et al. 1981c, 1983). This correspondence raises the possibility that one of the mechanisms through which basement membrane components could facilitate the proliferation of cells maintained on the membrane is through their stimulation or maintenance of an appropriate level of transferrin receptors on the cell surface. This appropriate level will in turn lead to active kidney epithelial cell proliferation, resulting from the cell's ability to internalize plasma transferrin through a transferrin receptor–mediated endocytotic pathway.

In addition to its role as a competence factor, the ECM could also affect the cytokinetics of cells. This possibility is supported by the observation of Yaoi and Kanaseki (1972) that low-density fibroblast cultures maintained on ECM (then called microexudate) have a high rate of DNA synthesis and a high mitotic index. In contrast, low-density cultures maintained on plastic have a high rate of DNA synthesis but a low mitotic index, thus suggesting that cells do not enter into mitosis. It is therefore likely that whereas plastic provides a foreign substrate upon which cells can attach tenaciously and spread in a vain attempt to phagocytose it, the ECM provides a natural substrate that cells recognize as their own and upon which they can undergo their characteristic changes in morphology (rounding up) that occur at mitosis. These morphological changes probably result from the hydrolysis of specific components of the ECM, leading to the disruption of the microfibrillar system and the rearrangement of the cellular cytoskeleton, so that cells can go through the cleavage steps, giving rise to progeny cells instead of undergoing endomitosis (Gospodarowicz et al. 1981a, b, 1982a).

PLASMA FACTORS INVOLVED IN THE PROLIFERATION OF NORMAL DIPLOID CELLS

Among the plasma factors that may be directly or indirectly responsible for the active proliferation of various cell types maintained on ECM are plasma lipoproteins (Gospodarowicz and Tauber 1980), insulin or somatomedin C, and epidermal or fibroblast growth factor. For all the cell types listed in Table 2 (Gospodarowicz 1983a, b, Gospodarowicz et al. 1982a), transferrin, the main iron-carrying protein in the bloodstream, had to be present if cells were to respond to these plasma factors. This absolute requirement for transferrin could either reflect its role in delivering iron to the cells or its ability to remove toxic traces of metals from the medium (Barnes and Sato 1980).

In addition to transferrin, the plasma factor requirements of the various normal diploid cell types grown on ECM fell into three broad categories. The first category, represented by vascular endothelial cells and kidney epithelial cells derived from the distal or collecting tubule, required only HDL to proliferate actively (Gospodarowicz 1983a, b, Gospodarowicz et al. 1982a). The second category, to which corneal endothelial cells, lens epithelial cells, and vascular smooth muscle cells belong, in addition to HDL, required insulin or somatomedin C and EGF or FGF (Gospodarowicz 1983b, Gospodarowicz et al. 1982a).

For both of these categories, HDL is the main mitogenic factor, since in its absence cells could not be triggered to divide by transferrin, insulin, somatomedin C, FGF, or EGF added alone or in combination. Although the mechanisms of action of insulin, somatomedin C, FGF, and EGF in triggering cell proliferation are at present unknown, the ways in which HDL could support cell proliferation have been shown to correlate with its ability to activate cellular HMG CoA reductase (Cohen et al. 1982a, b). This enzyme is responsible for the synthesis of mevalonic acid, which is later utilized for the synthesis of dolichol, ubiquinone, isopentenyl adenyl adenine, and cholesterol (Brown and Goldstein 1980). Of these compounds, the nonsterol products are probably required for cell proliferation (Gospodarowicz 1983b, Gospodarowicz et al. 1982a, Quesney-Huneeus et al. 1980). It should be emphasized that the effect of both FGF and EGF on cultured cells maintained on ECM and exposed to serum-free conditions is not so much on cell proliferation, but instead on cell senescence. In their presence the final cell density of the cultures is only 10% to 20% higher than that reached by cultures exposed to transferrin, HDL, and insulin. Both EGF and FGF, however, do sharply increase the longevity of cultures maintained in serum-free medium. This effect is best seen with corneal endothelial and lens epithelial cells; when maintained in synthetic medium supplemented with transferrin, HDL, and insulin they had lifespans of 20 and 28 generations, respectively. When FGF was present, the lifespans of the cultures became 85 and 100 generations, respectively (Giguere et al. 1982, Gospodarowicz and Massoglia 1982).

The third cell category is represented by steroid-producing cells such as adrenal cortex or ovarian granulosa cells. Both these cell types showed only a minor response to HDL, which at concentrations above 50 μg protein/ml became cytotoxic (Gospodarowicz 1983a, b, Gospodarowicz et al. 1982b). A major proliferative response of these two cell types to HDL would not have been expected, since these cells produce steroids and have an active HMG CoA reductase pathway. Instead, insulin is the main mitogenic factor for both granulosa and adrenal cortex cells, and although addition of HDL and EGF or FGF improved the final cell density of the cultures exposed to a defined medium, the improvements were slight (Gospodarowicz 1983b, Gospodarowicz et al. 1982b).

The substrate upon which cultures were maintained was found to be of crucial importance if a significant response to HDL or insulin (depending on cell type) was to be observed, since it was not observed with cells maintained on plastic dishes. This suggests that the integrity of the ECM upon which cells rest and migrate is an important factor in determining cell response to lipoproteins and insulinlike activity present in plasma.

ACKNOWLEDGMENTS

I wish to thank Mr. Dana Redhair for his expert assistance in the preparation of this manuscript. This work was supported by grants from the National Institutes of Health (HL 20197 and 23678, and EY 02186).

REFERENCES

Armelin HA. 1973. Pituitary extracts and steroid hormones in the control of 3T3 cell growth. Proc Natl Acad Sci USA 70:2702–2706.

Balk SC. 1971. Calcium as a regulator of the proliferation of normal, but not of transformed, chicken fibroblasts in a plasma-containing medium. Proc Natl Acad Sci USA 68:271–275.

Barnes D, Sato G. 1980. Serum-free cell culture: A unifying approach. Cell 22:649–655.

Benya PD, Padilla SR, Nimni ME. 1978. Independent regulation of collagen types by chondrocytes during the loss of differentiated function in culture. Cell 15:1313–1321.

Bernfield MR. 1978. The cell periphery in morphogenesis. *In* Littlefield JW, de Grouchy J (eds.), Birth Defects, Excerpta Medica, International Congress Series 432. Elsevier-North Holland, Amsterdam, New York, Oxford, pp. 111–125.

Bernfield MR, Banerjee SD, Cohn RH. 1972. Dependence of salivary epithelial morphology and branching morphogenesis upon acid mucopolysaccharide-protein (proteoglycan) at the epithelial surface. J Cell Biol 52:674–689.

Betsholtz C, Westermark B. 1984. Growth factor induced proliferation of human fibroblasts in serum-free culture depends on cell density and extracellular calcium concentration. J Cell Physiol 118:203–210.

Bishop JM. 1983. Cellular oncogenes and retroviruses. Ann Rev Biochem 52:301–354.

Bohlen P, Baird A, Esch F, Ling N, Gospodarowicz D. 1984a. Isolation and partial molecular characterization of pituitary fibroblast growth factor. Proc Natl Acad Sci USA 81:5364–5368.

Bohlen P, Blair A, Esch F, Ling N, Guillemin R, Gospodarowicz D, 1984b. Isolation from bovine pituitary and partial molecular characterization of a growth factor for mesoderm-derived cells. Lymphokine Res 3:82.

Broad TE, Ham RG. 1983. Growth and adipose differentiation of sheep preadipocyte fibroblasts in serum-free medium. Eur J Biochem 135:33–39.

Brown M, Goldstein J. 1980. Multivalent feedback regulation of HMG CoA reductase, a control mechanism coordinating isoprenoid synthesis on cell growth. J Lipid Res 21:505–517.

Buntrock P, Jentzsch KD, Heder G. 1982a. Stimulation of wound healing using brain extract with FGF activity. I. Quantitative and biochemical studies into formation of granulation tissue. Exp Pathol 21:46–53.

Buntrock P, Jentzsch KD, Heder G. 1982b. Stimulation of wound healing using brain extract with FGF activity. II. Histological and morphometric examination of cells and capillaries. Exp Pathol 21:62–67.

Carlone RL, Foret JE. 1979. Stimulation of mitosis in cultured limb blastemata of the newt, *Notophthalmus viridescens*. J Exp Zool 210:245–252.

Carlone RL, Ganagarajah M, Rathbone MP. 1981. Bovine pituitary fibroblast growth factor has neurotrophic activity for newt limb regenerates and skeletal muscle in vitro. Exp Cell Res 132:15–21.

Choo AF, Logan DM, Rathbone MP. 1981. Nerve trophic effects: partial purification from chick embryo brains of proteins that stimulate protein synthesis in cultured newt blastemata. Exp Neurol 73:558–570.

Clark JF, Jones KL, Gospodarowicz D, Sato GH. 1972. Hormone dependent growth response of a newly established ovarian cell line. Nature New Biol 236:180–182.

Cohen DC, Massoglia S, Gospodarowicz D. 1982a. Correlation between two effects of high density lipoproteins on vascular endothelial cells: The induction of 3-hydroxy-3-methylglutaryl coenzyme A reductase activity and the support of cellular proliferation. J Biol Chem 257:9429–9437.

Cohen DC, Massoglia S, Gospodarowicz D. 1982b. Feedback regulation of 3-hydroxy-3-methylglutaryl coenzyme A reductase in vascular endothelial cells: Separate sterol and non-sterol components. J Biol Chem 257:11106–11112.

Cohn RH, Banerjee SD, Bernfield MR. 1977. The basal lamina of embryonic salivary epithelia: nature of glycosaminoglycan and organization of extracellular materials. J Cell Biol 73:464–478.

Cooper GM. 1982. Cellular transforming genes. Science 217:801–806.

Corvol MT, Malemud CJ, Sokoloff L. 1972. A pituitary growth promoting factor for articular chondrocytes in monolayer culture. Endocrinology 90:262–271.

Dresden MH. 1969. Denervation effects on newt limb regeneration: DNA, RNA and protein synthesis. Dev. Biol. 19:311–320.

Dulbecco R, Elkington J. 1975. Induction of growth in resting fibroblastic cell cultures by Ca^{++}. Proc Natl Acad Sci USA 72:1584–1588.

Duthu GS, Smith JR. 1980. In vitro proliferation and life-span of bovine aorta endothelial cells: effect of culture conditions and fibroblast growth factor. J Cell Physiol 103:385–392.

Ekblom P. 1981. Formation of basement membranes in the embryonic kidney. An immunohistological study. J Cell Biol 91:1–10.

Ekblom P, Alitalo K, Vaheri A, Timpl R, Saxen L. 1980. Induction of a basement membrane glycoprotein in embryonic kidney: possible role of laminin in morphogenesis. Proc Natl Acad Sci USA 77:485–489.

Ekblom P, Lehtonen E, Saxen L, Timpl R. 1981a. Shift in collagen type is an early response to induction of the metanephric mesenchyme. J Cell Biol 89:276–283.

Ekblom P, Miettinen A, Virtanen I, Dawnay A, Wahlstrom T, Saxen L. 1981b. In vitro segregation of the metanephric nephron. Dev Biol 84:88–95.

Ekblom P, Thesleff I, Miettinen A, Saxen L. 1981c. Organogenesis in a defined medium supplemented with transferrin. Cell Differ 10:281–288.

Ekblom P, Thesleff I, Saxen L, Miettinen A, Timpl R. 1983. Transferrin as a fetal growth factor: Acquisition of responsiveness related to embryonic induction. Proc Natl Acad Sci USA 80:2651–2655.

Eva A, Robbins KC, Andersen PR, Srinivasan A, Tronick SR, Reddy EP, Nelson WE, Galen AT, Lautenberger JA, Papas TS, Westin EH, Wong-Staal F, Gallo RC, Aaronson SA. 1982. Cellular genes analogous to retroviral oncogenes are transcribed in human tumor cells. Nature 295:116–119.

Folkman J, Moscona A. 1978. Role of cell shape in growth control. Nature 273:345–348.

Giguere L, Cheng J, Gospodarowicz D. 1982. Factors involved in the control of proliferation of bovine corneal endothelial cells maintained in serum-free medium. J Cell Physiol 110:72–80.
Gospodarowicz D. 1974. Localization of a fibroblast growth factor and its effect alone and with hydrocortisone on 3T3 cell growth. Nature 249:123–127.
Gospodarowicz D. 1975. Purification of a fibroblast growth factor from bovine pituitary. J Biol Chem 250:2515–2519.
Gospodarowicz D. 1979. Fibroblast and epidermal growth factors. Their uses in vivo and in vitro in studies on cell functions and cell transplantation. Mol Cell Biochem 25:79–110.
Gospodarowicz D. 1983a. Control of endothelial cell proliferation and repair. *In* Cryer A (ed), Biochemical Interaction at the Endothelium. Elsevier-North Holland, New York, pp. 366–396.
Gospodarowicz D 1983b. The control of mammalian cell proliferation by growth factors, basement lamina, and lipoproteins. J Invest Dermatol 81:405–505.
Gospodarowicz D. 1984. Fibroblast growth factor. *In* Li CH (ed), Hormonal Proteins and Peptides, vol. 12. Academic Press, New York (in press).
Gospodarowicz D, Giguere L. 1982. Growth factors, effects on corneal tissue. *In* McDevitt DS (ed), Cell Biology of the Eye. Academic Press, New York, pp. 97–135.
Gospodarowicz, D, Greenburg G. 1981. Growth control of mammalian cells. Growth factors and extracellular matrix. *In* Ritzen M, Aperia A, Hall K, Larsson A, Zetterberg A, Zetterstrom R (eds), The Biology of Normal Human Growth. Raven Press, New York, pp. 1–21.
Gospodarowicz D, Ill CR. 1980. Do plasma and serum have different abilities to promote cell growth? Proc Natl Acad Sci USA 77:2726–2730.
Gospodarowicz D, Lui G-M. 1981. Effect of substrata and fibroblast growth factor on the proliferation in vitro of bovine aortic endothelial cells. J Cell Physiol 109:385–392.
Gospodarowicz D, Massoglia SL. 1982. Plasma factors involved in the in vitro control of proliferation of bovine lens cells grown in defined medium. Effect of fibroblast growth factor on cell longevity. Exp Eye Res 35:259–270.
Gospodarowicz D, Mescher AL. 1977. A comparison of the responses of cultured myoblasts and chondrocytes to fibroblast and epidermal growth factors. J Cell Physiol 93:117–128.
Gospodarowicz D, Mescher AL. 1981. Fibroblast growth factor and vertebrate regeneration. *In* Riccardi VM, Mulvihill JJ (eds), Advances in Neurology: Neurofibromatosis, vol 29. Raven Press, New York, pp. 149–171.
Gospodarowicz D, Moran JS. 1974. Effect of a fibroblast growth factor, insulin, dexamethasone and serum on the morphology of BALB/c 3T3 cells. Proc Natl Acad Sci USA 71:4648–4652.
Gospodarowicz D, Moran JS. 1975. Mitogenic effect of fibroblast factor on early passage cultures of human and murine fibroblasts. J Cell Biol 66:451–556.
Gospodarowicz D. Moran JS. 1976. Growth factors in mammalian cell culture. Ann Rev Biochem 45:531–558.
Gospodarowicz D, Tauber J-P. 1980. Growth factors. Extracellular matrix. Endocr Rev 1:201–227.
Gospodarowicz D, Zetter B. 1977. The use of fibroblast and epidermal growth factors to lower the serum requirement for growth of normal diploid cells in early passage: A new method for cloning. *In* Joint WHO/IABS Symposium on the Standardization of Cell Substrates for the Production of Virus Vaccine. S. Karger, Geneva, Switzerland, pp. 109–130.
Gospodarowicz D, Rudland P, Lindstrom J, Benirschke K. 1975. Fibroblast growth factor: localization, purification, mode of action, and physiological significance. Nobel symposium on growth factors. Adv Metab Disord 8:301–335.
Gospodarowicz D, Moran JS, Braun D, Birdwell CR. 1976a. Clonal growth of bovine endothelial cells in tissue culture: Fibroblast growth factor as a survival agent. Proc Natl Acad Sci USA 73:4120–4124.
Gospodarowicz D, Weseman J, Moran J, Lindstrom J. 1976b. Effect of fibroblast growth factor on the division and fusion of bovine myoblasts. J Cell Biol 70:395–407.
Gospodarowicz D, Moran JS, Braun D. 1977. Control of proliferation of bovine vascular endothelial cells. J Cell Physiol 91:377–388.
Gospodarowicz D, Bialecki H, Greenburg G. 1978a. Purification of the fibroblast growth factor activity from bovine brain. J Biol Chem 253:3736–3743.
Gospodarowicz D, Brown KS, Birdwell CR, Zetter B. 1978b. Control of proliferation of human vascular endothelial cells of human origin. I. Characterization of the response of human umbilical vein endothelial cells to fibroblast growth factor, epidermal growth factor, and thrombin. J Cell Biol 77:774–788.

Gospodarowicz D, Greenburg G, Bialecki H, Zetter B. 1978c. Factors involved in the modulation of cell proliferation in vivo and in vitro: the role of fibroblast and epidermal growth factors in the proliferative response of mammalian cells. In Vitro 14:85–118.
Gospodarowicz D, Mescher AL, Birdwell CR. 1978d. Control of cellular proliferation by the fibroblast and epidermal growth factors. National Cancer Inst Monogr 48:109–130.
Gospodarowicz D, Mescher AL, Moran JS. 1978e. Cellular specificities of fibroblast growth factor and epidermal growth factor. *In* Papaconstantinou J, Rutter WJ (eds), Molecular Control of Proliferation and Cytodifferentiation. 35th Symposium of the Society for Developmental Biology, Academic Press, New York, pp. 33–61.
Gospodarowicz D, Bialecki H, Thakral TK. 1979. The angiogenic activity of the fibroblast and epidermal growth factor. Exp Eye Res 28:501–514.
Gospodarowicz D, Vlodavsky I, Greenburg G, Alvarado J, Johnson LK, Moran J, 1979a. Studies on atherogenesis and corneal transplantation using cultured vascular and corneal endothelia. Recent Prog Horm Res 35:375–448.
Gospodarowicz D, Vlodavsky I, Greenburg G, Alvarado J, Johnson LK, Moran J. 1979b. Cellular shape is determined by the extracellular matrix and is responsible for the control of cellular growth and function. *In* Ross R, Sato G (eds), Cold Spring Harbor Conferences on Cell Proliferation, vol 9, Hormones and Cell Culture. Cold Spring Harbor Laboratory, Cold Spring Harbor, pp. 561–592.
Gospodarowicz D, Delgado D, Vlodavsky I. 1980a. Permissive effect of the extracellular matrix on cell proliferation in vitro. Proc Natl Acad Sci USA 77:4094–4098.
Gospodarowicz D, Vlodavsky I, Savion N. 1980b. The extracellular matrix and the control of proliferation of vascular endothelial and vascular smooth muscle cells. J Supramol Struct 13:339–372.
Gospodarowicz D, Cheng J, Hirabayashi K, Tauber J-P. 1981a. The extracellular matrix and the control of vascular endothelial and smooth muscle cell proliferation in cellular interactions. *In* Dingle, Gordon (eds), Research Monographs in Cell and Tissue Physiology, vol. 7. North Holland, pp. 135–165.
Gospodarowicz D, Giguere L, Savion N, Tauber J-P. 1981b. Effects of fibroblast growth factor and basal lamina on the growth of normal cells in serum-free medium. *In* Bizollon CA (ed), Physiological Peptides and New Trends in Radioimmunology. Elsevier-Horth Holland, New York, pp. 127–147.
Gospodarowicz D, Vlodavsky I, Savion N. 1981c. The role of fibroblast growth factor and the extracellular matrix in the control of proliferation and differentiation of corneal endothelial cells. Vision Res 21:87–103.
Gospodarowicz D, Vlodavsky I, Savion N. 1981d. The extracellular matrix and the control of proliferation of vascular endothelial and vascular smooth muscle cells. Prog Clin Biol Res 66A:53–86.
Gospodarowicz D, Cohen DC, Fujii DK. 1982a. Regulation of cell growth by the basal lamina and plasma factors: Relevance to embryonic control of cell proliferation and differentiation. *In* Cold Spring Harbor Conferences on Cell Proliferation: Growth of Cells in Hormonally Defined Media. Hormones and Cell Culture, vol. 9. Cold Spring Harbor Laboratory, Cold Spring Harbor, pp. 95–124.
Gospodarowicz D, Fujii DK, Vlodavsky I. 1982b. Basal lamina and the control of proliferation of malignant and normal cells. *In* Revoltella RP, Ponteri GM, Basilico C, Rovera G, Gallo RC, Subak-Sharpe JH (eds), Expression of Differentiated Functions in Cancer Cells. Raven Press, New York, pp. 121–139.
Gospodarowicz D, Lui G-M, Cheng JC. 1982c. Purification in high yield of brain fibroblast growth factor by preparative isoelectric focusing at pH 9.6. J Biol Chem 257:12,266–12,276.
Gospodarowicz D, Cheng J, Lirette M. 1983a. Bovine brain and pituitary fibroblast growth factors: comparison of their abilities to support the proliferation of human and bovine vascular endothelial cells. J Cell Biol 97:1677–1685.
Gospodarowicz D, Gonzalez R, Fujii DK. 1983b. Are factors originating from serum, plasma, or cultured cells involved in the growth-promoting effect of the extracellular matrix produced by cultured bovine corneal endothelial cells? J Cell Physiol 114:191–202.
Gospodarowicz D, Lui G-M, Cheng J, Baird A, Bohlen P. 1984a. Isolation by heparin Sepharose affinity chromatography of brain fibroblast growth factor: Identity with pituitary fibroblast growth factor. Proc Natl Acad Sci USA (in press).
Gospodarowicz D, Massoglia S, Cheng J, Lui G-M, Bohlen P. 1984b. Partial molecular characterization of bovine pituitary fibroblast growth factor purified by fast liquid chromatography (FPLC). J Cell Physiol (in press).
Gospodarowicz D, Lepine J, Massoglia S, Wood I. 1984c. Ability of various basement membranes to support differentiation in vitro of normal diploid bovine kidney tubule cells. J Cell Biol 99:947–961.

Gottesman MM. 1978. Transformation dependent secretion of low molecular weight protein by murine fibroblasts. Proc Natl Acad Sci USA 75:2767–2771.

Greenburg G, Gospodarowicz D. 1982. Inactivation of a basement membrane component responsible for cell proliferation but not cell attachment. Exp Cell Res 140:1–14.

Greenburg G, Vlodavsky I, Foidart J-M, Gospodarowicz D. 1980. Conditioned medium from endothelial cell cultures can restore the normal phenotypic expression of vascular endothelium maintained in vitro in the absence of fibroblast growth factor. J Cell Physiol 103:333–347.

Gutmann E. 1976. Neurotrophic relations. Annu Rev Physiol 38:177–216.

Hay ED. 1977. Interaction between the cell surface and extracellular matrix in corneal development. *In* Lash JW, Burger MM (eds), Cell and Tissue Interaction. Raven Press, New York, pp. 115–137.

Hay ED. 1980. Development of the vertebrate cornea. Int Rev Cytol 63:263–322.

Hay ED. 1981a. Embryonic induction and tissue interaction during morphogenesis. *In* Littlefield JW, de Grouchy J (eds), Birth Defects. Excerpta Medica International Congress Series 432. Elsevier-North Holland, New York, pp. 117–127.

Hay ED. 1981b. Extracellular matrix. J Cell Biol 91:205–223.

Hayward WS, Neel BG, Astrin SM. 1981. Activation of cellular oncogenes by promoter insertion in ALV-induced lymphoid leukosis. Nature 296:475–479.

Hogan BLM, Taylor A, Kurkkinen M, Couchman JR. 1982. Synthesis and localization of two sulphated glycoproteins associated with basement membranes and the extracellular matrix. J Cell Biol 95:197–204.

Holley RW, Kiernan JA. 1968. Contact inhibition of cell division in 3T3 cells. Proc Natl Acad Sci USA 70:300–304.

Jennings T, Jones RD, Lipton A. 1979. A growth factor from spinal cord. J Cell Physiol 100:273–278.

Jentzsch KD, Wellmitz GL, Heder G, Petzold E, Buntrock P, Gehme P. 1980. A bovine brain fraction with fibroblast growth factor activity inducing articular cartilage regeneration in vivo. Acta Biol Med Germ 39:967–971.

Jimenez de Asua L. 1980. An ordered sequence of temporal steps regulates the rate of initiation of DNA synthesis in cultured mouse cells. *In* Jimenez de Asua L, Levi-Montalcini R, Sheilds R, Iacobelli S (eds), Control Mechanisms in Animal Cells. Raven Press, New York, pp. 173–197.

Jimenez de Asua L, O'Farrell MK, Bennett D, Clinigan D, Rudland PS. 1977a. Interaction of two hormones and their effect on observed rate of initiation of DNA synthesis in 3T3 cells. Nature 265:151–153.

Jiminez de Asua L, O'Farrell MK, Clinigan D, Rudland PS. 1977b. Temporal sequence of hormonal interactions during the prereplicative phase of quiescent cultured 3T3 fibroblasts. Proc Natl Acad Sci USA 74:3845–3849.

Kato Y, Gospodarowicz D. 1984a. Effect of extracellular matrix produced by bovine corneal endothelial cells on proteoglycan biosynthesis by rabbit costal chondrocytes. J Cell Physiol 120:354–363.

Kato Y, Gospodarowicz D. 1984b. Effect of exogenous biomatrices on proteoglycan synthesis by cultured rabbit costal chondrocytes. J Cell Biol (in press).

Kato Y, Gospodarowicz D. 1984c. Sulfated proteoglycan synthesis by rabbit costal chondrocytes grown in the presence versus absence of fibroblast growth factor. J Cell Biol (in press).

Kelly K, Cochran BH, Stiles CD, Leder P. 1983. Cell specific regulation of the c myc by lymphocyte mitogens and platelet-derived growth factor. Cell 55:603–610.

Kleinman HK, Klebe FJ, Martin GR. 1981. Role of collagenous matrices in the adhesion and growth of cells. J Cell Biol 88:473–485.

Kraemer PM. 1971. Heparin sulfates of cultured cells. II. Acid-soluble and -precipitable species of different cell lines. Biochemistry 10:1445–1451.

Kratochwil K. 1972. Tissue interactions during embryonic development. *In* Tarin D (ed), General Properties in Tissue Interaction in Carcinogenesis. Academic Press, New York, pp. 1–48.

Land H, Parada LF, Weinberg RA. 1983. Tumorigenic conversion of primary embryo fibroblasts requires at least two cooperating oncogenes. Nature 304:596–602.

Leof EB, Wharton W, Van Wyk JJ, Pledger WJ. 1982. Epidermal growth factor (EGF) and somatomedin C regulate G1 progression in competent BALB/c 3T3 cells. Exp Cell Res 141:107–115.

Lewo I, Alitalo K, Riteli L, Vaheri A, Timpl R, Wartiovaara J. 1982. Basal lamina glycoproteins and type IV collagen are assembled into a fine fibered matrix in cultures of a teratocarcinoma-derived endoderm cell line. Exp Cell Res 137:15–23.

Longenecker JP, Kilty LA, Johnson LK. 1984. Glucocorticoid inhibition of vascular smooth muscle proliferation: Influence of homologous extracellular matrix and serum proteins. J Cell Biol 98:534–540.
Meier S, Hay ED. 1974. Control of corneal differentiation by extracellular materials. Collagen as a promoter and stabilizer of epithelial stroma production. Dev Biol 38:249–270.
Meier S, Hay ED. 1975. Stimulation of corneal differentiation by interaction between cell surface and extracellular matrix. I. Morphometric analysis of transfilter "induction." J Cell Biol 66:275–291.
Mescher AL, Gospodarowicz D. 1979. Mitogenic effect of a growth factor derived from myelin on denervated regenerates of newt forelimbs. J Exp Zool 207:497–503.
Mescher AL, Loh JJ. 1980. Newt forelimb regeneration blastemas in vitro: Cellular response to explantation and effects of various growth-promoting substances. J Exp Zool 216:235–245.
Nilsen-Hamilton M, Shapiro JM, Massoglia SL, Hamilton RT. 1980. Selective stimulation by mitogens of incorporation of ^{35}S-methionine into a family of proteins released into the medium by 3T3 cells. Cell 20:19–28.
Nilsen-Hamilton M, Hamilton RT, Allen WR, Massoglia SL. 1981. Stimulation of the release of two glycoproteins from mouse 3T3 cells bygrowth factors and by agents that increase intralysosomal pH. Biochem Biophys Res Commun 101:411–417.
Nilsen-Hamilton M, Hamilton RT, Adams GA. 1982. Rapid selective stimulation by growth factors of the incorporation by BALB/c 3T3 cells of (^{35}S)methionine into a glycoprotein and five superinducible proteins. Biochem Biophys Res Commun 101:158–166.
Ohno T, Kaneko T. 1981. Competitive growth stimulation by Ca^{++} and the platelet-derived growth factor in human diploid fibroblasts. Cell Structure and Function 6:83–86.
Ozawa E, Kimura I, Hasegawa T, Ii I, Saito K, Hagiwahara Y, Shimo-oka T. 1983. Iron-bound transferrin as a myotrophic factor. *In* Ebashi S, Ozawa E, (eds), Muscular Dystrophy: Biomedical Aspects. Japan Science Society Press, Tokyo/Springer-Verlag, Berlin, pp. 53–60.
Pardee AB. 1974. A restriction point for control of normal animal cell proliferation. Proc Natl Acad Sci USA 71:1286–1290.
Pledger WJ, Hart DA, Locatell, KL, Scher CD. 1981. Platelet-derived growth factor modulated protein constitutive synthesis by a transformed cell line. Proc Natl Acad Sci USA 78:4358–4361.
Powell JA. 1969. Analysis of histogenesis and regenerative ability of denervated forelimb regenerates of *Triturus viridescens*. J Exp Zool 170:125–148.
Quesney-Huneeus V, Wiley M, Siperstein M. 1980. Isopentenyladenine as a mediator of mevalonate-regulated DNA replication. Proc Natl Acad Sci USA 77:5842–5846.
Reddi AH. 1976. Collagen and cell differentiation. *In* Ramachadran GN, Reddi AH (eds), Biochemistry of Collagen. Plenum Press, New York, pp. 449–477.
Rudland PS, Jimenez de Asua L. 1979. Action of growth factors in the cell cycle. Biochim Biophys Acta 560:91–133.
Rudland PS, Seifert W, Gospodarowicz, D. 1974. Growth control in cultured mouse fibroblasts: induction of the pleiotypic and mitogenic responses by a purified growth factor. Proc Natl Acad Sci USA 71:2600–2604.
Saxen L, Ekblom P, Lehtonen E. 1982. The kidney as a model system for determination and differentiation. *In* Ritzen M (ed), The Biology of Normal Human Growth. Raven Press, New York, pp. 177–127.
Schlessinger J, Geiger B. 1981. Epidermal growth factor induces redistribution of actin and α actinin in human epidermal carcinoma cells. Exp Cell Res 134:273–279.
Serrero G, Khoo JC. 1982. An in vitro model to study adipose differentiation in serum-free medium. Anal Biochem 120:351–359.
Shing Y, Folkman J, Sullivan R, Butterfield C, Murray J, Klagsbrun M. 1984. Heparin affinity. Purification of a tumor-derived capillary endothelial cell growth factor. Science 223:1296–1299.
Singer M. 1952. The influence of the nerve in regeneration of the amphibian extremity. Q Rev Biol 27:169–200.
Singer M. 1954. Induction of regeneration of the forelimb of the post-metamorphic frog by augmentation of the nerve supply. J Exp Zool 126:419–471.
Singer M. 1974. Neurotrophic control of limb regeneration in the newt. Ann NY Acad Sci 228:308–322.
Singer M, Craven L. 1948. The growth and morphogenesis of the regenerating forelimb of adult triturus following denervation at various stages of development. J Exp Zool 108:279–308.

Stiles CD, Capne GT, Scher CD, Antoniades HM, Van Wyk JJ, Pledger WJ. 1979. Dual control of cell growth by somatomedins and "competence factors." Proc Natl Acad Sci USA 76:1279–1283.

Strickland S, Smith KS, Marotti KR. 1980. Hormonal induction of differentiation in teratocarcinoma stem cells: generation of parietal endoderm by retinoic acid and dibutyryl cAMP. Cell 21:347–355.

Thesleff I, Hurmerinta K. 1981. Changes in the distribution of type IV collagen, laminin, proteoglycan, and fibronectin during mouse tooth development. Differentiation 18:249–270.

Thesleff I, Barrach HJ, Foidart JM, Vaheri A, Pratt RM, Martin GR. 1981. Tissue interactions in tooth development. A review. Dev Biol 81:182–192.

Todd JT. 1823. On the process of reproduction of the members of the aquatic salamander. Q. J. Mic. Sci. Lit. Arts 16:84–96.

Togari A, Baker D, Dickens G, Guroff G. 1983. The neurite-promoting effect of fibroblast growth factor on PC 12 cells. Biochem Biophys Res Commun 114:1189–1193.

Tseng S, Savion N, Stern R, Gospodarowicz D. 1982. Fibroblast growth factor modulates synthesis of collagen in cultured vascular endothelial cells. Eur J Biochem 122:355–360.

Vlodavsky I, Gospodarowicz D. 1979. Structural and functional alterations in the surface of vascular endothelial cells associated with the formation of a confluent cell monolayer and with the withdrawal of fibroblast growth factor. J Supramol Struct 12:73–114.

Vlodavsky I, Johnson LK, Greenburg G, Gospodarowicz D. 1979. Vascular endothelial cells maintained in the absence of fibroblast growth factor undergo structural and functional alterations that are incompatible with their in vivo differentiated properties. J Cell Biol 83:468–486.

Vracko R. 1974. Basal lamina scaffold-anatomy and significance for maintenance of orderly tissue structure. Am J Pathol 77:314–329.

Wasteson A, Glimelius B, Busch C, Westermark B, Heldin C-H, Norling B. 1977. Effect of a platelet endoglycosidase on cell surface heparin sulfate of human cultured endothelial and glial cells. Thromb Res 11:309–321.

Wesels NK. 1977. Extracellular materials and tissue interaction. *In* Tissue Interaction and Development. W. A. Benjamin, Inc., New York, pp. 213–229.

Westall FC, Lennon VA, and Gospodarowicz D. 1978. Brain derived fibroblast growth factor: Identity with a fragment of the basic protein of myelin. Proc Natl Acad Sci USA 75:4675–4678.

Westermark B, Heldin CH, Ek B, Johnson A, Mellstrom K, Nister M, Wasteson A. 1982. Biochemistry and biology of platelet-derived growth factor. *In* Guroff G (ed), Growth and Maturation Factors, vol. I. John Wiley and Sons, New York, pp. 75–112.

Yamada K. 1983. Cell surface interaction with extracellular materials. Annu Rev Biochem 52:761–799.

Yaoi Y, Kanaseki T. 1972. Role of microexudate carpet in cell division. Nature 237:283–285.

Colony-Stimulating Factors, Stem Cells, and Hematopoiesis

Mediators in Cell Growth and Differentiation,
edited by Richard J. Ford and Abby L. Maizel.
Raven Press, New York © 1985.

Comparison of Different Assays for Multipotent Hematopoietic Stem Cells

Robert A. Phillips

Department of Medical Biophysics, University of Toronto and The Ontario Cancer Institute, Toronto, Canada M4X 1K9

The hematopoietic system offers many advantages for studies on cellular differentiation: (1) As a cell renewal system responsible for maintaining an enormous number of cells, it contains a full spectrum of differentiated cells, from almost immortal primitive stem cells to fully differentiated, nondividing cells with a limited life span. (2) Hematopoietic tissue exists in vivo as single cells, making preparation of cell suspensions possible without the use of agents such as enzymes that destroy cells or alter their properties. (3) Many cells in the hematopoietic system can be propagated in cell culture, and some immature cells can form colonies of differentiated cells in vitro. (4) Factors that regulate hematopoiesis and promote growth in vitro have been identified and characterized. (5) Detailed information about the major gene products such as hemoglobin and immunoglobulin offer unique opportunities for studies on the regulation of gene expression during differentiation. (6) Numerous mutations affecting hematopoiesis have arisen in humans and in experimental animals; these mutations facilitate the manipulation and characterization of this complex differentiation pathway.

STEM CELLS OF THE HEMATOPOIETIC SYSTEM

The continued production of differentiated elements with a limited life span requires the existence of stem cells. Although there is not an unequivocal definition for stem cells of normal tissue, they are usually considered to have the following four properties: potential for self-renewal, ability to make differentiated progeny, capacity to proliferate extensively, and responsiveness to regulatory signals (Siminovitch et al. 1963). Because there are no generally accepted minimum values for self-renewal or proliferative capacity, it is often difficult to determine whether or not a particular class of cells has stem cell properties. Thus, the characterization of putative stem cells depends on the properties of the assay used to detect it. In general, these assays can be classed into colony assays and reconstitution assays. The following sections will describe the various assays and discuss the possible relationships between the cells detected in the different assays.

Colony Assays

The modern era of research on hematopoiesis started when Till and McCulloch (1961) described the spleen colony assay. They observed that when a particular class of hematopoietic stem cells was injected into irradiated mice, the cells formed colonies in the spleen. These colonies contained between 1 million and 10 million cells from all lineages of the myeloid system and were derived from a single progenitor (Becker et al. 1963). In addition, transplantation of cells from an individual spleen colony produced secondary spleen colonies in another irradiated recipient (Siminovitch et al. 1963). Thus, the spleen colony-forming cell (CFU-S) has all of the properties of a stem cell, and the spleen colony assay has been used in hundreds of studies to enumerate hematopoietic stem cells. However, there are many indications that the CFU-S assay detects a heterogeneous population of stem cells. Numerous studies have shown heterogeneity of self-renewal potential (e.g., Schofield et al. 1980). In addition, Magli et al. (1982) have shown that the cells giving rise to colonies detected on day 8 after injection are different from the cells that produce colonies on day 12. The former appear restricted to erythroid differentiation, whereas the latter give rise to colonies of cells from several myeloid lineages.

Many of the immediate progenitors of the mature cells in the myeloid and lymphoid lineages can be detected by in vitro colony assays. Although the majority of these assays detect cells in only one or two lineages, there are examples of colonies derived from multipotent progenitors with limited self-renewal potential (Johnson and Metcalf 1977, Fauser and Messner 1979). Three groups have identified in vitro colony-forming cells in the mouse with potential to produce all of the cells in the myeloid lineage and with the capacity for self-renewal (Humphries et al. 1981, Keller and Phillips 1982, Nakahata and Ogawa 1982). The relationship of these multipotent colony-forming cells to CFU-S is not clear. Some colonies may contain cells able to form day 8 colonies in the spleens of irradiated recipients (Keller and Phillips 1982), but in general the in vitro colony-forming cells appear to have less self-renewal potential than CFU-S.

Reconstitution Assays

Injection of bone marrow stem cells gives long-term protection to irradiated recipients. Since Ford and his collaborators (1956) first showed that reconstitution of such recipients was the result of transplantation of cells and not factors, this technique has been used to study hematopoietic stem cells. The limitation of using irradiated recipients in the study of hematopoiesis is that the acute defect produced by irradiation creates a need for large numbers of differentiated progeny. This need in turn requires numerous hematopoietic stem cells. It is impossible, then, to study the long-term differentiative ability of a single hematopoietic stem cell, since it cannot produce sufficient differentiated cells to protect the recipient from the acute effects of irradiation. Similarly, such recipients cannot be used to study stem cells

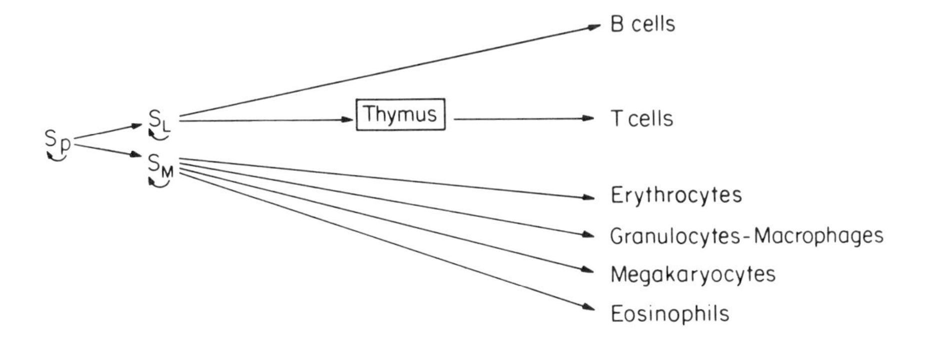

FIG. 1. Pluripotent and restricted stem cells detected by transplantation assays.

committed to only a single pathway of differentiation, since the irradiated recipient has an immediate need for cells in several lineages.

The use of mice with various mutations affecting specific lineages overcomes some of the disadvantages of using irradiated mice. Although numerous mutations affecting hematopoiesis in the mouse have been described (Russell 1979), three genotypes are of particular interest: W/W^v, scid, and xid. Mice homozygous for mutations at the w locus have a severe macrocytic anemia, but their lymphoid system is normal. These mice lack detectable CFU-S (McCulloch et al. 1964), but can be cured by grafts of normal, coisogenic bone marrow (Russell 1979). Mice homozygous for the recently described scid mutation (Bosma et al. 1983) have a normal myeloid system but lack B and T lymphocytes (Dorshkind et al. 1984). They cannot mount either cellular or humoral immune responses and cannot reject allografts or xenografts. Scid mice, unlike nude mice, have a thymus. Like W/W^v mice, scid mice can be cured of their lymphoid deficiency by grafts of normal bone marrow (Phillips et al. 1984). CBA/N mice carry the chromosome X-linked xid mutation affecting the ability of B lymphocytes to respond to some mitogens and antigens. In particular, they cannot respond to a mitogen in agar, and consequently, their lymphocytes are unable to form colonies in the standard B lymphocyte colony-forming assay (CFU-B) (Kincade 1977). The other components of their hematopoietic system are normal.

Whether or not the cells detected in the various colony assays are stem cells is unclear. However, most investigators will agree that a single cell with the capacity to regenerate the hematopoietic system in deficient animals can be called a stem cell. Indeed, this approach provides the most rigorous operational definition of a stem cell that is now available. Although reconstitution is not a necessary criterion to define a stem cell, it is a sufficient condition to define a cell as being a stem cell. Studies on the ability of single cells to reconstitute the hematopoietic system of mutant mice have provided evidence for at least three different stem cells in the system (Phillips et al. 1984). Their capacities for differentiation and their relationships to each other are shown schematically in Figure 1. The multipotent stem cell,

S_P, has the ability to produce both lymphoid and myeloid cells. The other two stem cells, S_M and S_L, are restricted to myeloid and lymphoid differentiation, respectively.

RECONSTITUTION OF W/W^v RECIPIENTS

W/W^v mice will survive without a graft of normal bone marrow cells. However, because of the genetic defect, normal stem cells have a selective advantage in these recipients (Harrison 1973, Mintz 1974). These properties allow analysis of the differentiation and self-renewal of grafted stem cells. In fact, Abramson et al. (1977) demonstrated that cures are apparently possible with a single stem cell. These investigators irradiated normal bone marrow cells to induce chromosome abnormalities in some of the stem cells. A few of the mice cured with these irradiated stem cells contained unique, radiation-induced chromosome abnormalities. Because of the uniqueness of the chromosome aberrations, it is likely that all cells with the same aberration were derived from a common stem cell. Three patterns of reconstitution were observed. In some recipients both myeloid and lymphoid cells had the same chromosome abnormality, thus providing evidence for the S_P type of stem cell; other recipients had a unique marker only in the myeloid pathway, suggesting the existence of a myeloid-restricted stem cell, S_M. Although these investigators also reported a T-lymphocyte-restricted stem cell in bone marrow, it is not possible to rule out the possibility that this marker was induced in a mature, postthymic lymphocyte rather than a prethymic stem cell, as originally suggested.

More recently, Boggs et al. (1982) used W/W^v recipients to establish a limiting dilution assay for hematopoietic stem cells. By determining the proportion of mice cured by various numbers of bone marrow cells, they were able to estimate the frequency of stem cells capable of curing W/W^v recipients—one stem cell per 10^4 bone marrow cells injected. They also estimated that during the 2 years over which the cures were stable, the stem cells responsible for this cure must have undergone approximately 8,000 doubling divisions. These experiments and those of others (Harrison 1973, Abramson et al. 1977, Fleischman et al. 1982) clearly demonstrate the existence of hematopoietic stem cells with extensive proliferative and self-renewal potential. However, the relationship of the stem cells to those detected by colony assays is unclear.

LONG-TERM CULTURE OF HEMATOPOIETIC STEM CELLS

Dexter and his colleagues (1977) have described an in vitro culture system that appears to maintain all classes of hematopoietic stem cells. The unique feature of this system is the creation of an adherent layer of epithelial cells, adipocytes, and macrophages. Within this adherent layer, stem cells can differentiate and self-renew for long periods of time, up to 7 months. Under appropriate conditions, all of the cells of the myeloid lineage, including CFU-S, can be detected in these cultures. In addition, cells from long-term culture can cure W/W^v recipients (Phillips RA, unpublished data); this result indicates that the cultures contain S_P or S_M stem cells, or both.

Mature B lymphocytes and pre-B cells, normal components of bone marrow, rapidly disappear from long-term cultures and do not reappear during the lifetime of the culture (Dorshkind and Phillips 1982). However, the lymphoid system recovers in irradiated recipients injected with cells from long-term culture. In experiments using chromosome markers to distinguish the cultured cells from those of the irradiated recipient, two patterns of reconstitution were observed (Jones-Villeneuve and Phillips 1980). In some mice, chromosomally marked myeloid and lymphoid cells were detected that indicated the presence in the cultures of some, if not all, of the stem cells indicated in Figure 1. In other recipients, marked cells from the long-term cultures were observed only in B and T lymphocytes, all of the myeloid cells in these recipients being derived from surviving stem cells in the irradiated recipient. This observation, together with the observation of a strong correlation between the numbers of B lymphocytes and T lymphocytes produced in individual mice, is consistent with the presence of lymphoid-restricted stem cells, S_L, in long-term cultures.

Results from two other experiments also indicate that long-term cultures contain stem cells capable of lymphoid differentiation. First, cells from long-term culture can cure scid mice of their immunological defect (Phillips et al. 1984). Second, cultured stem cells can also reconstitute CFU-B in CBA/N recipients (Dorshkind and Phillips 1982). These experiments are particularly interesting because they bear on the question of whether CFU-S have any potential for lymphocyte differentiation (Trentin et al. 1967). Lala and Johnson (1978) first reported that CFU-B are present in spleen colonies. However, Paige et al. (1979) found that most, if not all, CFU-B in spleen colonies were contaminants from the original bone marrow inoculum and were probably not produced by CFU-S. Because B lymphocytes rapidly disappear from long-term culture, it is possible to evaluate the ability of CFU-S from such cultures to produce CFU-B. In a detailed study of the ability of cells from long-term bone marrow cultures to regenerate CFU-B in CBA/N mice, Dorshkind and Phillips (1982) were unable to detect CFU-B above background levels in the spleens of these mice during the first 14 days after injection. Because the injected cells contained CFU-S and because spleen colonies must be examined before 14 days, this result is consistent with the results of Paige et al. (1979); namely, that CFU-S do not have the potential to make CFU-B.

Mauch et al. (1980) noted that the stem cells that float as a nonadherent population in long-term cultures have less proliferative potential than those firmly embedded in the adherent layer. Dorshkind and Phillips (1982) also found similar differences in the ability of these two stem cell populations to reconstitute CFU-B in CBA/N recipients. Cells from the adherent layer reconstituted much better than those from the nonadherent population. Again, there was no correlation between the reconstituting ability and the CFU-S content of the two populations. The conclusion from these experiments is that CFU-S are restricted to myeloid differentiation and cannot make lymphoid progeny.

FAILURE OF SPLEEN COLONIES TO CURE W/W^v MICE

As mentioned above, spleen colonies are derived from single progenitors, CFU-S, having all of the properties of stem cells. Since single stem cells appear capable of curing W/W^v mice of their anemia, one might predict that injection of single spleen colonies into W/W^v recipients will result in cures. Kitamura et al. (1981) detected only three cures following injection of single spleen colonies into 106 W/W^v recipients. We have repeated this experiment with similar results. Spleen colonies were produced by injection of 2×10^4 cells into irradiated recipients. Between 11 and 14 days later, individual spleen colonies were dissected free, made into cell suspensions, and injected into 60 W/W^v recipients. Only three cures were detected among these recipients. However, suspensions of similar spleens containing small numbers of spleen colonies were able to cure 12 of 24 recipients (Keller G, Dorshkind K, and Phillips RA, unpublished data). Since this procedure recovers all of the cells in the spleen, not just those in spleen colonies, our tentative conclusion from these experiments is that the spleens of irradiated mice do contain stem cells capable of curing W/W^v mice, but the majority of these stem cells are not in spleen colonies. The stem cells found in spleen colonies can form secondary spleen colonies upon injection into other irradiated recipients (Siminovitch et al. 1963), but they cannot cure W/W^v mice of their anemia.

DISCUSSION

The results described above indicate that CFU-S cannot produce lymphoid progeny and do not have the ability to cure W/W^v mice of their anemia. Thus, one can conclude that CFU-S must be distinct from the myeloid-restricted stem cell detected by Abramson et al. (1977) in their studies on the cure of W/W^v recipients. Hodgson and Bradley (1979) concluded from their studies on CFU-S from mice treated with 5-fluorouracil that CFU-S are derived from a more primitive progenitor in bone marrow; this conclusion was recently confirmed by van Zant (1984). Analogous data could be shown to indicate that the majority, if not all, of the in vitro colony assays detect progeny distinct from the CFU-S. No one has described an in vitro colony-forming cell with the proliferative and self-renewal potential of CFU-S. Thus, it is likely that all of the currently available in vitro assays detect progenitors that are more mature than CFU-S. A proposed model of the possible relationships between the various stem cells and colony-forming cells is shown in Figure 2. In this scheme, all of the progenitors capable of colony formation in vitro are shown as committed progenitors with little, if any, capacity for self-renewal. Intermediate stem cells are multipotent and have limited self-renewal potential; CFU-S is the only stem cell in this category. Primitive stem cells have extensive self-renewal capacity in addition to their ability to differentiate along several lineages. At present these stem cells can only be detected by their ability to reconstitute deficient recipients. There is little evidence that these primitive stem cells can be detected using in vivo or in vitro colony assays.

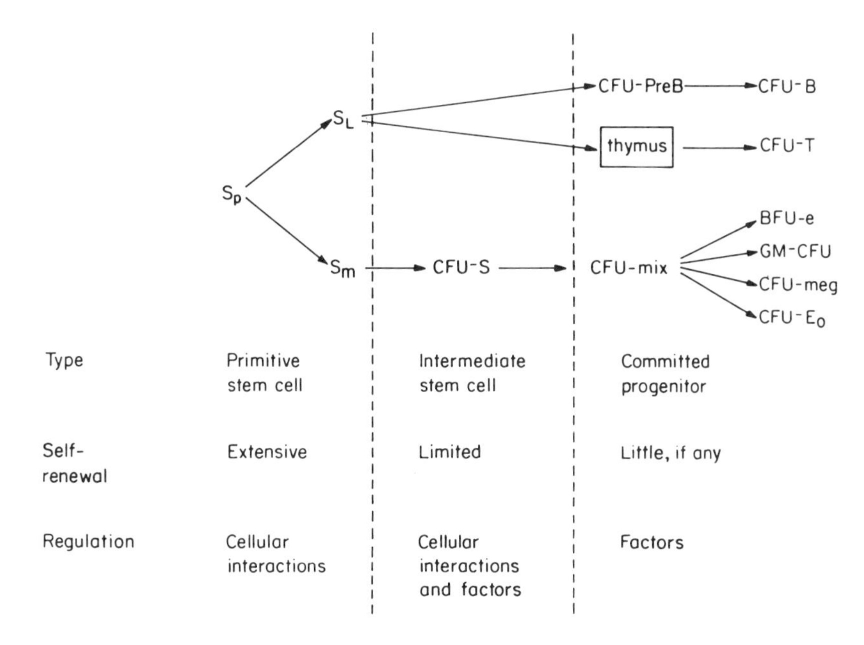

FIG. 2. Model showing proposed relationship between colony-forming stem cells and stem cells detected by transplantation.

It is interesting that the only in vitro system that appears to maintain primitive stem cells is the long-term culture system described by Dexter et al. (1977). It has been suggested that the adherent layer in these cultures is the in vitro counterpart of the hematopoietic microenvironment required for maintenance of primitive and intermediate stem cells (McCulloch et al. 1965, Patt et al. 1982). The observation that the adherent layer contains the cells with the most reconstituting ability indicates that intimate cellular interactions are essential for the maintenance and regulation of these stem cells. The long-term cultures established from hamsters may be an exception to the rule that primitive stem cells require an adherent layer. Long-term cultures from this animal do not have an adherent layer and still maintain hematopoiesis over long periods of time (Eastment et al. 1982). However, no one has yet done transplantation assays to determine whether or not these cultures contain primitive stem cells. Hematopoiesis in these cultures may be maintained by intermediate stem cells requiring only factors for their self-renewal and differentiation.

It is possible that the different growth requirements of the various progenitors shown in Figure 2 represent differences in their modes of regulation. Perhaps primitive stem cells require intimate contact with specific cells, loosely referred to as the hematopoietic microenvironment. Because such stem cells have enormous proliferative potential, they must be regulated precisely according to the needs of the host; their regulation is, therefore, likely to be complex and involve several components. At the other end of the spectrum, the regulation of committed progenitors with limited proliferative potential probably involves only soluble factors, because they allow rapid responses to environmental needs.

Since none of these progenitors have extensive proliferative potential, it may not be necessary to regulate individual progenitors in the same way as primitive stem cells. It is probably more important to regulate the population of progenitors rather than individual progenitors; such regulation could be achieved by soluble factors. The use of factors and not cellular interactions in the regulation of committed progenitors may explain why these progenitors grow readily in liquid and semisolid medium. CFU-S, which appear to have properties intermediate between primitive stem cells and committed progenitors, probably have a regulatory system intermediate between the two extremes, simpler than primitive stem cells and more complex than committed progenitors. The difficulty of maintaining CFU-S in liquid culture may indicate the requirement for cell interactions that are missing from all systems except the one described by Dexter et al. (1977).

The model outlined above is similar to age-structure hypotheses proposed by several investigators (Schofield 1978, Botnick et al. 1979). However, most of the previous models have assumed that the CFU-S is identical to either the S_M or the S_P primitive stem cells. These models usually propose that CFU-S retain immortality, i.e. are S_P, as long as they occupy a specific niche in the hematopoietic microenvironment (Schofield 1978). However, the inability of cells in spleen colonies to cure W/W^v recipients (Kitamura et al.1981, Keller GM, Dorshkind KA, and Phillips RA, unpublished data) appears to rule out this hypothesis. The majority of the CFU-S in spleen colonies do not have the capacity to cure, but the presence of curing units in the same spleens indicates that the niches are available and that the primitive stem cells required for cure are different from CFU-S.

The model shown in Figure 2 also explains an interesting curiosity. At any time in both mouse (Burton et al. 1982) and man (Fialkow 1973), the majority of the mature hematopoietic cells are derived from a small number of hematopoietic stem cells, much smaller than the number detected by transplantation assays. It is possible that when a primitive stem cell is stimulated to proliferate and produce intermediate stem cells, it produces only a small burst of such cells. In response to need, the first cells to be used will be those that respond to soluble factors, i.e., progenitors and then intermediate stem cells. Thus, with their large proliferative potential and their significant, but limited, potential for self-renewal, the small number of intermediate stem cells derived from a common primitive stem cell can generate large numbers of differentiated cells and will dominate the hematopoietic system for a significant period of time. The roles for the primitive stem cells are to maintain the pool of intermediate stem cells that are primarily responsible for the day-to-day maintenance of the production of hematopoietic cells and to serve as a reservoir for conditions of extreme demand. It is unlikely that an individual primitive stem cell will ever produce large numbers of progeny (intermediate stem cells) at a single time, and thus it may be difficult to develop a colony assay for this stem cell.

In addition to the efficient production of differentiated cells, there is another more subtle advantage to the host in having only a limited number of intermediate stem cells providing differentiated progeny at any given time. Because the induction of malignant disease requires multiple genetic changes in proliferating cells, the larger a clone becomes the much more likely it is to accumulate the changes required for malignant transformation. In the hematopoietic system, clones of intermediate stem cells will have only a limited life span because of their limited self-renewal potential (Siminovitch et al. 1964). Thus, as these clones die, to be replaced by new clones of intermediate stem cells, the somatic mutations of the old clones also disappear. By limiting the number of cells with extensive self-renewal potential, the individual limits the number of potential targets for malignant transformation. Indeed, there is evidence that clonal dominance is a prerequisite for development of leukemia (Till and McCulloch 1980). In this case, an initial mutation may alter self-renewal of an intermediate stem cell and allow it to produce progeny for an indefinite period. The excessive longevity of this mutant clone will increase the possibility that cells in this clone will acquire the additional mutations necessary for the development of malignant disease. Under normal conditions, the limited lifetime of intermediate stem cells makes the development of malignant disease an unlikely event.

The mathematical model developed by Goldie and Coldman (1983) to predict the probability of developing drug-resistance mutations in tumors is also applicable to predicting mutations in hematopoietic stem cells. These investigators concluded from an analysis of the various parameters in their model that the most important determinant for the accumulation of mutations in stem cells was not the size of the stem cell pool but its longevity. Thus, drug-resistance mutations were most likely to develop in slowly growing tumors in which many of the stem cells differentiate. The analogy with the hematopoietic system is obvious and again emphasizes the need to limit the number and life span of the intermediate stem cells that produce mature hematopoietic cells in order to reduce the probability of accumulating mutations that lead to malignant disease.

ACKNOWLEDGMENTS

This investigation was supported by the Medical Research Council and the National Cancer Institute of Canada.

REFERENCES

Abramson S, Miller RG, Phillips RA. 1977. Identification of pluripotent and restricted stem cells of the myeloid and lymphoid systems. J Exp Med 145:1567–1579.

Becker AJ, McCulloch EA, Till JE. 1963. Cytological demonstration of the clonal nature of spleen colonies derived from transplanted mouse marrow cells. Nature 197:452–454.

Boggs DR, Boggs SS, Saxe DF, Gross LA, Canfield DR. 1982. Hematopoietic stem cells with high proliferative potential. Assay of their concentration in marrow by the frequency and duration of cure of W/W^v mice. J Clin Invest 70:242–253.

Bosma GC, Custer RP, Bosma MJ. 1983. A severe combined immunodeficiency mutation in the mouse. Nature 301:527–530.

Botnick LE, Hannon EC, Hellman S. 1979. Nature of the hemopoietic stem cell compartment and its proliferative potential. Blood Cells 5:195–210.

Burton EI, Ansell JD, Gray RA, Micklem HS. 1982. A stem cell for stem cells in murine hemopoiesis. Nature 298:562–563.

Dexter TM, Allen TD, Lajtha LG. 1977. Conditions controlling the proliferation of hemopoietic stem cells in vitro. J Cell Physiol 91:335–344.

Dorshkind KA, Phillips RA. 1982. Maturational state of lymphoid cells in long-term bone marrow cultures. J Immunol 129:2444–2450.

Dorshkind K, Keller GM, Phillips RA, Miller RA, Bosma GC, O'Toole M, Bosma MJ. 1984. Functional status of cells from lymphomyeloid tissues in mice with severe combined immunodeficiency disease. J Immunol 132:1804–1808.

Eastment CE, Ruscetti FW, Denholm E, Katznelson J, Arnold E, Ts'o POP. 1982. Presence of mixed colony-forming cells in long-term hamster bone marrow suspension culture: Response to pokeweed spleen conditioned medium. Blood 60:495–502.

Fauser AA, Messner HA. 1979. Identification of megakaryocytes, macrophages, and esosinophils in colonics of human bone marrow containing neutrophilic granulocytes and erythroblasts. Blood 53:1023–1027.

Fialkow PJ. 1973. Primordial cell pool size and lineage relationships of five human cell types. Ann Hum Genet 37:39–48.

Fleischman RA, Custer RP, Mintz B. 1982. Totipotent hemopoietic stem cells: normal self-renewal and differentiation after transplantation between mouse fetuses. Cell 30:351–359.

Ford CE, Hamerton JH, Barnes WH, Loutit JF. 1956. Cytological identification of radiation chimeras. Nature 177:452–454.

Goldie JH, Coldman AJ. 1983. A quantitative model for multiple levels of drug resistance in clinical tumors. Cancer Treatment Reports 67:923.

Harrison DE. 1973. Normal production of erythrocytes by mouse bone marrow continuous for 73 months. Proc Natl Acad Sci USA 70:3184–3188.

Hodgson GS, Bradley TR. 1979. Properties of hematopoietic stem cells surviving 5-fluorouracil treatment: Evidence for a pre-CFU-S cell? Nature 281:381–382.

Humphries RK, Eaves AC, Eaves CJ. 1981. Self-renewal of hemopoietic stem cells during mixed colony formation in vitro. Proc Natl Acad Sci USA 78:3629–3633.

Johnson GR, Metcalf D. 1977. Pure and mixed erythroid colony formation in vitro stimulated by spleen conditioned medium with no detectable erythropoietin. Proc Natl Acad Sci USA 74:3879–3882.

Jones-Villeneuve EV, Phillips RA. 1980. Potentials for lymphoid differentiation by cells from long-term cultures of bone marrow. Exp Hematol 8:65–76.

Keller GM, Phillips RA. 1982. Detection in vitro of a unique, multipotent hemopoietic progenitor. J Cell Physiol (Suppl 1):31–36.

Kincade PW. 1977. Defective colony formation by B lymphocytes from CBA/N and C3H/HeJ mice. J Exp Med 145:249–263.

Kitamura Y, Yokoyama M, Matsunda H, Ohno T, Mori KJ. 1981. Spleen colony-forming cell as common precursor for tissue mast cells and granulocytes. Nature 291:159–160.

Lala PK, Johnson GR. 1978. Monoclonal origin of B lymphocyte colony-forming cells in spleen colonies formed by multipotential hemopoietic stem cells. J Exp Med 148:1468–1477.

Magli MC, Iscove NN, Odartchenko N. 1982. Transient nature of early hematopoietic spleen colonies. Nature 295:527–529.

Mauch P, Greenberger JS, Botnick L, Hannon E, Hellman S. 1980. Evidence for structured variation in self-renewal capacity within long-term bone marrow cultures. Proc Natl Acad Sci USA 77:2927–2930.

McCulloch EA, Siminovitch L, Till JE. 1964. Spleen colony formation in anemic mice of genotype W/W^v. Science 144:844–846.

McCulloch EA, Siminovitch L, Till JE, Russell ES, Bernstein SE. 1965. The cellular basis of the genetically determined hemopoietic defect in anemic mice of genotype Sl/Sl^d. Blood 26:399–410.

Mintz B. 1974. Gene control of mammalian differentiation. Ann Rev Genet 8:411–470.

Nakahata TL, Ogawa M. 1982. Identification in culture of a new class of hemopoietic colony-forming units with extensive capability to self-renew and generate multipotential hemopoietic colonies. Proc Natl Acad Sci USA 79:3843–3847.

Paige CJ, Kincade PW, Moore MAS, Lee G. 1979. The fate of fetal and adult B-cell progenitors grafted into immunodeficient CBA/N mice. J Exp Med 150:548–563.

Patt HM, Maloney MA, Flannery ML. 1982. Hematopoietic microenvironment transfer by stromal fibroblasts derived from bone marrow varying in cellularity. Exp Hematol 10:738–742.

Phillips RA, Bosma M, Dorshkind K. 1984. Reconstitution of immune deficient mice with cells from long-term bone marrow cultures. *In* Greenberger JS, Nemo GJ, Wright DG (eds), Long-Term Bone Marrow Culture. Alan R Liss, Inc, New York (in press).
Russell ES. 1979. Hereditary anemias of the mouse: A review for geneticists. Adv Genet 20:357–459.
Schofield R. 1978. The relationship between the spleen colony-forming cell and the hemopoietic stem cell. Blood Cells 4:7–25.
Schofield R, Lord BI, Kyffin S, Gilbert GW. 1980. Self-maintenance capacity of CFU-S. J Cell Physiol 103:355–362.
Siminovitch L, McCulloch EA, Till JE. 1963. The distribution of colony-forming cells among spleen colonies. J Cell Comp Physiol 62:327–336.
Siminovitch L, Till JE, McCulloch EA. 1964. Decline in colony-forming ability of marrow cells subjected to serial transplantation into irradiated mice. J Cell Comp Physiol 64:23–32.
Till JE, McCulloch EA. 1961. A direct measurement of the radiation sensitivity of normal mouse bone marrow. Radiat Res 14:213–222.
Till JE, McCulloch EA. 1980. Hemopoietic stem cell differentiation. Biochim Biophys Acta 605:431–459.
Trentin J, Wolf N, Cheng V, Fahlberg WL, Weiss D, Bonhag R. 1967. Antibody production by mice repopulated with limited numbers of clones of lymphoid cell precursors. J Immunol 98:1326–1337.
van Zant G. 1984. Studies of hematopoietic stem cells spared by 5-fluorouracil. J Exp Med 159:679–690.

Mediators in Cell Growth and Differentiation,
edited by Richard J. Ford and Abby L. Maizel.
Raven Press, New York © 1985.

Myeloid and Erythroid Stem Cells: Regulation in Normal and Neoplastic States

Malcolm A. S. Moore, Janice Gabrilove, Li Lu, and Yee Pang Yung

Department of Developmental Hematopoiesis, Memorial Sloan-Kettering Cancer Center, New York, New York 10021

A multiplicity of growth factors are implicated in the proliferation and differentiation of hematopoietic cells in vitro (Moore and Yung 1982). The best characterized of these are the colony-stimulating factors (CSFs), of which the granulocyte-specific (G-CSF), macrophage-specific CSF-1 (M-CSF), and the mixed granulocyte-macrophage-specific CSFs (GM-CSF) have been purified to homogeneity (Stanley and Guilbert 1981, Stanley 1979, Burgess and Metcalf 1980). CSFs also exist for megakaryocytes (Mk-CSF) (Metcalf et al. 1975), erythroid bursts (BPA) (Iscove et al. 1982), and multipotential stem cells (Multi-CSF) (Ruppert et al. 1983).

The relationship between these factors is much less clear-cut biochemically. A two-factor system was described for megakaryocyte differentiation (Williams et al. 1982). A similar two-factor model has been proposed by Iscove et al. (1982) and Hoang et al. (1983). These investigators, after using exhaustive purification procedures, found BPA to copurify with molecules stimulating pluripotential cells and early cells committed to granulocyte, macrophage, and megakaryocyte production. Thus, the existence of a single lineage-indifferent factor active on pluripotential hematopoietic precursor cells and their early committed progeny was proposed. According to this model, growth and maturation down any particular pathway will require the participation of at least two factors, one operating early in the pathway and the other late.

Meanwhile, various other investigators have reported on the copurification of factors having different functions. Schrader and Clark-Lewis (1982a,b) reported on the copurification of a T-cell–derived factor (CFUs-SF) that allowed for long-term (up to 4 weeks) maintenance of multipotential murine hematopoietic stem cells capable of forming spleen colonies in lethally irradiated mice. Nabel and co-workers (1981) reported on the purification of a growth factor from inducer T-cell clones that stimulated various types of hematopoietic cells as well as lymphoid cells. Bazill and co-workers (1983) reported on the copurification of a growth factor with activities stimulating multipotential stem cells, as well as megakaryocytic, erythroid, and granulocytic committed progenitor cells produced by the murine myelomonocytic leukemic cell line WEHI-3. Ihle and co-workers also reported on identification of a cytokine, interleukin 3 (IL-3), from WEHI-3CM likewise with multiple functions (Ihle et al. 1982a, b, c, 1983), including support of proliferation of several WEHI-3 conditioned-medium–dependent cell lines, CFUs-SF activities, BPA, mast-cell–stimulating activity, and histamine-producing cell–stimulating activities.

The murine IL-3 produced by the WEHI-3 cell line has also been reported to be active on the murine pre-B cell line Ea3 by increasing Ea3 proliferation and anaerobic glycolysis as determined by lactic acid production and by the ability of Ea3 to absorb partially purified IL-3 (Palacios et al. 1984). Ea3 cells are classified as pre-B cells on the basis of the presence of B-cell surface antigens, immunoglobulin gene rearrangement, and inducibility of IgM secretion. This intriguing observation suggests the possibility that the multilineage potential of pluripoietin (or IL-3) is related to its role in glycolysis, enabling both proliferation and possibly expression of binding sites for lineage-specific differentiation factors. This murine pluripoietin has been purified to homogeneity and its amino acid sequence reported. The protein from both the WEHI-3 myelomonocytic and T-cell leukemia lines is glycosylated, with the major species having a molecular weight of 28,000. IL-3 mRNA from WEHI-3 has recently been translated in *Xenopus* oocytes to give a biologically active factor (Fung et al. 1984). Sequence data from cloned cDNA has shown that it codes for a polypeptide of 166 amino acids including a signal peptide of possibly 27 amino acids. Comparison of the N-terminal sequence of cloned IL-3 is in complete agreement with that reported for earlier sequenced material from both WEHI-3 and LBRM-33 cell lines, but starts at residue 33, suggesting that an additional proteolytic cleavage has occurred apart from the cleavage of the signal peptide. It remains to be seen if the cloned IL-3 possesses all the activities of the hypothetical molecule.

The existence of a multipotential stimulating factor raises several questions: (1) Are all these investigators dealing with the same molecule? (2) If a single molecule is involved, how then is responsiveness of different cell types to this omnipotent factor regulated? (3) What then determines stem cell self-renewal versus differentiation along any particular pathway?

The ability to maintain multipotential stem cells in liquid cultures with no apparent differentiation in any particular pathway (Schrader and Clark-Lewis 1982b) and the ability of the same stem cells to form multilineage colonies (CFU-s) in vivo (Schrader and Clark-Lewis 1982a, b) and CFU-GEMM in vitro (Greenberger et al. 1983) would suggest that stem cell self-renewal and differentiation are independently regulated events. On the other hand, erythroid, granulocytic, macrophagic, and megakaryocytic cells, as well as mast cells and T and B lymphocytes, have been reported to be present in in vitro human multilineage GEMM colonies (Hara et al. 1982, Messner et al. 1981, Johnson and Metcalf 1977, Messner and Fauser 1980). Although the nature of the stimulating factor is as yet not well defined, it appears that an IL-3-type molecule is involved (Ruppert et al. 1983).

Thus, the question is raised as to whether other levels of regulation of hematopoiesis exist.

Several inbred strains of mice are known to suffer from various types of genetically controlled hematopoietic cell defects. Of special interest are the W/W^v and $S1/S1^d$ mice (Metcalf and Moore 1971). Both strains exhibit macrocytic anemia, which in the case of W/W^v has been ascribed to a stem cell defect and in the case of Sl/Sl^d to a microenvironmental defect. Besides the defect in erythropoiesis, these mice further exhibit defective myelopoiesis and mast cell deficiency (Kitamura et al. 1979a,b). It was recently reported (Bartlett 1982) that pluripotential hematopoietic stem cells, detectable in spleen colony (CFU-s) assay, though deficient in the marrow of W/W^v anemic mice, were nevertheless present in the brain in large numbers. It was also reported by Hara et al. (1982) that despite lack of microscopically evident colony formation in the spleen of irradiated mice, hematopoietic cells of W/W^v mice did produce macroscopically evident mixed colonies containing erythroid, macrophage, and megakaryocytic cells (CFU-mix) in vitro whose number, size, and constitution were comparable to those derived from + / + mice. Thus, in vivo hematopoiesis in the W/W^v mice, though obviously deficient, is nevertheless sufficient for survival. Deficiency is thus more quantitative than qualitative.

In contrast to the microenvironmental and stem cell defects of S1 and W mice, autoimmune disease-prone NZB mice display B-cell abnormality and lack of responsiveness of myeloid progenitors to G-CSF (Kincade et al. 1979). Although independent multigenic systems have been shown to be involved in the control of B-cell hyperactivity and autoimmunity in NZB mice, the cause of unresponsiveness to G-CSF is as yet not known, nor is it clear if unresponsiveness also affects hematopoiesis at the stem cell level. It is thus of great interest and importance to further analyze hematopoietic development in these genetically defective mice.

RELATIONSHIP OF MURINE MAST CELL GROWTH FACTORS—IL-3 AND PLURIPOIETIN

We first reported (Tertian et al. 1980, Yung et al. 1981) on the growth from precursors in long-term marrow cultures of a Thy-1-negative cell bearing basophilic cytoplasmic granules in mitogen-activated murine splenic leukocyte–conditioned medium. These cells were later identified as mast cells (Tertian et al. 1981) based on their characteristic morphology, the ultrastructure of their cytoplasmic granules, positive reactions with toluidine blue and alcian blue, the presence of IgE receptors, and the monoamines and sulfated products contained in the cytoplasm. Pure mast cell lines can be generated from spleen and from fresh or Dexter-type cultured bone marrow cells of normal as well as athymic mice. Such mast cells can be maintained in exponential growth indefinitely in the presence of growth factor supplied by mitogen-activated spleen–conditioned medium or medium conditioned by the myelomonocytic WEHI-3 cells or mitogen-stimulated murine T lymphoma LBRM-33 cells. The continuous presence of the growth factor is a prerequisite for viability. Cell death ensues within 24–48 hr in the absence of CM-derived factor.

The production of mast cell growth factor (MCGF) by murine spleen cells was T-cell dependent and relatively macrophage independent (Yung and Moore 1982). Spleen cells depleted of T cells failed to produce MCGF. The genetic deficiency of T cells in athymic nu/nu mice likewise compromised their ability to produce MCGF. MCGF biochemically derived by spleen-cell-conditioned medium (CM) has been shown to be a neuraminidase-sensitive and trypsin-sensitive glycoprotein of MW 35,000 (by Sephadex G-100 gel filtration). MCGF is relatively heat stable and has low affinity for DEAE-cellulose. It is readily separated from interleukin-2 (IL-2) and M-CSF by DEAE ion exchange chromatography. MCGF purified from spleen CM lacked G-CSF activity. Peritoneal macrophage–conditioned medium that contained GM-CSF activities failed to stimulate mast cell growth. Thus MCGF appeared to be distinct from GM-CSF.

Sequential purification steps involving ammonium sulfate precipitation, DEAE-cellulose ion exchange chromatography, ultragel AcA54 gel filtration chromatography, isoelectric focusing (Yung and Moore 1984), and reverse phase high performance liquid chromatography (RP-HPLC) were used to purify MCGF from serum-free WEHI-CM and LBRM-CM. As a variety of hematopoietic factors were known to be present in these conditioned media, attempts were made to dissociate them. Individual fractions collected at each stage of purification were assayed for GM-CSF, multi-CSF, and MCGF. These studies showed MCGF to be distinct from GM-CSF, but multi-CSF activity as detected in CFU-GEMM assay could not be separated from MCGF. Moreover, RP-HPLC revealed fractions containing multi-CSF but not MCGF activity; thus, the common identity of multi-CSF and MCGF remains to be resolved.

It was reported by Ihle et al. (1983) that IL-3 purified to homogeneity had MCGF activity. We had the opportunity to test a sample of purified IL-3 (courtesy of J. Ihle). IL-3 did stimulate murine mast cell proliferation; however, IL-3 was also found to be enriched for G-CSF (Yung and Moore 1984). These results are thus at variance with our findings that MCGF can be distinguished from G-CSF by relative sensitivity to neuraminidase and inhibition of production by lactoferrin (Yung and Moore 1982). Recombinant IL-3 may allow us to resolve the issue of the identity of IL-3, MCGF, G-CSF and the variety of activities ascribed to IL-3 (Ihle et al. 1983).

PURIFICATION AND CHARACTERIZATION OF A REGULATORY PROTEIN CAPABLE OF STIMULATING PROLIFERATION OF HUMAN PLURIPOTENTIAL HEMATOPOIETIC STEM CELLS AND INDUCING MYELOID LEUKEMIC CELL DIFFERENTIATION

Pure murine IL-3 was shown by us to possess pluripotential activity on murine bone marrow and murine long-term normal or Friend virus–infected marrow cultures. It lacked any activity on human bone marrow and did not induce differentiation of the murine myelomonocytic leukemia WEHI-3 nor the human promyelocytic leukemia HL-60. This species restriction prompted us to identify a human source of an IL-3–like activity. In this regard, after extensive screening of the human tumor cell line bank developed at Sloan-Kettering Institute, we identified two epithelial tumor lines, one derived from a bladder carcinoma and the other from a hepatoma, that produced a number of human-active factors. Biological activity was produced constitutively under low-serum or serum-free conditions, and we identified factors stimulating human granulocyte-macrophage progenitors (CFU-GM), erythroid progenitors (erythroid burst-forming units; BFU-E), and pluripotential progenitors (CFU-GEMM). In addition, the cell lines produced an activity that induced terminal granulocyte-macrophage differentiation of both the murine and human myeloid leukemic cell lines and associated loss of leukemic growth potential. Extensive purification of these factors has been undertaken from cell line–conditioned medium using a sequence of steps involving ammonium sulfate precipitation, ion-exchange chromatography, gel filtration and RP-HPLC. Purification by a factor of 10^6 has been obtained, allowing us to identify a protein (probably glycosylated) with a molecular weight of 32,000, 18-20,000 under reducing conditions, with biological activity at in vitro concentrations of 10^{-12} M. This material can be eluted as a single activity from SDS-PAGE and contains 10^9 units of human GM-CSF activity, is a potent stimulus of erythropoiesis, and stimulates human pluripotent cells (CFU-GEMM). This highly purified activity is probably homogeneous, and though comparable in many biological respects with murine IL-3, it appears to possess the potential for inducing human myeloid leukemic differentiation. This last feature is not found in murine IL-3.

The clinical and biological importance of this unique human pluripoietin resides in its potential to accelerate in vivo regeneration of bone marrow function, which is so frequently compromised by chemotherapy or radiation therapy. A possible role in stimulation of marrow function is also evident in bone marrow transplant situations and in genetic or acquired marrow stem cell defects such as aplastic anemia. Perhaps the most interesting feature of this pluripoietin is its potential for inducing normal maturation of leukemic cells while stimulating normal stem cell regeneration. Such a combination of biological functions suggests a role for pluripoietin-differentiation factor as an adjunct to conventional therapy for acute and chronic myeloid leukemias.

Sufficient pure pluripoietin has been obtained to permit amino acid sequencing, and this is currently under way with the goal of using molecular cloning to obtain sufficient quantities of homogeneous material for in vivo testing in murine and primate model systems as a prelude to clinical trials.

DEMONSTRATION OF PHYSIOLOGICAL AGENTS WITH CYTODIFFERENTIATING POTENTIAL IN ACUTE MYELOID LEUKEMIA

Fibach and Sachs (1976) reported that serum from mice injected with endotoxin induced granulocyte and macrophage differentiation of the mouse myeloid leukemic cell line M1. At that time it was unclear whether the differentiation-inducing factor was CSF (MG1). Metcalf (1979) also reported that a pure preparation of GM-CSF has some capacity to induce differentiation of murine WEHI-3 myelomonocytic leukemic cells. More recently, biochemical characterization of postendotoxin serum has shown that the differentiation factor could be separated from GM-CSF (Lotem et al. 1980) and was termed MGI-2 or could be separated from the bulk of serum CSF, but coeluted with a minor species of CSF that stimulated only granulocyte colony formation (Burgess and Metcalf 1980).

The cellular origin of the differentiation factor (GM-DF) for leukemic cells is diverse. Spleen lymphocytes and macrophages have been shown to produce the activity when treated with various mitogens including endotoxin (Yamamoto et al. 1981). Conditioned medium from a variety of murine organs also contained GM-DF, but much higher levels of activity were reported following endotoxin treatment (Nicola et al. 1982). GM-DF produced by the different organs in vitro and found in vivo in endotoxin serum was distinguishable from the majority of granulocyte-macrophage (GM) and macrophage (M) CSF.

Induction of GM-DF is radioresistant and independent of T cells, since nude mice respond normally (Metcalf 1982). Repeated injections of endotoxin lead to depressed responsiveness as measured by serum DF and CSF levels, and sustained serum GM-DF levels have not been obtained (Metcalf 1982). This is in part due to the short serum half-life of GM-DF (1.5–3 hr) and the development of endotoxin "tolerance," which provide possible major obstacles to the clinical use of endotoxin induction of endogenous mediators.

It has become clear that many of the diverse biological effects of endotoxin are not caused by endotoxin itself but by endogenous mediators released by host cells in response to endotoxin. Three mediators of possible relevance to the antileukemic effects of endotoxin are tumor necrosis factor (TNF), interleukin-1 (IL-1), and leukemia differentiating factor (GM-DF); in addition, endotoxin is a potent inducer of prostaglandins, interferons, and various species of myeloid CSF.

Induction of Serum GM-DF Activity and an Antineoplastic Cytotoxin (Tumor Necrosis Factor; TNF) in Mice and in Patients with Advanced Malignancy Receiving Purified Lipopolysaccharide

We examined the ability of mouse serum obtained 3 hr after intravenous injection of 5 μg of endotoxin to induce differentiation of WEHI-3B(D+) cells cloned in semisolid agar. When exposed to 5–10% postendotoxin serum, 100% of leukemic colonies converted from a tight, undifferentiated pattern to diffuse, differentiated colonies composed of neutrophil and mature macrophages. Differentiation was still noted at 1:64 dilution of serum. Significantly, colony inhibition was not observed even at high serum concentrations. TNF serum (unfractionated serum from *Corynebacterium parvum* + endotoxin-treated mice) inhibited 90% of leukemic colony growth when used at 10% concentration, but this presumed TNF-dependent inhibition rapidly titrated out. Significant levels of differentiation-inducing factor activity, comparable to levels in postendotoxin serum, were also noted.

To determine further the independent actions of TNF and GM-DF on myeloid leukemic cell growth, TNF was subjected to batch elution on DEAE-A-50 gel. The serum fraction eluting with 0.2 M NaCl was found to contain very little TNF activity, but did contain a major proportion of the differentiation-inducing factor activity. In contrast, the serum fraction eluted with 1.0 M NaCl contained most of the TNF, which was further purified on Sephadex G-100 columns. The TNF-rich peak was strongly inhibitory to WEHI-3B(D+) leukemic cloning, but contained no significant differentiation-inducing activity.

In view of the evidence that potent macromolecules, such as GM-DF, can be induced following endotoxin administration in the mouse, serum from patients with advanced nonhematopoietic malignancies treated with Novo-Pyrexal, a highly purified preparation of endotoxin extracted from *Salmonella abortus-equi*, was analyzed in order to determine if such a differentiation-inducing protein could also be detected in man.

Serum was collected at 0, 1/2 hr, 2 hr, 4 hr, 6 hr, 8 hr, 12 hr, and 24 hr after a patient had received an intravenous injection of Novo-Pyrexal, beginning with a dose of 0.1 $\mu g/m^2$ and escalating at weekly or biweekly intervals. The serum was assayed at 10% and 1% in 1-ml agar cultures containing 300 WEHI-3B(D+) murine myelomonocytic leukemic cells. Time 0 serum had no significant effect on cloning efficiency or colony differentiation, as measured by tight versus diffuse colonies. A serum activity that induced up to 85% of leukemic colonies to differentiate to mature granulocytes or macrophages reached its peak at 2–4 hr after Novo-Pyrexal administration, and disappeared after 12–24 hr. An acute decrease in white cell count was observed in every instance, as early as 30 min after the completion of Novo-Pyrexal infusion, followed by a return to normal or a rebound leukocytosis by 24 hr. In seven patients studied so far, induction of this differentiation factor has occurred with similar kinetics. In addition, serum collected 30 min after Novo-Pyrexal administration inhibited leukemic colony formation, but was not active in inducing differentiation. The relationship of this inhibitory activity to TNF is not known at this time.

The ability to repeatedly induce serum activities such as GM-DF following repeated endotoxin administration may be influenced by the development of tolerance. The mechanism of tolerance induction is not clear; it is not simply due to the appearance of antibodies directed against the polysaccharide component of endotoxin. In an attempt to prevent endotoxin tolerance in patients receiving repeated dosages of highly purified endotoxin, clinical protocols were developed that involved escalating doses and biweekly or weekly schedules of drug administration. Sera obtained before and at various intervals after endotoxin administration were assayed for their ability to induce differentiation of WEHI-3B(D+) cells. Leukemia differentiation-inducing activity was elevated from essentially undetectable levels to a peak between 2–6 hr following endotoxin, and the escalation of endotoxin at twice-weekly intervals to a maximum of 30 $\mu g/m^2$ produced consistent reinduction of this response. With repeated administration of a given dose of endotoxin, a degree of tolerance appears to occur with regard to the induction of GM-DF.

The ability of endotoxin to induce human active CSF in the serum of these patients was confirmed by the assay of the patients' own sera on the patients' own bone marrow obtained prior to and 24 hr following treatment with Novo-Pyrexal. CSF was observed in the serum at 4–8 hr after endotoxin administration. In addition, the pretreatment marrow exhibited a greater responsiveness, i.e., contained a greater number of CFU-GM, than did the marrow obtained 24 hr after treatment. This might suggest, as one possibility, an in vivo depletion of marrow CFU-GM, which have moved further along the myeloid differentiation pathway as a result of biological regulators of hematopoiesis released following endotoxin treatment.

Endotoxin-Induced Leukemia Differentiation Factor in the Sera of Patients with Preleukemia, Chronic Myelogenous Leukemia, or Acute Myelogenous Leukemia (ANLL)

Interest in host resistance factors in leukemia has continued for some decades, and for this reason incidents of spontaneous regression or remission have excited considerable attention. The majority of spontaneous regressions have occurred following streptococcal or prolonged staphylococcal infections. At Memorial hospital, patients with ANLL who received *Pseudomonas* vaccine to protect against *Pseudomonas* infection were found to have improved survival compared with patients who did not receive the vaccine (there was no difference in the incidence of *Pseudomonas* morbidity). The therapeutic role of bacterial vaccine or endotoxins has been seen in the context of various mechanisms of antineoplastic defense with augmentation of host immune mechanisms. Our present work suggests that such therapy might be effective by virtue of the acute induction of factors promoting leukemic cell differentiation or of an antiproliferative factor selective for transformed cells, and suggests a possible therapeutic role in human myeloid leukemia. In view of the evidence that potent macromolecules such as GM-DF and TNF are found in postendotoxin therapy serum in man, and in light of the work by Lotem and Sachs, who have shown that injection of either endotoxin or murine-derived MGI-1 containing significant amounts of MGI-2 (GM-DF) can inhibit the development of myeloid leukemia in vivo and stimulate normal hematopoiesis, we have begun to treat patients with Novo-Pyrexal who have ANLL, all FAB classifications, in first or second relapse, CML patients in blast crisis or accelerated phase of the disease, and patients with preleukemia, whether or not they received prior treatment.

At present, we have treated three leukemia patients with Novo-Pyrexal. The diagnoses were myelodysplastic syndrome (patient 1), chronic myelogenous leukemia in lymphoblastic crisis (patient 2), and erythroleukemia (patient 3). Patients 2 and 3 had received serveral courses of chemotherapy before treatment with endotoxin; patient 1 had received none. Patients were treated either with a weekly schedule of Novo-Pyrexal beginning at a dose of 0.2 $\mu g/m^2$, and escalating to a dose that produced significant clinical toxicity, i.e., fever and chills, or with a daily schedule designed to induce physiological tolerance to fever and chills, commencing with the same dose and escalating once tolerance was induced at a given dose level. So far patients 1 and 2 are evaluable for production of GM-DF; patient 3 is just being evaluated. The results demonstrate the presence of 70-100 units of GM-DF/0.1 ml of undiluted serum (50 units equals the activity that renders 50% of WEHI-3B(D +) colonies differentiated in the clonal agar system). In patient 1, this activity could be repeatedly induced when treated on a weekly schedule with 0.6 $\mu g/m^2$; no factor production, however, could be detected in the patient's undiluted serum when endotoxin was given daily at a dose comparable to that of the weekly regimen.

Although this is a phase I study of Novo-Pyrexal in this group of patients, and not designed to address the therapeutic benefit of such treatment, we have evaluated the evidence of therapeutic effect. Patient 1's initial bone marrow revealed 29% Auer rod–positive blasts plus promyelocytes; subsequently this level dropped to 12% blasts plus promyelocytes, as documented on successive bone marrow aspirations. The number of Auer rod–positive blasts also decreased, as did the number of micromegakaryocytes. Patient 2 received two treatments with Novo-Pyrexal and soon thereafter was treated with high-dose ara-C. His disease subsequently reverted to the chronic phase (in our series at Memorial hospital, in only 20% of patients has CML in blast crisis reverted to the chronic phase after high-dose ara-C alone). Whether or not treatment with endotoxin directly or indirectly affects the number of cells entering S-phase, possibly rendering them more susceptible to a cycle-specific agent such as ara-C, is of considerable interest. We will study this question in the future.

CONCLUSIONS

Significant advances in the therapy of ANLL have occurred over the past two decades. The frequency of complete clinical remission of these diseases induced by chemotherapy has exceeded 50%. Despite this improvement, further advances in the rate of remission induction have been slow and the duration of remission for the average patient has not been significantly improved. These clinical findings have prompted the consideration of alternative chemotherapeutic approaches to the treatment of leukemia. One possibility would be a therapy based on the induction of differentiation of the leukemia cells to end-stage forms that are no longer capable of division or its combination with marrow transplantation. For patients with ANLL who receive transplants in late remission or relapse, relapse after transplantation remains a significant problem. Of 29 patients transplanted for ANLL in tertiary remission or relapse, 14 patients died early of infectious complications. Of 15 patients surviving, 11 ultimately suffered relapse. Including ALL in tertiary or greater remission or relapse in relatively chemotherapy-refractory patients, the recurrence of disease exceeds 60%. Thus, better antileukemic approaches must be devised to address the problem of relapse in these high-risk groups.

Inducing leukemic cells to become mature cells with little or no proliferative potential may represent an effective alternative or adjunct to a cytoreductive modality. The dosages of agents employed as inducers of differentiation are probably lower than required for effective cytoreductive therapy and may well synergize with or complement more-conventional agents. Anthracyclines have a major role in treatment of acute leukemias and lymphomas and have proven to be potent inducers of differentiation of HL-60 leukemic cells in vitro and in vivo (Schwartz and Sartorelli 1982). Differences in the ability of various compounds to induce differentiation of myeloid leukemia cell lines or primary human myeloid leukemia can be detected in vitro. This may also apply to the effectiveness of various compounds in inducing differentiation of leukemic cells from different leukemia patients, either individually or in combination with other agents. Actinomycin-D (Hozumi 1983) has proved to be a particularly effective agent in induction of myeloid leukemic cell differentiation, and at low concentrations, it sensitizes "undifferentiated" leukemic cells to the differentiation-inducing action of various biological response modifiers (Hozumi 1983).

ACKNOWLEDGMENTS

This investigation was supported by grants number HL-31780, CA 20194, and CA 23766 awarded by the National Institutes of Health, ACS CH 3G awarded by the American Cancer Society, and the Gar Reichman Foundation.

REFERENCES

Barlett PF. 1982. Pluripotential hemopoietic stem cells in adult mouse brain. Proc Natl Acad Sci USA 79:2722.

Bazill GW, Haynes M, Garland J, Dexter TM. 1983. Characterization and partial purification of a haemopoietic cell growth factor in WEHI-3 cell conditioned medium. Biochem J 210:747–759.

Burgess AW, Metcalf D. 1980. The nature and action of granulocyte-macrophage colony stimulating factors. Blood 56:947–958.

Fibach E, Sachs L. 1976. Control of normal differentiation of myeloid leukemic cells. VIII. Induction of differentiation to mature granulocytes in mass culture. J Cell Physiol 86:221–230.

Fung MC, Hapel AJ, Ymer S, Cohen DR, Johnson RM, Campbell HD, Young IG. 1984. Molecular cloning of cDNA for murine interleukin-3. Nature 307:233.

Greenberger JS, Eckner RJ, Sakakeeny M, Marks P, Reid D, Nabel G, Hapel A, Ihle JN, Humphries KC. 1983. Interleukin 3-dependent hematopoietic progenitor cell lines. Fed Proc 42:2762–2771.

Hoang T, Iscove NN, Odartchenko N. 1983. Macromolecules stimulating human granulocytic colony-forming cells, precursors of these cells, and primitive erythroid progenitors: Some apparent non-identities. Blood (in press).

Hozumi M. 1983. Fundamentals of chemotherapy of myeloid leukemia by induction of leukemia cell differentiation. Adv Cancer Res 38:121–164.

Hara H, Ohe Y, Noguchi K, Tsuyama K, Kitamura Y. 1982. Presence of pluripotent haemopoietic precursors in vitro (CFU-mix) in haemopoietic tissues from mice of W/W^v genotype. Cell Tissue Kinet 15:25.

Ihle JN, Keller J, Greenberger JS, Henderson L, Yetter RA, Morse HC. 1982a. Phenotypic characteristics of cell lines requiring interleukin 3 for growth. J Immunol 129:1377.

Ihle J, Keller J, Henderson L, Klein F, Palaszynski E. 1982b. Procedures for the purification of interleukin 3 to homogeneity. J Immunol 129:2431.

Ihle JN, Rebar L, Keller J, Lee JC, Hapel A. 1982c. Interleukin 3: Possible roles in the regulation of lymphocyte differentiation and growth. Immunol Rev 63:5.

Ihle JN, Keller J, Oroszlan S, Henderson LE, Copeland TD, Fitch F, Prystowsky F, Goldwasser E, Schrader JW, Palaszynski E, Dy M, Lebel B. 1983. Biological properties of homogeneous interleukin 3. I. Demonstration of WEHI-3 growth factor activity, mast cell growth factor activity, P cell

stimulating factor activity, colony-stimulating factor activity, and histamine-producing cell-stimulating factor activity. J Immunol 131:282–287.
Iscove NN, Roitsch CA, Williams N, Guilbert LJ. 1982. Molecules stimulating early red cells, granulocyte, macrophage, and megakaryocyte precursors in culture: similarity in size, hydrophobicity and charge. J Cell Physiol Suppl 1:65.
Johnson GR, Metcalf D. 1977. Pure and mixed erythroid colony formation in vitro stimulated by spleen conditioned medium with no detectable erythropoietin. Proc Natl Acad Sci USA 74:3879–3882.
Kincade P, Lee G, Fernandes G, Moore MAS, Williams N, Good RA. 1979. Abnormalities in clonable B lymphocytes and myeloid progenitors in autoimmune NZB mice. Proc Natl Acad Sci USA 76:3464.
Kitamura Y, Matsuda H, Hatanaka K. 1979a. Clonal nature of mast cell clusters formed in W/W^{v} mice after bone marrow transplantation. Nature 281:154.
Kitamura Y, Shimada M, Go S, Matsuda H, Hatanaka K, Seki M. 1979b. Distribution of mast cell precursors in hematopoietic and lymphopoietic tissues in mice. J Exp Med 150:482.
Lotem J, Lipton JH, Sachs L. 1980. Separation of different molecular forms of macrophage- and granulocyte-inducing proteins for normal and leukemic myeloid cells. Int J Cancer 25:763–771.
Messner HA, Fauser AA. 1980. Culture studies of human pluripotent hemopoietic progenitors. Blut 41:327–333.
Messner HA, Izaguirre CA, Jamal N. 1981. Identification of T lymphocytes in human mixed hemopoietic colonies. Blood 58:402–405.
Metcalf D. 1979. Clonal analysis of the action of GM-CSF on the proliferation and differentiation of myelomonocytic leukemic cells. Int J Cancer 24:616–623.
Metcalf D. 1982. Regulator-induced suppression of myelomonocytic leukemic cells: clonal analysis of early cellular events. Int J Cancer 30:203–210.
Metcalf D, MacDonald HR, Odartchenko N, Sordat B. 1975. Growth of mouse megakaryocyte colonies in vitro. Proc Natl Acad Sci USA 72:1744–1748.
Metcalf D, Moore MAS. 1971. Hematopoietic Cells. Amsterdam, North Holland.
Moore MAS, Yung Y-P. 1982. Hemopoietic growth factor production and differentiation of a murine myelomonocytic leukemia cell line. *In* Revoltella RP, Pontieri GM, Basilico C, Rovera G, Gallo RC, Subak-Sharpe JH (eds), Expression of Differentiated Functions in Cancer Cells, Raven Press, New York, pp. 213–221.
Nabel G, Galli SJ, Dvorak AM, Drorak HF, Cantor H. 1981. Inducer T lymphocytes synthesize a factor that stimulates proliferation of cloned mast cells. Nature 291:332.
Nicola NA, Matsumoto M, Metcalf D, Johnson GR. 1982. Molecular properties of a factor inducing differentiation in murine myelomonocytic leukemia cells. *In* Modern Trends in Human Leukemia V, June 21-23, 1982, Wilsede, West Germany Deutsche Gesellschaft fur Hamatologie und Oncology, New York, Springer Verlag.
Palacios R, Henson G, Steinmetz M, McKearn JR. 1984. Interleukin-3 supports growth of mouse pre-B-cell clones in vitro. Nature 309:126.
Ruppert S, Lohr GW, Fauser AA. 1983. Characterization of stimulatory activity for human pluripotent stem cells (CFU_{GEMM}). Exp Hematol 11:54–161.
Schrader JW, Clark-Lewis I. 1982a. A T cell-derived factor stimulating multipotential hemopoietic stem cells: Molecular weight and distinction from T cell growth factor and T cell-derived granulocyte-macrophage colony-stimulating factor. J Immunol 129:30.
Schrader JW, Clark-Lewis I. 1982b. The use of T cell hybridomas in the biochemical and biological characterization of multiple regulatory factors produced by T cells. Curr Top Microbiol Immunol 100:221–229.
Schwartz E, Sartorelli A. 1982. Maturation activity of azacytidine in leukemia. Cancer Res 42:2651.
Stanley ER. 1979. Colony-stimulating factor (CSF) radioimmunoassay: Detection of a CSF subclass stimulating macrophage production. Proc Natl Acad Sci USA 76:2969–2973.
Stanley ER, Guilbert LF. 1981. Methods for the purification, assay, characterization and target cell binding of a colony stimulating factor (CSF-1). J Immunol Methods 42:253–284.
Tertian G, Yung Y-P, Guy-Grand D, Moore MAS. 1981. Long-term in vitro culture of murine mast cells. I. Description of a growth factor dependent culture technique. J Immunol 127:788.
Tertian G, Yung Y-P, Moore MAS. 1980. Induction and long-term maintenance of Thy-1-positive lymphocyte: Derivation from continuous marrow cultures. J Supramol Struct Cell Biochem 13:533.
Williams N, Eger RR, Jackson HM, Nelson DJ. 1982. Two-factor requirement for murine megakaryocyte colony formation. J Cell Physiol 110:101–104.

Yamamoto Y, Tomida M, Hozumi M, Azuma I. 1981. Enhancement by immunostimulants of the production by mouse spleen cells of factor(s) stimulating differentiation of mouse myeloid leukemic cells. Gann 72:828–833.

Yung Y-P, Eger R, Tertian G, Moore MAS. 1981. Long-term in vitro culture of murine mast cells: II. Purification of a mast cell growth factor and its dissociation from TCGF. J Immunol 127:794.

Yung Y-P, Moore MAS. 1982. Long-term in vitro culture of murine mast cells: III. Discrimination of mast cell growth factor and granulocyte-CSF. J Immunol 129:1256.

Mediators in Cell Growth and Differentiation,
edited by Richard J. Ford and Abby L. Maizel.
Raven Press, New York © 1985.

Hematopoietic Growth Factors

Antony W. Burgess

Ludwig Institute for Cancer Research, Victoria, 3050 Australia

The production of blood cells in mammalian organisms is regulated by a dynamic interaction between a set of progenitor cells and glycoprotein growth factors. Several regulatory glycoproteins have been purified (Burgess et al. 1977, Stanley and Heard 1977, Miyake et al. 1977, Gillis et al. 1980, Ihle et al. 1982), but most of these are still poorly characterized at the molecular level. We have little knowledge about the way in which these glycoproteins regulate the proliferation or differentiation of either normal or neoplastic hematopoietic progenitor cells.

As in other cellular systems, the hematopoietic growth factors (HGFs) (e.g., colony-stimulating factors (Burgess and Metcalf 1980a) and erythropoietin (Miyake et al. 1977)) almost certainly function via a cell surface receptor. It is important, however, to compare the apparent biological roles of the HGFs and other growth factors such as epidermal growth factor (EGF) (Cohen 1982), nerve growth factor (NGF) (Thoenen and Barde 1980), platelet-derived growth factor (PDGF) (Vogel et al. 1978), and fibroblast growth factor (FGF) (Thomas et al. 1980). Although fibroblast "survival" appears to be independent of EGF, FGF, and PDGF, normal hematopoietic progenitor cells (and macrophages; Tushinski et al. 1982) are dependent on specific growth factors for their survival (Nicola et al. 1981, Burgess et al. 1982, Ihle et al. 1983, Gillis et al. 1980). When bone marrow colony-forming cells are cultured in the absence of colony-stimulating factor (CSF), they die at a constant rate (approximately 5% per hour) (Nicola and Metcalf 1982). Addition of granulocyte/macrophage CSF (GM-CSF) prevents cell death and allows the rapid proliferation of the progenitor cells to continue. Indeed, in the presence of GM-CSF the hematopoietic progenitor cells proliferate and differentiate for 2 to 10 days, finally producing the mature cells capable of circulating in the blood. Thus, as distinct from EGF, PDGF, and FGF, which are usually considered to be stimulators of cell division, the HGFs appear to be essential for the survival of the progenitor cells. This distinction becomes particularly important when attempting to understand the intracellular biochemical pathways that are regulated by the HGFs, as most studies on other growth factors aim at identifying signals that initiate DNA synthesis, whereas studies with the HGFs need to identify the critical molecular processes that are shut down when the growth factor is removed.

Although there are suggestions that more than one growth regulator may be involved in the control of production of individual cell lineages in other tissues (e.g., EGF and the tissue-derived growth factors (Nexø et al. 1980, Roberts et al. 1981)), there appears to be a plethora of regulators controlling hematopoietic cell production (Table 1) (Axelrad et al. 1981, Nicola and Vadas 1984, Burgess and Nicola 1983). Undoubtedly many molecules are involved, but the disorganized state of the nomenclature for the HGFs and the absence of standard assay systems are now causing severe problems in understanding the significance of many current experiments (Lotem et al. 1980, Burgess and Metcalf 1980b, Burgess et al. 1980, Ihle et al. 1983, Lotem and Sachs 1983).

Five compartments of normal hematopoietic differentiation can be defined (Figure 1) in which there appears to be preferential activity of the different hematopoietic growth factors. Since there are more than eight distinct hematopoietic lineages and each has a different biological and biochemical profile (i.e., number of divisions, dividing- or non-dividing-end cells, nucleate- or enucleate-end cells), the compartments described in Figure 1 are not unique, but simply guides to the balance between proliferation, commitment, and differentiation. Similarly, some of the HGFs act in several of these compartments (e.g., GM-CSF, M-CSF, multi-CSF). However, within the functional framework of this model, the role of the different HGFs at the specific stages of cell lineage differentiation can be compared. It should be noted that in some cases, HGFs have been defined according to their action on neoplastic cells (e.g., differentiation factor (Burgess and Metcalf 1980b), macrophage-granulocyte inducer-2 (Lotem et al. 1980)). However, the role of these factors during normal hematopoiesis (fetal or adult) needs to be identified (Nicola et al. 1983) as it is unlikely that HGFs specific for leukemic cells will be produced by normal tissues.

ARE THE HGFs PRODUCED IN VIVO?

It must be emphasized that much of our knowledge of the HGFs is derived from culture studies in which unfractionated bone marrow cells, thymocytes, or spleen cells have been stimulated with growth factors produced by cell lines or tissues in vitro. Only two of the HGF molecules, erythropoietin (Epo) (see Goldwasser 1975) and GM-CSF (Metcalf 1971), have been detected in the serum or urine of animals. Erythropoietin is known to control the production of red blood cells: the correlation of Epo concentration and subsequent erythropoiesis is excellent (Sherwood and Goldwasser 1979), and the increase in erythropoiesis as a result of Epo injection is also directly related to the concentration of the growth factor in vivo. The relationship between Epo and multi-CSF in the regulation of erythropoiesis in vivo has not yet been elucidated (Eaves and Eaves 1978, Burgess et al. 1980, Johnson and Metcalf 1977). Although multi-CSF is able to stimulate erythroid colonies in vitro, even under conditions of severe anemia in which the serum levels of Epo are raised over 1000-fold, no multi-CSF has been detected (Burgess AW and Johnson GR, unpublished results). Similarly, attempts to modulate erythropoiesis or eosinophilopoiesis in vivo, by injections of multi-CSF (or by growing tumor cells that secrete multi-CSF, i.e., WEHI-3B(D^-)), have not been successful.

TABLE 1. *Hematopoietic growth factors*

Name	Alternative names	Biological action	Compartment*	Reference
M-CSF	MGF, MGI-1M, CSF-1	Stimulates macrophages and macrophage progenitors	II-V	Stanley and Heard 1977
GM-CSF	CSF-2, MGI-1GM	Stimulates a subset of multi-potential progenitors, all GM-progenitors, macrophages and granulocytes	I†-V	Burgess et al. 1977, Burgess and Metcalf 1980a, Gough et al. 1981
Multi-CSF	BPA, IL-3, PSF, MGI-1G, HCGF, MCGF, PSF, CFU-S-SA (MEG-CSF, EO-CSF, E-CSF)	Stimulates all mixed colony forming cells, factor-dependent cell lines and lymphocyte functions	I-IV	Burgess et al. 1980, Ihle et al. 1983, Fung et al. 1984, Axelrad et al. 1981
T cell GF	TCGF, IL-2	Stimulates the proliferation of subsets of T lymphocytes	III	Robb et al. 1981, Gillis et al. 1980, Taniguchi et al. 1983
B cell GF	BCGF, IL-1	Stimulates the proliferation of subsets of B lymphocytes	III	Conlon 1983
B cell maturation	BMF	Stimulates maturation of B lymphocytes		Sidman et al. 1984
G-CSF	GM-DF, DF, MGI-2	Stimulates subset of multi-potential progenitors, G progenitors, and the differentiation of some myeloid leukemic cells	I†-IV	Burgess and Metcalf 1980b, Lotem et al. 1980, Nicola et al. 1983
SAF	SHSF, CFU-S-SA	Stimulates self-renewal and commitment of multi-potential hematopoietic progenitor cells	I	Axelrad et al. 1981, Staber et al. 1984
Erythropoietin	Epo	Stimulates proliferation of mature erythroid precursors and production of hemoglobin	II†-IV	Miyake et al. 1977

*See Fig. 1.

†Stimulates only a subset of the cells in this compartment.

Abbreviations: CSF, colony-stimulating factor; MGF, macrophage growth factor; MGI, macrophage granulocyte inducer; M, macrophage; G, granulocyte; BPA, burst-promoting activity; IL, interleukin; PSF, P-cell–stimulating factor; HCGF, hematopoietic cell growth factor; CFU-S-SA, colony-forming unit–spleen-stimulating activity; MEG, megakaryocyte; EO, eosinophil; E, erythroid; BMF, B-cell maturation factor; DF, differentiation factor; SAF, stem cell activating factor; SHSF, splenic hematopoiesis stimulating factor.

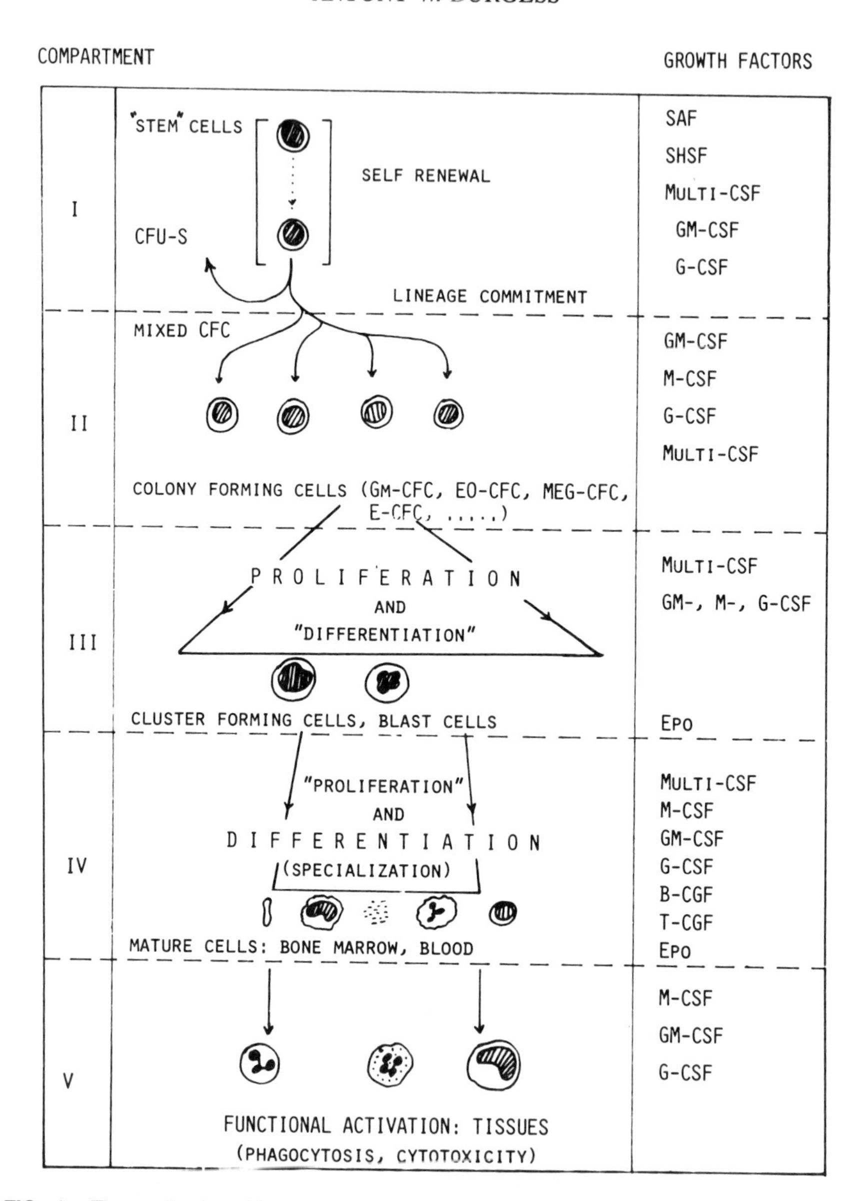

FIG. 1. The production of hematopoietic cells is usefully considered in the five compartments outlined in this diagram. Each of the compartments is preferentially controlled by a set of growth factors (see footnote to Table 1 for abbreviations). Most of the growth factors appear to function in several compartments, but are not able to stimulate all of the steps needed to produce mature blood cells. One of these molecules, multi-CSF, is able to stimulate the production of many different types of blood cells; the other regulators all appear to be specific for particular lineages. There appear to be minor populations of cells in the first three compartments capable of being stimulated by both G-CSF and GM-CSF, but these cells require either multi-CSF or their lineage-specific regulators to enter compartment IV (i.e., the final phases of proliferation and specialization).

Although GM-CSF has been detected in the serum of mice, it has been much more difficult to determine the relationship between its serum concentration and subsequent production of granulocytes and macrophages (Lange 1983, Asano et al. 1977). It has not been possible to inject sufficient quantities of GM-CSF into mice or humans to study directly its effects on myelopoiesis. Induction of neutrophil deficiencies (e.g. antineutrophil sera) are invariably accompanied by other physiological changes (e.g., infection, increased turnover of other cells, marrow cell depletion) that make it difficult to attribute any change in the production of neutrophils or macrophages directly to GM-CSF. The HGFs associated with endotoxin serum have not been characterized in detail. However, there are at least three forms of CSF (G – , GM – , and M –) in postendotoxin serum (Burgess and Metcalf 1980a, Lotem et al. 1980) as well as a separate activity capable of stimulating the self-renewal and proliferation of multipotential hematopoietic progenitor (and stem) cells (Staber et al. 1984). Preliminary characterization suggests that the CSFs in endotoxin serum have molecular weights of approximately 23,000 (Nicola et al. 1979); thus, this GM-CSF appears to be related to the molecule purified from mouse lung–conditioned medium. However, the biological activity of the GM-CSF from endotoxin serum can be completely neutralized by an antibody raised against CSF prepared from L cells (which has different biological and molecular properties (Stanley and Heard 1977)), whereas the GM-CSF from mouse lung–conditioned medium is not inactivated by this antiserum (Stanley 1979).

A similar state of knowledge exists for the molecules regulating the production of lymphocytes. Although it has been possible to purify T-cell growth factor (TCGF: also called IL-2) (Gillis et al. 1980), the distribution, function, and significance of this molecule in vivo have proved difficult to study. In recent years it has become a routine laboratory procedure to grow subsets of murine T lymphocytes using TCGF (Kelso et al. 1982). These continuous cultures of T lymphocytes are completely dependent on the presence of TCGF: removal results in the death of these cell lines. As valuable as TCGF will prove to be for increasing our knowledge of T-cell proliferation, it is difficult to understand why TCGF has been so difficult to detect in vivo. Perhaps both TCGF (Muhlradt and Opitz 1982) and multi-CSF are short-range GF that act only in the microenvironment in which they are produced. Indeed, in order to prevent widespread effects of a local antigenic challenge, it may be advantageous for organisms to clear TCGF from the circulation rapidly. Measurement of the production and clearance rates of TCGF from the murine circulation indicates that the rate of clearance from the serum exceeds the maximum rate of splenic production by a factor of at least 500 (Donohue and Rosenberg 1983). If this is the case, endogenous TCGF may never be detected in the serum, and the sites of TCGF production may need to be identified using in situ hybridization with molecular probes for mRNA.

IN VITRO PRODUCTION OF HGFs

The in vitro production and assay of the HGFs has greatly facilitated the identification and characterization of these proteins (Metcalf 1977). Although most cellular and tissue sources of the HGFs produce only small quantities of these factors (10-100 pg/10^6 cells), several laboratories have succeeded in purifying small amounts of at least five of the HGFs produced by cells in culture: TCGF, multi-CSF, M-CSF, GM-CSF, and G-CSF (Table 1). All of the HGFs are glycoproteins with some heterogeneity associated with the carbohydrate side chains (Burgess et al. 1980, Ihle et al. 1983). Initially, the heterogeneity in the sialic acid moieties and interference from contaminating proteins led to the belief that there were many different GM-CSFs (i.e., that each tissue produced a distinct molecular form). However, careful analysis of the asialo-GM-CSF molecules under dissociating conditions indicated that the apparently different, murine tissue–derived GM-CSFs are composed of the same protein chain with different carbohydrate moieties (Nicola et al. 1979).

The purification of these HGFs has allowed limited amino acid sequence information to be obtained for at least four of these molecules, and cDNA clones corresponding to TCGF (Taniguchi et al. 1983), multi-CSF (Fung et al. 1984), and GM-CSF (Gough et al. 1984) have now been obtained. Comparison of the predicted amino acid sequences for these regulators reveals no direct sequence homologies nor any predicted structural homologies between these molecules (Garnier et al. 1978) (Figure 2). TCGF is predicted to have a considerable proportion of α-helical secondary structure, and the short α-helical segments predicted for GM-CSF and multi-CSF do not occupy similar positions in the polypeptide chain. These initial data indicate that despite the functional relationships and gross biochemical similarities (molecular weights and isoelectric points) exhibited by the HGFs, there are unlikely to be close nucleotide or amino acid sequence homologies between the different HGFs. At present there is only limited information on the structure of the genes encoding these growth factors, and it will be fascinating to follow expression, copy number, and context of these genes in the different forms of leukemia.

MECHANISM OF ACTION OF THE HGFs

The biological specificity of the HGFs resides not only in their own production and structure, but in the distribution of the appropriate cell surface receptors. An understanding of the initial events in both myelopoiesis and lymphopoiesis will require a detailed knowledge of the induction of growth factor synthesis and release, as well as the receptor status of the hematopoietic progenitor cells and their intracellular responses to different concentrations of the growth factors (Metcalf and Burgess 1982). Until recently, it has not been possible to characterize the nature and distribution of the HGF receptors. In most cases only the overall receptor number and tissue distribution have been reported (Guilbert and Stanley 1980, Robb et al. 1981, Pigoli et al. 1982).

The specific cellular binding and the tissue distribution of receptors for M-CSF have been studied in two laboratories (Guilbert and Stanley 1980, Pigoli et al. 1982). The initial binding studies on cell lines and mononuclear phagocytes indi-

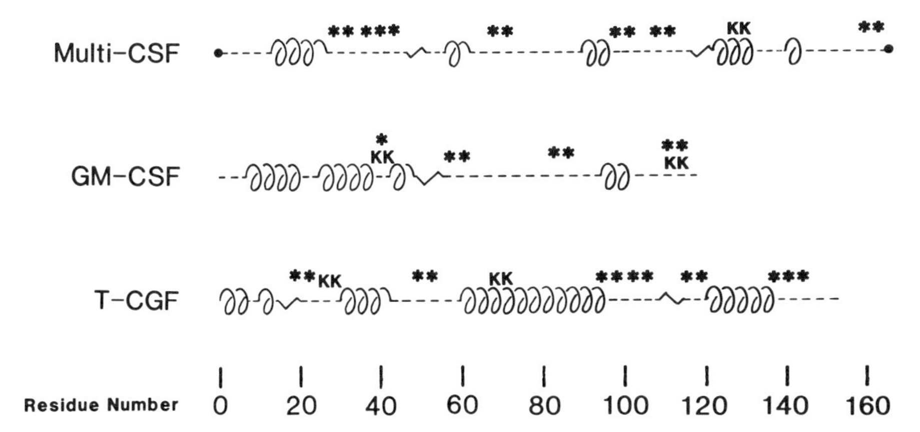

FIG. 2. Predicted regions of α-helical (ᘓᘓᘓ), extended structure (--⌒--) and reverse turns (**) for three of the hematopoietic growth factors GM-CSF, multi-CSF, and TCGF. The predictions were made using an empirical algorithm (Garnier et al. 1978). The amino acid sequences used for these predictions were derived from the nucleotide sequences of the cDNA clones of TCGF (Taniguchi et al. 1983), multi-CSF (or IL-3) (Fung et al. 1984), and GM-CSF (Gough et al. 1984).

cated that the binding was saturable, but at 37°C M-CSF was rapidly degraded. Peritoneal exudate macrophages have 20,000-50,000 receptors/cell (Table 2), but nucleated bone marrow cells and many macrophage cell lines have less than 1,000 receptors/cell. There is also some indication that bone marrow blast cells and neutrophils are able to bind ^{125}I-labeled M-CSF, but only after an initial incubation period (up to 24 hr) in vitro (Pigoli et al. 1982). However, it should be noted that the purification procedures for the M-CSF in these two studies are quite different, and recent results (Burgess AW, Metcalf D, Nice EC, Kozka I, and Hamilton JA, unpublished results) suggest that the affinity-purified CSF from mouse L-cell conditioned medium could also contain a GM-CSF.

Murine G-CSF has been purified (Nicola et al. 1983) and radioiodinated without loss of biological activity (Nicola and Metcalf 1984). ^{125}I-labeled G-CSF binds to bone marrow cells and to at least one myelomonocytic leukemic cell line (Nicola and Metcalf 1984); however, in both cases the number of receptors per cell is extremely low (Table 2). There appear to be at least two classes of G-CSF receptor, only one of which is competed for by GM-CSF. Similar studies have been initiated with ^{125}I-labeled GM-CSF (Walker F and Burgess AW, unpublished data). Specific binding is detected only in bone marrow cells, peritoneal neutrophils, and some myeloid cell lines. The number of GM-CSF receptors appears to be similar (but fewer) than for G-CSF receptors, and again there appear to be at least two classes of receptor, only one of which is competed for by G-CSF. At present, the receptor number is so low that characterization of the receptor and its distribution on progenitor cells has not been achieved. Competitive binding studies between these molecules and multi-CSF will undoubtedly help us understand the control signals regulating both the production and activation of neutrophils and macrophages.

TABLE 2. *Receptors for hematopoietic growth factors (HGF)*

HGF*	Dissociation constant (pm)	Number of sites/cell				Reference
		Normal cells		Cell lines		
M-CSF	10	Peritoneal exudate macrophages	50,000	P388D macrophages	6,000	Guilbert and Stanley 1980
TCGF	5–25	ConA–stimulated splenocytes	11,000	HUT-102B2 T-lymphoma	15,000	Robb et al. 1981
G-CSF	50–100	Bone marrow neutrophil	800	WEHI-3B(D^+) (myeloid leukemia)	700	Nicola and Metcalf 1984
GM-CSF	50–500	Peritoneal neutrophil	400	WEHI-3B(D^+)	250	Walker and Burgess 1984

*See Table 1 for abbreviations.

The receptor for TCGF has been detected on activated TCGF-responsive T lymphocytes. Unstimulated lymphocytes (either on T or B) display few or no TCGF receptors (Robb et al. 1981). TCGF binds with similar affinity to continuous T-cell lines (Robb et al. 1981), mitogen- (or antigen-) activated T cells, and several T lymphomas (Table 2). Myeloid cells and several TCGF-independent cell lines have no detectable TCGF receptors. The binding of TCGF is specific, and no other polypeptide hormones appear to compete for TCGF binding. The biological response of T cells to TCGF has been correlated with the fraction of TCGF receptors occupied; thus, T-cell (and myeloid cell—see earlier discussion) proliferation is presumably regulated by both receptor number and the concentration of TCGF.

The critical biochemical events activated as a result of polypeptide growth factors binding to the cell surface are still poorly understood. Several systems indicate that hormone binding is associated with the activation of a receptor kinase activity (Cohen et al. 1979, Kasuga et al. 1982, Nishimura et al. 1982), which is assumed to be the trigger for subsequent control of DNA synthesis (Sefton et al. 1980). Even in the most widely studied systems (e.g., EGF modulation of the EGF receptor tyrosine kinase in the A431 carcinoma and PDGF modulation of fibroblast DNA synthesis), it has been difficult to define which molecular changes are critical for the action of the hormone and which changes are by-products of a general (but secondary) change in cellular metabolism (Sefton et al. 1980).

The HGFs are essential for the survival and proliferation of their target cells, but there is little or no information about the molecular basis of this dependence. A recent observation, that ATP levels change rapidly when a factor-dependent cell line is deprived of multi-CSF (Whetton and Dexter 1983), may provide a clue to the function of HGF receptors. The activation of the HGF receptors may control cell metabolism at the most basic of levels: the supply of ATP. Interestingly, Whetton and Dexter (1983) also reported that an exogenous source of ATP was able to substitute for multi-CSF. Our own studies on the action of GM-CSF or hematopoietic progenitor have shown a rapid decrease (10-fold in 3 hr) in the incorporation of ^{32}P into proteins when hematopoietic progenitor cells are deprived of GM-CSF (Stanley et al. 1984). This overall change in the protein phosphorylation pattern will effect a set of proteins critical for the control of DNA synthesis.

There are already indications as to the nature of some of the intracellular proteins that are intimately involved in the action of the growth factors (e.g., the oncogene products (Downward et al. 1984), the cytoskeletal protein (Birchmeier 1981), and even peptide hormones that may act intracellularly (Baldwin 1982)). However, identifying the molecular modifications involved when the hematopoietic progenitor cells become committed to a particular lineage will also require a considerable improvement in our current preparations of purified progenitor cells and a thorough understanding of the early molecular markers of cell lineage–specific differentiation.

REFERENCES

Asano S, Urate A, Okabe T, Sato N, Kondo Y, Ueyama Y, Chiba S, Ohsawa N, Kosaka K. 1977. Demonstration of granulopoietic factor(s) in the plasma of nude mice transplanted with a human lung cancer in the tumor tissue. Blood 49:845–852.

Axelrad A, Burgess A, Iscove N, Wagemaker G. 1981. *In* Control of Cellular Division and Development, Part B. Alan R. Liss, New York, pp. 421–425.
Baldwin G. 1982. Gastrin and the transforming protein of polyoma virus have evolved from a common ancestor. FEBS Lett 137:1–5.
Birchmeier W. 1981. Fibroblasts focal contacts. Trends Biochem Sci 6:234–237.
Burgess AW, Camakaris J, Metcalf D. 1977. Purification and properties of colony stimulating factor from mouse lung conditioned medium. J Biol Chem 252:1998–2003.
Burgess AW, Metcalf D. 1980a. The nature and action of granulocyte-macrophage colony stimulating factor. Blood 56:1947–1958.
Burgess AW, Metcalf D. 1980b. Characterization of a serum factor stimulating the differentiation of myelomonocytic leukemic cells. Int J Cancer 26:647–654.
Burgess AW, Metcalf D, Russell SHM, Nicola NA. 1980. Granulocyte/macrophage- megakaryocyte-, eosinophil and erythroid colony stimulating factors produced by mouse spleen cells. Biochem J 185:301–314.
Burgess AW, Nicola NA. 1983. Growth Factors and Stem Cells. Academic Press, Sydney.
Burgess AW, Nicola NA, Johnson GR, Nice EC. 1982. Colony forming cell proliferation: a rapid and sensitive assay system for murine granulocyte and macrophage colony stimulating factors. Blood 60:1219–1223.
Cohen S. 1982. The epidermal growth factor (EGF). *In* Accomplishments in Cancer Research 1982. J. B. Lippincott Co., Philadelphia, pp. 76–83.
Cohen S, Carpenter G, King L. 1979. Epidermal growth factor-receptor-protein kinase interactions. J Biol Chem 255:4834–4842.
Conlon PJ. 1983. A rapid biologic assay for interleukin-1. J Immunol 131:1280–1282.
Donohue JH, Rosenberg SA. 1983. The fate of interleukin-2 after in vivo administration. J Immunol 130:2203–2208.
Downward J, Yarden Y, Mayes E, Scrace G, Totty N, Stockwell P, Ullrich A, Schlessinger J, Waterfield MD. 1984. Close similarity of epidermal growth factor receptor and v-erb-B oncogene protein sequences. Nature 307:521–557.
Eaves CJ, Eaves AC. 1978. Erythropoietin (Ep) dose-response curves for three classes of erythroid progenitors in normal human marrow and in patients with polycythemia vera. Blood 52:1196–1210.
Fung MC, Hapel AJ, Yoner S, Cohen DR, Johnson RM, Campbell HD, Young IG. 1984. Molecular cloning of cDNA for murine interleukin-3. Nature 307:233–237.
Garnier J, Osguthorpe DJ, Robson B. 1978. Analysis of the accuracy and implications of simple methods for predicting the secondary structure of globular proteins. J Mol Biol 120:97–120.
Gillis S, Smith KA, Watson JD. 1980. Biochemical and biological characterization of lymphocyte regulatory molecules. II. Purification of a class of rat and human lymphokines. J Immunol 124:1954.
Goldwasser E. 1975. Erythropoietin and the differentiation of red blood cells. Fed Proc 34:2285–2291.
Gough NM, Gough J, Metcalf D, Kelso A, Grail D, Nicola NA, Burgess AW, Dunn AR. 1984. Molecular cloning of cDNA encoding the murine haemopoietic growth regulator, GM-CSF. Nature (in press).
Guilbert LJ, Stanley ER. 1980. Specific interaction of murine colony-stimulating factor with mononuclear phagocytic cells. J Cell Biol 85:153–159.
Ihle JN, Keller J, Henderson L, Klein F, Palaszynski EW. 1982. Procedures for the purification of interleukin 3 to homogeneity. J Immunol 129:2431.
Ihle JN, Keller J, Oroszlan S, Henderson LE, Copeland TD, Fitch F, Prystowsky MB, Goldwasser E, Schrader JW, Palaszynski E, Dy M, Lebel B. 1983. Biologic properties of homogeneous interleukin 3. J Immunol 131:282–287.
Johnson GR, Metcalf D. 1977. Pure and mixed erythroid colony formation in vitro stimulated by spleen conditioned medium with no detectable erythropoietin. Proc Natl Acad Sci USA 74:3879–3882.
Kasuga M, Zick Y, Blithe DG, Crettaz M, Kahn CR. 1982. Insulin stimulates tyrosine phosphorylation of the insulin receptor in a cell-free system. Nature 298:667–669.
Kelso A, Glasebrook AL, Kanagawa O, Brunner KT. 1982. Production of macrophage-activating factor by T lymphocyte clones and correlation with other lymphokine activities. J Immunol 129:550–556.
Lange RD. 1983. Cyclic hematopoiesis. Exp Hematol 11:435–451.
Lotem J, Lipton JH, Sachs L. 1980. Separation of different molecular forms of macrophage and granulocyte inducing proteins for normal and leukemic myeloid cells. Int J Cancer 25:763–771.
Lotem J, Sachs L. 1983. Coupling of growth and differentiation in normal myeloid precursors and the breakdown of this coupling in leukemia. Int J Cancer 32:127–134.
Metcalf D. 1971. Acute antigen-induced elevation of serum colony stimulating factor (CSF) levels. Immunology 21:427–436.

Metcalf D. 1977. Hemopoietic Colonies: In Vitro Cloning of Normal and Leukemic Cells. Springer Verlag, Berlin.
Metcalf D, Burgess AW. 1982. Clonal analysis of progenitor cell commitment to granulocyte-macrophage production. J Cell Physiol 111:275–283.
Miyake T, King CKH, Goldwasser E. 1977. Purification of human erythropoietin. J Biol Chem 252:5558–5564.
Muhlradt PF, Opitz HG. 1982. Clearance of interleukin 2 from the blood of normal and T cell-depleted mice. Eur J Immunol 12:983–985.
Nexø E, Hollenberg MD, Figueroa A, Batt RM. 1980. Detection of epidermal growth factor and its receptor during fetal mouse development. Proc Natl Acad Sci USA 77:2782–2785.
Nicola NA, Burgess AW, Metcalf D. 1979. Similar molecular properties of granulocyte-macrophage colony stimulating factors produced by different mouse organs in vitro and in vivo. J Biol Chem 254:5290–5299.
Nicola NA, Metcalf D. 1982. Analysis of purified fetal liver hemopoietic progenitor cells in liquid culture. J Cell Physiol 112:257–264.
Nicola NA, Metcalf D. 1984. Binding of the differentiation-inducer, granulocyte-colony-stimulating factor, to responsive but not unresponsive leukemic cell lines. Proc Natl Acad Sci USA (in press).
Nicola NA, Metcalf D, Matsumoto M, Johnson GR. 1983. Purification of a factor inducing differentiation in murine myelomonocytic leukemia cells: Identification as granulocyte colony-stimulating factor (G-CSF). J Biol Chem 258:5290–5299.
Nicola NA, Metcalf D, von Melchner H, Burgess AW. 1981. Isolation of murine fetal hemopoietic progenitor cells and selective fractionation of various erythroid precursors. Blood 58:376–386.
Nicola NA, Vadas M. 1984. Hemopoietic colony-stimulating factors. Immunology Today (in press).
Nishimura J, Huang JS, Deuel TF. 1982. Platelet-derived growth factor stimulates tyrosine-specific protein kinase activity in Swiss mouse 3T3 cell membranes. Proc Natl Acad Sci USA 79:4303–4307.
Pigoli G, Waheed A, Shadduck RK. 1982. Observations on the binding and interaction of radioiodinated colony-stimulating factor with murine bone marrow cells in vitro. Blood 52:408–420.
Robb, RJ, Munck A, Smith KA. 1981. T-cell growth factor receptors. J Exp Med 154:1455–1474.
Roberts AB, Anzano MA, Lamb LC, Smith JM, Sporn MB. 1981. New class of transforming growth factors potentiated by epidermal growth factor: Isolation from non-neoplastic tissues. Proc Natl Acad Sci USA 78:5339–5343.
Sefton BM, Hunter T, Beemon K, Eckhart W. 1980. Evidence that the phosphorylation of tyrosine is essential for cellular transformation by Rous sarcoma virus. Cell 20:807–816.
Sherwood JB, Goldwasser E. 1979. A sensitive radioimmunoassay for erythropoietin. Blood 54:885–893.
Sidman CL, Paige CJ, Schreier MH. 1984. B cell maturation factor (BMF): A lymphokine or family of lymphokines promoting the maturation of B lymphocytes. J Immunol 132:209–222.
Staber FG, Burgess AW, Nicola NA, Metcalf D. 1984. Biological and biochemical properties of a serum factor which stimulates splenic hemopoiesis in mice. Exp Hematol (in press).
Stanley ER. 1979. Colony stimulating factor (CSF) radioimmunoassay: Detection of a CSF subclass stimulating macrophage production. Proc Natl Acad Sci USA 76:2969–2973.
Stanley ER, Heard BM. 1977. Factors regulating macrophage growth and production. J Biol Chem 252:4305–4312.
Stanley IJ, Nicola NA, Burgess AW. 1984. Expression of a phosphorylated c-ras p21 in normal hemopoietic progenitor cells (in preparation).
Taniguchi T, Matsui H, Fujita T, Takaoka C, Kashima N, Yoshimoto R, Hamuro J. 1983. Structure and expression of a cloned cDNA for human interleukin-2. Nature 302:305–310.
Thoenen H, Barde YA. 1980. Physiology of nerve growth factor. Physiol Rev 60: 1284–1335.
Thomas KA, Riley MC, Lemmon SK, Baglan NC, Bradshaw RA. 1980. Brain fibroblast growth factor. J Biol Chem 255:5517–5520.
Tushinsky RJ, Oliver IT, Guilbert LJ, Tynan PW, Warner JR, Stanley ER. 1982. Survival of mononuclear phagocytes depends on a lineage-specific growth factor that the differentiated cells selectively destroy. Cell 28:71–81.
Vogel A, Raines E, Kariya B, Rivest MJ, Ross R. 1978. Co-ordinate control of 3T3 cell proliferation by platelet-derived growth factor and plasma components. Proc Natl Acad Sci USA 75:2810–2814.
Whetton AD, Dexter TM. 1983. Effect of haematopoietic cell growth factor on intracellular ATP levels. Nature 303:629–631.

Lymphoid Growth Factors

Mediators in Cell Growth and Differentiation,
edited by Richard J. Ford and Abby L. Maizel.
Raven Press, New York © 1985.

The Purification and Biological Properties of Human Interleukin 1

Lawrence B. Lachman

Department of Cell Biology, The University of Texas M. D. Anderson Hospital and Tumor Institute at Houston, Houston, Texas 77030

Interleukin 1 (IL-1), previously known as lymphocyte-activating factor (Gery and Waksman 1972), is released by macrophages undergoing an immune response. Although one of many macrophage factors (monokines), IL-1 has demonstrated the most potential as an immunostimulant. In some in vitro immunological assay systems, IL-1 can completely fulfill the requirement for macrophages (Maizel et al. 1981a). In others, IL-1 can greatly reduce the number of macrophages required for a measurable immunological response (Bobak and Whisler 1980, Rosenberg and Lipsky 1981). IL-1 cannot, however, replace the functional properties of antigen processing and antigen presentation performed by macrophages (Lipsky et al. 1983). Likewise, IL-1 cannot subsitute for histocompatibility antigens found on the surface of macrophages (Gilman et al. 1983). IL-1 can, however, stimulate [^{3}H]thymidine incorporation by mouse and human T lymphocytes (Lachman et al. 1979, Maizel et al. 1981a), regulate B-lymphocyte differentiation (Ford et al. 1981, Hoffman and Watson 1979, Howard et al. 1982, Wood 1979), control the growth of bone marrow cells (Mishell et al. 1982), and effect the generation of cytotoxic T lymphocytes (Farrar and Hilfiker 1982). Some of these properties of IL-1, but clearly not all, may be due to the ability of IL-1 to initiate the release of interleukin 2 (IL-2), previously known as T-cell growth factor, from helper T lymphocytes (Larsson and Coutinho 1980, Maizel et al. 1981b, Smith et al. 1980). IL-2 can be readily distinguished from IL-1 by its ability to initiate and maintain T lymphocytes in continuous culture (Smith 1980).

IL-1 also regulates many nonimmunological responses involving macrophages. Specifically, IL-1 may be identical to the macrophage-derived fever-producing factor endogenous pyrogen (Murphy et al. 1980, Rosenwasser et al. 1979). IL-1 stimulates the release of acute phase reactants by hepatocytes (Selinger et al. 1980), the release of prostaglandin and collagenase from synovial cells (Mizel et al. 1981), and the growth of fibroblasts (Schmidt et al. 1982).

Of course, a mediator with such a broad spectrum of biological properties has been the subject of intense research. Many researchers, working in diverse areas of macrophage function, have rapidly reached agreement that they have all been studying the same molecule. This chapter will discuss the preparation of homogeneous human IL-1, as well as some of its biological properties.

MATERIALS AND METHODS

Most materials and methods have been previously described (Lachman et al. 1979, 1980). Human IL-1 was prepared from peripheral blood monocytes and leukemic cells as described below. The assay for IL-1 activity (Lachman et al. 1980) measures the increased incorporation of [^{3}H]thymidine by CD-1 strain (Charles River Breeding Laboratories) mouse thymocytes. The IL-1 assay measures the effect of IL-1 on [^{3}H]thymidine incorporation by thymocytes in the *absence* of lectin. Although IL-1 can greatly potentiate the response of mouse thymocytes to concanavalin A (ConA) or phytohemagglutinin (PHA), human IL-1 is directly mitogenic for murine thymocytes. A unit of IL-1 activity was calculated according to the procedure of Gillis et al. (1978). (A unit of IL-1 activity is arbitrarily measured by individual investigators because no international standard exists.)

RESULTS

Production of Human IL-1

Human IL-1 can be prepared from several cell sources. IL-1 is a monocyte- or macrophage-derived factor and is released into the culture medium by peripheral blood monocytes or leukemic monoblasts after treatment with endotoxin or phagocytic stimuli. Peripheral blood monocytes, the most common source of IL-1, can be difficult to obtain in large quantity, but plasmapheresis residue bags can be an excellent source. Leukemic cells of monocytic origin are also an excellent source of IL-1 (Lachman et al. 1979). Leukemic cells obtained from acute monocytic leukemia (AMOL) or acute myelomonocytic leukemia (AMML) patients receiving therapeutic leukapheresis can be used to prepare large quantities of IL-1 from a single donor. Leukemic patients with peripheral white blood cell counts of 300,000/mm^3 may yield 10^{11}–10^{12} leukemic cells after leukapheresis. Leukemic cells may be frozen without losing the ability to produce IL-1. (The serum and urine of one recent AMOL patient were found to lack IL-1 activity when measured in the mouse thymocyte assay. Since the biological assay is not ideal for body fluids, it should be replaced by radioimmunoassays and enzyme-linked immunosorbent assays as these become available.) As previously determined from peripheral blood monocytes, AMOL and AMML cells produce maximum IL-1 in the presence of serum. Human and fetal calf serum work equally well, but a minimum of 1% is required, with maximum IL-1 release occurring in the presence of 5% human serum (Figure 1).

For the purification described below, IL-1 was prepared from AMOL cells frozen at -60°C in 50-ml tubes. The cells (2×10^{10}) were slowly thawed and added to a spinner flask containing 6 liters of minimal essential medium, 5% human serum, and 10 μg/ml of *Escherichia coli* lipopolysaccharide (to stimulate IL-1 release). The flask was incubated at 37°C for 48 hr, at which time the IL-1–containing

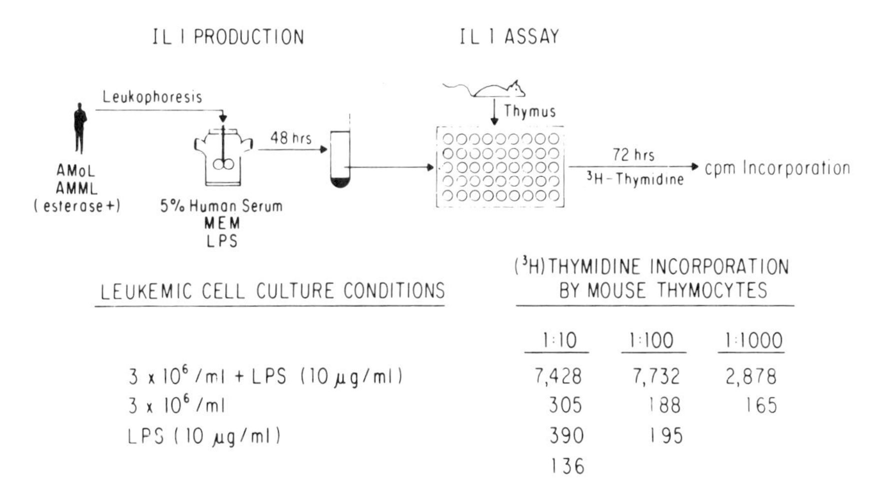

LEUKEMIC CELL CULTURE CONDITIONS	(³H)THYMIDINE INCORPORATION BY MOUSE THYMOCYTES		
	1:10	1:100	1:1000
3 x 10^6/ml + LPS (10 µg/ml)	7,428	7,732	2,878
3 x 10^6/ml	305	188	165
LPS (10 µg/ml)	390	195	
	136		

FIG. 1. Preparation of human IL-1 from leukemic cells and the mouse thymocyte assay for IL-1.

supernatant was separated from the cells by centrifugation. The conditioned medium was frozen in 500-ml aliquots for later purification. The conditioned medium exhibited IL-1 activity in the mouse thymocyte assay at a final dilution of 1:1000 (Figure 1).

Purification of AMOL Cell IL-1

Hollow Fiber Diafiltration and Ultrafiltration

This procedure has been described in detail (Lachman et al. 1980). Briefly, 1 liter of conditioned medium was extracted with 10 liters of 0.85% NaCl by using a 50,000-Da cut-off hollow fiber device. IL-1 activity was found in both the diafiltrate with molecular weight below 50,000 and the extracted medium with molecular weight above 50,000. This observation agrees with previous ones from gel filtration chromatography, which indicate that IL-1 activity in conditioned medium is a low-molecular-weight activity of 13,000 and a high-molecular-weight activity of 70,000 (Blyden and Handschumacher 1977). The purification procedure described in this report is for the 13,000-Da activity only. The diafiltrate (10 liters) was concentrated overnight at 4°C by using three ultrafiltrate cells with YM10 membranes, a self-feeding reservoir, and an automatic shut-off device. In the morning, the 10,000- to 50,000-Da pool (500 ml) was ultrafiltered by using the 50,000-Da hollow fiber device used in the first step of the purification. The purpose of this ultrafiltration procedure was to remove the small quantity of 50,000-Da serum protein that passed through the hollow fiber cartridge during diafiltration. The 50,000-Da ultrafiltrate (450 ml) was concentrated to 40 ml by using a single stirred cell with a YM10 membrane.

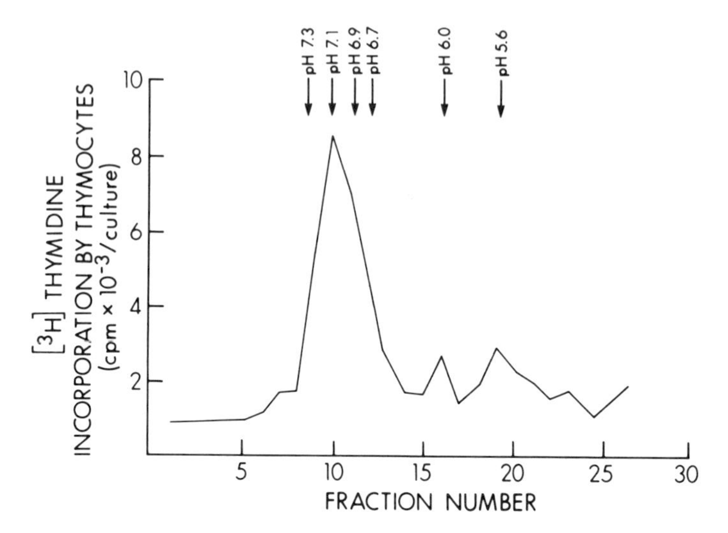

FIG. 2. Isoelectric focusing of the 10–50,000 MW IL-1 activity. The 10–50,000 MW IL-1 fraction from the hollow fiber fractionation was further purified by sucrose gradient IEF in a pH 4-8 Ampholine gradient. When a stable pH gradient had been obtained (18 hr at 7°C, 16,000V constant power), 1.5-ml fractions were collected, the pH determined, and each fraction dialyzed to remove the sucrose and Ampholine. The amount of IL-1 in each fraction was determined in the [^{3}H]thymidine incorporation assay using mouse thymocytes. The profile of activity shown above was obtained by testing the IEF fractions at a final concentration of 1% (v/v) in the thymocyte assay.

Isoelectric Focusing (IEF) of the 10,000- to 50,000-Da Fraction

The 40-ml sample was isoelectrically focused in a pH 4–8 Ampholine and sucrose gradient for 20 hr at 4°C (Lachman et al. 1980). The pH 6.8–7.2 region was found to contain the IL-1 activity (Figure 2).

Preparative Gel Purification of the IEF-Purified IL-1 Activity

Previous experiments (Lachman et al. 1977) have demonstrated that IEF-purified IL-1 could be further purified by nondenaturing polyacrylamide gel electrophoresis (PAGE). Unfortunately, this procedure does not result in homogeneous IL-1. Analytical PAGE of the gel-purified activity demonstrated the presence of small quantities of serum proteins as well as IL-1. Therefore, we decided to further purify IL-1 by taking advantage of its low molecular weight. Sodium dodecyl sulfate (SDS)-PAGE was performed on IEF-purified IL-1 that had been dialyzed against water for 1 day and lyophilized. The lyophilized sample was hydrated in 25 μl of sample buffer (not containing reducing agents) and warmed for 2 hr at 37°C. SDS-PAGE of the sample was performed at 10°C by using a 1% tube gel of 100×5.5 mm with a 5 mm 3% stacking gel (Laemmli 1970). Identically prepared gels containing prestained molecular weight standards (Bethesda Research Laboratories) and molecular weight standards to be stained after electrophoresis were used to calibrate the gels. After electrophoresis, the IL-1–containing gel was sliced

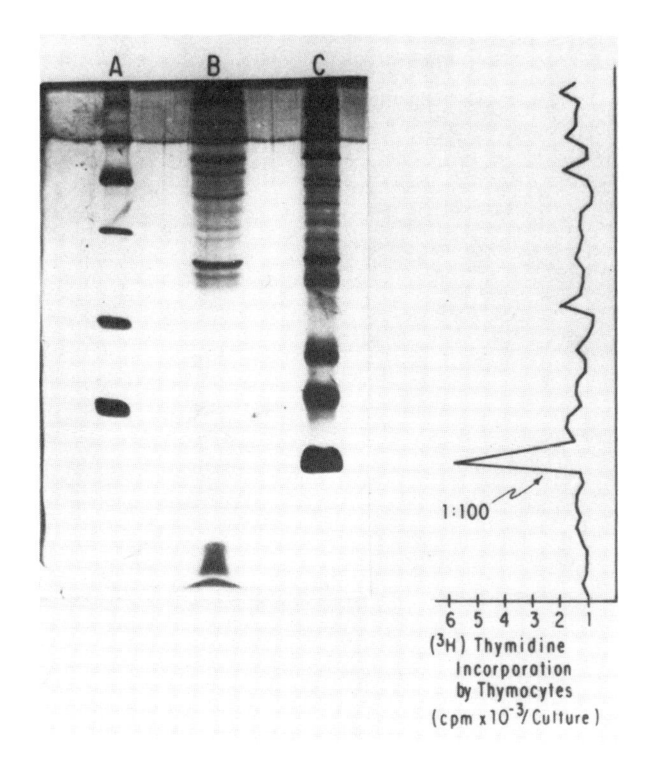

FIG. 3. Sodium dodecyl sulfate-polyacrylamide gel electrophoresis of IEF-purified IL-1. Two 1.0-ml samples of the IEF-purified IL-1 were dialyzed against water and lyophilized. One sample was applied to a 15% analytical slab gel and the other to a 15% tube gel; each gel contained a 3% stacking gel. The IEF-purified IL-1 (C) contained many protein bands, but in particular three bands of 20,000 MW that were not present in an identical purified sample of control medium (B), which contained medium and serum, but no cells. The MW standards (A) are 94,000, 43,000, 30,000, 20,100, and 14,400. A [^{3}H]thymidine incorporation assay of the tube gel slices demonstrated a single peak of IL-1 activity that corresponded with the darkly staining band of 11,000 MW. The recovered IL-1 was active at a final concentration of 1% (v/v) in the thymocyte culture.

into fifty 2-mm slices and eluted overnight in 0.5 ml of 0.85% NaCl to recover the IL-1 activity. The eluted fractions were dialyzed against 0.85% NaCl for 48 hr to remove as much SDS as possible, and the fractions were assayed for IL-1 activity in the thymocyte assay.

Figure 3 shows the results of such an assay compared with analytical slab gel of the IEF-purified activity. The IL-1 activity was found to correspond with a darkly staining band of approximately 11,000 Da evident after $AgNO_3$ staining (Oakley et al. 1980). To determine if the 11,000-Da protein was homogeneous, analytical SDS-PAGE was performed in a second dimension. Four 0.5-mm slices from the 11,000-Da region of a 15% SDS tube gel containing the IL-1 activity were placed in the sample wells of an 18% SDS slab gel (1.5 × 150 mm), and electrophoresis was performed at 10°C using the same buffer system as that used for the preparative

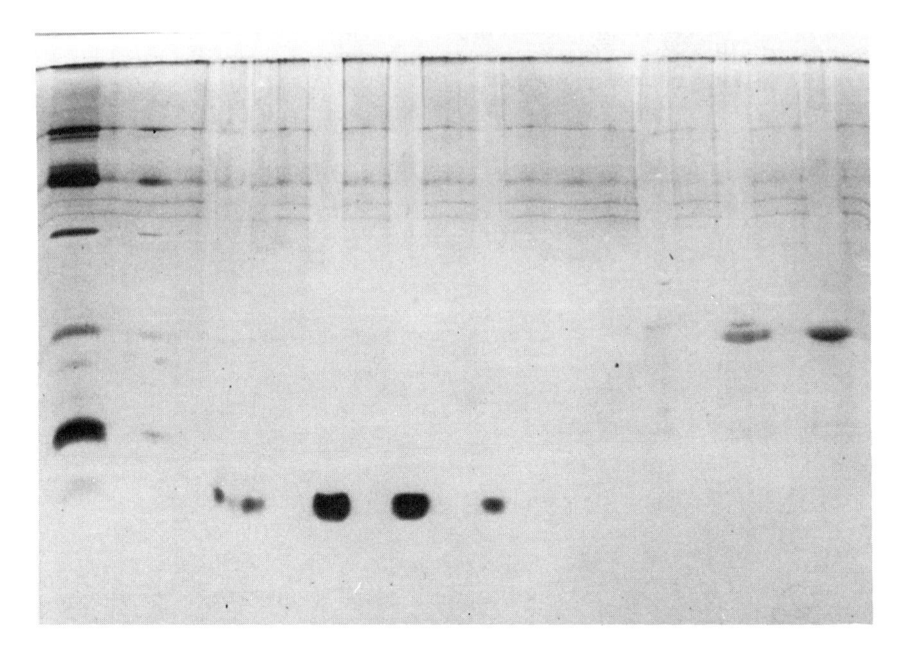

FIG. 4. Evaluation of purity of SDS-PAGE-purified IL-1. To evaluate the purity of the tube gel–purified IL-1, analytical SDS-PAGE with $AgNO_3$ staining was performed. Four 0.5-mm gel slices in the 11,000 MW range of the tube gel were placed in the sample wells of an 18% slab gel and electrophoresis was performed. MW standards (see Figure 2) were placed in a separate well (left lane). $AgNO_3$ staining of the slab gel revealed a single protein band of ~11,000 MW in the gel slices. Very faintly staining lines seen at the top of the gel are buffer contaminants since they are visible between wells.

tube gel (Figure 3). $AgNO_3$ staining of the analytical gel demonstrated that the protein band that corresponded to the IL-1 activity was contained in the four gel slices taken from the tube gel and that only a single protein band was visible (Figure 4).

The recovery of IL-1 biological activity after preparative SDS-PAGE was quite low. Although the gel fractions had been dialyzed to reduce the levels of SDS, the samples were toxic to the mouse thymocytes used for the IL-1 assay. The gel fractions required a 100-fold dilution to reduce the toxicity of the samples. They rapidly lost activity when diluted much more than 100-fold, and thus contained a very small percentage of the biological activity that had been applied. The loss of biological activity during SDS-PAGE has been addressed for IL-2. A newly devised buffer system has resulted in recovery of highly active IL-2 after SDS-PAGE (Gillis et al. 1982).

Recovery of IL-1 after the Complete Purification

The overall recovery of IL-1 activity after IEF was usually about 4%. This poor recovery of activity exemplifies the necessity for beginning purifications with large quantities of extremely active medium. The major loss of activity occurs during the IEF step. This large loss of activity is probably due to the very hydrophobic nature of IL-1: IL-1 binds to many types of membranes (not the YM series of hydrophobic membranes) and chromatographic media. IL-1 also binds to most types of sterile filters and must be sterile filtered with hydrophobic filters (Gelman Acrodiscs) or in the presence of serum.

As noted above, preparative SDS-PAGE of the IEF-purified activity results in poor recovery of biological activity, but significant recovery of homogeneous protein. The amount of protein recovered from the preparative gel can be estimated by comparing the intensity of silver staining with known amounts of protein. Conventional protein determinations such as Lowry or Bradford would require that the entire SDS-PAGE–purified sample be used for a single protein determination. The amount of protein recovered from the preparative gel was estimated to be 3 μg/liter of starting conditioned medium. Because the amount of IL-1 in the IEF-purified sample from 1 liter of conditioned medium was about 3 μg and the yield was 4%, the implication is that 1 liter of conditioned medium used for the purification originally contained approximately 75 μg of IL-1 or 75 ng of IL-1 per milliliter. That the original conditioned medium exhibits strong biological activity when added to thymocytes at a final concentration of 1% (v/v) implies that IL-1 exhibits biological activity at about 10^{-11} M. It should also be noted that although the SDS-PAGE–purified IL-1 retains little biological activity, the material is suitable for amino acid sequence determination and the preparation of monoclonal antibody.

Biochemical Properties of Human IL-1

Biochemical characterization of IL-1 has progressed through the indirect route of treatment with various enzymes, chemicals, or heat. Chymotrypsin and trypsin were able to inactivate IEF-purified IL-1, whereas neuraminidase and ribonuclease did not affect the biological activity. Treatment of IEF-purified IL-1 with cyanogen bromide (BrCN) under acidic conditions also destroyed the biological activity. IL-1 activity was fairly stable in the strongly acidic condition (0.3 N HCl) used to perform the BrCN hydrolysis. Reduction and alkylation of IL-1 with iodoacetamide did not affect the biological activity of IL-1. Heat treatment at 100°C for 10 min inactivated IL-1 biological activity, whereas 56°C for 30 min reduced the activity by only one half. The conclusion from these experiments is that IL-1 is a nonglycosylated protein that is quite stable under heat and extremely acidic conditions. IEF-purified IL-1 has been screened for protease activity in the very sensitive casein hydrolysate assay (Levin et al. 1976). It did not hydrolyze casein to a measurable extent during a 1-hr period. This assay has been demonstrated to be sensitive to 1–2 ng of tyrpsin per microliter of sample.

Biological Properties of Human IL-1

The introduction of this paper contained a brief summary of the general biological properties of IL-1. This section will discuss in greater detail some recently determined biological properties of IEF and nondenaturing PAGE-purified human IL-1.

Effect of IL-1 on Mouse Thymocyte Subpopulations

Stimulation of [^{3}H]thymidine incorporation by mouse thymocytes was the assay used to monitor for IL-1 activity through purification. Human IL-1 (Lachman et al. 1980) and rabbit IL-1 (Simon and Willoughby 1981) are directly mitogenic for mouse thymocytes. Alternative IL-1 assays include the stimulation of [^{3}H]thymidine incorporation by ConA- or PHA-stimulated thymocytes. The mode of action of IL-1 in all these assays is probably to stimulate release of the truly mitogenic substance IL-2 (Larsson and Coutinho, 1980, Smith et al. 1980). Mature and immature mouse thymocyte subpopulations, separated by peanut lectin agglutination (PNA), have been tested for [^{3}H]thymidine incorporation in the presence of lectin and IL-1 (Conlon et al. 1982, Oppenheim et al. 1982, Otterness et al. 1981). The mature PNA subpopulation responds maximally to IL-1. The PNA-positive immature subpopulation may show a small stimulation by IL-1 and lectin, but it is due to slight contamination by mature cells in the PNA-positive population (Oppenheim et al. 1982). The inability of PNA-positive thymocytes to respond mitogenically to the same signals that stimulate PNA-negative cells reflects both poor production and poor response to IL-1 and IL-2 by immature thymocytes (Conlon et al. 1982, Otterness et al. 1981).

Effect of IL-1 on Human Thymocytes and T Lymphocytes

Maizel et al. (1981b) demonstrated that IEF- and PAGE-purified human IL-1 could stimulate [^{3}H]thymidine incorporation by human thymocytes and T lymphocytes. Unlike mouse thymocytes, human IL-1 could not stimulate significant [^{3}H]thymidine incorporation in the absence of ConA or PHA. However, in the presence of lectin, IL-1 could stimulate significant [^{3}H]thymidine incorporation by macrophage-depleted T lymphocytes and thymocytes. In fact, the levels of [^{3}H]thymidine incorporation in the presence of IL-1 are equal to the levels of [^{3}H]thymidine incorporation found when 5% human monocytes were added to the T lymphocyte cultures. These results demonstrated the ability of IL-1 to completely replace macrophages in this system.

Of further interest was a subsequent report by Maizel et al. (1981b) that demonstrated that entry of lectin-stimulated lymphocytes into the S phase of the cell cycle was regulated by either monocytes or IL-1. Although lectin-stimulated T lymphocytes are morphologically lymphoblasts, these cells do not progress through the G_1 phase of the cell cycle. Entry of these lymphoblasts into the S phase, with concomitant [^{3}H]thymidine incorporation, is dependent on the presence of IL-1 or monocytes.

Effect of IL-1 on B Lymphocytes

As greater understanding of the role of IL-1 and IL-2 in T-cell mitogenesis developed, it was obvious that attention would turn to the role of these factors or comparable factors in stimulation of B-cell growth, differentiation, and release of antibody. Many lines of evidence indicate that a B-cell growth factor, distinct from IL-2, is produced by lymphocytes (Farrar and Hilfiker 1982). A macrophage factor, which may be identical to IL-1, has been demonstrated to effect differentiation of B cells grown in culture (Ford et al. 1981, Howard et al. 1982, Swain et al. 1981). It is interesting to note that several earlier papers (Hoffman and Watson 1979, Wood 1979) demonstrate the involvement of a macrophage factor (possibly IL-1) in B-cell activation.

Effect of IL-1 on Fibroblasts and Synovial Cells

IL-1 is very likely a growth factor for fibroblasts. The role of macrophages in wound healing and fibroblast growth has been well documented since the original report by Leibovitch and Ross (1975). Definite proof does not exist that IL-1 is the fibroblast growth factor, but highly purified mouse IL-1 (Schmidt et al. 1982) and high-pressure liquid chromatography–purified human IL-1 (Postlethwaite et al. 1983) are mitogenic for nonconfluent fibroblasts.

IL-1 may exacerbate inflammation during arthritis. A recent paper by Hamilton et al. (1972) has demonstrated that PAGE-purified IL-1 stimulates the release of plasminogen activator from primary cultures of human fibroblasts isolated from arthritic joints. In a similar report (Mizel et al. 1981), murine IL-1 was demonstrated to stimulate prostaglandin and collagenase from synovial cells isolated from rheumatoid joints.

In Vivo Effects of Human IL-1

The biological effects of IL-1 can be scientifically evaluated only by using endotoxin-free samples. Fortunately, IEF-purified human IL-1 is endotoxin free as determined by the *Limulus* lysate clot assay. This is a fortuitous result, because no special precautions were taken during the purification to ensure endotoxin-free conditions. The explanation for this finding is that endotoxin has an isoelectric point of 3, and is thus separated from IL-1, which has an isoelectric point of 7.

Human IEF-purified IL-1 was administrered i.v. to rabbits by Ackins, and when the rabbits were monitored for an increase in body temperature (Bernheim et al. 1980), it was raised 1°C by 0.5 ml of IEF-purified IL-1. The observed fever was monophasic, and occurred within 30 min after injection and subsided within 1 hr. An identical febrile response was observed in animals that had recently been made tolerant to endotoxin, which confirms that IL-1–induced fever was not due to endotoxin.

These results agree with the findings of Murphy et al. (1980) and Rosenwasser et al. (1979): that IL-1 may be the same factor as the previously described endogenous pyrogen. Unfortunately, all these experiments suffer from the same basic problem; they have not been performed with homogeneous IL-1. These experiments will no doubt be repeated as homogeneous IL-1 becomes available.

TABLE 1. *The effect of intravenously injected IEF-purified IL-1 on the level of circulating neutrophil**

Number of trials	Dose of IL-1 injected (ml/animal)	Number of blood neutrophils (cells/mm³ ± SE)	Difference from normal
7	—	1,334 ± 154	0
7	0.010	1,335 ± 184	0
5	0.020	3,405 ± 481	2,070
4	0.050	5,557 ± 442	4,222
3	0.075	6,881 ± 1,384	5,476
4	0.100	8,727 ± 1,439	7,392

*200-g female Holtzman rats were injected with IEF-purified IL-1, or saline, in the amount indicated. One hour after injection, the number of circulating neutrophils was determined on a sample of heart blood by counting the number of leukocytes and by differential counting of Wright's stained smear.

Kampschmidt and Upchurch administered IEF-purified IL-1 to rats and monitored the animals for (1) rapid increase in the circulating levels of peripheral blood neutrophils and (2) the 24-hr levels of plasma fibrinogen. The rationale for these experiments was that IL-1 exhibited many of the same properties as the previously described monokine leukocyte endogenous mediator (LEM) (Kampschmidt and Upchurch 1980). Injection i.v. of 0.10 ml of IEF-purified IL-1 was found by Kampschmidt and Upchurch to increase circulating neutrophil levels by 8,000 cells/mm³ 1 hr after injection (Table 1). Also, 0.25 ml of IL-1 was found to increase 24-hr fibrinogen levels by 45 mg/100 ml of plasma (Table 2). These properties are identical to those described for LEM and again confirmed that IEF-purified IL-1 was free of contaminating endotoxin. As mentioned previously, the shortcoming of these experiments is that IEF-purified IL-1 is not homogeneous. In any event, these experiments may indicate that IL-1 is a central mediator of inflammation that can exert biological effects on the nervous system (fever), the bone marrow (margination of neutrophils), and the liver (secretion of acute phase reactants such as fibrinogen).

TABLE 2. *Qualitative dose-response of IL-1*

Units of IL-1 injected*	Response
31.0	1°C Fever
6.3	Increase of 8,000 neutrophils/mm³ of blood
15.8	Fibrinogen increase of 44 mg/100 ml of plasma

*1.0 ml of IEF-purified IL-1 contained 63.0 units of thymocyte-stimulating activity. An aliquot of 0.01 units can stimulate [³H]TdR incorporation by mouse thymocytes equal to five times the level of unstimulated cells.

DISCUSSION

Macrophages stimulated with endotoxin or phagocytic stimulants release IL-1 into the culture medium. Very highly purified, but not yet homogeneous, preparations of human IL-1 have been demonstrated to stimulate mouse and human T lymphocytes, to induce fever in rabbits, and to stimulate margination of neutrophils and the release of acute phase proteins in rats. IL-1 has been demonstrated to enhance the antibody response of B cells, stimulate the generation of cytotoxic T cells, and stimulate synovial cells and fibroblasts to secrete plasminogen activator, prostaglandin, and collagenase. IL-1 may affect the total response to bacterial antigens or tumors by simultaneously stimulating several lines of host defense. The availability of homogeneous preparations of human IL-1 with high specific biological activity is of paramount importance. Homogeneous IL-1 will lead to amino acid sequence data that can be used to search for peptide fragments that may have IL-1 activity. Also, amino acid sequence data can be used to prepare nucleic acid probes for eventual gene cloning. Hybridoma antibody to IL-1 has not yet been fully described (Swain et al. 1981). Simplified assays with monoclonal antibodies can be used to determine IL-1 levels in serum, and thus allow exploration of the possibility that IL-1 release may be reduced in tumor-bearing patients and immunodeficient patients. This avenue of research may indicate that IL-1 could be used as an immunotherapeutic agent, similar to interferon. Also, the primary amino acid sequence data could be used to design novel peptides that would inhibit the biological activity of IL-1. These antagonists could be valuable therapeutic agents for the treatment of chronic inflammatory diseases such as arthritis and sclerosis.

ACKNOWLEDGMENTS

I wish to thank Ms. Shelia Buckner for the careful preparation of this paper. This research was supported by American Cancer Society grant IM 296.

REFERENCES

Bernheim HA, Block LH, Francis L, Atkins E. 1980. Release of endogenous pyrogen-activating factor from concanavalin A-stimulated human lymphocytes. J Exp Med 152:1811–1816.
Blyden G, Handschumacher RE. 1977. Purification and properties of human lymphocyte activating factor (LAF). J Immunol 188:1631–1637.
Bobak D, Whisler R. 1980. Human B lymphocyte colony responses. I. General characteristics and modulation by monocytes. J Immunol 125:2764–2768.
Conlon PJ, Henney CS, Gillis S. 1982. Cytokine dependent thymocyte responses: Characterization of IL 1 and IL 2 target subpopulations and mechanism of action. J Immunol 128:797–801.
Farrar JJ, Hilfiker ML. 1982. Antigen-nonspecific helper factors in the antibody response. Fed Proc 41:263–268.
Ford RJ, Mehta SR, Franzini D, Montagna R, Lachman LB, Maizel AL. 1981. Soluble factor activation of human B lymphocytes. Nature 294:261–263.
Gery I, Waksman BH. 1972. Potentiation of T-lymphocyte response to mitogens: II. The cellular source of potentiating mediators. J Exp Med 136:143–148.
Gilman SC, Rosenberg JS, Feldman JD. 1983. Inhibition of interleukin synthesis and T cell proliferation by a monoclonal anti-Ia antibody. J Immunol 130:1236.

Gillis S, Ferm MM, Ou W, Smith KA. 1978. T cell growth factor; parameters of production and a quantitative microassay for activity. J Immunol 120:2027–2037.
Gillis S, Mochizuki D, Conlon PJ, Hefeneider SJ, Ramthun CA, Gillis AE, Frank MB, Henney C, Watson JD. 1982. Molecular characterization of interleukin 2. Immunol Rev 63:167.
Hamilton JA, Zabriskie JD, Lachman LB, Chen Y-S. 1972. Streptococcal cell walls with synovial cell activation. Stimulation of synovial fibroblast plasminogen activator activity by monocytes treated with group A streptococcal cell wall sonicates and muramyl dipeptide. J Exp Med 155:1702–1709.
Hoffman MK, Watson J. 1979. Helper T cell-replacing factors secreted by thymus-derived cells and macrophages: Cellular requirements for B cell activation and synergistic properties. J Immunol 122:1371–1375.
Howard M, Farrar J, Hilfiker M, Johnson B, Takatsu K, Hamaoka T, Paul W. 1982. Identification of a T cell derived B cell growth factor distinct from interleukin 2. J Exp Med 155:914–923.
Kampschmidt RF, Upchurch HF. 1980. Neutrophil release after injection of endotoxin or leukocytic endogenous mediator into rats. J Reticuloendothel Soc 28:191–198.
Lachman LB, Hacker MP, Handschumacher RE. 1977. Partial purification of human lymphocyte-activating factor (LAF) by ultrafiltration and electrophoretic techniques. J Immunol 119:2019–2023.
Lachman LB, Moore JO, Metzgar RS. 1979. Preparation and characterization of lymphocyte activating factor (LAF) from acute monocytic and myelomonocytic leukemia cells. Cell Immunol 41:199–207.
Lachman LB, Page SO, Metzgar RS. 1980. Purification of human interleukin 1. J Supramol Struct 13:457–462.
Laemmli VK. 1970. Determination of protein molecular weight in polyacrylamide gels. Nature (London) 277:680–684.
Larsson E-L, Coutinho A. 1980. Mechanism of T cell activation. I. A screening of "step one" ligands. Eur J Immunol 10:93–98.
Leibovich SJ, Ross R. 1975. The role of macrophages in wound repair. A study of hydrocortisone and antimacrophage serum. Am J Pathol 78:71–91.
Levin N, Hatcher VB, Lazarus GS. 1976. Proteinases of human epidermis: A possible mechanism for polymorphonuclear leukocyte chemotaxis. Biochim Biophys Acta 452:458–465.
Lipsky PE, Thompson PA, Rosenwasser LJ, Dinarello CA. 1983. The role of interleukin 1 in human B cell activation: Inhibition of B cell proliferation and the generation of immunoglobulin-secreting cells by an antibody against human leukocytic pyrogen. J Immunol 130:2708.
Maizel AL, Mehta SR, Ford RJ, Lachman LB. 1981a. Effect of interleukin 1 on human thymocytes and purified human T cells. J Exp Med 153:470–474.
Maizel AL, Mehta SR, Hauft S, Franzini D, Lachman LB, Ford RJ. 1981b. Human T lymphocyte/monocyte interaction in response to lectin: Kinetics of entry into S-phase. J Immunol 127:1058–1064.
Mishell RI, Lee DA, Grabstein KH, Lachman LB. 1982. Prevention of the in vitro myelosuppressive effects of glucocorticosteroids by interleukin 1 (IL-1). J Immunol 128:1614–1619.
Mizel SB, Dayer J-M, Krane SM, Mergenhagen SE. 1981. Stimulation of rheumatoid synovial cell collagenase and prostaglandin production by partially purified lymphocyte-activating factor (interleukin 1). Proc Natl Acad Sci USA 78:2474–2478.
Murphy PA, Simon PL, Willoughby WF. 1980. Endogenous pyrogens made by rabbit alveolar macrophages. J Immunol 124:2498–2503.
Oakley BR, Kirsch DR, Morris NR. 1980. A simple ultra-sensitive silver stain for detecting proteins in polyacrylamide gels. Anal Biochem 105:361–364.
Oppenheim JJ, Stadler BM, Siraganian RP, Mage M, Mathieson B. 1982. Lymphokines: Their role in lymphocyte responses. Properties of interleukin 1. Fed Proc 41:257–262.
Otterness IG, Lachman LB, Bliven ML. 1981. Effects of levamisole on the proliferation of thymic lymphocyte subpopulations. Immunopharmacology 3:61–66.
Postlethwaite AE, Lachman LB, Mainardi CL, Kang AH. 1983. Interleukin 1 stimulation of collagenase production by cultured fibroblasts. J Exp Med 157:801–806.
Rosenberg AS, Lipsky PE. 1981. The role of monocytes in pokeweed mitogen-stimulated human B cell activation: Separate requirements for intact monocytes and a soluble monocyte factor. J Immunol 126:1341–1346.
Rosenwasser LJ, Dinarello CA, Rosenthal AS. 1979. Adherent cell function in murine T-lymphocyte antigen recognition. IV. Enhancement of murine T-cell antigen recognition by human leukocytic pyrogen. J Exp Med 150:709–713.

Schmidt JA, Mizel SB, Cohen D, Green I. 1982. Interleukin 1, a potential regulator of fibroblast proliferation. J Immunol 128:2177.
Selinger MJ, McAdam KP, Kaplan MM, Sipe JD, Vogel SN, Rosenstreich DL. 1980. Monokine induced synthesis of serum amyloid A protein by hepatocytes. Nature 285:498–501.
Simon PL, Willoughby WF. 1981. The role of subcellular factors in pulmonary immune function: Physiochemical characterization of two distinct species of lymphocyte-activating factor produced by rabbit alveolar macrophages. J Immunol 126:1534–1540.
Smith KA. 1980. Interleukin 2. Immunol Rev 51:337–352.
Smith KA, Lachman LB, Oppenheim JJ, Favata MF. 1980. The functional relationship of the interleukins. J Exp Med 151:1551–1553.
Swain SL, Dennert G, Warner JF, Dutton RW. 1981. Culture supernatants of a stimulated T-cell line have helper activity that acts synergistically with interleukin 2 in response of B cells to antigen. Proc Natl Acad Sci USA 78:2517–2521.
Wood DD. 1979. Mechanism of action of human B cell-activating factor. I. Comparison of the plaque-stimulating activity with thymocyte-stimulating activity. J Immunol 123:2400–2407.

Mediators in Cell Growth and Differentiation,
edited by Richard J. Ford and Abby L. Maizel.
Raven Press, New York © 1985.

The Determinants of T-Cell Growth

Kendall A. Smith

The Department of Medicine, Dartmouth Medical School,
Hanover, New Hampshire 03756

The study of the in vitro growth characteristics of T-lymphocytes (T-cells) has contributed immeasurably to the understanding of the critical elements necessary for DNA replication and mitosis (Morgan et al. 1976, Ruscetti et al. 1977). When considered in the light of other cell systems available (e.g., fibroblasts and epithelial cells), the in vitro T-cell system is unique, since a liquid suspension environment closely approximates the surrounding milieu that lymphocytes encounter in the peripheral blood. In contrast, cultured fibroblasts and epithelial cells must be placed into an environment that lacks the normal surrounding supporting cells and structures.

Thus, normal anchorage-dependent cells, by definition, are altered by the experimental situation necessary for their study. Consequently, information relevant to physiological growth controls that affect these cell types may not be forthcoming. In addition to serum, which is absent in vivo, fibroblasts have been shown to be influenced by a variety of polypeptide "growth factors" (examples include platelet-derived growth factor, epidermal growth factor, fibroblast growth factor, insulin, and insulin-like growth factors). Therefore, it is obscure as to which of these polypeptides is responsible for initiation of cell growth in vivo, and how cell growth is regulated.

However, it is apparent that fibroblast growth in vivo is not a continuous phenomenon: rather, fibroblasts are normally quiescent unless stimulated by tissue injury. Moreover, upon completion of the tissue repair process, fibroblasts again become quiescent. That this transient growth phenomenon even occurs suggests that an exquisitely sensitive and specific signaling mechanism exists that initiates and retards normal fibroblast proliferation.

Since the immune system depends upon a specific and intermittent proliferative clonal expansion of lymphocytes, there are unique advantages of this system as a model for the tightly regulated control of cell growth common to all tissues and cell types. Lymphocytes that populate the peripheral lymphoid tissue after a maturation process in the bone marrow and thymus persist in the resting (G_0) phase of the cell cycle and are incapable of proliferating or performing differentiating functions (e.g., antibody secretion, cytolysis, lymphokine production), unless the antigen receptor is triggered appropriately. Thus, lymphocytes can be considered analogous to fibroblasts, since an external or environmental signal initiates a series of molecular events that ultimately culminates in DNA replication and mitosis.

Particularly noteworthy is the understanding that although antigen-receptor interaction initiates cell activation, there is a discrete molecular mechanism whereby the process of cell division is carried forward. Thus, as a consequence of T-cell antigen receptor triggering, there is a rapid induction of the T-cell lymphocytotrophic growth hormone, interleukin 2 (IL-2), and IL-2 receptor gene expression (for review, see Smith 1984). The subsequent IL-2–receptor interaction then initiates a G_1-S phase transition and the cell becomes irreversibly committed to mitosis.

Accordingly, in the T-cell system, only one polypeptide hormone-receptor system is involved in promoting cell division. Among other things, the involvement of only a single class of high-affinity, specific, and saturable receptors simplifies considerably studies directed toward understanding the DNA replication signaling pathway that is initiated by IL-2 receptor triggering. In contrast, as already mentioned, not one but several hormone-receptor systems promote fibroblast proliferation; which system promotes cell division physiologically is obscure.

The evidence that the IL-2–receptor interaction is the key and *only* signal capable of promoting T-cell DNA synthesis comes from studies with IL-2 and IL-2 receptor–specific monoclonal antibodies (Smith et al. 1983, Leonard et al. 1982). Despite the presence of serum, antigen, or lectin, both antibodies completely abrogate DNA synthesis and cellular division by preventing the IL-2–receptor interaction. However, it is worthy of mention that serum-derived nutritional or more-specific cofactors, or both, may be necessary for either the antigen receptor triggering of IL-2 and IL-2–receptor gene expression or events subsequent to IL-2–receptor triggering. For example, it has already been shown that the IL-2–receptor interaction induces transferrin receptor expression (Neckers and Cossman 1983) and that transferrin is necessary for DNA synthesis and cellular division. However, since transferrin-receptor interaction has been shown to be required for many proliferating cell types (Barnes and Sato 1980), it must be concluded that transferrin serves a nutritive, facilitatory function in cell division, rather than an obligatory step in the DNA replication signaling pathway triggered by IL-2. In a similar fashion, insulin receptors are known to appear on antigen- or lectin-triggered T-cells (Strom and Bangs 1982). However, the role that an insulin-receptor interaction might play in T-cell mitosis remains ill defined.

THE IL-2 HORMONE-RECEPTOR SYSTEM

The continuing dissection of the elements essential for T-cell proliferation has resulted in several developments within the past 2 years that have finally permitted a detailed analysis of the determinants of T-cell growth. Between 1965, when mitogenic activities were first discovered in cultured leukocyte media (Gordon and MacLean 1965, Kasakura and Lowenstein 1965), and 1983, when a cDNA encoding the T-cell growth factor activity was isolated (Tanaguchi et al. 1983), the study of lymphokine-mediated T-cell mitosis was hampered by inadequate quantities of lymphokines and a consequent lack of understanding of the functions of the molecular moieties in question. In the past 3 years, the T-cell growth factor, IL-2, has been isolated (Robb and Smith 1981) and purified to homogeneity (Smith et al. 1983), and the activity has been ascribed to a single, variable glycosylated polypeptide. Moreover, monoclonal antibodies have been prepared that react specifically with IL-2 (Smith et al. 1983), which have greatly facilitated its purification so that quantities sufficient for further biologic experiments could be realized.

These developments subsequently facilitated the preparation of immunoaffinity-purified, homogeneous, biosynthetically radiolabeled IL-2 so that the mechanism of IL-2 interaction with T-cells could be explored (Smith et al. 1983). Accordingly, it has been enlightening to discover that the IL-2–T-cell interaction displays all of the characteristics ascribed to classic hormone receptor interactions (i.e., high affinity, saturability, and ligand and tissue specificity (Robb et al. 1981)). In addition to the insight this concept has provided for the design of functional experiments exploring the correlation of IL-2–receptor interaction and the T-cell proliferative response, the IL-2 radioreceptor assay was instrumental in identifying a monoclonal antibody (anti-Tac) reactive with IL-2 receptors (Leonard et al. 1982). Now, very recent experiments utilizing anti-Tac have resulted in the identification of cDNA clones that encode the IL-2 receptor (Nikaido et al. unpublished data, 1984, Leonard et al. unpublished data, 1984).

Thus, in contrast to every other cell system available for studies of eukaryotic cell growth, the T-cell–IL-2 hormone-receptor system is unique: IL-2 has been shown to be the *only* molecule that is capable of promoting DNA duplication and cellular division, the mechanism of interaction of IL-2 with target T-cells is known, monoclonal antibodies are available that react both with IL-2 and the IL-2 receptor, and the genes encoding IL-2 and the IL-2 receptor have been cloned.

IL-2 AND LEUKEMIC CELL GROWTH

Collectively, the technical and intellectual progress in our understanding of the determinants of normal T-cell growth have allowed, for the first time, a rational approach to potential abnormalities in the IL-2–receptor T-cell growth regulatory system that may be responsible for neoplastic transformation. As a consequence of the availability of sensitive, unambiguous assays for IL-2, IL-2 mRNA, and the IL-2 receptor, T-cell leukemias could be screened for these crucial determinants of T-cell growth. The resulting information has been quite definitive: leukemic cells that express an immature T-cell surface phenotype isolated from patients with childhood acute lymphoblastic leukemic (T-ALL) do not produce IL-2, nor do they express IL-2 receptors (Smith 1982, Holbrook et al. 1984). In striking contrast, leukemic cells and cell lines isolated from patients with adult T-cell leukemia (ATL), a disease endemic to southern Japan and first described by Yodoi and co-workers (1974), express IL-2 receptors 50-fold in excess of those found on normal activated T-cells (Wang and Smith 1984). Moreover, the ATL cell IL-2 receptor expression is abnormal in an additional aspect, in that ATL cell–IL-2 receptors are continuously expressed, rather than transiently expressed, a property characteristic of normal T-cells.

Initially, evidence was presented by Gootenberg and co-workers (1981) that indicated that ATL cells also produced IL-2, thus suggesting that the leukemic cells proliferate by virtue of an IL-2–receptor-mediated autocrine mechanism. However, a more extensive survey of ATL cell lines by Arya et al. (1984) has recently shown that ATL cell lines do not express the IL-2 gene. Thus, an autocrine mechanism involving the IL-2–receptor system is not responsible for the neoplastic proliferative characteristics of ATL cells. Whether the IL-2 receptor itself somehow promotes the growth of these cells remains to be explored.

In this regard, recent experiments (unpublished) from our own laboratory point toward the idea that the abnormally expressed IL-2 receptors present on ATL cells behave as though they are already triggered in the absence of IL-2. Detailed IL-2 binding studies on four ATL cell lines (HUT-102B2, ATL-2, MT-2, and C91-P1) reveal two classes of IL-2 binding sites. One relatively high affinity class of receptors binds IL-2 with a 5-fold lower affinity than do normal T-cells (K_d, 50 pM) and there are 4- to 5-fold greater numbers of these receptors ($4–5 \times 10^4$ sites/cell compared to 1×10^4 sites/cell on normal T-cells). In addition, another class of receptors has even a lower affinity for IL-2 (K_d, 10–50 nM), and there are even more of these sites ($1–2 \times 10^6$ sites/cell). Thus, not only do ATL cells constitutively express IL-2 receptors that are present in greater numbers, there are two classes of binding sites, and these sites have a lower affinity for IL-2 than do normal IL-2 receptors.

Other experiments indicate that IL-2 binding to ATL cell IL-2 receptors mediates no effect (stimulatory or inhibitory) on the growth characteristics of these cell lines: IL-2 concentrations 10,000-fold in excess of those required to promote normal T-cell growth have no effect on ATL cells, and both IL-2 and IL-2–receptor monoclonal antibodies, which completely inhibit normal T-cell growth, also have no effect. Accordingly, by inference, the IL-2 receptors expressed by ATL cells are either completely nonfunctional or already maximally triggered in the absence of IL-2. Future studies directed at the molecular mechanism whereby the IL-2–triggered receptor signals DNA replication may discriminate between these alternatives.

IL-2–DEPENDENT NORMAL T-CELL GROWTH

Normal IL-2–dependent T-cells are available to compare and contrast with the ATL cells. In this regard, it is important to realize that two types of cells exist. Freshly isolated T-cells placed into culture with agents that stimulate the T-cell antigen receptor (e.g., antigen, mitogenic lectins, and clonotypic monoclonal antibodies) only transiently express IL-2 receptors; the receptors appear within the cell population asynchronously and also disappear asynchronously, so that the IL-2–dependent proliferative response slowly reaches a maximum over 2–3 days and then declines over the succeeding 7–10 days of culture (Cantrell and Smith 1983,

Meuer et al. 1984). In contrast, long-term IL-2–dependent murine cytolytic T-cell lines (CTLL) (Gillis and Smith 1977) and clones (Baker et al. 1979) continuously express IL-2 receptors; consequently, CTLL grow indefinitely provided an adequate supply of IL-2 is furnished to the cell population. In many ways, CTLL resemble neoplastic ATL cells. However, the CTLL clearly are not transformed, since they are dependent upon an exogenous source of IL-2 and cease to proliferate if IL-2 is withdrawn. Moreover, the CTLL are not tumorigenic when inoculated into syngeneic normal or immunodeficient animals (Smith and Ruscetti 1981). Thus, by analogy to fibroblasts, the CTLL resemble murine embryonic 3T3 cells, which also have an unlimited in vitro proliferative capacity but are dependent upon serum-derived polypeptide growth factors for their proliferative potential.

Because of the unique characteristics of the normal T-cells, i.e., their transient expression of IL-2 receptors after antigen-receptor triggering, T-cells can be synchronized in the resting phase (G_0/G_1) of the cell cycle (Cantrell and Smith 1983, 1984). It is then possible to render the cell population IL-2 receptor-positive by a reexposure to the initial immunostimulatory signal. The proliferative effects of IL-2 can then be studied in a kinetic fashion. As a result, a series of recently published experiments provide information that explain fully the variability of cell cycle times, a peculiar characteristic of cell growth that had remained perplexing for more than 50 years (Cantrell and Smith 1984).

When an asynchronously growing, genetically homogeneous population of cells is observed, and the time required for cell division is plotted according to the number of cells, a striking feature, true for all living cells that have been examined (i.e., bacteria, yeast, protozoa, avian, and mammalian cells) is that the time required for one cell division is highly variable (for review, see Pardee et al. 1979). Quite enlightening are data plotted according to the *rate* of cell division, which is the reciprocal of the time required for each cell division. Division rate converts the abscissa to a logarithmic plot and yields a normal distribution of the individual cells within the population (the rate-normal distribution). This plot is of considerable interest, since it reveals that some cells transit the cell cycle very quickly, while others display a markedly retarded rate of cell cycle progression.

Many investigators have sought to explain this phenomenon and several observations have become axioms: (1) cell cycle variability cannot be explained by clonal heterogeneity since cloned cell populations display the same cell cycle variability as uncloned populations; (2) cell cycle times are not genetically determined, since there is a poor correlation between the division rates of mother cells and the resulting daughter cells; (3) nongenetic (or epigenetic) influences seem to be important since the cell cycle times of sibling daughter cells are quite similar; (4) a cell population retains the same mean division rate over many generations and there is no selection for more rapidly growing cells (Pardee et al. 1979).

As a result of these observations, two models have been presented to explain cell cycle variability. The deterministic theory, proposed 20 years ago by Koch and Schaechter (1962), states that genetically identical cells are *functionally different*, and that variability in cell cycle times results from the cumulative effects of many small differences. Thus, this theory implies that the system is so complex that it may be impossible to dissect. The probabalistic theory, proposed 15 years ago by Burns and Tannock (1970), and subsequently popularized and extended by Smith and Martin (1973), states that cells are *functionally identical* and that cell cycle variability results from an unpredictable Poissonian event that occurs quite independently of other events or properties of the cell population. This theory implies that it is impossible to account for cell cycle variability because it is a matter of chance.

In this regard, it is noteworthy that the debate between determinism and indeterminism continues in the field of quantum physics. Einstein, in a letter to Max Born on December 12, 1926, voiced in favor of determinism in this debate with a now quite famous quote: "Quantum mechanics is certainly imposing. But an inner voice tells me that it is not yet the real thing. The theory says a lot, but does not really bring us any closer to the secret of the Old One. I, at any rate, am convinced that He does not throw dice." Moreover, Einstein went on to propose, along with de Broglie, that perhaps all information concerning quantum mechanical events was not known, and that an undiscovered or "hidden variable" might be responsible for the absence of predictability observed in quantum systems.

With regard to cell cycle variability among T-cells, we are in a unique position to investigate the elements critical for determining cell growth, since the signal for DNA replication is known. As well, the system is simple and has been reduced to the thermodynamic principles governing the interaction of a single ligand and a single class of high-affinity receptors. Furthermore, unique cellular and molecular reagents have been developed to aid the analysis, including synchronized G_0/G_1 IL-2 receptor-positive T-cell populations, homogeneous, immunoaffinity purified IL-2, and methods to determine IL-2 receptor expression both quantitatively (through the IL-2 radioreceptor assay) and qualitatively (by means of cytofluorometric analysis using the IL-2 receptor monoclonal antibody). Consequently, it is possible to proceed beyond a mathematical analysis of kinetic data derived from proliferating cell populations to a critical study of the parameters known to be responsible for T-cell cycle progression.

In contrast to those investigators who had studied other cell systems, the identification of the IL-2–receptor interaction as the sole event involved in T-cell cycle progression aided our efforts considerably and focused our attention on the interplay between IL-2 concentration and IL-2 receptor density. Moreover, the application of an immunologic approach to cell surface molecules, i.e., the cytofluorometric analysis of T cells labeled with IL-2 receptor monoclonal antibodies, for the first time allowed the visualization of the broad heterogeneity of the IL-2 receptor densities among individual cells of both uncloned and cloned human T-cell populations.

Quite striking is the finding that IL-2 receptors are distributed lognormally within the populations. Consequently, it became natural to propose that perhaps the cellular IL-2 receptor density is important for T-cell cycle progression and responsible for the rate-normal distribution of cell cycle times of proliferating T-cell populations. The results of several experimental approaches are decisive: Cells with high IL-2 receptor density transit the cell cycle more rapidly than those with a low IL-2 receptor density. Moreover, only three parameters are critical for the rate of cell division: IL-2 concentration, IL-2 receptor density, and the duration of the IL-2–receptor interaction (Cantrell and Smith 1984). Thus, it is easy to understand that so long as these variables remained "hidden" the variability of T-cell cycle times continued to be perplexing.

Furthermore, of even greater intellectual and emotional significance for the biologic sciences, the variability of cell cycle times is not a result of a throw of the dice. T-cell cycle progression is deterministic and absolutely predictable, provided the three parameters responsible for cell division are known. Finally, the element of time, in this system expressed by the duration of the IL-2–receptor interaction, indicates that a critical threshold of intracellular signals are generated that determine the quantal response, namely DNA replication and cell division. Accordingly, the beacon that will guide future investigators searching to probe the events controlling both normal and neoplastic T-cell growth is bright, and easily discernible among the kaleidoscope of possibilities.

REFERENCES

Arya SK, Wong-Staal F, Gallo RC. 1984. T-cell growth factor gene: Lack of expression in human T-cell leukemia-lymphoma virus-infected cells. Science 223:1086–1087.

Baker PE, Gillis S, Smith KA. 1979. Monoclonal cytolytic T-cell lines. J Exp Med 149:273–278.

Barnes D, Sato G. 1980. Serum-free culture: A unifying approach. Cell 22:649–655.

Burns FJ, Tannock IF. 1970. On the existence of a G_0-phase in the cell cycle. Cell Tissue Kinet 3:321–334.

Cantrell DA, Smith KA. 1983. Transient expression of interleukin 2 receptors: Consequences for T-cell growth. J Exp Med 158:1895–1911.

Cantrell DA, Smith KA. 1984. The interleukin 2-T-cell system: A new cell growth model. Science 224:1312–1316.

Gillis S, Smith KA. 1977. Long-term culture of cytotoxic T-lymphocytes. Nature 268:154–156.

Gootenberg JE, Ruscetti FW, Mier JW, Gazdar A, Gallo RC. 1981. Human cutaneous T cell lymphoma and leukemia cell lines produce and respond to T cell growth factor. J Exp Med 154:1403–1418.

Gordon J, MacLean LD. 1965. A lymphocyte-stimulating factor produced in vitro. Nature 208:795–796.

Holbrook NJ, Smith KA, Fornace AJ, Comeau CM, Wiskocil RL, Crabtree GR. 1984. T-cell growth factor: Complete nucleotide sequence and organization of the gene in normal and malignant cells. Proc Natl Acad Sci USA 81:1634–1638.

Kasakura S, Lowenstein L. 1965. A factor stimulating DNA synthesis derived from the medium of leucocyte cultures. Nature 208:794–795.

Koch AL, Schaechter M. 1962. A model for statistics of the cell division process. J Gen Microbiol 29:435–454.

Leonard WJ, Depper JM, Uchiyama T, Smith KA, Waldmann TA, Greene WC. 1982. A monoclonal antibody, anti-Tac, blocks the membrane binding and action of human T-cell growth factor. Nature 300:267–269.

Meuer SC, Hussey RE, Cantrell DA, Hodgdon JC, Schlossman SF, Smith KA, Reinherz EL. 1984. Triggering of the T3-Ti antigen-receptor complex results in clonal T-cell proliferation through an interleukin 2–dependent autocrine pathway. Proc Natl Acad Sci USA 81:1509–1513.

Morgan DA, Ruscetti FW, Gallo R. 1976. Selective in vitro growth of T lymphocytes from normal human bone marrows. Science 193:1007–1008.

Neckers LM, Cossman J. 1983. Transferrin receptor induction in mitogen-stimulated human T lymphocytes is required for DNA synthesis and cell division and is regulated by interleukin 2. Proc Natl Acad Sci USA 80:3493–3498.

Pardee AB, Shilo BZ, Koch AL. 1979. Variability of the cell cycle. *In* Sato GH, Ross R (ed), Hormones and Cell Culture. Cold Spring Harbor Laboratory, Cold Spring Harbor, New York, pp. 373–392.

Robb RJ, Smith KA. 1981. Heterogeneity of human T-cell growth factor due to glycosylation. Mol Immunol 18:1087–1094.
Robb RJ, Munck A, Smith KA. 1981. T-cell growth factor receptors: Quantitation, specificity and biological relevance. J Exp Med 154:1455–1474.
Ruscetti FW, Morgan DA, Gallo RC. 1977. Functional and morphologic characterization of human T cells continuously grown in vitro. J Immunol 119:131–138.
Smith JA, Martin L. 1973. Do cells cycle? Proc Natl Acad Sci USA 70:1263–1267.
Smith KA. 1982. T-cell growth factor-dependent leukemic cell growth: Therapeutic implications. *In* Bloomfield CD (ed), Adult Leukemias. Martinus Nijhoff, The Hague, Netherlands, pp. 43–61.
Smith KA. 1984. Interleukin 2. Ann Rev Immunol 2:319–333.
Smith KA, Favata MF, Oroszlan S. 1983. Production and characterization of monoclonal antibodies to human interleukin-2: Strategy and tactics. J Immunol 131:1808–1815.
Smith KA, Ruscetti FW. 1981. T-cell growth factor and the culture of cloned functional T-cells. Adv Immunol 31:137–175.
Strom TB, Bangs JD. 1982. Human serum-free mixed lymphocyte response: The stereospecific effect of insulin and its potentiation by transferrin. J Immunol 128:1555–1559.
Yodoi J, Takatsuki K, Masuda T. 1974. Two cases of T-cell chronic leukemia in Japan. N Engl J Med 290:572–576.
Wang HM, Smith KA. 1984. Interleukin 2 receptor expression by human leukemia cells (Abstract). Fed Proc 43:1607.

Mediators in Cell Growth and Differentiation,
edited by Richard J. Ford and Abby L. Maizel.
Raven Press, New York © 1985.

Heterogeneity of Macrophage-Activating Factors (MAFs) and Their Effects In Vivo

Peter H. Krammer, Diethard Gemsa,* Ute Hamann,
Brigitte Kaltmann, Claire Kubelka,† and Wolfgang Müller

*Institute for Immunology and Genetics, German Cancer Research Center, Heidelberg, *Department of Molecular Pharmacology, University of Hannover, and †Institute for Immunology and Serology, University of Heidelberg, Federal Republic of Germany*

Antigen or mitogen stimulation of T-cells leads to secretion of soluble mediators called lymphokines. Lymphokines are hormone-like substances that act on cells of the immune and other biological systems and induce growth, differentiation, and activation. One T-cell clone has the capacity to secrete several lymphokines with specificity for cells from various organs. Among the many lymphokines secreted by T-cells are macrophage-activating factors (MAFs), which stimulate several distinct macrophage functions (Gemsa et al. 1983). The heterogeneity of MAFs and initial investigations of their in vivo effects are the subject of this chapter.

HETEROGENEITY OF MAFs SECRETED BY T-CELL CLONES

Lymphokine secretion by T-cells has previously been studied in supernatants from bulk cultures. A more refined analysis of lymphokine secretion and of the secretory cells became possible with the establishment of long-term T-cell clones. Three T-cell clones and the lymphokine activities found in their supernatants are shown in Table 1. Clone 29 and 96 grow in the presence of concanavalin A (ConA)-induced supernatant (CAS) and secrete immune interferon (IFN-γ), MAF, and variable quantities of colony-stimulating factor for granulocytes and macrophages (GM-CSF) following ConA induction. The alloreactive AKR anti-C57B1/6 T-cell clone PK 7.1.2 E8 grows in the presence of CAS and irradiated C57B1/6 stimulator cells and secretes colony-stimulating factors (CSF), MAF, histamine-producing-cell–stimulating factor (HCSF), B-cell growth factors (BCGF), and B-cell differentiation factors for IgM and IgG (BCDF μ, BCDF γ) following T-cell mitogen or alloantigen stimulation (Krammer et al. 1982).

Both types of T-cell clones secrete MAFs. We therefore studied whether MAFs are a heterogeneous group of lymphokines. Supernatants from each clone were incubated with resident peritoneal macrophages, and several parameters of macrophage activation were measured (Gemsa et al. 1983).

TABLE 1. *Lymphokines secreted by the T-cell clones 29, 96 and PK 7.1.2 E8*

T-cell clones	Lymphokines	References
29, 96	IFN-γ, MAF, GM-CSF	Gemsa et al. 1983 Krammer et al. 1982, 1983
PK 7.1.2 E8	GM-CSF, Meg-CSF, Eo-CSF, E-CSF MAF, HCSF, LDCF, BCGF, BCDF μ, BCDF γ	Gemsa et al. 1983 Krammer et al. 1982, 1983

TABLE 2. *Different macrophage activities induced by supernatants from clone 96 and PK 7.1.2 E8*

Supernatants derived from	Macrophage activities		
	No effect	Increased	Decreased
Clone 96	Phagocytosis	Glucosamine incorporation HMPS O_2^- release H_2O_2 release PGE_2 release Schistosomula killing Tumor cytostasis Tumor cytolysis	RNA synthesis Protein synthesis Pinocytosis
Clone PK 7.1.2 E8	PGE_2 release Schistosomula killing Tumor cytolysis	RNA synthesis Protein synthesis Glucosamine incorporation HMPS O_2^- release H_2O_2 release Pinocytosis Phagocytosis Tumor cytostasis	

Table 2 summarizes results obtained from such experiments. Stimulation or depression of macrophage activities showed different patterns with the supernatant from each clone. Incubation with the supernatant of clone 7.1.2 E8 increased RNA, protein, and glycoprotein synthesis, hexosemonophosphate shunt (HMPS) activity, release of oxygen metabolites (O_2^-, H_2O_2), pinocytosis, phagocytosis, and tumor-cytostasis, wnereas no effect on prostaglandin E_2 (PGE_2) release, schistosomula killing, and tumor cytolysis could be observed. In contrast, incubation with supernatant of clone 96 increased glycoprotein synthesis, HMPS activity, release of oxygen metabolites and PGE_2, schistosomula killing, and tumor cytostasis and cytolysis, whereas RNA and protein synthesis and pinocytosis were decreased and phagocytosis remained uneffected. These results suggested that different lymphokines were present in the supernatant of both clones, and each induced distinct macrophage functions.

This assumption was further supported by the following lines of evidence: (1) A rabbit antiserum raised against lymphokine-rich fractions of a clone secreting the same lymphokines as clone 96 inhibited MAFs only in the supernatant of clone 96. MAFs in the supernatant of clone PK 7.1.2 E8 were not affected (Gemsa et al. 1983, Krammer et al. 1983). (2) In supernatants of T-cell hybridomas constructed by fusing clone 96 with the AKR/J T-cell tumor BW 5147, we could separate MAFs further and found indication that induction of macrophage tumor cytotoxicity (MAF-C), tumorcytostasis (MAF-G), and schistosomula killing (MAF-S) were stimulated by separate lymphokines. The data are shown in Table 3. Table 3 further shows that MAFs in the supernatants of the T-cell hybridomas also segregated from IFN-γ activity (Krammer et al. 1983). (3) These data reinforce data from the literature indicating that IFN-γ alone is insufficient to activate the tumor cytotoxicity of resident macrophages (Pace et al. 1983, Schultz and Kleinschmidt 1983). This is in line with our analysis of IFN-γ and MAF-C in the supernatant of ConA-activated T-cell clones grown in limiting dilution microcultures and induced to secrete lymphokines with ConA. Supernatants of microwells containing T-cells with a high probability of being derived from one precurser cell showed a pattern of MAF-C and IFN-γ activity outlined in Table 4. Several supernatants contained IFN-γ and MAF-C activity; other supernatants contained either IFN-γ or MAF-C activity (U. Hamann and P. Krammer, Eur J Immunol, in press).

Taken together, these results lead to two conclusions: (1) T-cell–derived MAFs are a heterogeneous group of lymphokines comprising MAFs that induce tumor cytolysis, tumor cytostasis, schistosomula killing, etc.; (2) IFN-γ alone cannot

TABLE 3. *Segregation of lymphokine activities in T-cell hybridoma supernatants*

Hybridoma number	IFN-γ (U/ml)	MAF-C (% ^{51}Cr release)	MAF-G (% Growth inhibition)	MAF-S (% Dead schistosomula)
15	96	28	90	44
1	0	25	93	48
9	0	16	99	1
21	0	0	63	2
32	0	34	34	2

TABLE 4. *Segregation of IFN-γ and MAF-C in supernatants of T-cell clones grown in limiting dilution microcultures*

Pattern of lymphokine activity		Number of cultures showing pattern of lymphokine activity
IFN-γ	MAF-C	
+	+	14
−	+	21
+	−	3
−	−	56

induce tumor cytolysis and may act together with a second signal. This signal may be provided by lipopolysaccharide (LPS). Alternatively, IFN-γ may act in concert with another lymphokine. The heterogeneous macrophage functions induced by supernatants of different T-cell clones may ensue from the activity of single molecules or from the combined action of several factors. In addition, it cannot be excluded that the same macrophage functions are stimulated by different factors binding to different receptors. Future work will determine whether these conclusions based on biological assay systems can be substantiated by molecular analysis on the protein or gene level.

IN VIVO ACTIVATION OF MACROPHAGES BY LYMPHOKINES

Our data on macrophage activation with T-cell clone–derived lymphokines showed that such lymphokines were active in vitro at high dilution. For example, some batches of clone 29 or 96 supernatant could induce macrophage tumor cytostasis of the Eb T-cell lymphoma in vitro at a dilution of $>1/20{,}000$. The potency of this in vitro effect prompted experiments to test whether T-cell clone–derived lymphokines could also activate macrophages in vivo.

To analyze whether the clone 29 supernatant (containing the same lymphokines as clone 96 supernatant) was capable of activating resident macrophages in vivo, we performed the following experiments. Aliquots of clone 29 supernatant were i.p. injected into DBA/2 mice. After the last injection, macrophages from such mice or from control mice injected with culture medium devoid of MAFs were tested for their capacity to kill P815 tumor cells or schistosomula of *Schistosoma mansoni* in the presence or absence of LPS. In the presence of LPS, only macrophages from mice injected with clone 29 supernatant were effective, showing low but significant tumor cell and substantial schistosomula killing (Table 5). The most likely interpretation of these results is that resident macrophages can be preactivated in vivo to kill tumor cells or schistosomula in vitro in the presence of a second signal. It cannot be excluded, however, that lymphokine injection into the peritoneal cavity resulted in recruitment of an active macrophage population or stimulated an intermediate cell that in turn activated macrophages (C. Kubelka, unpublished data).

Preliminary experiments in our laboratory demonstrated that DBA/2 mice that had received 10^5 Eb T-lymphoma cells i.p. and were treated with T-cell clone 96

TABLE 5. *Preactivation of macrophages in vivo with T-cell clone–derived lymphokines to kill tumor cells and schistosomula in vitro*

Injection in vivo	Addition of LPS in vitro	% Dead schistosomula	Lysis of P815 tumor cells (%)
Medium	+	2	4
Clone 29 supernatant	+	47	17

supernatant daily for a period of 14 days beginning 3 days after tumor inoculation were tumor free for over 2 months.

These findings show that lymphokines are active under in vivo conditions and may be used in therapy of cancer and infectious diseases. Future work in the investigation of macrophage activation, therefore, will focus on the characterization of the signals that induce competent antitumor and antimicrobial macrophages in vivo.

REFERENCES

Gemsa D, Debatin K-M, Kubelka C, Kramer W, Deimann W, Kees U, Krammer PH. 1983. Macrophage activating factors from different T cell clones induce distinct macrophage functions. J Immunol 131: 833–844.

Krammer PH, Dy M, Hültner L, Isakson P, Kees U, Lohmann-Mattes M-L, Marcucci F, Michnay A, Puré E, Schimpl A, Staber F, Vitetta ES, Waller M. 1982. Production of lymphokines by murine T cells grown in limiting dilution and long-term cultures. *In* Fathman G, Fitch F (eds), Isolation, Characterization, and Utilization of T-lymphocyte Clones. Academic Press, New York, pp. 253–262.

Krammer PH, Echtenacher B, Gemsa D, Hamann U, Hültner L, Kaltmann B, Kees U, Kubelka C, Marcucci F. 1983. Immune interferon (IFN-γ), macrophage-activating factors (MAFs), and colony-stimulating factors (CSFS) secreted by T-cell clones in limiting dilution microcultures, long-term cultures, and by T-cell hybridomas. Immunol Rev 76:5–28.

Pace JL, Russel SW, Torres BA, Johnson HM, Gray PW. 1983. Recombinant mouse interferon induces the priming step in macrophage activation for tumor cell killing. J Immunol 130:2011–2013.

Schultz RM, Kleinschmidt WJ. 1983. Functional identity between murine interferon and macrophage activating factor. Nature 305:239–240.

Mediators in Cell Growth and Differentiation,
edited by Richard J. Ford and Abby L. Maizel.
Raven Press, New York © 1985.

Interleukins and Inhibitors in Human Lymphocyte Regulation

Peter C. Nowell, Gary A. Koretzky, John C. Reed, and Anne C. Hannam-Harris

Department of Pathology and Laboratory Medicine, University of Pennsylvania School of Medicine, Philadelphia, Pennsylvania 19104

The biology and biochemistry of interleukin-1 (IL-1) and interleukin-2 (IL-2), as well as other immunologically nonspecific mediators of lymphocyte function, have been reviewed in detail elsewhere in this volume (Lachman 1985, see pages 171–183, Smith 1985, pages 185–192, this volume). Our laboratory has been interested in the functions of a number of these stimulatory factors in the regulation of human lymphocytes, in vitro and, to a lesser extent, in vivo. We have also recently begun to investigate several reagents that inhibit human lymphocyte proliferation, hoping to determine the physiological relevance of such suppressive effects and their specific relationship to the stimulatory action of the interleukins.

In the following sections are summarized briefly some recent findings relevant to these topics.

THE INTERLEUKINS IN HUMAN LYMPHOCYTE STIMULATION

The importance of immunologically nonspecific, stimulatory mediators in the activation of T-lymphocytes has been well documented in the last decade. Since Gery and his colleagues first described IL-1 (then called lymphocyte-activating factor) in 1971, there has been considerable investigation of the role of macrophages and macrophage products in the initiation of lymphocytic responses (Gery and Waksman 1972, Gery and Handschumacher 1974, Mizel et al. 1978, Mizel and Ben-Zvi 1980, Oppenheim et al. 1979, Mizel 1982). Similarly, the demonstration in 1976 by Morgan et al. that a soluble factor in lymphocyte culture supernatants could support the continued growth of T-cells in vitro led to a large number of studies that have clearly established the central role of IL-2 in the proliferation of T-cells from various species (Farrar et al. 1980, Gillis et al. 1978, 1980, Gillis and Smith 1977, Smith et al. 1979, 1980, Watson et al. 1979).

In general, less work has been done with freshly isolated human cells than with established cell lines and material from experimental animals. Although many questions concerning the functions of the interleukins have been answered, a number also remain incompletely resolved.

The Role of Interleukin-1

The key role of IL-1 in triggering the release of IL-2 by lymphocytes has been well documented in several laboratories (Smith et al. 1979, 1980, Gillis and Mizel 1981, Ruscetti and Gallo 1981, Lachman 1984). Less clear is the possible role of IL-1 in inducing the expression of IL-2 receptors on potentially responsive T-cells. Several workers have reported that T-cells respond to IL-2 when stimulated by lectin mitogen alone (Larsson et al. 1980). Others, however, have suggested that mitogen alone is insufficient to make lymphocytes responsive to IL-2 (Hunig et al. 1983). We have used the monoclonal antibody anti-Tac, which appears to identify the IL-2 receptor on human T-cells (Leonard et al. 1982, Robb and Greene 1983), to investigate this question (Koretzky et al. 1983a).

In Table 1 are presented the results of seven experiments in which we compared human peripheral blood mononuclear cells (PBM) with the same cell population after monocyte depletion (MDC). The monocyte-depleted cells (MDC) were routinely <0.5% esterase-positive (Koretzky et al. 1982). These cells were also functionally depleted of monocytes, as evidenced by their poor proliferative response to phytohemagglutinin (PHA) when compared to the PBM controls.

The experiments in Table 1 demonstrate the expression of the IL-2 receptor induced by PHA in PBM and MDC cultures. Consistent with the report of other investigators (Uchiyama et al. 1981), we generally observed that 40–60% of T-cells expressed Tac antigen after PHA stimulation of PBM. Monocyte depletion resulted in markedly fewer T-cells' expressing the IL-2 receptor in each experiment ($p<.01$). When cells were cultured without mitogen, less than 5% of the T-cells were Tac-positive for both MDC and PBM (data not shown).

Table 1 also illustrates the ability of a macrophage culture supernatant (which was shown to contain IL-1 activity in a mouse thymocyte assay) to restore partially the proliferative response of MDC cultures to PHA (Koretzky et al. 1982). Interestingly, although the supernatant used in this series of experiments had a relatively small effect on proliferation, it was able to restore the expression of the IL-2 receptor in response to the lectin mitogen to levels statistically indistinguishable from those of undepleted cells. The supernatant alone did not induce Tac-positivity above background levels in either PBM or MDC cultures (data not shown). These findings indicate that although low levels of proliferation and Tac expression are seen in our MDC cultures after exposure to PHA alone, monocytes or their products are required, in addition to mitogen, for levels approaching those of PHA-stimulated PBM.

We have also found that the phorbol ester 12-0-tetradecanoylphorbol-13-acetate (TPA) can replace monocytes in this culture system, to restore not only proliferation but also Tac-positivity (Koretzky et al. 1982, 1983a). Table 2 illustrates experiments with TPA done with the same monocyte-depleted cells used for the experiments in Table 1.

TABLE 1. *Monocyte-depleted cells respond poorly to PHA by proliferation or the expression of the IL-2 receptor*

Exp.	Proliferation*			% Tac-positivity†		
	PBM + PHA	MDC + PHA	MDC + IL-1 + PHA	PBM + PHA‡	MDC + PHA§	MDC + IL-1 + PHA‖
1	68,737	3,634	13,192	80	14	n.d.
2	61,954	6,474	17,232	55	26	33
3	65,046	6,462	14,581	36	26	56
4	72,821	10,255	16,987	42	17	31
5	34,355	4,364	7,101	40	22	42
6	43,892	1,803	8,892	43	13	29
7	101,780	7,103	n.d.	66	15	n.d.

*[^{3}H]thymidine incorporation (cpm) after 48 hr of culture. Values given are the means of triplicate cultures.

†Tac-positivity = (% Tac-positive cells)/(% OKT11A-positive cells) × 100. The determination was made after 48 hr of culture. Without mitogen, both PBM and MDC were ≤5% Tac-positive at 48 hr.

‡$p < 0.01$ comparing PBM + PHA to MDC + PHA by the Wilcoxon signed rank test.

§$p < 0.03$ comparing MDC + PHA to MDC + IL-1 + PHA by Wilcoxon signed rank test.

‖$p < 0.31$ comparing PBM + PHA to MDC + IL-1 + PHA by Wilcoxon signed rank test.

n.d., not done.

TABLE 2. *TPA replaces monocytes in stimulation of proliferation and in the induction of IL-2 receptor expression in lymphocyte cultures exposed to PHA*

	Proliferation*		% Tac-positivity†	
Exp.	PBM + PHA + TPA	MDC + PHA + TPA	PBM + PHA + TPA	MDC + PHA + TPA
1	97,111	93,067	91	97
2	80,810	99,538	n.d.	92
3	89,422	78,149	91	100
4	110,240	119,516	88	100
5	66,839	52,505	n.d.	99
6	n.d.	104,750	n.d.	100
7	128,079	88,610	93	99

*[^{3}H]thymidine incorporation (cpm) after 48 hr of culture. Values given are the means of triplicate cultures.

†Tac-positivity = (% Tac-positive cells)/(% OKT11A-positive cells) × 100. The determination was made after 48 hr of culture. Without mitogen, both PBM and MDC were ≤5% Tac-positive at 48 hr.

n.d., not done.

The data discussed above are consistent with the view that for optimal expression of the IL-2 receptor by T-lymphocytes at least two signals are normally required: mitogen or antigen plus a monocyte-derived factor (IL-1) or a monocyte-replacing factor such as TPA. It also appears, however, that even with populations of cells rigorously depleted of monocytes, there is some degree of Tac-positivity developed in response to mitogen alone. These results confirm those of other investigators using anti-Tac antibody as a probe for the IL-2 receptor (Lipkowitz et al. 1984, Metah et al. 1984), as well as studies using the proliferative response to IL-2 as an indicator of IL-2 receptor expression (Larsson et al. 1980, Palacios 1982).

The data suggest that there may be subsets of T-cells that differ in their ability to express the IL-2 receptor. It is possible that the cells that express Tac after exposure to mitogen alone may be similar to those lymphocytes described by Cantrell and Smith (1983) that expressed and then lost the IL-2 receptor in vitro. These cells were more easily and rapidly stimulated to reexpress the IL-2 receptor than were Tac-negative lymphocytes freshly isolated from peripheral blood. These workers suggested that such cells might represent one aspect of T-cell memory, and the findings could certainly help to explain why lymphocytes from recently sensitized animals give an accelerated proliferative response when exposed to antigen in culture (Wilson and Nowell 1971). The data presented in this section indicate that one possible feature distinguishing between "sensitized" and "unsensitized" T-cells is the number of signals required for the expression of IL-2 receptors.

The Role of Interleukin-2

The central role of IL-2 in the proliferation of T-lymphocytes in many species has now been clearly demonstrated in several laboratories (Smith et al. 1979, Smith et al. 1980, Gillis and Mizel 1981, Ruscetti and Gallo 1981) and extensively reviewed elsewhere (Smith 1980, Gillis et al. 1982, Gillis 1983), including this volume (Smith 1985, see pages 185–192). A number of aspects remain to be clarified, such as the role of IL-2 in the release of other stimulatory lymphokines including B-cell growth factor (BCGF) and T-cell replacing factor (TRF) (Howard et al. 1983, Maizel et al. 1984, Nakanishi et al. 1983), but the limiting role of IL-2 in T-cell proliferation, at least in vitro, seems clearly established. This role has practical as well as theoretical importance in the laboratory. One problem incurred by investigators using lymphocyte culture techniques is the difficulty of obtaining significant proliferation when cells are cultured at low concentrations. The cumbersome use of autologous or heterologous, nondividing feeder cells has been the usual solution to this problem (Blomgren 1976, Winger and Nowell 1976, Winger et al. 1977a,b, 1978), but adding these cells also makes interpretation of results difficult. We have shown that by adding IL-2 instead of feeder cells to cultures of human PBM, we can obtain the same time course and proliferative response with as few as 1000 cells per microwell as had been previously observed in feeder cultures (Figure 1). We have found this approach useful in studying the cytogenetics and function of neoplastic T-cells in circumstances in which only small numbers of lymphocytes may be available for culture. For instance, we now have preliminary evidence that IL-2 supplementation can replace allogeneic feeder cells in the culture of lymphocytes from early skin lesions of patients with cutaneous T-cell lymphomas.

There are also now limited data indicating that IL-2 may play a significant role in immune responses in vivo. We have attempted to relate our in vitro studies of the interleukins to processes occurring in the body by collaborating with clinical investigators to study macrophages and lymphocytes recovered through bronchoalveolar lavage from healthy human volunteers. In addition to demonstrating spontaneous production of IL-1 by human alveolar macrophages (Koretzky et al. 1983c), we have also shown that T-cells recovered from the lung express IL-2 receptors (Davidson et al. 1983). With the exception of some neoplastic lymphocytes (Uchiyama et al. 1981), the IL-2 receptor is not normally demonstrable on human T-cells without in vitro stimulation. Our data, showing that cells obtained directly from human alveolar spaces express the receptor, suggest that IL-2 may be important in the triggering of human T-cells in vivo.

One of the more intriguing recent observations concerning the central function of IL-2 is that by Cantrell and Smith (1983) referred to above, concerning the loss and reexpression of the IL-2 receptor in vitro. Their data suggest that the continued presence of IL-2 is one of the signals necessary for both maximum expression of the IL-2 receptor and its maintenance, perhaps analogous to the interrelationship between insulin and its receptor (Robb et al. 1981, Cantrell and Smith 1983). If true, this dual role defines even more clearly the critical nature of IL-2 in immune responses both in vitro and in vivo.

There do appear to be some experimental circumstances, such as those we have recently demonstrated with the calcium ionophore A23187 (Koretzky et al. 1983b), in which initiation of lymphocyte proliferation may bypass the IL-1–IL-2 sequence,

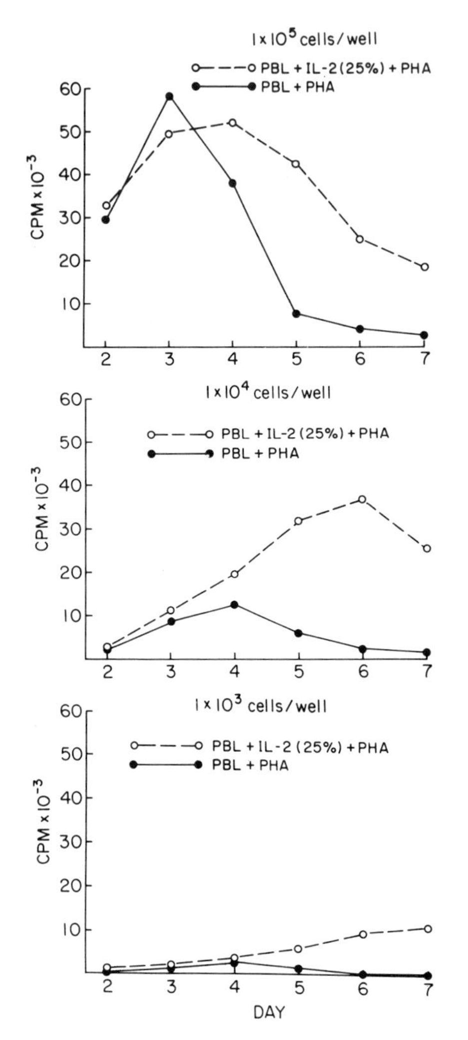

FIG. 1. Exogenous IL-2 greatly augments proliferation of PBM cultured with PHA at 10^4 or 10^3 cells/microwell. Cultures were pulsed with [^{3}H]thymidine for 8 hr and harvested on days 2, 3, 4, 5, 6, and 7. PHA was used at 50 μg/ml.

but it is not yet clear to what extent such interleukin-independent pathways may be physiologically relevant.

The increasing evidence concerning the central regulatory role of IL-2 in lymphocyte activation makes it of considerable interest to determine whether various inhibitory molecules act by interfering with the production or function of IL-2 or through mechanisms that do not involve the interleukins. We have begun several studies related to this question.

INHIBITORY REGULATORS OF HUMAN LYMPHOCYTES

The catalogue of suppressive mediators of T-cell proliferation thus far described is substantial and includes: interferon (Wheelock 1965), proliferation inhibiting factor (PIF) (Badger et al. 1974), inhibitor of DNA synthesis (IDS) (Namba et al. 1977, Jegasothy et al. 1978, Jegasothy and Battles 1981), immunoglobulin-binding factor (IBF) (Gisler and Friedman 1975), soluble immune response suppressor (SIRS) (Rich and Pierce 1974, Pierce et al. 1976), histamine-induced suppressor factor (HSF) (Rocklin 1977), soluble immune suppressor supernatant of T-cell proliferation (SISS-T) (Greene et al. 1981), soluble immune suppressor supernatant of B-cell proliferation (SISS-B) (Fleisher et al. 1981), and prostaglandins (Goodwin and Webb 1980, Tilden and Balch 1982). Unfortunately, few of these factors have been characterized in detail, and so to better understand possible mechanisms by which such suppressors might act, we and others have begun to study better-defined inhibitory agents. Some of our preliminary results, particularly as related to interleukin-mediated events, are summarized below.

Inhibitory Effects of Monoclonal Antibody OKT11A on Human Lymphocytes

Recently the monoclonal antibody OKT11A (with specificity for the E-rosette receptor on human T-cells) (Van Wauwe et al. 1981) has been shown to suppress lectin mitogen stimulation of human lymphocytes (Van Wauwe et al. 1981, Palacios and Martinez-Maza 1982), and it has been suggested that the interaction between OKT11A and its receptor might provide a "negative signal" for lymphocyte activation.

Several workers have also provided evidence that the addition of OKT11A to human lymphocyte cultures results in greatly decreased production of IL-2 (Palacios and Martinez-Maza 1982; M. Kamoun, personal communication). This observation has been confirmed in our laboratory, and recently extended with the demonstration that exogenous IL-2 can overcome the inhibitory effect of the antibody on proliferation. Figure 2 demonstrates one such experiment in which OKT11A depressed to a large degree the response of human PBM to PHA. The proliferative response was completely restored by supplementing the culture with a supernatant having human IL-2 activity.

In addition, we have demonstrated that OKT11A inhibits the expression of the IL-2 receptor on human lymphocytes. Table 3 shows the results of several experiments in which OKT11A reduced the proportion of cells expressing the IL-2 receptor (Tac antigen) after human PBM were stimulated with PHA. As also shown in Table 3, this inhibitory effect of OKT11A on IL-2 receptor acquisition was again largely reversed by the addition of exogenous IL-2.

We also have preliminary data indicating that OKT11A inhibits immunoglobulin production in human lymphocyte cultures stimulated with pokeweed mitogen (PWM) or with antigen (influenza virus, Hong Kong NT60 H3N2). PBM or tonsillar B-cells depleted of T-cells and then supplemented with irradiated T-cells were cultured with mitogen or antigen for 8–10 days in the presence of the antibody, and Ig secreted into the supernatants was quantified by ELISA.

Several representative experiments are illustrated in Figures 3 and 4. Induction of IgM secretion by PWM in peripheral blood cells and in tonsillar B-lymphocytes

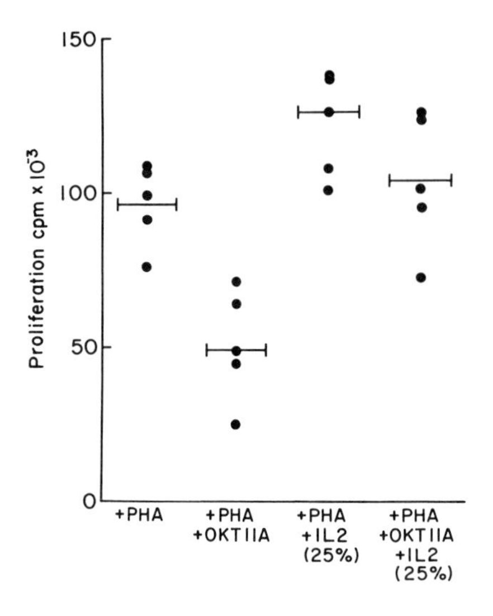

FIG. 2. OKT11A inhibits proliferation of PBM in response to PHA. The response can be restored by addition of exogenous IL-2. Cultures were pulsed with [^{3}H]thymidine for 8 hr. Points shown represent results from day of peak proliferation (days 3–4). The cpm in cultures with PBM alone were uniformly <1,000.

was inhibited by 40–94% (mean 73%) by OKT11A (Figure 3). Specific anti-influenza antibody production in antigen-stimulated cultures was similarly inhibited by >80% (Figure 4). Antibody against another T-cell surface antigen, Leu 1, had no inhibitory effect on Ig production. Attempts to overcome the inhibition of Ig production by addition of IL-2–containing supernatants have thus far produced variable results.

The observations that OKT11A does not bind to B-cells and that Ig synthesis by EBV-infected B-cells is not inhibited by the antibody (Palacios and Martinez-Maza 1982) make it likely that the suppression of Ig production by OKT11A is secondary to inhibition of T-cell–derived factors necessary for B-cell proliferation and differentiation, although a direct effect on B-cells has not yet been ruled out formally.

TABLE 3. *OKT11A antibody inhibits acquisition of IL-2 receptors**

Exp.		PBM	PHA	PHA + OKT11	PHA + IL-2 (25%)	PHA + OKT11 + IL-2 (25%)
1	cpm†	1,268	107,169	71,052	137,322	125,595
	% Tac +‡	0	40	22	43	38
2	cpm	797	79,322	40,870	—	—
	% Tac +	1	42	24		
3	cpm	1,454	112,711	27,015	—	—
	% Tac +	0	53	19		
4	cpm	1,003	56,765	28,587	99,811	89,693
	% Tac +	4	39	18	63	45

*PBM were cultured with PHA-P (50 μg/ml), OKT11A (150 ng/ml), IL-2–containing supernatant (25% vol/vol), or various combinations of these reagents as shown.

†cpm = [^{3}H]thymidine incorporation after 48 hr of culture. Values given are means of triplicate cultures.

‡Tac-positivity = (% Tac-positive cells)/(% OKT11A-positive cells [T-cells]) × 100. The determination was made after 24 to 36 hr of culture.

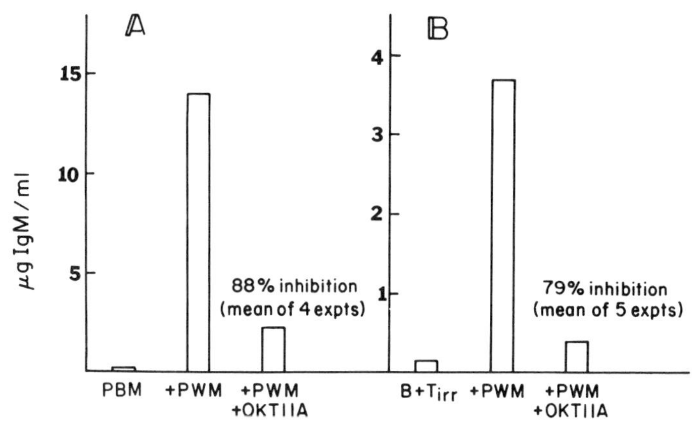

FIG. 3. OKT11A inhibits PWM-stimulated immunoglobulin production in **(A)** peripneral blood mononuclear cells and **(B)** tonsil T-depleted lymphocytes with irradiated T-cells (T_{irr}).

Taken together, these findings with human lymphocytes suggest that the inhibitory action of OKT11A is mediated through its effect on both IL-2 production and the acquisition of IL-2 receptors by human T-cells. The fact that the addition of IL-2 to cultures containing OKT11A completely reverses the suppression indicates that the T-cells are capable of responding to IL-2 in the presence of OKT11A if sufficient quantities are present. It is not yet clear how the addition of exogenous IL-2 leads to restoration of IL-2 receptor expression, but if, as discussed above, a critical concentration of IL-2 is important for maximal IL-2 receptor acquisition, it may be that the only direct action of OKT11A antibody is to interfere with IL-2 production, with resulting inhibition of IL-2 receptor expression.

This apparent relationship between the inhibitory action of OKT11A and its effect on IL-2 production further emphasizes the essential role of this lymphokine

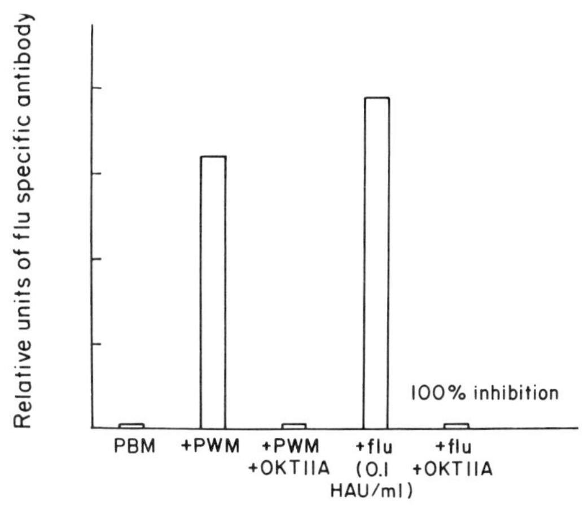

FIG. 4. OKT11A inhibits production of specific anti-influenza antibody by PBM stimulated with PWM or influenza antigen.

in T-cell activation and suggests that at least some endogenous soluble inhibitors may act via the interleukin pathway. It has recently been demonstrated that prostaglandin E_2 (PGE_2), a monocyte product, may inhibit PHA-induced T-cell proliferative responses by suppressing the production of IL-2 (Tilden and Balch 1982, Chouaib and Fradelizi 1983). Other suppressive factors, however, may act through different mechanisms (see below).

Inhibitory Effects of Wheat Germ Agglutinin (WGA) on Human Lymphocytes

WGA is a tetravalent lectin that, at appropriate concentrations, suppresses human T-lymphocyte proliferation and inhibits many of the molecular processes triggered by lectin mitogens such as PHA (Greene et al. 1976, Parker et al. 1976, Karsenti et al. 1975). Interestingly, it also shares many characteristics with SISS-T described by Greene and Waldmann (Greene and Waldmann 1980, Greene et al. 1981). Both WGA and SISS-T have a saccharide specificity for N-acetyl-D-glucosamine (NAG), and inclusion of NAG in lymphocyte cultures abrogates suppression by either WGA or SISS-T, presumably by preventing binding. We have extended previous studies by others on the mechanism of WGA-induced suppression of PHA-stimulated human lymphocyte cultures, with particular emphasis on its relationship to the IL-1–IL-2 pathway.

Our findings to date (Reed et al. 1984) have indicated that WGA does not reduce IL-2 activity in mitogen-stimulated cultures, and in most instances the levels of IL-2 are significantly increased over PHA-stimulated cultures not containing WGA. This observation suggested a failure by the WGA-suppressed lymphocytes to utilize IL-2. This possibility was further supported by the finding that addition of exogenous IL-2 did not overcome the inhibition.

We determined that WGA does bind to detergent-solubilized IL-2 receptors from PHA-activated human T-cells, but it does not bind human IL-2. We then employed radiolabeled IL-2 to investigate the effect of WGA on the IL-2 receptor as expressed on intact cells. In these studies, PBM cultured for 48 hr in the presence of PHA and WGA bound markedly less radiolabeled IL-2 than cells stimulated with PHA alone. This decrease in radiolabeled IL-2 binding in WGA-suppressed cultures was not attributable to blocking by WGA of the binding of IL-2 to its cellular receptor. These observations suggested that WGA prevents expression of receptors for IL-2. We also found that WGA failed to suppress proliferation in several established neoplastic lymphoid cell lines (JURKAT, CEM, MOLT-4), although it did have a suppressive effect on several IL-2–dependent T-cell lines (Reed et al. 1984).

While a number of these findings suggest that WGA may inhibit human lymphocyte cultures by binding to IL-2 receptors and impairing their expression on activated human T-cells, we cannot yet rule out the possibility that its inhibitory action is mediated indirectly through suppressor cells or their soluble products. In this regard, WGA has a small direct stimulatory effect on PBM in which an OKT8-positive subpopulation has been implicated (Boldt and Dorsey 1983). In any event,

WGA appears to act in a different fashion from OKT11A or PGE_2, as described above.

Another soluble inhibitor we have studied is the T-cell product IDS (Jegasothy and Battles 1981). We have preliminary data indicating that human IDS inhibits mitogen-induced lymphocyte proliferation without impairing the elaboration of IL-2 or the expression of either IL-2 receptors or transferrin receptors (J. C. Reed, P. C. Nowell, and B. V. Jegasothy, unpublished data). Thus, the inhibitory action of IDS may be mediated through a mechanism entirely independent of IL-2. It would appear that further studies of the sort that we have initiated with WGA, OKT11A, and IDS may shed light on immunologically nonspecific inhibitory pathways that are physiologically relevant.

SUMMARY AND CONCLUSIONS

It seems clear, from studies in many laboratories, that both IL-1 and IL-2 play an important role in lymphocyte activation and are thus significant in the initiation and amplification of an optimal immune response. These conclusions have been extended to human lymphocytes, and preliminary data with macrophages and T-cells from alveolar spaces suggest that the interleukins are also important in vivo. A number of details remain to be worked out concerning these complex cellular and humoral interactions involved in triggering the immune response, and the increasing availability of purified reagents should help to speed the clarification of remaining issues.

To date, much less work has been done with inhibitory factors produced by macrophages and lymphocytes than with the stimulatory interleukins. The preliminary studies by our laboratory and by others suggest that some suppressive agents may act through interference with the IL-1–IL-2 pathway and that others may be interleukin independent. Such investigations are only beginning, however, and much more investigation is needed to dissect in detail the mechanisms by which immunologically nonspecific inhibitory factors may contribute to the regulation of early steps in the immune response.

REFERENCES

Badger AM, Cooperband SR, Greene JA. 1974. Studies on the mechanism of action of proliferation inhibitory factor (PIF). Cell Immunol 13:355.

Blomgren H. 1976. Modification of the response of human T lymphocytes to phytomitogens by cocultivation with unresponsive non-T leukocytes. Scand J Immunol 5:467.

Boldt DH, Dorsey SA. 1983. Interactions of lectins and monoclonal antibodies with human mononuclear cells. J Immunol 130:1646.

Cantrell DA, Smith KA. 1983. Transient expression of interleukin-2 receptors. J Exp Med 158:1895–1911.

Chouaib S, Fradelizi D. 1983. The mechanism of inhibition of human IL2 production. J Immunol 129:2463.

Davidson B, Rossman M, Greene W, Koretzky G, Nowell P, Daniele R. 1983. Activated lung lymphocytes in normal subjects and in sarcoidosis: Identification of IA and TAC antigens on lung T cells (Abstract). Am Rev Resp Dis 127:62.

Farrar J, Mizel S, Fuller-Farrar J, Farrar W, Milfiker M. 1980. Macrophage-independent activation of helper T cells. I. Production of interleukin 2. J Immunol 125:793.

Fleisher TA, Greene WC, Blaese RM, Waldmann TA. 1981. Soluble suppressor supernatants elaborated by concanavalin A-activated human mononuclear cells. II. Characterization of a soluble suppressor of B cell immunoglobulin production. J Immunol 126:192.

Gery I, Handschumacher RE. 1974. Potentiation of the T lymphocyte response to mitogen. III. Properties of the mediator(s) from adherent cells. Cell Immunol 11:162.

Gery I, Waksman BH. 1972. Potentiation of the T lymphocyte response to mitogen. II. The cellular source of potentiating mediator(s). J Exp Med 136:143.

Gery I, Gershon RK, Waksman BH. 1971. Potentiation of cultured mouse thymocyte responses by factors released by peripheral leukocytes. J Immunol 107:1778.

Gillis S. 1983. Interleukin 2: Biology and biochemistry. J Clin Immunol 3:1.

Gillis S, Mizel SB. 1981. T cell lymphoma model for the analysis of interleukin-1 mediated T cell activation. Proc Natl Acad Sci USA 78:1133.

Gillis S, Smith KA. 1977. Long-term culture of tumor-specific cytotoxic T cells. Nature 268:254.

Gillis S, Fern MM, Ou W, Smith KA. 1978. T cell growth factor: Parameters of production and a quantitative microassay for activity. J Immunol 120:2027.

Gillis S, Shield M, Watson J. 1980. Biochemical and biological characterization of lymphocyte replacing molecules. III. Isolation and phenotypic characterization of interleukin-2 producing lymphomas. J Immunol 125:250.

Gillis S, Mochizuki DY, Conlon PJ, Hefeneider SH, Ranthun CA, Gillis AE, Frank MB, Henney CS, Watson JD. 1982. Molecular characterization of interleukin-2. Immunol Rev 63:167.

Gisler RH, Friedman WH. 1975. Suppression of in vitro antibody synthesis by immunoglobulin binding factor. J Exp Med 142:507.

Goodwin JS, Webb DR. 1980. Regulation of the immune response by prostaglandins. Clin Immunol Immunopathol 15:106.

Greene WC, Waldmann TA. 1980. Inhibition of human lymphocyte proliferation by the non-mitogenic lectin wheat germ agglutinin. J Immunol 124:1979.

Greene WC, Parker CM, Parker CW. 1976. Opposing effects of mitogenic and non-mitogenic lectins of lymphocyte activation: Evidence that wheat germ agglutinin produces a negative signal. J Biol Chem 251:4017.

Greene WC, Fleisher TA, Waldmann TA. 1981. Soluble suppressor supernatants elaborated by concanavalin A-activated human mononuclear cells. I. Characterization of a soluble suppressor of T cell proliferation. J Immunol 126:1185.

Howard M, Matis L, Malek TR, Shevach E, Kell W, Cohen D, Nakanishi K, Paul WE. 1983. Interleukin 2 induces antigen-reactive T cell lines to secrete BCGF-1. J Exp Med 158:2024.

Hünig T, Loos M, Schimpl A. 1983. The role of accessory cells in polyclonal T cell activation. I. Both activation of IL2 production and of IL2 responsiveness by Con A are accessory cell dependent. Eur J Immunol 13:1.

Jegasothy BV, Battles DR. 1981. Immunosuppressive lymphocyte factors. III. Complete purification and partial characterization of human inhibitor of DNA synthesis. Molec Immunol 18:395.

Jegasothy BV, Namba Y, Waksman BH. 1978. Regulatory substances produced by lymphocytes. VII. IDS (inhibitor of DNA synthesis) inhibits stimulated lymphocyte proliferation by activation of membrane adenylate cyclase at a restriction point in late G1. Immunochem 15:551.

Karsenti E, Bornens M, Avrameas S. 1975. Stimulation and inhibition of DNA synthesis in rat thymocytes: Action of concanavalin A and wheat germ agglutinin. Eur J Immunol 5:73.

Koretzky GA, Daniele RP, Nowell PC. 1982. A phorbol ester (TPA) can replace macrophages in human lymphocyte cultures stimulated with a mitogen but not with an antigen. J Immunol 128:1776.

Koretzky GA, Nowell PC, Greene WC. 1983a. The role of monocytes or monocyte replacing factors in the expression of the IL2 receptor in human T lymphocytes (Abstract). Fed Proc 42:452.

Koretzky GA, Daniele RP, Greene WC, Nowell PC. 1983b. Evidence for an interleukin-independent pathway for human lymphocyte activation. Proc Natl Acad Sci USA 80:3444.

Koretzky GA, Elias JA, Kay SL, Rossman MD, Nowell PC, Daniele RP. 1983c. Spontaneous production of interleukin-1 by human alveolar macrophages. Clin Immunol Immunopathol 29:443.

Lachman LB. 1984. Human interleukin 1: Central mediator of macrophage function (Abstract). Lymphokine Res 3:68.

Larsson EL, Iscove NN, Coutinho A. 1980. Two distinct factors are required for induction of T cell growth. Nature 283:664.

Leonard W, Depper J, Uchiyama T, Smith K, Waldmann T, Greene W. 1982. A monoclonal antibody, anti-Tac, blocks the action and binding of human T cell growth factor. Nature 300:267.

Lipkowitz S, Greene WC, Rubin AL, Novogrodsky A, Stenzel KH. 1984. Expression of receptors for interleukin 2: Role in the commitment of T lymphocytes to proliferate. J Immunol 132:31.
Maizel A, Morgan J, Mehta S, Kouttab N, Sahasrabuddhe CG. 1984. Human B cell growth factor: Biological and biochemical characterization (Abstract). Lymphokine Res 3:68.
Mehta S, Koretzky G, Nowell P, Sandler R, Maizel A. 1984. Kinetics of human T cell interleukin 2 receptor acquisition (Abstract). Lymphokine Res 3:94.
Mizel SB. 1982. Interleukin-1 and T cell activation. Immunol Rev 63:51.
Mizel SB, Ben-Zvi A. 1980. Studies on the role of lymphocyte activating factor (interleukin 1) in antigen-induced lymph node lymphocyte proliferation. Cell Immunol 54:382.
Mizel SB, Oppenheim JJ, Rosenstreich DL. 1978. Characterization of lymphocyte activating factor (LAF) produced by the macrophage cell line P338D. I. Enhancement of LAF production by activated T lymphocytes. J Immunol 120:1497.
Morgan DA, Ruscetti FW, Gallo R. 1976. Selective in vitro growth of T lymphocytes from normal human bone marrow. Science 193:1007.
Nakanishi K, Howard M, Muraguchi A, Farrar J, Takatsu K, Hamaoka T, Paul WE. 1983. Soluble factors involved in B cell differentiation: Identification of two distinct T cell replacing factors (TRF). J Immunol 130:2219.
Namba Y, Jegasothy BV, Waksman BH. 1977. Regulatory substances produced by lymphocytes. V. Production of an inhibitor of DNA synthesis (IDS) by proliferating T lymphocytes. J Immunol 118:379.
Oppenheim JJ, Mizel SB, Meltzer MS. 1979. Biological effects of lymphocyte and macrophage-derived mitogenic "amplification" factors. *In* Cohen S, Pick E, Oppenheim JJ (eds), Biology of the Lymphokines. Academic Press, New York, p. 291.
Palacios R. 1982. Mechanism of T cell activation: Role and functional relationship of HLA DR antigens and interleukins. Immunol Rev 63:73.
Palacios R, Martinez-Maza O. 1982. Is the E receptor on human T lymphocytes a "negative signal receptor"? J Immunol 129:2479.
Parker CW, Dankner RE, Falkenheim SF, Greene WC. 1976. Suggestive evidence for both stimulatory and inhibitory domains on human lymphocytes, as indicated by phospholipid turnover studies with wheat germ agglutinin and other lectins. Immunol Commun 5:134.
Pierce CW, Tadakuma T, Kuhner AL. 1976. Characterization of a soluble immune response suppressor (SIRS) produced by concanavalin A-activated spleen cells. *In* Oppenheim JJ, Rosenstreich DL (eds), Mitogens in Immunobiology. Academic Press, New York.
Reed, JC, Robb RJ, Greene WC, Nowell PC. 1984. Effect of wheat germ agglutinin on the interleukin pathway of human T lymphocyte activation (Abstract). Fed Proc 43:1504.
Rich RR, Pierce CW. 1974. Biological expression of lymphocyte activation. III. Suppression of plaque forming cell responses in vitro by supernatant fluids from concanavalin A-activated spleen cell cultures. J Immunol 112:1630.
Robb RJ, Greene WC. 1983. Direct demonstration of the identity of T-cell growth factor binding protein and the Tac antigen. J Exp Med 158:1332.
Robb RJ, Munck A, Smith KA. 1981. T cell growth factor receptors. J Exp Med 154:1445.
Rocklin RE. 1977. Histamine-induced suppressor factor (HSF): Effect on migration inhibitory factor (MIF). J Immunol 118:1734.
Ruscetti FW, Gallo RC. 1981. Human T lymphocyte growth factor: Regulation of growth and function of T lymphocytes. Blood 57:379.
Smith KA. 1980. T cell growth factor. Immunol Rev 51:338.
Smith KA. 1984. The interleukin-2 hormone-receptor system (Abstract). Lymphokine Res 3:75.
Smith KA, Gillis S, Baker P, McKenzie D, Ruscetti FW. 1979. T-cell growth factor mediated T-cell proliferation. Ann NY Acad Sci 332:423.
Smith KA, Lachman LB, Oppenheim JJ, Favata MF. 1980. The functional relationship of the interleukins. J Exp Med 151:1551.
Tilden AB, Balch CM. 1982. A comparison of PGE_2 effects on human suppressor cell function and on interleukin-2 function. J Immunol 129:2469.
Uchiyama T, Nelson D, Fleisher T, Waldmann T. 1981. A monoclonal antibody (anti-Tac) reactive with activated and functionally mature human T cells. II. Expression of Tac antigen on activated cytotoxic killer T cells, suppressor cells, and on one of two types of helper T cells. J Immunol 126:1398.
Van Wauwe J, Goossens J, Decouk W, Kung P, Goldstein G. 1981. Suppression of human T cell mitogenesis and E-rosette formation by the monoclonal antibody OKT11A. Immunology 44:865.

Watson J, Gillis S, Marbrook J, Mochizuki D, Smith KA. 1979. Biochemical and biological characteristics of lymphocyte regulatory molecules. I. Purification of a class of murine lymphokines. J Exp Med 150:849.

Wheelock EF. 1965. Interferon-like virus inhibitor induced in human leukocytes by phytohemagglutinin. Science 149:310.

Wilson DB, Nowell PC. 1971. Quantitative studies on the mixed lymphocyte interaction in rats. V. Tempo and specificity of the proliferative response and the number of reactive cells from immunized donors. J Exp Med 133:442.

Winger LA, Nowell PC. 1976. Suppressor activity in PHA cultures of human peripheral blood lymphocytes (Abstract). Fed Proc 35:435.

Winger LA, Nowell PC, Daniele RP. 1977a. Sequential proliferation induced in human peripheral blood lymphocytes by mitogen. I. Growth of one thousand lymphocytes in feeder layer cultures. J Immunol 118:1763.

Winger LA, Nowell PC, Daniele RP. 1977b. Sequential proliferation induced in human peripheral blood lymphocytes by mitogen. II. Suppression by PHA-activated cells. J Immunol 118:1768.

Winger LA, Nowell PC, Daniele RP. 1978. Sequential proliferation induced in human peripheral blood lymphocytes by mitogen. III. Possible mechanisms of suppression by PHA-activated cells. J Immunol 121:2422.

Growth Factors in Neoplasia

Mediators in Cell Growth and Differentiation,
edited by Richard J. Ford and Abby L. Maizel.
Raven Press, New York © 1985.

Endocrine and Autocrine Estromedins for Mammary and Pituitary Tumor Cells

David A. Sirbasku, Tatsuhiko Ikeda,* and David Danielpour

*The Department of Biochemistry and Molecular Biology, University of Texas Medical School at Houston, Houston, Texas, 77030, and *The Faculty of Nutrition, Kobe-Gakuin University, Igawadani-cho Arise, Nishi-ku, Kobe, Japan 673*

As is evident from the topic of this symposium, the role of growth factors in malignant cell growth has become an area of intense research interest. Generally, growth factors have been characterized as mitogenic agents whose actions are mediated by specific, high-affinity cell surface (and possibly internal) receptors and whose actions are demonstrable at nanomolar concentrations or less, which precludes a simple cell nutrient role. Of the polypeptide growth factors currently being studied, several types have been distinguished. One of the first characterized was epidermal growth factor (EGF), isolated by Savage and Cohen (1972) in high milligram amounts from mouse submaxillary gland. This 6,045-Da polypeptide was shown to promote growth of mesenchymal and epithelial origin cells (see review by Carpenter and Cohen 1979) and to act via binding to a 170,000-Da cell surface receptor (Cohen et al. 1982). More recently, another related group of growth factors, the alpha transforming growth factors (α-TGF), have been isolated in low microgram amounts from large volumes of conditioned medium of cells in culture (Roberts et al. 1980) and shown to interact with the EGF receptor (Todaro et al. 1980). These polypeptides were approximately the same size as EGF, but were different structurally (Marquardt et al. 1983). Using the same type of assay (i.e., stimulation of normal rat kidney (NRK) cell colony formation in soft agar) another higher molecular weight transforming activity has been purified in low microgram quantities from kilogram amounts of normal kidney. This beta-transforming growth factor (β-TGF) has an approximate molecular weight of 25,000 (Roberts et al. 1983).

Growth factors of another type have been identified and classified based on their insulinlike action on target cells. Human insulinlike growth factor I (IGF-I) and insulinlike growth factor II (IGF-II) have both been isolated in low milligram amounts from several hundred pounds of Cohn's plasma fractions. The amino acid sequence of IGF-I has been determined (Rinderknecht and Humbel 1978a) and shown to be equivalent to somatomedin C (Klapper et al. 1983). The sequence of human IGF-II has been determined (Rinderknecht and Humbel 1978b) and shown to be nearly equivalent to that of a multiplication-stimulating activity (MSA) isolated from the conditioned medium of rat liver cells in culture (Marquardt et al. 1981). From the data available, it appears that these growth factors bind to different receptors and that insulin (at high concentrations) binds to the IGF-I receptor, but not to the IGF-II receptor (King et al. 1980, Kasuga et al. 1981, Massague and Czech 1982).

One other very important mitogenic agent under study is platelet-derived growth factor (PDGF). This factor has been purified in microgram quantities from several hundred units of outdated human platelets (Antoniades et al. 1979) and shown to have a molecular weight of 29,000 to 31,000. On reducing agent treatment, this factor loses activity. Polyacrylamide gel analysis after reduction shows it to be composed of two dissimilar lower molecular weight subunits (Raines and Ross 1982). The amino acid sequence of PDGF has been determined and shown to resemble the expected product of a common viral oncogene (Deuel et al. 1981, Doolittle et al. 1983, Stiles 1983, Waterfield et al. 1983). This observation has led to the now-prevalent concept that unregulated production of growth factors is one of the primary lesions that occurs on a cell's pathway to malignancy.

In this report, we review our studies of the role of estrogen-inducible growth factors in estrogen-responsive growth of mammary and pituitary tumor cells. In 1978, we originally proposed that estrogens may induce some target tissues to synthesize or secrete polypeptide growth factors that enter the circulation and promote the growth of distant target tissue tumors (Sirbasku 1978a). Potential source tissues for such factors were identified as uterus, kidney, and pituitary (Sirbasku 1978a, Sirbasku et al. 1982) for mammary cells, and uterus, kidney, and hypothalamus for pituitary cells (Sirbasku 1978a, Sirbasku and Moo 1982). We tentatively designated these growth factors estromedins to indicate their role as mediators of the estrogenic growth effects in vivo.

With the report by Sporn and Todaro (1980) suggesting that, in addition to endocrine effects, paracrine and autocrine tumor-promoting growth factors also were important in malignant cell growth, we examined MTW9/PL rat mammary tumors for estrogen-inducible autostimulatory activity. A steroid hormone–related autostimulatory activity was found in extracts of these tumors growing in rats (Sirbasku 1981). Since that time, we have reported that both the MTW9/PL (Danielpour and Sirbasku 1983a) and the human MCF-7 (Danielpour and Sirbasku 1983b) mammary tumor cells have low molecular weight autostimulatory growth factors inducible by estrogen and associated with the cells in serum-free culture. These mitogens have also been tentatively named estromedins since they may equally serve as mediators of the estrogenic response.

We will now describe some of the properties of the purified mammary/pituitary tumor cell estromedins from uterus, kidney, and pituitary, and describe briefly the data supporting the possibility of autostimulatory (autocrine) estromedins in rat and human mammary cells in culture.

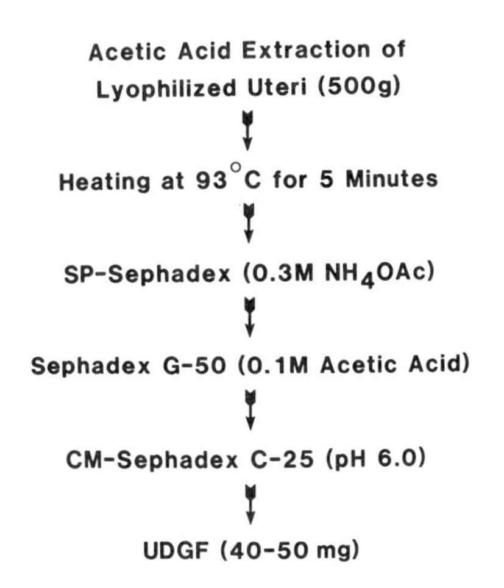

FIG. 1. A schematic representation of the method of purification of sheep UDGF as described in detail in Ikeda and Sirbasku (1984).

PROPERTIES OF ENDOCRINE ESTROMEDINS

Uterine-Derived Growth Factors From Sheep

As cited in the introduction, of the many polypeptide growth factors now isolated, only EGF is available in quantities that allow relatively unlimited characterization of many of the aspects of physical-chemical properties, the mechanism of action, the isolation of receptor, and the study of growth factor action in vivo. For this range of studies, high milligram amounts of mitogen must be available.

Our first attempts to isolate large amounts of the uterine-derived growth factor (UDGF) were done with acetone powders of pregnant sheep uteri (Iio and Sirbasku 1982). At near neutral pH, we partially purified a 25,000-Da activity in microgram quantities from 100-g amounts of acetone powder. The state of homogeneity was not estimated directly (Iio and Sirbasku 1982), but from our recent data appeared to be $<3\%$ UDGF (Ikeda and Sirbasku 1984). Overall yields from neutral methods were $<1\%$ after six fractionation steps. These data suggested that kilogram amounts of acetone powder would have been required to prepare less than a milligram of purified (homogeneous) sheep UDGF.

Since many of the growth factors purified to date were treated early in their isolation with heat or acid dissociating conditions, or both, the combination was attempted with the sheep UDGF. To summarize our results presented in greater detail elsewhere (Ikeda and Sirbasku 1984), we have purified 40 to 50 mg of UDGF from 500 g of lyophilized powder of early pregnant sheep uteri. The five-step isolation method is summarized in Figure 1. These steps were acetic acid extraction, heat treatment at 95°C, SP Sephadex cation exchange chromatography, Sephadex G-50 molecular sieve chromatography, and finally CM Sephadex C-25 cation exchange chromatography at pH 6.0. The overall activity yield was 33%.

The purification of the UDGF activity was monitored by an assay that measured the stimulation of incorporation of tritium-labeled thymidine into MTW9/PL rat mammary tumor cell DNA under serum-free conditions, and in the absence of

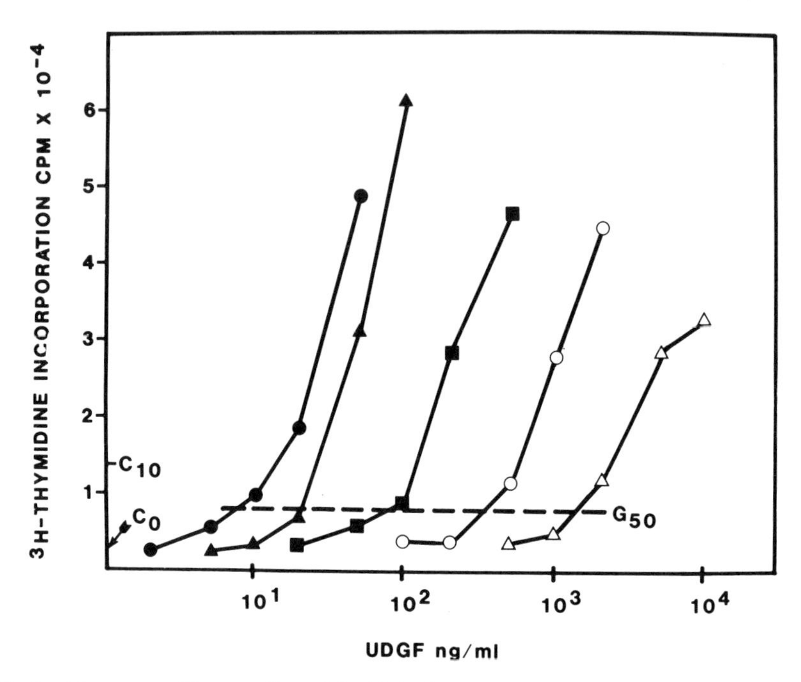

FIG. 2. The assay of the specific activity (G_{50}) of UDGF after the various stages of preparation outlined in Figure 1 is shown. Growth was measured in response to serum-free Dulbecco's modified Eagle's medium (DME) (C_0), 10% (vol/vol) fetal calf serum (C_{10}), the initial acetic acid extract (open triangles), the heated supernatant (open circles), the pooled active fractions from SP-Sephadex (closed squares), the pooled active fractions from Sephadex G-50 (closed triangles), and finally the activity from the CM-Sephadex chromatography (closed circles). The assay used was described previously in detail (Ikeda and Sirbasku 1984), and was based on measuring the growth factor stimulation of tritium-labeled thymidine into MTW9/PL cell DNA in serum-free culture.

other growth and attachment factors that are known to be required for mammary tumor cell growth in serum-free culture (Barnes and Sato 1979). (See Ikeda and Sirbasku (1984) for complete details of this assay method.) As shown in Figure 2, the concentration of protein required to replace half the MTW9/PL cell response to 10% (vol/vol) fetal calf serum decreased with each step of purification. This value, designated the G_{50}, was the measure of specific activity. The G_{50} of the UDGF from the final CM Sephadex step was 8 ng/ml or 1.9×10^{-9}M based on a molecular weight of 4200 calculated from the urea-sodium dodecyl sulfate (SDS) 12.5% polyacrylamide gel analysis shown in Figure 3. As can be seen from these data, the final preparation shows only one Coomassie blue stained band; this is the case even when considerable amounts (i.e., 100 μg) of protein were applied to gels. Analysis by non-SDS polyacrylamide gels at acid and basic pH, hydrophobic chromatography on octyl Sepharose, and high-performance liquid chromatography (HPLC) on reverse phase (C_8 and C_{18}) columns, and a molecular sieve TSK-125 column all confirmed >95% homogeneity (Ikeda and Sirbasku 1984). The pI of purified sheep UDGF was estimated by standard methods to be 7.3 (Figure 4).

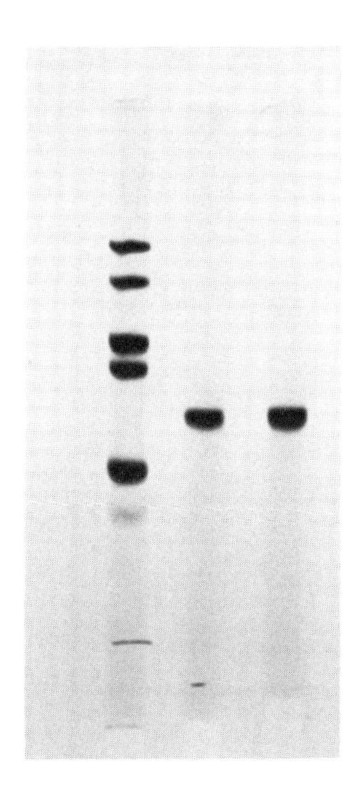

FIG. 3. The 8 M urea, 0.1% SDS polyacrylamide gel electrophoresis of the final UDGF preparation was done as described in Ikeda and Sirbasku (1984). The gel at the left received a mixture of myoglobin (M_r = 16,947), and sequenced cyanogen bromide-generated fragments of myoglobin of masses (top to bottom) 14,404, 8,159, 6,214, and 2,512 daltons. The middle gel shows the migration position of 25 μg of UDGF, and the right gel the position of migration of 50 μg of the same sample. The gels were stained with Coomassie Blue.

Several types of cell specificity studies have been done with this factor including a demonstration that UDGF promotes MTW9/PL growth as measured by cell number increase and that this factor promotes GH3/C14 rat pituitary tumor cell growth in serum-free medium. This activity does not promote growth of normal diploid fibroblasts and is a relatively weak mitogen for 3T3 mouse fibroblasts (Ikeda and Sirbasku 1984).

In addition to cell type specificity studies, we have shown (Ikeda and Sirbasku 1984) that the other known growth factors, EGF, PDGF, IGF-I, IGF-II, MSA, and and extensive list of steroid and polypeptide hormones do not substitute for the mitogenic action of sheep UDGF on MTW9/PL rat mammary tumor cells. However, samples of the TGFs were not available for direct testing, and purification simply for comparison purposes was impractical. To determine whether the sheep UDGF was equivalent to any of the other characterized TGFs, we attempted reduction of the UDGF activity using 10 mM dithiothreitol or 5% (vol/vol) 2-mercaptoethanol. Under any conditions used, UDGF mobility in urea-SDS polyacrylamide gels did not change (i.e., no evidence for subunits covalently attached by intermolecular disulfide bonds), and no change in UDGF specific activity was found (Ikeda and Sirbasku 1984). Similar treatments of the TGFs result in changes in electrophoretic mobility and loss of activity (Roberts et al. 1983, Frolik et al. 1983).

From the data available, we conclude that sheep UDGF is a new growth factor. Final evidence to support this conclusion is being sought by determination of the complete amino acid sequence.

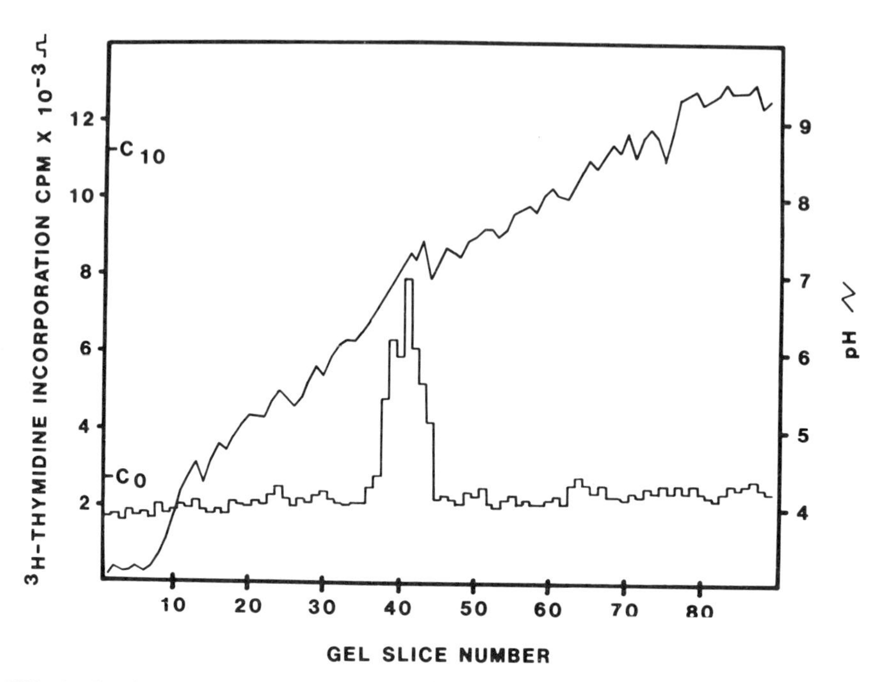

FIG. 4. Isoelectric focusing of the purified UDGF in 4.5% polyacrylamide gels exactly as described in Ikeda and Sirbasku (1984). The pH gradient ranged between 3 and 9. Three identical gels were run, with one being stained with Coomassie Blue to localize the protein, and the other two being sliced into 1-mm thick discs for measurement of pH and elution of growth factor activity for assay by the tritium-labeled thymidine incorporation method. The stained protein band and the eluted UDGF activity were coincident.

Kidney-Derived Growth Factors From Sheep

Purification of a mammary/pituitary/kidney tumor cell growth factor has been accomplished from lyophilized powder of mature ewe kidney. The isolation methods employed are outlined in Figure 5. Throughout the purification, the activity was monitored by following the stimulation of incorporation of tritium-labeled thymidine into MTW9/PL rat mammary cell DNA under serum-free conditions as described before (Ikeda and Sirbasku 1984). Beginning with 500 g of powder, the kidney-derived growth factor (KDGF) was extracted with acetic acid, and the active supernatant treated at 93°C to remove major impurities; the activity was further purified by successive applications of BioRad AG5OW ×8 ion exchange chromatography, two DEAE Sepharose CL-6B separations at pH 5.8 and 6.2, and finally Sephadex G-50 chromatography in acetic acid. Between 8 and 14 mg of KDGF are obtained from each preparation. The increase in specific activity (G_{50}) after each step of the isolation is shown in Figure 6.

Estimations of the degree of homogeneity of the KDGF preparations were made by several methods. As shown in Figure 7, sheep KDGF gave only one Coomassie blue stained band when analyzed by urea-SDS 12.5% polyacrylamide gel electro-

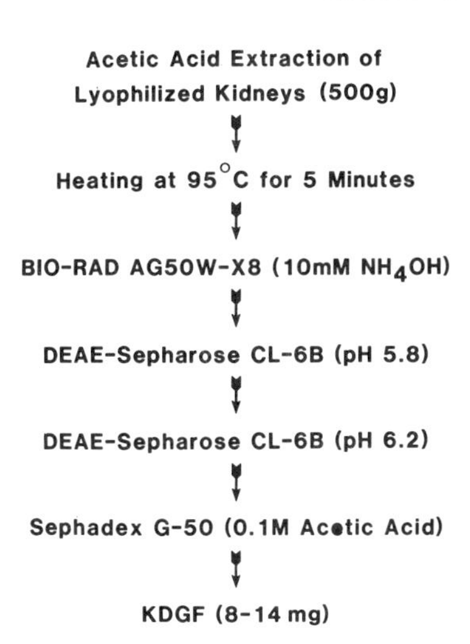

FIG. 5. Schematic representation of the purification of sheep KDGF. The methods of purification will be described in detail elsewhere (T. Ikeda and D. A. Sirbasku, manuscript in preparation).

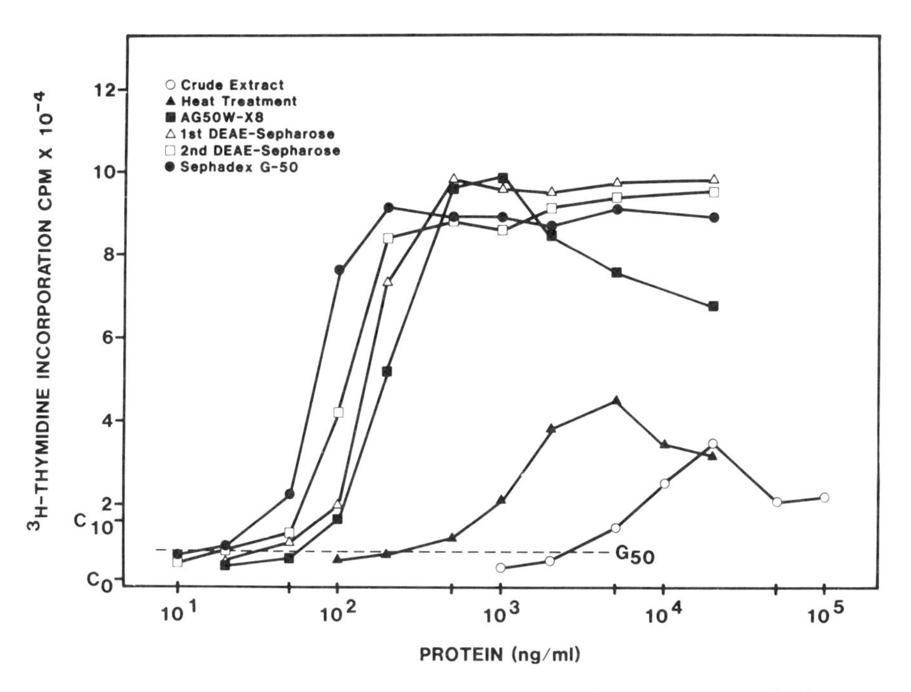

FIG. 6. Assay of the specific activity (G_{50}) of sheep KDGF after the various purification steps outlined in Figure 5. The assay used was the stimulation of incorporation of tritium-labeled thymidine into MTW9/PL mammary tumor cell DNA as described in Ikeda and Sirbasku (1984). The C_o and C_{10} designations indicate growth in the absence of serum and in 10% fetal calf serum, respectively.

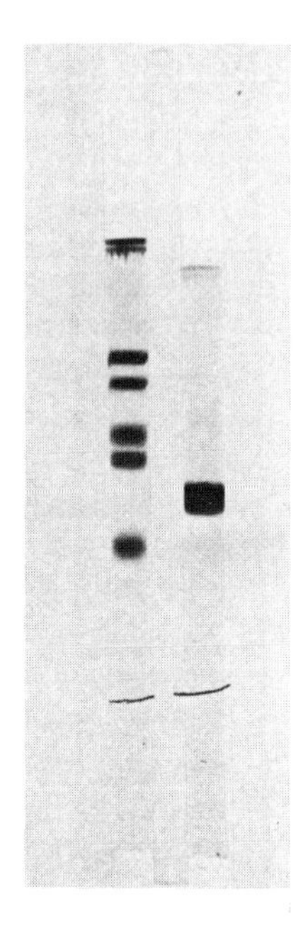

FIG. 7. The 8 M urea, 0.1% SDS polyacrylamide gel electrophoresis, and Coomassie Blue staining analysis of the KDGF preparation from the Sephadex G-50 chromatography step performed as shown in Figure 5. The migration positions of the known molecular weight myoglobin and myoglobin fragment markers (see legend of Figure 3) is shown on the left, and the migration position of 50 μg of KDGF is shown on the right.

phoresis. Other analyses by a variety of methods including polyacrylamide gel electrophoresis under nonreducing conditions, reverse phase HPLC, and molecular sieve HPLC (GS2000SW), all confirmed >90% homogeneity. As found for sheep UDGF, the estimated molecular weight of sheep KDGF was 4200 (see Figure 8).

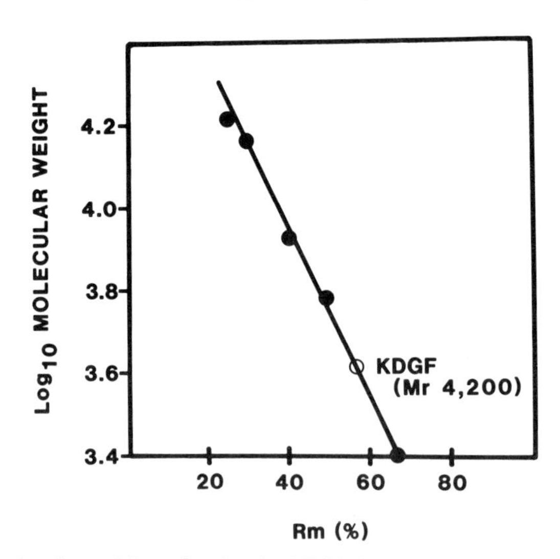

FIG. 8. The molecular weight estimation for KDGF from the data presented in Figure 7. The calibration curve was constructed from the relative migration positions of the myoglobin standards versus the migration position of bromophenol blue indicator (Rm%).

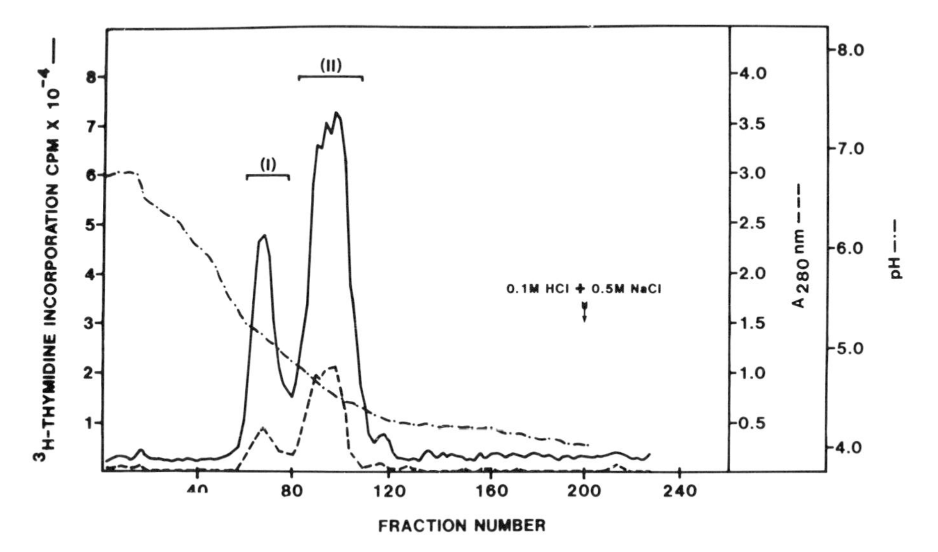

FIG. 9. The determination of the isoelectric point (pI) of purified KDGF was accomplished by the chromatofocusing procedure described in the Pharmacia Fine Chemicals, Inc., procedures manual. A column (1.5 × 30 cm) of PBE94 chromatofocusing exchanger was equilibrated with 25 mM histidine-HCl (pH 6.3) prior to application of the sample, which had already been equilibrated in the same buffer. The column was washed with the histidine buffer, followed by a 1:7 dilution of polybuffer 74 (pH 4.0). The fractions collected were assayed with the MTW9/PL cells as described before (Ikeda and Sirbasku 1984).

As calculated from the data in Figure 6, the G_{50} of the final KDGF preparation from the Sephadex G-50 column was 19 ng/ml, or 4.5×10^{-9} M. This concentration of KDGF was two to three times the G_{50} of UDGF (see Figure 2). Thus, even though both factors shared the same apparent molecular weight, the specific activities of each were different. When the pI of KDGF was estimated by standard chromatofocusing methods (Pharmacia Fine Chemicals, Inc.), the KDGF resolved into two peaks of activity designated KDGF-I and KDGF-II of pIs 5.2 and 4.8, respectively (see Figure 9). KDGF-II was 70% to 80% of the total activity, and showed G_{50} of 19 ng/ml, whereas the G_{50} of KDGF-I was 10 ng/ml under the same assay conditions. At present, the relationship between the two forms of KDGF is unknown. Studies are in progress to establish their molecular identities by protein chemistry and immunological methods.

In another series of studies, the purified KDGF was shown to be a potent mitogen for the GH3/C14 rat pituitary tumor cell line in serum-free medium (T. Ikeda and D. A. Sirbasku, unpublished results). In addition, KDGF is not mitogenic for normal diploid fibroblasts. Another very important study (to be reported elsewhere) was the determination of the possible mitogenic effect of KDGF on NRK cells in culture. We have shown that the 4200-Da mixture of KDGF-I and KDGF-II promotes the incorporation of tritium-labeled thymidine into NRK cell DNA under completely serum-free culture conditions. It must be noted that this effect

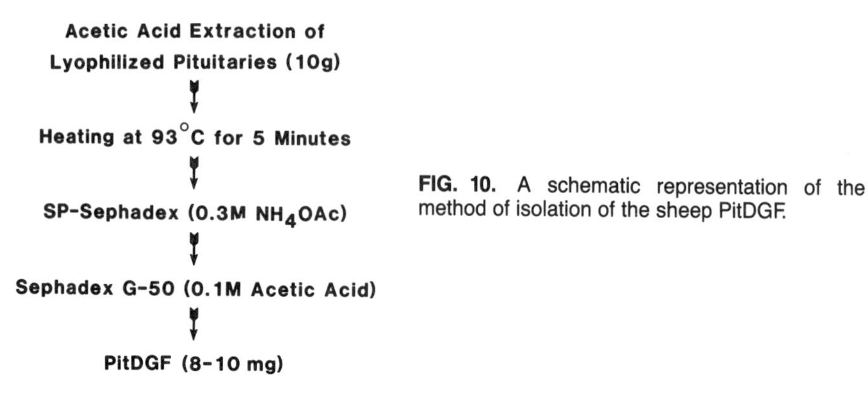

FIG. 10. A schematic representation of the method of isolation of the sheep PitDGF.

of KDGF was demonstrated in the absence of the serum and EGF supplements required for the usual assay of the TGF activities. These data, coupled with the observations that dithiothreitol and 2-mercaptoethanol had no effect upon KDGF mobility in urea-SDS gels or upon KDGF activity, again were important indications of major differences between this mammary/pituitary mitogen and the 25,000-Da form of TGF that was isolated from kidney (Roberts et al. 1983).

Pituitary-Derived Growth Factor from Sheep

Using both assay and isolation methods similar to those applied to the purification of sheep UDGF, we have identified a potent pituitary-derived growth factor (PitDGF) in the acetic acid extracts of whole sheep pituitaries. As summarized in Figure 10, as little as 10 g of lyophilized powder gave 8 to 10 mg of a potent growth factor for the MTW9/PL mammary and the GH3/C14 pituitary tumor cells. The methods of isolation employed were appropriately reduced scale procedures of Ikeda and Sirbasku (1984), which included acetic acid extraction, heat treatment at 95°C, SP Sephadex chromatography, and finally Sephadex G-50 chromatography in acetic acid. The pooled active fractions from the Sephadex G-50 column were analyzed by urea-SDS polyacrylamide gel electrophoresis and shown to be one major stained protein band (Figure 11) of estimated molecular weight 3900 Da (calculated from the data presented in Figure 11). Further analysis of the state of purity of this preparation is in progress.

Even in the absence of further characterizations of homogeneity, it is important to note that the G_{50} of this preparation with MTW9/PL cells was 29 ng/ml (Figure 12), which compares favorably to the G_{50} values of 8, 10, and 19 ng/ml for UDGF, KDGF-I, and KDGF-II, respectively. Another important point to note is that the molecular weight of this activity is far below those of the other pituitary hormones often cited (see review by Welsch and Nagasawa 1977) as potential mitogens for mammary cells. As shown in Figure 12, prolactin (24,000 Da) and growth hormone (22,000 Da) are not mitogenic for MTW9/PL cells at concentrations of up to 5 μg/ml, and as cited from our other report (Ikeda and Sirbasku 1984) several other

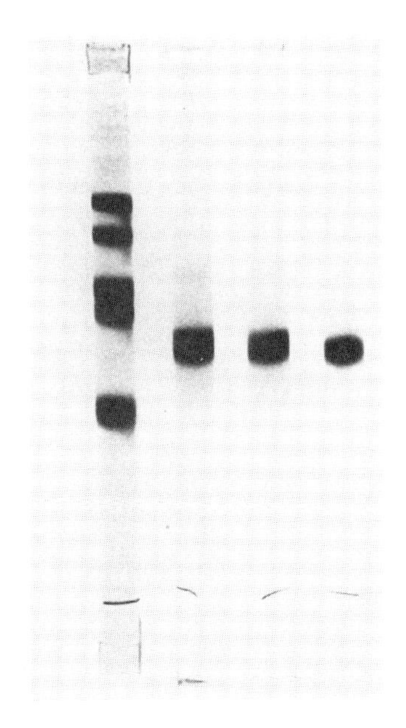

FIG. 11. The 8 M urea, 0.1% SDS polyacrylamide gel, and Coomassie Blue staining analysis of the sheep PitDGF preparation from the Sephadex G-50 chromatography step shown above. The gel to the far left received the mixture of myoglobin and sequenced fragments of myoglobin as standards used to calculate the molecular weight of PitDGF. The other gels (from right to left) received 75, 100, and 150 μg of PitDGF, respectively.

hormones and growth factors also do not substitute for PitDGF (or UDGF) with the MTW9/PL cells in serum-free culture.

When the mitogenic effects of PitDGF were assayed on other cell lines, it was demonstrated that GH3/C14 pituitary tumor cells responded to low concentrations of factor 24 hr after PitDGF addition under DNA synthesis assay conditions; when growth was measured 48 hr after addition of PitDGF, only a very modest response was seen (Figure 13), indicating that the growth factor may induce only a single round of replication under these assay conditions.

Lastly, one of the other important properties of PitDGF was shown to be its potent activity toward human breast cancer cell lines in culture. For example, the human MCF-7 mammary tumor cells grew in response to low concentrations of PitDGF in completely serum-free medium; under these same conditions prolactin and growth hormone showed no mitogenic effects at concentrations of up to 5 μg/ml (Figure 14). Thus, a potentially important new mitogen has been identified in pituitary gland, and further studies are planned to characterize its physiological significance. In addition, both immunological and protein chemistry methods will be employed to determine what, if any, relationship PitDGF has to UDGF, KDGF-I, and KDGF-II.

Along the lines of attempting to establish whether PitDGF is an endocrine estromedin, we have shown before that the estrogen-responsive growth of the MTW9/PL cell line in W/Fu rats was stimulated by the prolactin- and growth hormone–secreting GH3/C14 pituitary tumor (Sirbasku et al. 1982a). The data presented in Figure 12 strongly suggested that prolactin or growth hormone, or both, were not the active agents responsible for this stimulation and that there might

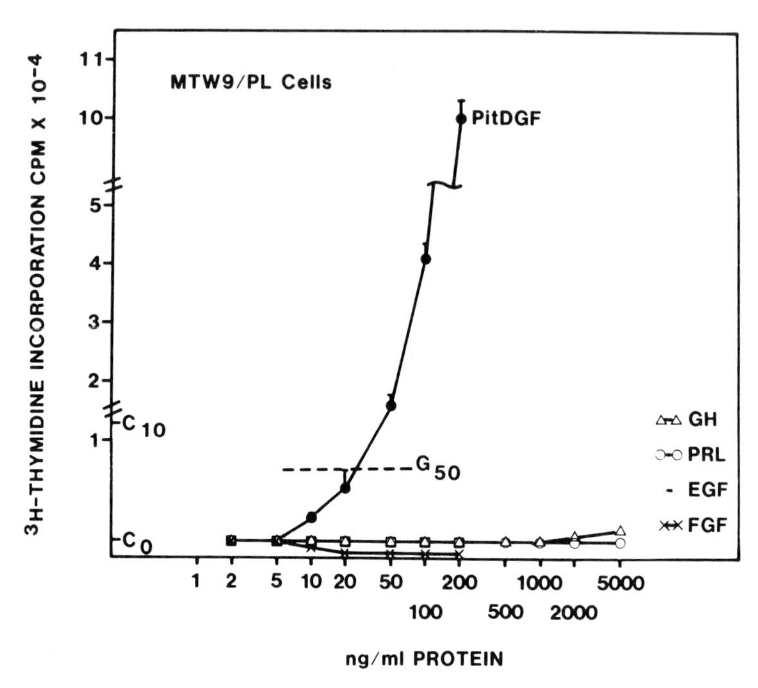

FIG. 12. Growth of MTW9/PL rat mammary tumor cells measured in response to PitDGF, growth hormone (GH), prolactin (PRL), fibroblast growth factor (FGF), and EGF. The assays were conducted in serum-free medium by the tritium-labeling method described in Ikeda and Sirbasku (1984).

be another, previously unrecognized factor (hormone) secreted from the GH3/C14 cells that was mitogenic for mammary origin cells. We have now identified and partially purified a mammary tumor cell growth factor from the GH3/C14 rat pituitary tumors, and by all criteria applied to date conclude that this rat-derived activity is very similar to the PitDGF obtained from normal sheep pituitaries. If such proves to be the case, the pituitary cells in culture will offer a system for study of the possible estrogen regulation of pituitary estromedin production.

PROPERTIES OF AUTOCRINE ESTROMEDINS

Estrogen-Inducible Rat Mammary Tumor Autostimulatory Growth Factors

At the 1981 Banbury Conferences on Hormones and Breast Cancer, we reported (Sirbasku 1981) the presence of a potent autostimulatory growth factor in neutral buffer extracts of the estrogen-responsive (Sirbasku 1978b) MWT9/PL rat mammary tumor growing in W/Fu rats. This activity had properties of a high molecular weight (i.e., >60,000) protein when determined under nondissociating conditions. The specific activity appeared related to the estrogen status of the host. Extracts of tumors growing in response to physiological levels of estrogen had greater growth

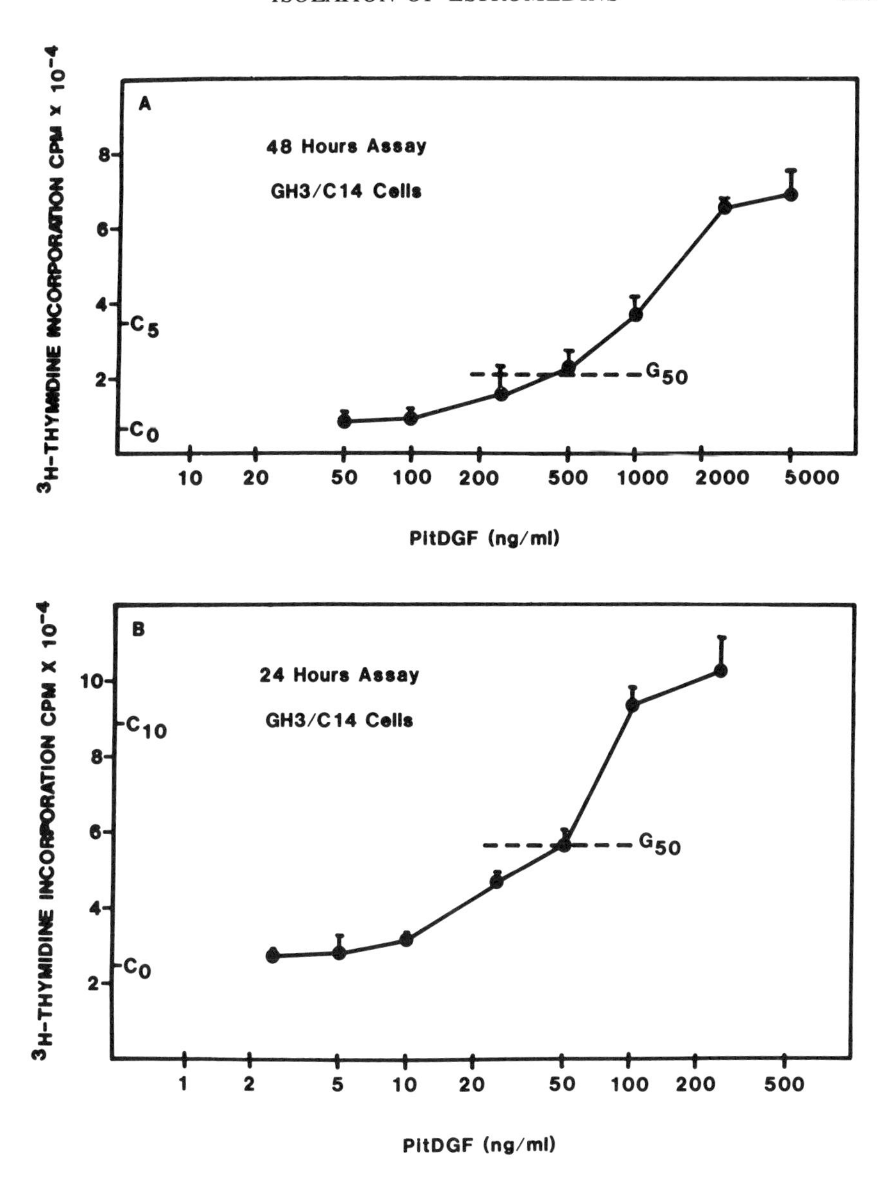

FIG. 13. Effect of sheep PitDGF on GH3/C14 rat pituitary tumor cell growth in serum-free medium was assayed by two methods described before (Ikeda and Sirbasku 1984). The growth-promoting potency of PitDGF was estimated 48 hr after addition of the growth factor to the serum-free cultures of cells **(A)**, or at 24 hr after addition of the growth factor **(B).** The specific activity (G_{50}) of PitDGF in each assay was estimated as the concentration of factor that half-replaced the mitogenic response to either 5% or 10% serum addition. The different serum concentrations were required in each assay to optimize the effect on labeling index at that time.

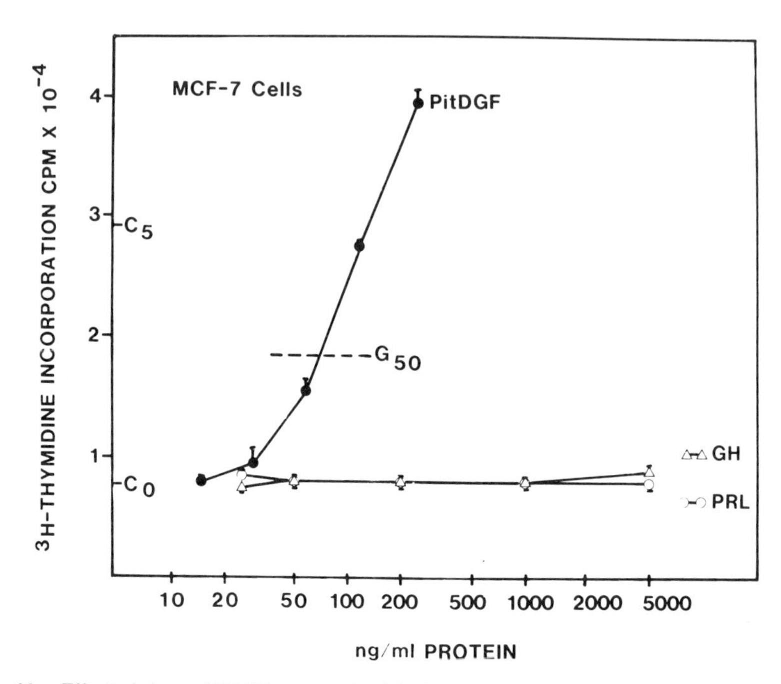

FIG. 14. Effect of sheep PitDGF on growth of the human MCF-7 mammary tumor cell line was assayed as described before (Danielpour and Sirbasku 1983b). For comparison purposes, the effects of the growth factor were assayed in parallel with the designated concentrations of growth hormone (GH) and prolactin (PRL).

factor activity than tumors that were regressing after oophorectomy of the rats. These observations led us to suggest that estrogens may have local effects on cell growth by inducing autocrine or paracrinelike growth factors or both (Sirbasku 1981, Leland et al. 1981, 1982, Sirbasku and Leland 1982a, 1982b, Ikeda et al. 1982, 1984, Ikeda and Sirbasku 1984).

In this report, we have applied the methods used to identify and isolate the endocrine estromedins to the problem of the further characterization of autocrine estromedins. Using acetic acid extraction and heating at 95°C, potent autocrine growth factors for the MTW9/PL cells have been identified in their serum-free conditioned medium and from the extracts of these cells growing in serum-free medium (Figure 15). In another series of experiments (Figure 16), we have shown that the rat mammary tumor–derived growth factor (r-MTDGF) was induced two-fold in the cells in culture by physiological concentrations (1 to 10×10^{-9} M) estradiol. The inductions of activity required 3 to 5 days incubation in serum-free medium before activity above that of the no-estrogen controls was found. We have described these results in a preliminary report (Danielpour and Sirbasku 1983a).

From the data available, r-MTDGF from the MTW9/PL cells in culture appears to be a 2,000- to 6,000-Da mitogen that cannot be replaced by other known growth factors such as EGF, PDGF, fibroblast growth factor (FGF), or IGF-I and IGF-II.

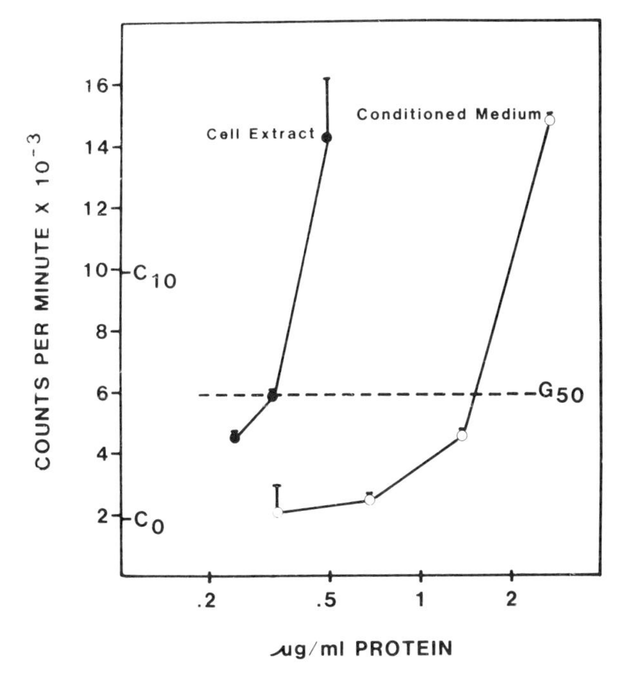

FIG. 15. Identification of autostimulatory (autocrine) growth factors in extracts of the MTW9/PL cells growing in serum-free medium and identification in serum-free conditioned medium done as described before (Danielpour and Sirbasku 1983a). The acetic acid-treated and -heated extracts of the cells were compared to serum-free conditioned medium prepared under the same conditions as the cells. The assays were done with the same MTW9/PL cells as described before (Ikeda and Sirbasku 1984).

The data obtained (D. Danielpour and D. A. Sirbasku, manuscript in preparation) show that conditioned medium from MTW9/PL cells in culture is a relatively poor source of growth factor for large-scale isolations (also see data in Figure 15). Alternately, 100 g of MTW9/PL cells from monolayers might yield as much as 1 to 5 mg of r-MTDGF, making this tumor cell–derived activity approximately 100 to 1000 times more abundant than the transforming activities isolated from conditioned medium as described by others (Todaro et al. 1980). However, even growth of 100 g of cells represents a considerable effort, which could easily be circumvented by use of fresh MTW9/PL tumors, which can be obtained in kilogram amounts.

We have reported elsewhere (Ikeda et al. 1984) the partial purification of the MTW9/PL autostimulatory activity from MTW9/PL tumors. The r-MTDGF was partially purified by the same first four steps used to purify sheep UDGF (Ikeda and Sirbasku 1984). From 46 g of tumor, 24 mg of partially purified factor was obtained, which showed a G_{50} of 84 ng/ml. Assuming a final G_{50} of r-MTDGF equivalent to that of sheep UDGF (i.e., $G_{50} = 8$ ng/ml) approximately another tenfold

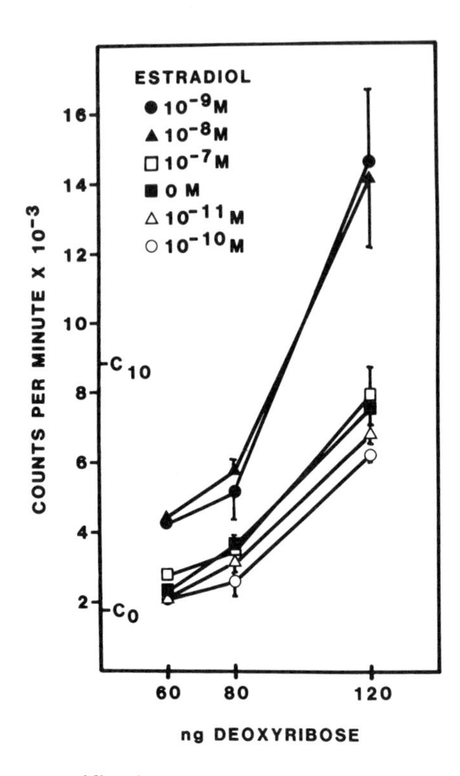

FIG. 16. Assays were conducted to determine whether the autocrine activity associated with the MTW9/PL cell monolayers was estrogen inducible. The MTW9/PL cells were grown in serum-free medium for 4 days with either no addition of estradiol, or the addition of estradiol at the designated concentrations. Cell extracts were prepared as described before (Danielpour and Sirbasku 1983a) using acetic acid treatment and heating at 95°C. The assays were conducted as before with the exception that the specific activity was based on DNA to correct for differences in cellular protein content in estradiol-treated cultures.

purification will be required. If this is the case, then as much as 50 mg amounts of r-MTDGF might be expected from a kilogram of tumor.

Estrogen-Inducible Human Mammary Tumor Autostimulatory Growth Factors

We have examined three human origin mammary tumor cell lines for acid-extractable, heat-stable autocrine growth factor activities. An example of the results obtained is shown in Figure 17, and it demonstrates that acetic acid and heated extracts of the MCF-7 cells were potent growth-promoting preparations for the same cells maintained in serum-free culture. Further, similar extracts of the human T-47D and ZR-75-1 mammary tumor cells also promoted the growth of the MCF-7 cells in culture (Figure 17). Thus, human-origin cells appear to possess an MTDGF similar in some respects to that of the rat origin MTW9/PL cells. In a final series of experiments, we have reported preliminary evidence (Danielpour and Sirbasku 1983b) that the acetic acid- and heat-stable MTDGF of MCF-7 cells was estradiol inducible in serum-free culture. Further characterization and purification of the human MTDGF is in progress.

DISCUSSION

As summarized in Table 1, we have purified three possible endocrine estromedins from uterus, kidney, and pituitary, and have begun characterizations of their cell

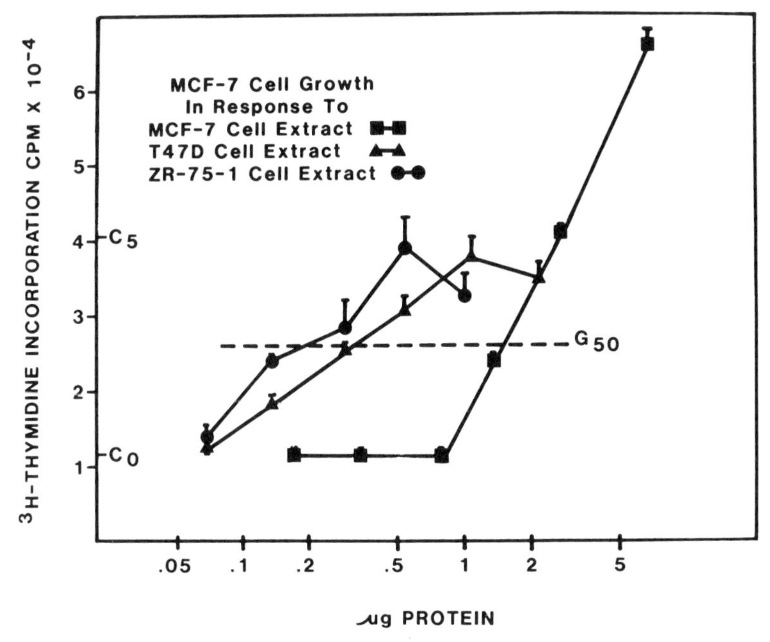

FIG. 17. The presence of autocrine growth factors in the human mammary tumor cell lines MCF-7, T47D, and ZR-75-1 were assayed. The cell lines were grown for at least 2 days in serum-free DME before being extracted with acetic acid and heated as described before (Danielpour and Sirbasku 1983b). The heated extracts from each cell line were assayed for growth promotion of the MCF-7 line in serum-free DME. The specific activity of each extract was estimated from the G_{50} values calculated from above.

type specificities and biochemical properties. The next obvious steps required to establish their role as estromedins is to first determine their presence in plasma and then to evaluate the effects of estrogens on their circulating concentrations. Currently, we are raising specific polyclonal antibodies to each factor with the

TABLE 1. *Properties and targets of four growth factors*

Purified factors	MW	Possible endocrine targets	Possible autocrine targets
Pituitary-derived growth factor (PitDGF)	3,900	Mammary, uterus, kidney	Pituitary
Uterine-derived growth factor (UDGF)	4,200	Mammary, uterus, pituitary	Uterus
Kidney-derived growth factor (KDGF)	4,200	Mammary, pituitary, (uterus)	Kidney
Mammary tumor–derived growth factor (MTDGF)	2,000–6,000	?	Mammary

intention of establishing their immunological similarities or differences and developing sensitive radioimmunoassay methods of measurement.

One unexpected possibility raised by our most recent data (T. Ikeda and D. A. Sirbasku, manuscript in preparation; D. Danielpour and D. A. Sirbasku, manuscript in preparation) is that all three of the endocrine estromedins thus far isolated might serve dual functions in vivo. We have shown that in each case (i.e., uterus, kidney, and pituitary) the same purified growth factor shows both strong endocrine activity toward mammary and pituitary cells and potent autocrine activity toward normal or neoplastic cells, or both, from the tissue of origin. For example, we have shown that the sheep UDGF is a potent mitogen for mammary cells and for cells of uterine origin (Ikeda and Sirbasku 1984). Similarly, the sheep PitDGF is mitogenic for both the MTW9/PL cells (possible endocrine role) and for the GH3/C14 pituitary cells (autocrine role). Further studies are required to define the dual role of these factors, but if this should prove to be the case, an entirely new approach to estrogen regulation of growth will become testable.

Finally, the relative roles of the endocrine versus autocrine activities in estrogen-responsive and -autonomous tumor growth deserve special comment. It is well known that the presence of the estrogen receptor correlates positively with the state of hormone responsiveness of target tissue tumors. We believe this is the case because those same tumors produce autostimulatory growth factors in response to the steroid hormone and that this effect is mediated by the receptor. It is equally well known that autonomous tumors (i.e., those that do not regress or stop growth after oophorectomy) are both receptor positive and receptor negative. We propose that the state of autonomy is independent of the presence of receptor because autonomous cells have achieved this state by genetic changes that lead to continuous and unregulated production of the autocrine factors. This means that in normal tissues and highly estrogen-responsive tumors both endocrine and autocrine growth factors may be important in growth regulation, and during the usually gradual conversion to autonomy the production of autocrine activity increases and growth becomes independent of the former controls.

ACKNOWLEDGMENTS

We thank Ms. Judy M. Roscoe for expert technical assistance and preparation of the photographs presented in this report. This work was supported by National Cancer Institute grants RO1-CA26617, and RO1-CA-38024 and a grant BC-255 from the American Cancer Society. DD is the recipient of a Predoctoral Fellowship in Cancer Research from the Rosalie B. Hite Foundation, Houston, Texas. DAS is a recipient of an American Cancer Society Faculty Research Award (FRA-212).

REFERENCES

Antoniades HN, Scher CD, Stiles CD. 1979. Purification of human platelet-derived growth factor. Proc Natl Acad Sci USA 76:1809–1813.

Barnes D, Sato G. 1979. Growth of a human mammary tumor cell line in serum free medium. Nature 281:388–389.

Carpenter G, Cohen S. 1979. Epidermal growth factor. Ann Rev Biochem 48:193–216.
Cohen S, Fava RA, Sawyer ST. 1982. Purification and characterization of epidermal growth factor receptor/protein kinase from normal mouse liver. Proc Natl Acad Sci USA 79:6237–6241.
Danielpour D, Sirbasku DA. 1983a. Autocrine control of estrogen-responsive mammary tumor cell growth. (Abstract) In Vitro 19:252.
Danielpour D, Sirbasku DA. 1983b. Estrogen-inducible autostimulatory growth factors found in human breast cancer cells. (Abstract) J Cell Biol 95 (part 2):394a.
Deuel TF, Huang JS, Proffitt RT, Baenziger JU, Chang D, Kennedy BB. 1981. Human platelet derived growth factor: Purification and resolution into two active protein fractions. J Biol Chem 256:8896–8899.
Doolittle RF, Hunkapiller MW, Hood LE, Devare SG, Robbins KC, Aaronson SA, Antoniades HN. 1983. Simian sarcoma virus onc gene, v-*sis*, is derived from the gene (or genes) encoding a platelet-derived growth factor. Science 221:275–277.
Frolik CA, Dart LL, Meyers CA, Smith DM, Lehrman SR, Sporn MB. 1983. Purification and initial characterization of a human placental type β transforming growth factor. (Abstract) Fed Proc 42:1833.
Iio M, Sirbasku DA. 1982. Partial purification of mammary tumor cell estromedins from the uterus of a pregnant sheep. *In* Sato GH, Pardee AB, and Sirbasku DA (eds), Cold Spring Harbor Conferences on Cell Proliferation, Vol 9, Growth of Cells in Hormonally Defined Media. Cold Spring Harbor Laboratory, Cold Spring Harbor, New York, pp. 751–761.
Ikeda T, Sirbasku DA. 1984. Purification and properties of a mammary/uterine/pituitary tumor cell growth factor from pregnant sheep uterus. J Biol Chem 259:4049–4064.
Ikeda T, Liu QF, Danielpour D, Officer JB, Iio M, Leland FE, Sirbasku DA. 1982. Identification of estrogen-inducible growth factors (estromedins) for rat and human mammary tumor cells in culture. In Vitro 18:961–979.
Ikeda T, Danielpour D, Galle PR, Sirbasku DA. 1984. General methods for isolation of acetic acid and heat stable polypeptide growth factors for mammary and pituitary tumor cells. *In* Barnes DW, Sirbasku DA, Sato GH (eds), Cell Culture Methods for Molecular and Cell Biology, Vol 2, Methods of Serum Free Culture of Cells of the Endocrine System. Allen R. Liss, Inc., New York (in press).
Kasuga M, van Obberghen E, Nissley SP, Rechler MM. 1981. Demonstration of two subtypes of insulin like growth factor receptors by affinity crosslinking. J Biol Chem 256:5305–5308.
Klapper DG, Svoboda ME, Van Wyk JJ. 1983. Sequence analysis of somatomedin C: Confirmation of identity with insulin like growth factor I. Endocrinology 112:2215–2217.
King GL, Kahn CR, Rechler MM, Nissley SP. 1980. Direct demonstration of separate receptors for growth and metabolic multiplication stimulating activity (an insulin like growth factor) using antibodies to the insulin receptor. J Clin Invest 66:130–140.
Leland FE, Iio M, Sirbasku DA. 1981. Hormone dependent cell lines. *In* Sato GH (ed), Functionally Differentiated Cell Lines. Allen R. Liss, Inc., New York, pp. 1–46.
Leland FE, Danielpour D, Sirbasku DA. 1982. Studies of the endocrine, paracrine, and autocrine control of mammary tumor cell growth. *In* Sato GH, Pardee AB, Sirbasku DA (eds), Cold Spring Harbor Conferences on Cell Proliferation, Vol 9, Growth of Cells in Hormonally Defined Media. Cold Spring Harbor Laboratory, Cold Spring Harbor, New York, pp. 741–750.
Marquardt H, Todaro GJ, Henderson LE, Oroszlan S. 1981. Purification and primary structure of a polypeptide with multiplication stimulating activity from rat liver cell cultures: Homology with insulin like growth factor. II. J Biol Chem 256:6859–6865.
Marquardt H, Hunkapiller MW, Hood LE, Twardzik DR, DeLarco JE, Stephenson JR, Todaro GJ. 1983. Transforming growth factors produced by retrovirus transformed rodent fibroblasts and human melanoma cells: Amino acid sequence homology with epidermal growth factor. Proc Natl Acad Sci USA 80:4684–4688.
Massague J, Czech MP. 1982. The subunit structure of two distinct receptors for insulin like growth factors I and II and their relationships to the insulin receptor. J Biol Chem 257:5038–5045.
Raines EW, Ross R. 1982. Platelet derived growth factor. I. High yield purification and evidence for multiple forms. J Biol Chem 257:5154–5160.
Rinderknecht E, Humbel RE. 1978a. The amino acid sequence of human insulin like growth factor I and its structural homology with proinsulin. J Biol Chem 253:2769–2776.
Rinderknecht E, Humbel RE. 1978b. Primary structure of human insulin like growth factor II. FEBS Lett, 89:283–286.

Roberts AB, Lamb LC, Newton DL, Sporn MB, DeLarco JE, Todaro GJ. 1980. Transforming growth factors: Isolation of polypeptides from virally and chemically transformed cells by acid/ethanol extraction. Proc Natl Acad Sci USA 77:3494–3498.

Roberts AB, Anzano MA, Meyers CA, Wideman J, Blacher R, Pan YCE, Stein S, Lehrman SR, Smith JM, Lamb LC, Sporn MB. 1983. Purification and properties of a type β transforming growth factor from bovine kidney. Biochemistry 22:5692–5698.

Savage CR Jr, Cohen S. 1972. Epidermal growth factor and a new derivative: Rapid isolation procedures and biological and chemical characterization. J Biol Chem 247:7609–7611.

Sirbasku DA. 1978a. Estrogen induction of growth factors specific for hormone responsive mammary, pituitary, and kidney tumor cells. Proc Natl Acad Sci USA 75:3786–3790.

Sirbasku DA. 1978b. Hormone responsive growth in vivo of a tissue culture cell line established from the MT-W9A rat mammary tumor. Cancer Res 38:1154–1165.

Sirbasku DA. 1981. New concepts in control of estrogen responsive tumor growth. Banbury Rep 8:425–443.

Sirbasku DA, Leland FE. 1982a. Estrogen inducible growth factors: Proposal of new mechanisms of estrogen promoted tumor growth. *In* Litwack G (ed), Biochemical Action of Hormones. Academic Press, New York, pp. 115–140.

Sirbasku DA, Leland FE. 1982b. Growth factors for hormone sensitive tumor cells. *In* Leung BS (ed), Hormonal Regulation of Mammary Tumors, Vol 2, Peptides and Other Hormones. Eden Press, Montreal, pp. 88–112.

Sirbasku DA, Moo JM. 1982. Characterization of hypothalamic origin growth factor(s) for rat pituitary tumor cells in culture. (Abstract) Fed Proc 41:489.

Sirbasku DA, Officer JM, Leland FE, Iio M. 1982. Evidence of a new role for pituitary derived hormones and growth factors in mammary tumor cell growth in vivo and in vitro. *In* Sato GH, Pardee AB, Sirbasku DA (eds), Cold Spring Harbor Conferences on Cell Proliferation, Vol 9, Growth of Cells in Hormonally Defined Media. Cold Spring Harbor Laboratory, Cold Spring Harbor, New York, pp. 763–778.

Sporn MB, Todaro GJ. 1980. Autocrine secretion and malignant transformation of cells. N Engl J Med 303:878–880.

Stiles CD. 1983. The molecular biology of platelet-derived growth factor. Cell 33:653–655.

Todaro GJ, Fryling C, DeLarco JE. 1980. Transforming growth factors produced by certain human tumor cells: Polypeptides that interact with epidermal growth factor receptors. Proc Natl Acad Sci USA 77:5258–5262.

Waterfield MD, Scrace GT, Whittle N, Stroobant P, Johnsson A, Wasteson A, Westermark B, Heldin CH, Huang JS, Deuel TF. 1983. Platelet derived growth factor is structurally related to the putative transforming protein $p28^{sis}$ of simian sarcoma virus. Nature 304:35–39.

Welsch CW, Nagasawa H. 1977. Prolactin and murine mammary tumorigenesis: A review. Cancer Res 37:951–963.

Mediators in Cell Growth and Differentiation,
edited by Richard J. Ford and Abby L. Maizel.
Raven Press, New York © 1985.

Growth Factors for Human Lymphoid Neoplasms

Richard J. Ford, Frances Davis,* and Irma Ramirez†

Departments of Pathology, Chemotherapy Research, and Pediatrics,† The University of Texas, M. D. Anderson Hospital and Tumor Institute at Houston, Houston, Texas 77030*

Human lymphoid neoplasms are a heterogeneous group of tumors derived from the T- and B-cell lineages in the so-called non-Hodgkin's lymphomas (NHL) (Ford and Maizel 1982) and most probably from the monocyte-macrophage lineage in Hodgkin's disease (HD), although this latter designation has not been conclusively proved (Ford et al. 1982). In addition to these "solid tumors" of the lymphoreticular (lymphoid + reticuloendothelial) system, there are varieties of lymphoid leukemias, derived from both T- and B-cell precursors. These leukemias are usually primary neoplastic diseases of the bone marrow that extend into the peripheral blood, but occasionally also infiltrate the tissues of the lymphatic or, less often, other organ systems, giving rise to lesions usually morphologically indistinguishable from lymphomas. The reverse is also occasionally found, in which NHL cells enter the peripheral circulation, giving rise to a leukemic form of the lymphomas.

Lymphoid neoplasms are common tumors that can occur at any time in the human life cycle from infancy to senescence. The realization that lymphoid neoplasms represent tumors of the immune system (Hansen and Good 1974) has stimulated particular interest in these tumors, largely as a result of the impact of cellular immunology and the ability to apply recently developed phenotyping reagents such as monoclonal antibodies to identify the putative T- or B-cell lineage of the neoplasm (Anderson et al. 1984). In addition to cell identification and phenotyping, recent advances in immunology and molecular biology have led to the development of experimental technology that should provide a better understanding of the biology of human lymphoid neoplasms. This information also provides the prospect of new treatment modalities, probably through the use of immunologically active molecules (lymphokines, monoclonal antibodies, etc.) as biological response modifiers, adjuvants, or even replacements for standard conventional chemotherapy or radiotherapy. In the following sections, several experimental approaches to studying the immunobiology of human lymphoid neoplasms, with particular focus on lymphoid growth factors (interleukins) will be discussed.

PHENOTYPIC CHARACTERISTICS OF HUMAN LYMPHOID NEOPLASMS

Human lymphoid tumors are usually diagnosed by pathologists on the basis of effacement of the normal lymphoid tissue architecture by a rather monomorphic population of lymphoid cells. This results in the breakdown of the usual compartmentalized lymphoid cellular areas, such as the B-cell predominant cortex, which contains lymphoid follicles and germinal centers, or the T-cell predominant paracortex, its postcapillary or high endothelial venules (which are so important in lymphocyte trafficking (Butcher et al. 1979)), and the presence of interdigitating reticulum (IDC) or dendritic cells involved in T-cell activation (van Valk et al. 1984). Cellular atypia is present in some lymphoid malignancies and is a helpful diagnostic parameter if present, but isolated neoplastic T and B cells often show little or no obvious cytologic abnormality.

The spectrum of NHL includes virtually all the recognized stages in human T- and B-cell ontogeny (Aisenberg 1983). NHL cells are thought to represent the neoplastic counterparts of the various stages in T- and B-cell differentiation, in which the tumor cell has been "frozen" (i.e., cannot proceed further in the differentiative sequence) (Salmon and Seligmann 1974). Lymphoid cells that have been neoplastically transformed (e.g., by retrovirus, oncogene expression, etc.) proliferate but do not concomitantly differentiate, thus giving rise to a lymphoid neoplasm composed of monoclonal lymphoma cells.

More than a decade ago, after the advent of conventional immunologic cell surface markers for normal T- and B-lymphocytes (Seligmann et al. 1975), these were applied to human lymphoid tumors (Lukes et al. 1978). Such techniques established the putative T- or B-cell lineage of the NHL as well as other human lymphoid neoplasms (Ford 1981). Phenotyping of human lymphoid neoplasms was further refined when hybridoma technology developed T- and B-cell–specific monoclonal antibodies that identified antigens characteristic of differentiation stages in both T- and B-cell ontogeny (Bernard et al. 1984). These developments together have provided a rather detailed definition of the phenotypic characteristics of the differentiative cascade in T- and B-cell maturation and a framework for identifying where the various NHL fit into this scheme (Nadler et al. 1981).

Even though the aforementioned batteries of conventional and monoclonal antibody markers are available for cell surface phenotyping, the problem of discriminating normal from neoplastic human lymphoid cells still exists. This is an important problem, as virtually all human lymphoid neoplasms are infiltrated with varying numbers of normal or reactive lymphoid cells (Dvoretsky et al. 1982). The immunologic markers identified are usually present on both normal and neoplastic cells of a particular lymphoid lineage and thus provide little help in the discrimination of normal from neoplastic cells.

Recently, we have used heteroantibodies to the human malignancy–associated nucleolar antigen (HMNA) to identify neoplastic human lymphoid cells and to distinguish the neoplastic cells from the reactive lymphoid cells present in cell

TABLE 1. *Nucleolar antigen (HMNA) expression in neoplastic and normal human lymphoid cell populations*

Lymphoid cell type*	No. tested†	No. positive‡
I. Non-Hodgkin's lymphomas		
Well-differentiated lymphocytic	18	18
Nodular poorly differentiated lymphocytic	15	15
Diffuse poorly differentiated lymphocytic	6	6
Nodular large cell	4	4
Diffuse large cell	16	16
Undifferentiated (Burkitt's, non-Burkitt's)	7	7
II. Lymphocytic leukemias		
Chronic lymphocytic	6	6
Hairy cell	5	5
III. Normal lymphoid cells		
Peripheral mononuclear cells	6	0
Mitogen-stimulated T or B cells	6	0
Long-term T-cell lines (HTLV-)	4	0
Long-term B-cell lines (EBV-)	2	0
Mixed lymphocyte cultures	5	0

*Modified Rappaport classification.

†Frozen sections or cytocentrifuge preparations were fixed in cold methanol and assayed by indirect immunofluorescence.

‡Positivity was defined by greater than 50% of the morphologically malignant or phenotypically malignant (i.e., monoclonal) cells (or both) showed nucleolar fluorescence.

suspensions of lymphomatous lesions (Ford et al. 1984a). This antibody is made by injecting rabbits with sonicated HeLa cell nucleoli and subsequently absorbing with normal cell nuclei until the activity on normal cells is removed (Ford et al. 1984a). The HMNA has been shown to be present in virtually all T- and B-cell NHL, as well as various types of lymphocytic leukemia. The antigen identified by these antisera does not, however, appear to be present in normal or reactive lymphoid cells, in T- or B-cell lymphoid populations stimulated by lectins, or in growth factor–dependent long-term cell lines of human T or B lymphocytes (Table 1). This antibody is therefore a very useful marker for assessing cell suspensions that have been enriched for malignant cells by various cell separation procedures (Ford et al. 1984b). One can, for instance, assess a cell suspension enriched for neoplastic B cells by sheep red blood cell rosetting and adherent cell depletion, by assaying for immunoglobulin light chains as a measure of monoclonality (Levy et al. 1977) and HMNA (Ford et al. 1984b) as a measure of malignancy (Table 2). We have very recently developed a two-color immunoenzyme technique that allows us to demonstrate these two markers, as well as others, in conjunction with the HMNA (Lu et al. 1984). The use of such phenotyping technology in concert with cell separation procedures allows considerable confidence that the populations being assayed in in vitro experimental systems is indeed representative of the malignant cells present in the lymphoid neoplasm.

TABLE 2. *Neoplastic human B-cell populations assessed for monoclonality and human malignancy-associated nucleolar antigen (HMNA)*

Lymphoma type*	SIg light chain†	HMNA‡
WDL	κ(98)λ(1)	96
WDL	κ(96)λ(0)	98
N-PDL	κ(95)λ(2)	93
N-PDL	κ(3)λ(96)	96
N-PDL	κ(89)λ(3)	99
Undiff Burkitt's	κ(95)λ(2)	93
N-mixed	κ(91)λ(1)	89
D-LCL	κ(90)λ(3)	85
D-LCL	κ(92)λ(2)	88
D-LCL	κ(3)λ(89)	80

*Modified Rappaport classification, B-cell lymphoma cell populations were prepared by sheep red blood cell rosette depletion and adherent cell depletion.

†SIg light chain typing was performed by direct immunofluorescence.

‡HMNA staining was performed by indirect immunofluorescence described in Ford et al. 1984a.

WDL, well-differentiated lymphocytic; N-PDL, nodular poorly differentiated lymphocytic; N-mixed, nodular mixed; D-LCL, diffuse large cell lymphoma.

GROWTH FACTORS FOR NON-HODGKIN'S LYMPHOMAS

One of the most important discoveries of the last decade in immunology has been the lymphoid growth factors, or the interleukins as they are usually referred to (Morgan et al. 1976). The lymphoid growth factors include T-cell growth factor (TCGF) (Smith 1984) and B-cell growth factor (BCGF) (Ford et al. 1981). Immunologists were actually rather tardy in discovering these very important molecules, which control cell proliferation, as cell biologists in a number of other systems had long since described similar molecules (for review, see Cohen 1983). It has become quite clear, however, that growth factors such as TCGF (or IL-2) and BCGF are lymphocytotrophic hormones (Smith 1984), which drive normal activated T and B lymphocytes, respectively, from the resting, G_0 stage of the cell cycle into the proliferative, S, G_2, and M phases. In the T-cell system, the monokine interleukin 1 plays an initial, crucial role as an amplification factor in the concatenation of events leading to T-cell activation (Smith et al. 1980).

Growth control in NHL, like most other human neoplasms, is a little-explored area. Aside from the older intuitive notions about tumors representing "uncontrolled" growth and the empiric in vivo clinical observations that untreated or treatment-resistant tumor growth appears to be unrestrained, we know very little about what controls cell proliferation in lymphoid tumors. One thing that we do know is that when most lymphoid tumor cells are placed in cell culture, under standard in vitro conditions, they do not proliferate and usually die out in 7–10 days.

Since lymphoid tumors retain many of the cell surface phenotypic characteristics of their putative normal counterparts (Ford and Maizel 1982, Hansen and Good 1974), we were interested in ascertaining if neoplastic human lymphoid cells could respond to a corresponding interleukin growth factor (i.e., TCGF or BCGF). In these experiments, fresh lymphoma cells obtained from untreated patients at diagnostic biopsy were separated into tumor cells and accessory cells by rosetting and adherence. In Table 2 it can be seen that the homologous lymphoid growth factor stimulated thymidine incorporation in both small T-cell and small B-cell NHL. These data indicate that at least some neoplastic human lymphoid cells can respond to normal immunoregulatory factors, in this case, lymphoid growth factors. This ability implies that these neoplasms represent functional T- and B-cell tumors rather than completely autonomous tumor cells that have escaped immunoregulatory control by mutation.

Large-cell NHL (LCL), on the other hand, are quite different tumors both clinically (Stein 1981) and biologically. Both the T- and B-cell LCL tend to be rather aggressive tumors, but many are now cured with combination chemotherapy. When we studied LCL of both T and B lineages in vitro, we found the freshly prepared tumor cells to be refractory to both growth factors and mitogenic lectins, as shown in Table 3. Low background [^{3}H]thymidine incorporation was also noted, indicating that the tumor cells were not spontaneously proliferating to any appreciable extent.

The results indicate that the small-cell NHL are sensitive to immunoregulatory control at least as regards proliferative stimulation by growth factor, whereas the LCL appear to be refractory. These findings raise several important questions regarding the nature of the responsiveness of lymphoma cells to growth factor. Growth factor–mediated proliferation suggests that at least the small-cell NHL have receptors for the appropriate interleukin, whereas the LCL either lack the receptors or have had the receptors down-regulated to the extent that the tumor cells appear to be unreactive.

From the thymidine incorporation data it appears that, even in the small-cell NHL, less than 10% of the tumor cells are proliferating. In fact, we were unable to stimulate [^{3}H]thymidine incorporation in the nonproliferating lymphoma cells when we treated them with various mitogens, including staphylococcal protein A (soluble and insoluble), lipopolysaccharide, phytohemagglutinin, etc. When other nonimmunologic stimulants, such as phorbol ester (TPA) or calcium ionophore (A23187), which appear to be mitogenic in many systems (Bertoglio 1983, Luckasen et al. 1974), were tried, no further [^{3}H]thymidine incorporation was observed. These studies suggest that the majority of cells present in a NHL lesion are nondividing cells and, furthermore, may not be capable of further cell division regardless of the stimulus.

At present, our working hypothesis regarding lymphoma cell proliferation is that most of the small cell lymphomas of both B- and T-cell type possess receptors for

TABLE 3. *Growth factor-mediated stimulation of non-Hodgkin's lymphoma cells*

Lymphoma phenotype*	$\alpha - \mu$†	BCGF/TCGF‡	$\alpha\mu$ + BCGF	PA − S§
B-cell type				
D-WDL B1B4HMNA+T11−	1,250	4,282	12,220	10,640
D-WDL B1B2B4HMNA+T11−	1,688	5,990	10,620	9,620
D-WDL B1B2B4HMNA+T11−	2,118	4,530	4,876	2,292
N-PDL B1B2B4HMNA+T11−	1,990	2,527	6,794	3,950
N-PDL B1HMNA+T11−	1,708	3,296	6,880	3,620
Undiff Burkitts B1B2HMNA+T11−	2,290	4,886	10,660	4,220
D-LCL B1B4HMNA+T11,3−	290	320	766	ND
D-LCL B1B4HMNA+T11−	660	420	484	320
D-LCL B1DR+T3−	906	1,100	985	466
T-cell type				
D-PDL T3T4HMNA+B1−	ND	4,210	ND	1,640
D-PDL T11T4HMNA+B1−	ND	8,380	ND	890
D-PDL T11DRHMNA+SIg−	ND	648	ND	220
D-PDL T3T4HMNA+SIg−	ND	290	ND	346

*Modified Rappaport classification, monoclonal antibody analysis was done by direct and/or indirect immunofluorescence; antigens were considered positive if present on >70 of the T- and adherent cell-depleted B-cell tumors or if present on >80 of the E-rosette–positive T-cell lymphomas.

†Anti-mu beads (Biorad) were used at a final concentration of 15 μg/ml.

‡BCGF and TCGF (IL-2) were prepared as described previously (Maizel et al. 1982) by purification of lymphocyte conditioned media.

§Protein A-Sepharose was added at a final concentration of 5% (V/V).

See Table 2 for abbreviations.

their homologous interleukin growth factors. Most of the tumor cells present in a lymphomatous lesion, however, appear to have either lost or down-regulated their growth factor receptors and therefore are unresponsive to the growth factor. A small percentage of the tumor cells, less than 10% by cytofluorometry and autoradiography, do respond to specific growth factor, implying the presence of the homologous growth factor receptors.

The presence of growth factor receptors on the growth fraction or "proliferating pool" of tumor cells in a lymphoma is interesting in that it implies that at least this small fraction is stimulated by the same type of growth factor-mediated mechanisms as seen in normal lymphocytes. This type of mechanism also suggests possible strategies for attacking the tumor cells by blocking growth factor receptors or other biological response modifications.

Growth factor sensitivity also raises the question of where the factor that drives tumor proliferation originates. A likely source is the abundant complement of T accessory cells that is present in virtually all human lymphoid tumors (Harris et al. 1982). These accessory cells may be involved in tumor cell proliferation in either a positive or possibly even a negative (suppressor) sense. Another possible explanation is that the tumor cells themselves produce and respond to their own growth factor (autocrine stimulation) (Sporn and Todaro 1980). There have been several reports that Burkitt's lymphoma cell lines, as well as other B lymphoblastoid cell lines, make autostimulatory factors in vitro (Blazar et al. 1983, Gordon et al. 1984). We have preliminary data that suggest that this may also be the case for at least some B-cell type NHL, in cells freshly explanted in vitro from biopsy specimens (Ford et al. 1984c).

An interesting question to be posed if this autocrine stimulation mechanism is found to be a general characteristic of human lymphoid tumors is whether the growth factor made by the tumors is identical to or related to the normal lymphoid growth factor for the putative tumor cell lineage. Alternatively, it might be envisioned that a lymphoid tumor would make a "tumor growth factor" that only the autochthonous tumor or related tumors could respond to: the normal lymphocyte counterparts of the lymphoma or leukemia cell would not respond to it.

Whatever the molecular nature of the stimulus behind human lymphoid tumor cell growth, it is apparent that the discovery and delineation of interleukin growth factors for normal lymphoid cells has paved the way for major strides in advancing our understanding of the basic immunobiology of lymphoid tumors.

ACKNOWLEDGMENTS

This investigation was supported by Grant Number CA31479 awarded by the National Cancer Institute, U.S. Department of Health and Human Services. We would like to thank Tammy Hazelrigs and Linda Kimbrough for preparation of this manuscript.

REFERENCES

Aisenberg AC. 1983. Cell lineage in lymphoproliferative disease. Am J Med 74:679–685.

Anderson K, Bates MP, Slaughenhaupt BL, Pinkus GS, Schlossman SF, Nadler LM. 1984. Expression of human B-cell associated antigens on leukemias and lymphomas. Blood 63:1424–1430.

Bernard A, Boumsell L, Dausset J, Milstein C, Schlossman SF. 1984. Leucocyte Typing: Human Leucocyte Differentiation Antigens Detected by Monoclonal Antibodies. Springer-Verlag, New York.

Bertoglio JH. 1983. Monocyte independent stimulation of human B lymphocytes by phorbol myristic acetate. J Immunol 131:2279–2281.

Blazar BA, Sutton LM, Strome M. 1983. Self-stimulating growth factor production by B cell lines derived from Burkitt's lymphomas and other lines transformed in vitro by Epstein-Barr virus. Cancer Res 43:4562–4568.

Butcher EC, Scollay RG, Weissman IL. 1979. Lymphocyte adherence to high endothelial venules. J Immunol 123:1996-2002.

Cohen S. 1983. Epidermal growth factor. Annu Rev Biochem 52:1-27.

Dvoretsky P, Wood GS, Levy R, Warnke R. 1982. T-lymphocyte subsets in follicular lymphomas. Hum Pathol 13:618–625.

Ford RJ. 1981. Characterization of normal and malignant human lymphoid cell populations. Cancer Bulletin 33:255–260.

Ford RJ and Maizel AL. 1982. Immunobiology of lymphoreticular neoplasms. *In* Twomey J (ed), The Pathophysiology of Human Immunologic Disorders. Urban and Schwartzenberg, Baltimore, pp. 199-217.

Ford RJ, Mehta SR, Franzini D, Montagna R, Maizel A. 1981. Soluble factor activation of human B lymphocytes. Nature 294:161-163.

Ford RJ, Mehta SR, Maizel AL. 1982. Growth factors in Hodgkin's disease. Cancer Treat Rep 66:633–638.

Ford RJ, Cramer M, Davis F. 1984a. Identification of human lymphoma cells by antisera to malignancy associated nucleolar antigens. Blood 63:559–563.

Ford RJ, Kouttab N, Davis F. 1984b. Growth factor-mediated proliferation in non-Hodgkin's lymphomas. Submitted for publication.

Ford RJ, Ramirez I, Sahasrabuddhe CG. 1984c. Autocrine stimulation in human lymphoid neoplasms. In preparation.

Gordon J, Ley SC, Melamed MD, Hughes-Jones NC. 1984. Soluble factor requirements for the autostimulatory growth of B lymphoblasts immortalized by Epstein-Barr virus. J Exp Med 159:1554–1559.

Hansen JA, Good RA. 1974. Malignant disease of the lymphoid system in immunological perspective. Hum Pathol 5:567–599.

Harris NL, Data RE. 1982. The distribution of neoplastic and normal B lymphoid cells in nodular lymphomas. Hum Pathol 13:610–617.

Levy R, Warnke R, Dorfman R, Haimovich J. 1977. The monoclonality of human B cell lymphomas. J Exp Med 145:1014–1028.

Lu M, Davis F, Ford RJ. 1984. Simultaneous determination of lymphoid lineage and malignancy on human lymphoma cells. Submitted for publication.

Luckasen JR, White JG, Kersey JH. 1974. Mitogenic properties of calcium ionophore, A 23187. Proc Natl Acad Sci USA 71:5088–5090.

Lukes RJ, Taylor CR, Parker JW, Pattengale P, Tindle BH. 1978. A morphologic and immunologic surface marker study of 299 non-Hodgkin lymphomas. Am J Pathol 90:461–468.

Maizel AL, Sahasrabuddhe C, Mehta S, Morgan J, Ford RJ. 1982. Isolation of a human B cell mitogenic factor. Proc Natl Acad Sci USA 79:5998–6002.

Morgan DA, Ruscetti FW, Gallo RC. 1976. Selective in vitro growth of T lymphocytes from normal human bone marrows. Science 193:1007–1008.

Nadler LM, Ritz J, Reinherz EL, Schlossman SF. 1981. Cellular origins of human leukemia and lymphomas. *In* Knapp W (ed), Leukemic Markers. Academic Press, New York, pp. 1–17.

Salmon S, Seligmann M. 1974. B cell neoplasm in man. Lancet 2:1229–1231.

Seligmann M, Preud'Homme J-L, Kourlisky. 1975. Membrane Receptors on Lymphocytes. Elsevier, New York.

Smith KA. 1984. Interleukin 2. Ann Rev Immunol 2:319–333.

Smith KA, Lachman LB, Oppenheim JJ, Favata M. 1980. The functional relationship of the interleukins. J Exp Med 151:1551–1555.

Sporn MB, Todaro G. 1980. Autocrine stimulation and malignant transformation of cells. N Engl J Med 303:878–883.

Stein RS. 1981. Clinical features and clinical evaluation of Hodgkin's disease and the non-Hodgkin's lymphomas. *In* Bennett JM (ed), Lymphoma 1. Nijhoff, Boston, pp. 129-175.

van Valk P, van der Loo EM, Daha MR, Meijer C. 1984. Analysis of lymphoid and dendritic cells in human lymph node, tonsil, and spleen. Virchows Arch [Pathol Anat] 43:169–178.

Mediators in Cell Growth and Differentiation,
edited by Richard J. Ford and Abby L. Maizel.
Published by Raven Press, New York 1985.

Human Proto-oncogenes, Growth Factors, and Cancer

Stuart A. Aaronson, Keith C. Robbins, and Steven R. Tronick

National Cancer Institute, Bethesda, Maryland 20205

Investigations of acute transforming retroviruses have led to important insights concerning a small group of cellular genes with transforming potential. Acute retroviruses have arisen in nature by recombination of replication-competent type C retroviruses with evolutionarily well conserved cellular genes, termed proto-oncogenes. The transduced cellular *(onc)* sequences confer on the virus properties essential for the induction and maintenance of transformation (Weiss et al. 1982, Duesberg 1983, Bishop 1983).

Recent studies have provided evidence that proto-oncogenes can be activated as oncogenes by other mechanisms as well. Of particular importance in this regard is accumulating evidence that proto-oncogenes may be frequent targets for genetic alterations involved in neoplastic processes affecting human cells. In this review, we summarize studies that have led to identification of the first normal cellular function of a proto-oncogene and present evidence suggesting how this gene may be involved in certain kinds of human cancer. We also describe investigations of another family of proto-oncogenes that have been more firmly implicated as human transforming genes but whose normal function remains to be determined.

THE *sis* TRANSFORMING GENE

The *sis* oncogene was first identified as the transforming gene of the simian sarcoma virus (SSV). SSV is the only primate-derived acutely transforming retrovirus (Theilen et al. 1971). SSV, like other sarcoma viruses, transforms cultured fibroblasts (Aaronson 1973) and induces fibrosarcomas and glioblastomas in appropriate host animals (Wolfe et al. 1971, 1972).

The cloning of biologically active SSV DNA made possible a detailed analysis of this transforming viral genome (Robbins et al. 1981). When the cloned 5.8 kbp SSV genome was compared to the genome of its helper virus, SSAV, by restriction enzyme mapping and heteroduplex analyses, a restriction fragment approximately 1 kbp long that was not detectably related to SSAV sequences could be localized toward the 3′ end of the SSV genome. In accordance with convention, this segment was designated v-*sis*.

In order to determine the origin of the *sis* sequence, retroviral DNAs and normal cellular DNAs were analyzed by Southern blotting using a molecularly cloned v-*sis*-specific probe. Retroviral genomes failed to hybridize with *sis*; however, *sis*-related restriction fragments were found at low copy number in the DNAs of species as diverse as humans and quails. These findings suggested that recombination between SSAV and a normal cellular gene led to the creation of SSV (Robbins et al. 1981).

In order to determine the species of origin of *sis*, we measured the extent to which a v-*sis* probe could anneal to DNAs of various species. DNAs derived from New World primates annealed with this probe to a significantly greater degree than the DNAs isolated from any other species tested. Furthermore, hybrids formed between *sis* and woolly monkey cellular DNA exhibited the same thermal stability as that of the homologous v-*sis* DNA duplex. These results strongly implied that v-*sis* arose from the woolly monkey genome (Robbins et al. 1982a).

v-*sis* Is Required for the Oncogenic Activity of SSV

In order to determine whether v-*sis* is required for SSV to exhibit transforming activity, a series of deletion mutants was constructed from molecularly cloned SSV DNA. These DNAs were then tested for their ability to transform NIH/3T3 cells in transfection assays. All constructions that resulted in the deletion of *sis* sequences were unable to transform NIH/3T3 cells (Robbins et al. 1982b). For example, deletion of the 3′ LTR had no effect on transforming efficiency. However, when the deletion was extended 82 base pairs into *sis*, no activity could be detected. Similarly, a mutant that lacked 250 bp at the 5′ end of *sis* also was inactive. By this analysis, the transforming gene of SSV could be localized to v-*sis*.

Nucleotide Sequence Analysis of SSV

The potential coding capacity of v-*sis* was revealed by the determination of the complete nucleotide sequence of SSV (Devare et al. 1983). An open reading frame 271 codons long encompassed v-*sis* and also included helper sequences (254 bp) 5′ to v-*sis*. The termination codon was located 347 bp upstream from the 3′ end of *sis*. By sequence comparison with Moloney murine leukemia virus (MuLV) and SSAV, the 5′ end of the open reading frame was identified as the amino terminus of the SSAV *env* gene. The carboxyl terminus of the SSAV *env* gene encoding p15E was found to flank *sis* at its 3′ end. Thus, creation of SSV involved a large deletion of the SSAV *env* gene and substitution of v-*sis*.

How might a *sis* gene product or products be synthesized? In addition to the first methionine of the open reading frame at position 3657, two methionines are predicted to occur prior to the start of v-*sis* sequences (positions 3969 and 3792). RNA splicing acceptor sites reside near each of these methionines. One of the ATG codons corresponds to that proposed as the initiator of the Moloney MuLV *env* gene product (Shinnick et al. 1981). Thus, an intriguing possibility is that the *sis* gene uses the same sequences for the initiation of its transcription and translation

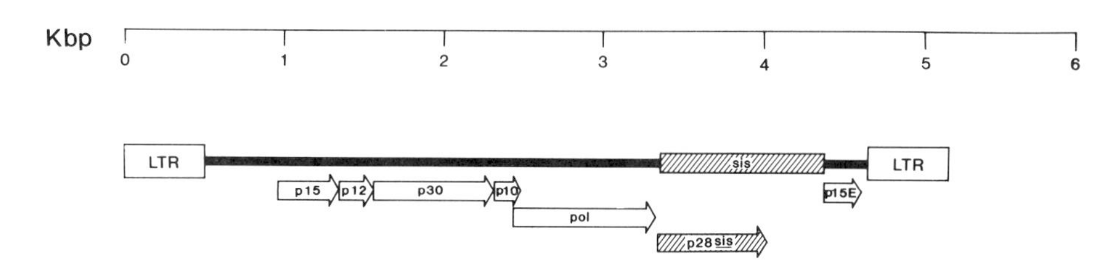

FIG. 1. The genome of SSV. The figure depicts the structural organization of biologically active SSV DNA and also indicates its gene products.

as the *env* gene it replaced. In fact, the SSV genome codes for a 2.7 kbp mRNA that contains LTR and *sis* sequences, but does not hybridize with *gag*-specific probes (Igarashi, H., Aaronson, S. A., and Robbins, K. C., unpublished observations). A message of this size could accommodate enough information required to code for proteins of 33,000, 30,000, and 28,000 MW predicted to initiate from the first three methionines, in order, in the v-*sis* open reading frame.

Identification of the SSV Transforming Protein

The task of identifying a *sis*-encoded protein could now be readily attempted by synthesizing peptides predicted by the nucleotide sequence and preparing antibodies to them (Robbins et al. 1982b). One such antibody, directed toward amino terminal sequences of the predicted *sis* protein, specifically precipitated a 28,000 MW polypeptide from extracts of SSV-transformed cells. In contrast, antibodies against the SSAV *env* gene product did not precipitate this protein. These findings suggested that the translation of the *sis* protein commenced at the third ATG of the open reading frame. Further confirmation that the 28,000 MW polypeptide ($p28^{sis}$) is coded for by the v-*sis* open reading frame was obtained by the demonstration that cyanogen bromide fragments predicted by the sequence could be detected with the anti-*sis* peptide serum. Additionally, a carboxyl terminal peptide antiserum specifically precipitated $p28^{sis}$ from SSV-transformed cells. The structural features of the SSV genome and its gene products are summarized in Figure 1.

Having identified the SSV-transforming protein, we attempted to identify its function. Studies on the origin of oncogenes carried out over the last few years have shown in every case that oncogene sequences are highly conserved among eukaryotic species. Thus, mammalian v-*onc* probes detect specific restriction fragments in the genomes of evolutionarily distant species such as flies and worms (Shilo and Weinberg 1981), and most recently, even in yeasts (DeFeo-Jones et al. 1983). Such evolutionary conservation must imply that proto-oncogenes function in critical cellular processes such as growth and development.

Elucidation of the molecular structure of retroviral *onc* genes has led to the detection, in many instances, of their protein products. The prototype of one group of *onc* genes, the *src* gene of Rous sarcoma virus, possesses tyrosine-specific protein kinase activity (Collett and Erikson 1978, Levinson et al. 1978). Several other v-*onc*-encoded proteins, although lacking in detectable protein kinase activity, nevertheless share partial amino acid sequence similarities with the known retroviral tyrosine kinases (Bishop 1983). The *ras* oncogene family encodes proteins possessing GTP-binding activity (Scolnick et al. 1979), and other *onc* proteins are thought to interact with substrates in the cell nucleus, due to their localization in this cell compartment (Alitalo et al. 1983, Hann et al. 1983, Curran et al. 1984). The transforming protein of SSV displayed none of the known properties of other characterized transforming gene products.

The SSV-Transforming Protein is Related to Human Platelet–Derived Growth Factor

The rapid proliferation of nucleotide sequence data, and thus of predicted amino acid sequences for numerous proteins, in combination with the development of sequence data banks and computer programs for rapidly searching for similarities among these sequences (Wilbur and Lipman 1983), has recently led to a number of striking observations. One such discovery was that the amino acid sequence of a peptide chain of platelet-derived growth factor (PDGF), a potent human connective tissue cell mitogen, highly matched a segment of the predicted v-*sis* sequence (Doolittle et al. 1983, Waterfield et al. 1983). Thus, for the first time, a direct link between an *onc* gene and a known biological function had been established.

In active PDGF preparations, two peptides have been identified, designated PDGF-1 and PDGF-2 by Antoniades and Hunkapiller (1983). At their amino terminal ends, these peptides share 8 of 19 residues without the introduction of gaps in their sequences. The predicted *sis* coding sequence, starting at residue 67, demonstrated an 84% match to PDGF-2 over this same stretch. Furthermore, in the total of 70 PDGF-2 residues identified, 87.1% corresponded to the $p28^{sis}$ sequence. Taking into account the New World primate origin of v-*sis* (Robbins et al. 1982a), it was concluded that the v-*sis* transforming gene arose by recombination between the SSAV genome and a host cell gene for PDGF or a very highly related protein. Thus, all the observed amino acid differences between human PDGF-2 and v-*sis* could be accounted for by the known degree of divergence between the genomes of humans and cebids (Wilson et al. 1977).

Efforts were undertaken to establish directly that the v-*sis* gene product and human PDGF shared structural, immunological, and biological properties (Robbins et al. 1983). We utilized antibodies directed against peptides synthesized on the basis of the predicted N- and C-termini of $p28^{sis}$ (Devare et a. 1983) to study its biogenesis. Marmoset cells infected with SSV were pulse-labeled with [^{35}S]methionine for various times, extracted, and then immunoprecipitated with the appropriate antisera. The immunoprecipitates were then analyzed on sodium dodecyl sulfate-polyacrylamide gels in the presence or absence of reducing agent. The following events in the biogenesis and posttranslational modification of the SSV-transforming protein were shown to occur. First, a single 28,000-Da peptide chain is synthesized that rapidly undergoes disulfide bond–mediated dimerization to a 56,000 Da species. The 56 kDa dimer is then cleaved at its amino terminal end. The products of these proteolytic events detected under reducing conditions are single chains of 11 kDa (amino terminal) and 20 kDa (C-terminal) peptides (designated p11^{sis} and p20^{sis}, respectively).

Other steps involved in the processing of p28^{sis} were uncovered by studies utilizing anti-PDGF serum. Under nonreducing conditions, the same sized species as were precipitated by the anti-*sis* C-terminal serum were detected. In addition, the PDGF antiserum detected a 24-kDa protein that was not recognized by the *sis* N- and C-terminal specific antisera. This species appeared to be the most stable processed form of the *sis* gene product. These results demonstrated that the *sis* protein also undergoes cleavage at its 3′ end.

In addition to the close structural homology between p28^{sis} and PDGF demonstrated by the antibody studies, the intracellular forms of the *sis* protein bear striking similarities to protein species in biologically active PDGF preparations. Thus, PDGF is composed of molecules ranging from 28,000 to 35,000 Da that upon reduction are converted to peptides ranging from 12,000 to 18,000 Da (Antoniades et al. 1979, Heldin et al. 1981, Deuel et al. 1981, Raines and Ross 1982). In fact, a proteolytic cleavage signal (Lys-Arg) is present at residues 65–66 in the v-*sis* sequence, and the next residue commences the homology between v-*sis* and PDGF-2. Cleavage here results in a peptide of approximately 20,000 Da, which corresponds closely in size to PDGF-2.

The Human *sis* Proto-oncogene

The demonstration of striking structural and immunological similarities between a known human growth factor and a viral oncogene product presented the opportunity to define, at the molecular level, the role of the *sis* proto-oncogene in human malignancies. *Sis*-related mRNAs have been detected in human tumor cells originating from connective tissue, but not in epithelial tumor cells (Eva et al. 1982). Moreover, human glioblastoma and osteosarcoma cell lines have been reported to produce PDGF-like polypeptides (Heldin et al. 1980, Nister et al. 1982, Graves et al. 1983). Genetic alterations affecting the transcription or translation of the *sis* proto-oncogene might induce sustained, inappropriate cell division of human cells responsive to the mitogenic activity of a PDGF-like molecule. Thus, activation of the *sis* locus might be implicated in the malignant transformation of certain types of human cells.

In order to characterize the *sis*/PDGF-2 locus, we isolated v-*sis*-related sequences from a bacteriophage library of normal human DNA (Chiu et al. 1984). These clones represented a continuous stretch of approximately 30 kbp. By Southern blotting analysis and hybridization with a v-*sis* probe, six v-*sis*-homologous restriction fragments were identified that could be localized within a 15-kbp region. Nucleotide sequence analysis of the v-*sis*-related regions was performed, and it demonstrated that an open reading frame was contained within the first five c-*sis* (human) exons. The 5′-most exon lacked a translation initiation codon in its open reading frame (Josephs et al. 1984, Chiu et al. 1984). Thus, c-*sis* coding sequences are incompletely represented in SSV. This conclusion is further supported by the observation that a 4.2-kbp *sis*-related transcript is present in human cells (Eva et al. 1982). When the predicted c-*sis* (human) coding sequence was compared with that of the polypeptides representing PDGF-2, there was complete homology except at two of a total of 104 residues (Chiu et al. 1984) (Figure 2). These findings strongly imply that c-*sis* (human) is the structural gene for PDGF-2.

Expression of the Normal Coding Sequence for Human PDGF-2/*sis* Causes Cellular Transformation

The transcription of c-*sis* (human) was studied by using probes representing 30 kbp of the human *sis*/PDGF-2 locus. A 4.2-kb mRNA in the A2781 human tumor cell line was detected by a v-*sis* probe and by c-*sis* probes representing each of the c-*sis* exons (Igarashi, H., Aaronson, S. A., and Robbins, K.C., unpublished data). The only other probe that hybridized to the same sized message was derived from the region from 0 to 0.6 kbp on the c-*sis* map. This probe (pc-*sis*l) did not detect v-*sis* RNA and could represent an exon of the c-*sis* transcriptional unit unrelated to v-*sis* (Gazit et al. 1984).

Nucleotide sequence analysis of pc-*sis*l was performed and revealed three open reading frames initiated by a methionine codon. Donor splice sites were found at positions within pc-*sis*l that would allow for in-phase translation of each reading frame when spliced to the acceptor splice site in the first v-*sis*-related exon.

The DNA of a c-*sis* (human) phage clone (λ-c-*sis* clone 8) that contained all of the PDGF-2 coding sequences (Chiu et al. 1984) was introduced into NIH/3T3 cells via transfection in order to study the structural requirements for c-*sis* gene expression. The sequences contained within clone 8 were not capable of transforming cells, nor did they synthesize transcripts as determined by cotransfection experiments using a selectable marker gene (pSV2-gpt). Positioning a retroviral long terminal repeat (LTR) upstream of the λ-c-*sis* clone 8 coding regions failed to confer transforming activity.

Since the putative upstream exon of c-*sis* (human) contained a potential amino terminal sequence for a PDGF/*sis* precursor, but not identifiable promotor signals, this segment was ligated in the proper orientation between the retroviral LTR and first exon of c-*sis* (human) clone 8. Upon transfection of NIH/3T3 cells with this molecularly cloned construct, transforming activity comparable to that observed for SSV DNA was observed (10^5 ffu/pmol). The foci induced by the human construct and by SSV DNA were morphologically indistinguishable (Gazit et al. 1984). These findings established that the normal human gene encoding PDGF-2/*sis* is capable of acquiring transforming activity when expressed in a cell susceptible to the effects of PDGF. As such, they provide direct proof for the accuracy of a model in which the constitutive expression of the gene for a normal growth factor in a cell susceptible to its growth-promoting actions can result in cell transformation.

```
                  ┌→(v-sis-helper virus junction)      ┌→(exon 2)               60          ┌→(exon 3)
p28sis   MTLTWOGDPI PEELYKMLSG HSIRSFDDLQ RLLQGDSGKE DGAELDLNMT RSHSGGELES LAPGKRSLGS LSVAEPAMIA
c-sis         ***** *****E***D ********** ***H**P*E* ********** ********** ****R***** *TI*******
PDGF-2                                                                             **** *TI*******

                                          120                   ┌→(exon 4)
p28sis   ECKTRTEVFE ISRRLIDRTN ANFLVWPPCV EVQRCSGCCN NRNVQCRPTQ VQLRPVQVRK IEIVRKKPIF KKATVTLEDH
c-sis    ********** ********** ********** ********** ********** ********** ********** **********
PDGF-2   *****E***C *C***?**?? ??????**** **K**T**** ****K***S* *****?**** ******

                    180     ┌→(exon 5)                                   226
p28sis   LACKCEIVAA ARAVTRSPGT SQEQRAKTTQ SRVTIRTVRV RRPPKGKHRK CKHTHDKTAL KETLGAtrm
c-sis    ******T*** **P******G ********P* T********* ********** F********* ******trm
PDGF-2
```

FIG. 2. Comparison of amino acid sequences of p28[sis], c-*sis* (human), and PDGF-2. The predicted amino acid sequence of p28[sis], shown in the one-letter amino acid code, is given in the top line. Shared residues are indicated by *. The location of c-*sis* (human) exons relative to the v-*sis* nucleotide sequence is indicated, as is the 5′ helper virus v-*sis* junction. The PDGF-2 sequence is taken from Antoniades and Hunkapiller (1983). One-letter code: A = alanine, R = arginine, N = asparagine, D = aspartic acid, C = cysteine, Q = glutamine, E = glutamic acid, G = glycine, H = histidine, I = isoleucine, L = leucine, K = lysine, M = methionine, F = phenylalanine, P = proline, S = serine, T = threonine, W = tryptophan, Y = tyrosine, V = valine.

THE *ras* PROTO-ONCOGENES

Unlike the *sis* proto-oncogene product, whose normal function has been identified, the physiological role of the *ras* family gene product, p21, is unknown. However, much is known about the involvement of *ras* genes in the steps leading to malignant transformation of human cells.

Identification of *ras* Genes

Studies on the Kirsten and Harvey strains of murine sarcoma viruses (Kirsten-MSV, Harvey-MSV) led eventually to the identification of a small group of highly conserved cellular oncogenes, designated the *ras* family. Early research on Harvey-MSV and Kirsten-MSV, isolated from rats infected with Moloney and Kirsten leukemia viruses, respectively, demonstrated that each was genetically related to the other in non-helper virus-derived regions of the genome (Shih et al. 1978). With the advent of molecular cloning techniques, it was possible to demonstrate that most of the non-helper-derived sequences were contributed by endogenous rat retroviral information (Ellis et al. 1981). However, an additional set of sequences in each viral genome, designated *ras*, was shown to be derived from the cellular genome and was not homologous to any known viral sequences. The *ras* sequences in Kirsten-MSV and Harvey-MSV were found to be only partially related and thus represent different members of a gene family (Ellis et al. 1981).

Subsequent to the isolation of Kirsten-MSV and Harvey-MSV, two other murine sarcoma viruses, BALB-MSV and Rasheed-MSV, were shown to contain *ras* oncogenes derived from the BALB/c mouse and Fisher rat, respectively (Andersen et al. 1981, Gonda et al. 1982). Another *ras* family member, designated N-*ras*, has been identified in mammalian cells, but has not been identified to date as a transforming gene of any known retrovirus (Shimizu et al. 1983a, Eva et al. 1983).

ras Proto-oncogenes of Human Cells

Three *ras* genes of human cells have been molecularly cloned and characterized in detail. The organization of their coding sequences is similar in that each gene contains four exons; however, the exons are distributed over a region anywhere from 4.8 kbp (H-*ras*) to 45 kbp (K-*ras*) in length (Chang et al. 1982, Santos et al. 1982, Shimizu et al. 1983b). Pseudogenes of H-*ras* and K-*ras* have also been identified (designated H-*ras*-2 and K-*ras*-1) (Chang et al. 1982). Each *ras* locus has been shown to reside on a separate human chromosome as shown in Table 1. The molecular cloning and nucleotide sequence analysis of *ras* genes of yeast has demonstrated a remarkable degree of evolutionary conservation of *ras* gene structure (Powers et al. 1984).

ras Genes Are Frequently Detected as Human Transforming Genes

The involvement of the *ras* gene family in naturally occurring malignancies came to light in studies in which investigators asked whether DNAs of animal or human

TABLE 1. *Detection of* ras *oncogenes in human tumors*

Tumor source	Percent positive	*ras* Oncogene activated		
		H-*ras*	K-*ras*	N-*ras*
Carcinoma (lung, gastrointestinal, genitourinary)	10–30	4/12	6/12	2/12
Sarcoma (fibrosarcoma, rhabdomyosarcoma)	~10	0/2	0/2	2/2
Hematopoietic (AML, CML, ALL, CLL)	10–50	0/9	1/9	8/9

AML, acute myelogenous leukemia; CML, chronic myelogenous leukemia; ALL, acute lymphocytic leukemia; CLL, chronic lymphocytic leukemia.

tumor cells possessed the capacity to directly confer the neoplastic phenotype on a susceptible assay cell. Some human tumor DNAs were shown to induce transformed foci in the continuous NIH/3T3 mouse cell line, which is highly susceptible to the uptake and stable incorporation of exogenous DNA (Cooper 1982, Weinberg 1982).

The first molecularly cloned human transforming gene, whose source was the T24 bladder carcinoma cell line, was demonstrated to be the activated homologue of the normal H-*ras* gene (Goldfarb et al. 1982, Pulciani et al. 1982a, Shih and Weinberg 1982). Subsequent analysis of oncogenes detected by transfection assays has established that the majority of such transforming genes belong to the *ras* family. For example, K-*ras* oncogenes have been detected at high frequency in lung and colon carcinomas (Pulciani et al. 1982b). Carcinomas of the digestive tract, including pancreas and gall bladder, genitourinary tract tumors, and sarcomas have also been shown to contain *ras* oncogenes. N-*ras* appears to be the most frequently activated *ras* transforming gene in human hematopoietic neoplasms (Eva et al. 1983). These results are summarized in Table 1.

Not only can a variety of tumor types contain the same activated *ras* oncogene, but the same tumor type can contain different activated *ras* oncogenes. Thus, in hematopoietic tumors, we have observed different *ras* oncogenes (K-*ras*, N-*ras*) activated in lymphoid tumors at the same stage of hematopoietic cell differentiation, as well as N-*ras* genes activated in tumors as diverse in origin as acute and chronic myelogenous leukemia (Eva et al. 1983) (Table 1). These findings suggest that *ras* oncogenes detected in the NIH/3T3 transfection assay are not specific to a given stage of cell differentiation or tissue type.

Retroviruses that contain *ras*-related *onc* genes are known to possess a wide spectrum of target cells for transformation in vivo and in vitro. In addition to inducing sarcomas and transforming fibroblasts (Gross 1970), these viruses are capable of inducing tumors of immature lymphoid cells (Pierce and Aaronson 1982). They also can stimulate the proliferation of erythroblasts (Hankins and Scolnick 1981) and monocyte/macrophages (Greenberger 1979) and can even induce alterations in the growth and differentiation of epithelial cells (Weissman and

Aaronson 1983). Thus, the wide array of tissue types that can be induced to proliferate abnormally by these *onc* genes may help to explain the high frequency of detection of their activated human homologues in diverse human tumors.

It should be noted that not all oncogenes detected by NIH/3T3 transfection analysis are related to known retroviral oncogenes. For example, Cooper and coworkers have identified and molecularly cloned an oncogene, B-*lym*, from a B-cell lymphoma (Goubin et al. 1983). B-*lym* appears to be activated in a large proportion of tumors at a specific stage of B-cell differentiation (Diamond et al. 1983). Studies to date indicate that this oncogene is relatively small (<600 bp), and sequence analysis indicates that it possesses distant homology to transferrin (Goubin et al. 1983). These investigators have also detected oncogenes that appear to be specifically activated in tumors at other stages of lymphoid differentiation (Lane et al. 1982) or in mammary carcinomas (Lane et al. 1981). None of these transforming genes appear to possess detectable homology with known retroviral *onc* sequences.

Mechanism of Activation of *ras* Oncogenes

The availability of molecular clones of the normal and activated alleles of human *ras* proto-oncogenes made it possible to determine the molecular mechanisms responsible for the malignant conversion of these genes. The genetic lesions responsible for activation of a number of *ras* oncogenes have been localized to single-base changes in their p21 coding sequences. In the T24/EJ bladder carcinoma oncogene, a transversion of a G to a T causes a valine residue to be incorporated instead of a glycine in the 12th position of the predicted p21 primary structure (66-69). During our analyses of human cells for transforming DNA sequences, we were able to isolate and molecularly clone an oncogene from a human lung carcinoma, designated Hs242 (Yuasa et al. 1983). This gene was also identified as an activated H-*ras* (human) proto-oncogene, making it possible to compare the mechanisms by which the same human proto-oncogene has been independently activated in human tumor-derived cells.

The Hs242 transforming sequence was isolated and subjected to restriction enzyme analysis in order to compare its physical map with that of the previously reported T24/EJ bladder tumor oncogene (Yuasa et al. 1983), as well as c-H-*ras* cloned from a normal human fetal liver library (Santos et al. 1982). The restriction map of the Hs242 oncogene closely corresponded with both, diverging only outside the region previously shown to be required for the transforming activity of the T24 oncogene.

To map the position of the genetic lesion in Hs242 leading to its malignant activation, recombinants were constructed in which homologous sequences of H-*ras* (human) were substituted for fragments of the Hs242 oncogene. By this analysis the genetic alteration that activated the Hs242 oncogene was localized to a 0.45-kbp region that encompassed its second coding exon. Nucleotide sequence analysis of this region revealed that the Hs242 oncogene and H-*ras* (human) differed at a single base within codon 61 (Figure 2B). The change of an A to a T resulted in

TABLE 2. *Genetic lesions that activate* ras *oncogenes of human tumors*

ras Oncogene	Tumor	Base/amino acid change	Codon no.	Reference
H-*ras*				
T24	Bladder carcinoma	Gly→Val (GGC→GTC)	12	Tabin et al. 1982, Reddy et al. 1982, Taparowsky et al. 1983b, Capon et al. 1983a
Hs0578	Mammary carcinoma	Gly→Asp (GGC→GAC)	12	Kraus et al. 1984
Hs 242	Lung carcinoma	Gln→Leu (CAG→CTG)	61	Yuasa et al. 1983
K-*ras*				
Calu-1	Lung carcinoma	Gly→Cys (GGT→TGT)	12	Shimizu et al. 1983b, Capon et al. 1983b
SW480	Colon carcinoma	Gly→Val (GGT→GTT)	12	Capon et al. 1983b
N-*ras*				
SK-N-SH	Neuroblastoma	Gln→Lys (CAA→AAA)	61	Taparowsky et al. 1983a
SW1271	Lung carcinoma	Gln→Arg (GGT→GTT)	61	Yuasa et al. 1984

the replacement of glutamine by leucine in this codon. Thus, a single amino acid substitution seems sufficient to confer transforming properties on the product of the Hs242 oncogene. These results also established that the site of activation in the Hs242 oncogene was totally different from that of the T24/EJ oncogene (Yuasa et al. 1983).

In subsequent studies, we have assessed the generality of point mutations as the basis for acquisition of malignant properties by *ras* proto-oncogenes by molecularly cloning and analyzing other activated *ras* oncogenes (Table 2). Activation of an H-*ras* transforming gene of the Hs578T human breast carcinosarcoma line has been localized to a point mutation at position 12 that changes glycine to aspartic acid in the amino acid sequence (Kraus et al. 1984). Recently Wigler and co-workers (Taparowsky et al. 1983a) reported that the lesion leading to activation of the N-*ras* oncogene in a neuroblastoma line was due to the alteration of codon 61 from CAA to AAA, causing the substitution in this case of lysine for glutamine. Another N-*ras* transforming gene, this one isolated from a human lung carcinoma cell line, SW1271, has been shown to result from a single point mutation of an A to a G at position 61 in the coding sequence, resulting in the substitution of arginine for glutamine (Yuasa et al. 1984).

Investigators analyzing K-*ras* oncogenes (Shimuzu et al. 1983b, Capon et al. 1983b) have achieved strikingly similar results. In the two K-*ras* transforming genes so far analyzed, single-point mutations in the 12th codon have been shown to be responsible for acquisition of malignant properties. Thus, mutations at positions 12 or 61 appear to be the genetic lesions most commonly responsible for activation of *ras* oncogenes under natural conditions in human tumor cells (Table 2).

IMPLICATIONS

The pathways leading toward the malignant state are rapidly becoming better understood. The identification of the function of the *sis* proto-oncogene as a human growth factor has been followed by the discovery that amino acid sequences of human epidermal growth factor receptor are highly related to those predicted for the *erb*-B oncogene product (Downward et al. 1984). Thus, perturbations in at least two of the steps by which growth factors exert their proliferative effects may lead to transformation.

The number of steps in the neoplastic process may be relatively limited as indicated by studies on the effects of the simultaneous addition of oncogenes to cells either by transfection or virus infection. Rat embryo fibroblasts in culture can be successfully transformed when transfected by *ras* and *myc* or *ras* and the adenovirus Ela gene, but do not become malignant when transfected by only one (Land et al. 1983, Ruley 1983). These findings also suggest the existence of complementing groups of oncogenes. It is likely that more proto-oncogenes will be discovered. The identification of such genes, their normal functions, and their mechanisms of activation as oncogenes in human malignancies may be of critical importance in devising clinically applicable strategies for cancer diagnosis, detection, and prevention.

REFERENCES

Aaronson SA. 1973. Biologic characterization of mammalian cells transformed by a primate sarcoma virus. Virology 52:562–567.

Alitalo K, Ramsay G, Bishop JM, Pfeifer SO, Colby WW, Levinson AD. 1983. Identification of nuclear proteins encoded by viral and cellular *myc* oncogenes. Nature 306:274–277.

Andersen PR, Devare SG, Tronick SR, Ellis RW, Aaronson SA, Scolnick EM. 1981. Generation of BALB-MuSV and Ha-MuSV by type C virus transduction of homologous transforming genes from different species. Cell 26:129–134.

Antoniades HN, Hunkapiller MW. 1983. Human platelet-derived growth factor (PDGF): Amino terminal amino acid sequence. Science 220:963–965.

Antoniades HN, Scher CD, Stiles CD. 1979. Purification of human platelet-derived growth factor. Proc Natl Acad Sci USA 76:1809–1813.

Bishop JM. 1983. Cellular oncogenes and retroviruses. Ann Rev Biochem 52:301–354.

Capon DJ, Chen EY, Levinson AD, Seeburg PH, Goeddel DV. 1983a. Complete nucleotide sequences of the T24 human bladder carcinoma oncogene and its normal homologue. Nature 302:33–37.

Capon DJ, Seeburg PH, McGrath JP, Hayflick JS, Edman U, Levinson AD, Goeddel DV. 1983b. Activation of Ki-*ras* 2 gene in human colon and lung carcinomas by two different point mutations. Nature 304:507–513.

Chang EH, Gonda MA, Ellis RW, Scolnick EM, Lowy DR. 1982. Human genome contains four genes homologous to transforming genes of Harvey and Kirsten murine sarcoma viruses. Proc Natl Acad Sci USA 79:4848–4852.

Chiu I-M, Reddy EP, Givol D, Robbins KC, Tronick SR, Aaronson SA. 1984. Nucleotide sequence analysis identifies the human c-*sis* proto-oncogene as a structural gene for platelet-derived growth factor. Cell 37:123–129.

Collett MS, Erikson RL. 1978. Protein kinase activity associated with the avian sarcoma virus *src* gene product. Proc Natl Acad Sci USA 75:2021–2084.

Cooper GM. 1982. Cellular transforming genes. Science 218:801–806.
Curran T, Miller AD, Zokas L, Verma IM. 1984. Viral and cellular fos proteins: A comparative analysis. Cell 36:259–268.
DeFeo-Jones D, Scolnick EM, Koller R, Dhar R. 1983. Ras-related gene sequences identified and isolated from *Saccharomyces cerevisiae*. Nature 306:707–709.
Deuel TF, Huang JS, Proffit RT, Baenziger UU, Chang D, and Kennedey BB. 1981. Human platelet-derived growth factor. Purification and resolution into two active protein fractions. J Biol Chem 256:8896–8899.
Devare SG, Reddy EP, Law JD, Robbins KC, Aaronson SA. 1983. Nucleotide sequence of the siman sarcoma virus genome: Demonstration that its acquired cellular sequences encode the transforming gene product, $p28^{sis}$. Proc Natl Acad Sci USA 80:731–735.
Diamond A, Cooper GM, Ritz J, Lane MA. 1983. Identification and molecular cloning of the human *blym* transforming gene activated in Burkitt's lymphomas. Nature 305:112–116.
Doolittle RF, Hunkapiller MW, Hood LE, Devare SG, Robbins KC, Aaronson SA, Antoniades HN. 1983. Simian sarcoma virus *onc* gene, v-*sis*, is derived from the gene (or genes) encoding a platelet-derived growth factor. Science 221:275–277.
Downward J, Yarden Y, Mayes E, Scrace E, Totty N, Stockwell P, Ullrich A, Schlessinger J, Waterfield MD. 1984. Close similarity of epidermal growth factor receptor and v-*erb*-B oncogene protein sequences. Nature 307:521–526.
Duesberg PH. 1983. Retroviral transforming genes in normal cells (?). Nature 304:219–226.
Ellis RW, DeFeo D, Shih TY, Gonda MA, Young HA, Tsuchida N, Lowy DR, Scolnick EM. 1981. The p21 *src* genes of Harvey and Kirsten sarcoma viruses originate from divergent members of a family of normal vertebrate genes. Nature 292:506–511.
Eva A, Robbins KC, Andersen PR, Srinivasan A, Tronick SR, Reddy EP, Ellmore NW, Galen AT, Lautenberger JA, Papas TS, Westin EH, Wong-Staal F, Gallo RC, Aaronson SA. 1982. Cellular genes analogous to retroviral *onc* genes are transcribed in human tumor cells. Nature 295:116–119.
Eva A, Tronick SR, Gol RA, Pierce JH, Aaronson SA. 1983. Transforming genes of human hematopoietic tumors: Frequent detection of *ras*-related oncogenes whose activation appears to be independent of tumor phenotype. Proc Natl Acad Sci USA 80:4926–4930.
Gazit A, Igarashi H, Chiu I-M, Srinivasan A, Yaniv A, Tronick SR, Robbins KC, Aaronson SA. 1984. Expression of the normal human *sis*/PDGF-2 coding sequence induces cellular transformation. Cell 39 (in press).
Goldfarb M, Shimizu K, Perucho M, Wigler M. 1982. Isolation and preliminary characterization of a human transforming gene from T24 bladder carcinoma cells. Nature 296:404–409.
Gonda MA, Young HA, Elser JE, Rasheed S, Talmadge CB, Nagashima K, Li CC, Gilden RV. 1982. Molecular cloning, genomic analysis, and biological properties of rat leukemia virus and the onc sequences of Rasheed rat sarcoma virus. J Virol 44:520–529.
Goubin G, Goldman DS, Luce J, Neiman PE, Cooper GM. 1983. Molecular cloning and nucleotide sequence of a transforming gene detected by transfection of chicken B-cell lymphoma DNA. Nature 302:114–119.
Graves DT, Owen AJ, Antoniades HN. 1983. Evidence that a human osteosarcoma cell line which secretes a mitogen similar to platelet-derived growth factor requires growth factors present in platelet-poor plasma. Cancer Res 43:83–87.
Greenberger JS. 1979. Phenotypically distinct target cells for murine sarcoma virus and murine leukemia virus marrow transformation in vitro. JNCI 62:337–344.
Gross L. 1970. Oncogenic Viruses. 2nd ed. Pergamon Press, Oxford.
Hankins DW, Scolnick EM. 1981. Harvey and Kirsten sarcoma viruses promote the growth and differentiation of erythroid precursor cells in vitro. Cell 26:91–97.
Hann SR, Abrams HD, Rohrschneider LR, Eisenman RN. 1983. Proteins encoded by v-*myc* and c-*myc* oncogenes: Identification and localization in acute leukemia virus transformants and bursal lymphoma cell lines. Cell 34:789–798.
Heldin CH, Westermark B, Wasteson A. 1980. Chemical and biological properties of a growth factor from human-cultured osteosarcoma cells: Resemblance with platelet-derived growth factor. J Cell Physiol 105:235–246.
Heldin CH, Westermark B, Wasteson A. 1981. Demonstration of antibody against platelet-derived growth factor. Exp Cell Res 136:255–261.
Josephs SF, Guo C, Ratner L, Wong-Staal F. 1984. Human proto-oncogene nucleotide sequences corresponding to the transforming region of simian sarcoma virus. Science 223:487–491.

Kraus M, Yuasa Y, Aaronson SA. 1984. A position 12-activated H-*ras* oncogene in all HS578T mammary carcinosarcoma cells but not normal mammary cells of the same patient. Proc Natl Acad Sci USA 81:5384–5388.

Land H, Parada LF, Weinberg RA. 1983. Tumorigenic conversion of primary embryo fibroblasts requires at least two cooperating oncogenes. Nature 304:596–602.

Lane MA, Sainten A, Cooper GM. 1981. Activation of related transforming genes in mouse and human mammary carcinomas. Proc Natl Acad Sci USA 78:5185–5189.

Lane MA, Sainten A, Cooper GM. 1982. Stage-specific transforming genes of human and mouse B- and T-lymphocyte neoplasms. Cell 28:873–880.

Levinson AD, Opperman H, Levintow L, Varmus HE, Bishop JM. 1978. Evidence that the transforming gene of avian sarcoma virus encodes a protein kinase associated with a phosphoprotein. Cell 15:561–572.

Nister M, Heldin CH, Wasteson A, Westermark B. 1982. A platelet-derived growth factor analog produced by a human clonal glioma cell line. Ann NY Acad Sci 397:25–33.

Pierce JH, Aaronson SA. 1982. BALB- and Harvey-murine sarcoma virus transformation of a novel lymphoid progenitor cell. J Exp Med 156:873–877.

Powers S, Kataoka T, Fasano O, Goldfarb M, Strathern J, Broach J, Wigler M. 1984. Genes in *S. cerevisiae* encoding proteins with domains homologous to the mammalian *ras* proteins. Cell 36:607–612.

Pulciani S, Santos E, Lauver AV, Long LK, Robbins KC, Barbacid M. 1982a. Oncogenes in human tumor cell lines: Molecular cloning of a transforming gene from human bladder carcinoma cells. Proc Natl Acad Sci USA 79:2845–2849.

Pulciani S, Santos E, Lauver AV, Long LK, Aaronson SA, Barbacid M. 1982b. Oncogenes in solid human tumors. Nature 300:539–542.

Raines EW, Ross R. 1982. Platelet-derived growth factor. I. High yield purification and evidence for multiple forms. J Biol Chem 257:5154–5160.

Reddy EP, Reynolds RK, Santos E, Barbacid M. 1982. A point mutation is responsible for the acquisition of transforming properties by the T24 human bladder carcinoma oncogene. Nature 300:149–152.

Robbins KC, Devare SG, Aaronson SA. 1981. Molecular cloning of integrated simian sarcoma virus: Genome organization of infectious DNA clones. Proc Natl Acad Sci USA 78:2918–2922.

Robbins KC, Devare SG, Reddy EP, Aaronson SA. 1982a. In vivo identification of the transforming gene product of simian sarcoma virus. Science 218:1131–1133.

Robbins KC, Hill RL, Aaronson SA. 1982b. Primate origin of the cell-derived sequences of simian sarcoma virus. J Virol 41:721–725.

Robbins KC, Antoniades HN, Devare SG, Hunkapiller MW, Aaronson SA. 1983. Structural and immunological similarities between simian sarcoma virus gene product(s) and human platelet-derived growth factor. Nature 305:605–608.

Ruley HE. 1983. Adenovirus early region 1A enables viral and cellular transforming genes to transform primary cells in culture. Nature 304:602–606.

Santos E, Tronick SR, Aaronson SA, Pulciani S, Barbacid M. 1982. T24 human bladder carcinoma oncogene is an activated form of the normal human homologue of BALB- and Harvey-MSV transforming genes. Nature 298:343–347.

Scolnick EM, Papageorge AG, Shih TY. 1979. Guanine nucleotide-binding activity as an assay for *src* protein of rat-derived murine sarcoma viruses. Proc Natl Acad Sci USA 76:5355–5359.

Shih C, Weinberg RA. 1982. Isolation of a transforming sequence from a human bladder carcinoma cell line. Cell 29:161–169.

Shih TY, Williams DR, Weeks MO, Maryak JM, Vass WC, Scolnick EM. 1978. Comparison of the genomic organization of Kirsten and Harvey sarcoma viruses. J Virol 27:45–55.

Shilo BZ, Weinberg RA. 1981. DNA sequences homologous to vertebrate oncogenes are conserved in *Drosophila melanogaster*. Proc Natl Acad Sci USA 78:6789–6792.

Shimizu K, Goldfarb M, Perucho M, Wigler M. 1983a. Isolation and preliminary characterization of the transforming gene of a human neuroblastoma cell line. Proc Natl Acad Sci USA 80:383–387.

Shimizu K, Birnbaum D, Ruley MA, Fasano O, Suard Y, Edlund L, Taparowsky E, Goldfarb M, Wigler M. 1983b. Structure of the Ki-*ras* gene of the human lung carcinoma cell line Calu-1. Nature 304:497–500.

Shinnick TM, Lerner RA, Sutcliffe JG. 1981. Nucleotide sequence of Moloney murine leukemia virus. Nature 293:543–548.

Tabin CJ, Bradley SM, Bartmann CI, Weinberg RA, Papageorge AG, Scolnick EM, Dhar R, Long DR, Chang EH. 1982. Mechanism of activation of a human oncogene. Nature 300:143–149.
Taparowsky E, Shimizu K, Goldfarb M, Wigler M. 1983a. Structure and activation of the human N-*ras* gene. Cell 34:581–586.
Taparowsky E, Suard Y, Fasano O, Shimizu K, Goldfarb M, Wigler M. 1983b. Activation of the T24 bladder carcinoma transforming gene is linked to a single amino acid change. Nature 300:762–765.
Theilen GH, Gould D, Fowler M, Dungworth DL. 1971. C-type virus in tumor tissue of a woolly monkey *(Lagothrix* spp.) with fibrosarcoma. JNCI 47:881–899.
Waterfield MD, Scrace GT, Whittle N, Stroobant P, Johnsson A, Wasteson A, Westermark B, Heldin CH, Huang JS, Deuel TF. 1983. Platelet-derived growth factor is structurally related to the putative transforming protein $p28^{sis}$ of simian sarcoma virus. Nature 304:35–39.
Weinberg RA. 1982. Fewer and fewer oncogenes. Cell 30:3–4.
Weiss RA, Teich N, Varmus H, Coffin RJ (eds). 1982. Molecular Biology of Tumor Viruses, RNA Tumor Viruses, 2nd ed. Cold Spring Harbor Laboratories, Cold Spring Harbor, New York.
Weissman BE, Aaronson SA. 1983. BALB and Kirsten murine sarcoma viruses alter growth and differentiation of EGF-dependent BALB/c mouse epidermal keratinocyte lines. Cell 32:599–606.
Wilbur WJ, Lipman DJ. 1983. Rapid similarity searches of nucleic acid and protein data banks. Proc Natl Acad Sci USA 80:726–730.
Wilson AC, Carlson SS, White TJ. 1977. Biochemical evolution. Annu Rev Biochem 46:573–639.
Wolfe LG, Deinhardt F, Theilen GJ, Rabin H, Kawakami T, Bustad LK. 1971. Induction of tumors in marmoset monkeys by simian sarcoma virus, type 1 *(Lagothrix)*: A preliminary report. JNCI 47:1115–1120.
Wolfe LG, Smith RK, Dienhardt F. 1972. Simian sarcoma virus type 1 *(Lagothrix)*: Focus assay and demonstration of nontransforming associated virus. JNCI 48:1905–1907.
Yuasa Y, Srivastava SK, Dunn CY, Rhim JS, Reddy EP, Aaronson SA. 1983. Acquisition of transforming properties by alternative point mutations within c-*bas/has* human proto-oncogene. Nature 303:775–779.
Yuasa Y, Gol RA, Chang A, Chiu I-M, Reddy EP, Tronick SR, Aaronson SA. 1984. Mechanism of activation of an N-*ras* oncogene of SW-1271 human lung carcinoma cells. Proc Natl Acad Sci USA 81:3670–3674.

Interferon

Mediators in Cell Growth and Differentiation,
edited by Richard J. Ford and Abby L. Maizel.
Raven Press, New York © 1985.

Marked Cytolysis of Human Tumor Cells by Interferon Gamma and Leukocytes

Samuel Baron, Steven Tyring, Gary Klimpel, Sam Barranco,* Miriam Brysk,† Vicram Gupta,‡ and W. Robert Fleischmann, Jr.

Departments of Microbiology, Radiation Therapy, Dermatology†, and Internal Medicine‡, University of Texas Medical Branch, Galveston, Texas 77550*

Interferon can exert its antitumor action through a number of mechanisms (Gresser 1981–82, Baron 1981). These antitumor mechanisms include: (1) direct inhibition of cell multiplication; (2) direct cytolysis; (3) modulation of the immune response; (4) hormonal actions; (5) activation of effector cells such as sensitized T lymphocytes, natural killer (NK) cells, antibody-dependent cellular cytotoxicity (ADCC) effector cells, and macrophages.

We conducted a study to help determine whether some of these mechanisms might be more effective than others.

We made the assumption that cytolysis would be more likely to eliminate tumor cells rapidly than would cytostasis (such as the direct inhibition of cell division by interferon). Cytolysis by interferon alone has been reported under certain circumstances (Ito and Buffet 1981, Tyring et al. 1982, Rubin and Gupta 1980). Interferon-α may be cytolytic, but only against a few tumors. Interferon-γ, in high concentrations, is cytolytic against many tumor cells (Tyring et al. 1982). The tumors varied in their sensitivities to this cytolysis. Normal cells could also be lysed by interferon-γ.

The cytolytic activity in the preparations was shown to be due to interferon-γ by its copurification with antiviral activity, presence in an interferon-γ preparation with a specific activity of 10^8 units/mg protein, species specificity, a heat stability similar to interferon-γ and not to lymphotoxin, similar instability to low pH, and neutralization by antibodies specific to interferon-γ (Tyring et al. 1982), including an antiserum directed against an interferon-γ synthetic peptide (Johnson et al. 1982).

In further studies for interferon-induced cytolysis, we studied the three types of interferon in the presence and absence of effector leukocytes. Cultured human melanoma cells were studied first. Dramatic cytolysis of the NR melanoma cells occurred only in the presence of 200 units/ml of interferon-γ plus peripheral blood leukocytes (Figure 1). A comparison of a number of different cultured human tumors (human melanomas, adenomas, and an astrocytoma) and some normal cells

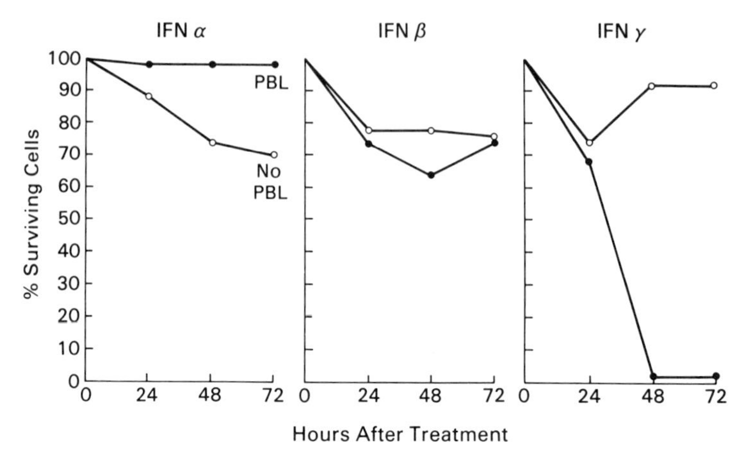

FIG. 1. Effect of human peripheral blood leukocytes (PBL) on the cytolytic action of interferons against human melanoma NR.

showed that: (1) complete elimination of about half the tumor types could be achieved in the presence of interferon-γ and peripheral blood leukocytes (Table 1); (2) interferons-α and -β were much less effective under the same conditions (data not shown); and (3) the normal cells were less affected by the cytolytic mechanism.

TABLE 1. *Effect of human peripheral blood leukocytes (PBL) on cytolysis by IFNγ*

Cells	Average % surviving cells 1–3 experiments* IFNγ†	
	No PBL	PBL‡
Transformed		
Melanoma, WM75, human	71	<1
Melanoma, NRD	66	<1
Adenocarcinoma, human	21	16
Adenocarcinoma I, human	5	<1
Astrocytoma, human	95	15
Wish, human	28	5
L929, mouse	39	<1
Normal		
Fibroblast, human	56	16
Embryo skin muscle, human	82	8

*% surviving cells at 72 hr relative to control without IFN and without PBL for the first column.

†The concentration of IFN was 200 reference units per ml.

‡A ratio of 100 PBL per target cell in each culture. The number of PBLs in actual contact with each target cell was estimated to be 3–6.

TABLE 2. *Effect of interferons and spleen cells on JB8 mouse tumor cells in culture and in mice*

Treatment		% Viable cells after 72 hours in culture	% Control mice with tumors after 9 weeks observation†
IFN	Spleen cells*		
Medium	0	100	100
IFNα/β‡	0	78	67
IFNγ§	0	84	63
IFNα/β + γ	0	9	17
Medium	+	52	100
IFNα/β	+	40	100
IFNγ	+	38	83
IFNα/β + γ	+	1	0

*Ratio of spleen cells to target cells = 100:1.
†Mice were treated at the local subcutaneous tumor site daily for the first 7 days.
‡IFNα/β used at 2×10^4 units/ml.
§IFNγ used at 2×10^3 units/ml.

Pretreatment of the effector leukocyte with interferon-γ followed by removal of interferon-γ gave the same results, indicating that the cytolysis was mediated by activated leukocytes. The leukocyte type responsible for the cytolytic activity has yet to be determined.

A parallel set of in vitro and in vivo experiments was done with mouse JB8 chemically transformed epidermal cells. As shown in Table 2, one of the best conditions for cytolysis of these mouse cells was found to be the synergistic combination of interferon-α/β with -γ (Fleischmann et al. 1984). That same combination was even more effective in the presence of mouse spleen cells. In an experiment representative of two such experiments, similar protection occurred in mice injected subcutaneously with 10^5 tumor cells and treated daily for 7 days at the local site of the tumor beginning 24 hr after tumor injection. It appears that the in vitro protection is paralleled by that in vivo for the JB8 tumor.

Present and previous studies indicate that specific conditions for tumor cell eradication must be carefully defined, perhaps for each tumor. Such an approach to the understanding and regulation of interferon action against human tumors may lead to more effective therapy.

REFERENCES

Baron S. 1981. Interferons: anti-tumor and immunoregulatory activities. *In* Fundamental Mechanisms of Human Cancer Immunology. Elsevier-North Holland, New York, pp. 125–134.

Fleischmann WR Jr, Newton RC, Fleischmann CM, Colburn NH, Brysk MM. 1984. Discrimination between nonmalignant and malignant cells by combinations of IFN-γ plus IFN-α/β. J Bio Res Mod (in press).

Gresser I. 1981–82. On the mechanisms of antitumor effects of interferon. Tex Rep Biol Med 41:582–589.

Ito M, Buffet RF. 1981. Cytocidal effect of purified human fibroblast interferon on tumor cells *in vitro*. JNCI 66:819.

Johnson HM, Langford MP, Lakhchaura B, Chan T, Stanton GJ. 1982. Neutralization of native human gamma interferon (HuIFNγ) by antibodies to a synthetic peptide encoded by the 5′ end of HuIFNγ cDNA. J Immunol 129:2357–2359.

Rubin BY, Gupta SL. 1980. Differential efficacies of human type I and type II interferons as antiviral and antiproliferative agents. Proc Natl Acad Sci USA 77:5928–5932.

Tyring S, Klimpel GR, Fleischmann WR Jr, Baron S. 1982. Direct cytolysis by partially-purified preparations of immune interferon. Int J Cancer 30:59–64.

Mediators in Cell Growth and Differentiation,
edited by Richard J. Ford and Abby L. Maizel.
Raven Press, New York © 1985.

The Human Interferons: From the Past and into the Future

Sidney Pestka, Jerome A. Langer, Paul B. Fisher,*
I. Bernard Weinstein,* John Ortaldo,†
and Ronald B. Herberman†

*Roche Institute of Molecular Biology, Roche Research Center, Nutley, New Jersey 07110; *Institute of Cancer Research, Columbia University College of Physicians & Surgeons, New York, New York 10032; †Biological Therapeutics Branch, Biological Response Modifiers Program, National Cancer Institute–Frederick Cancer Research Facility, Frederick, Maryland 21701*

Interferon was discovered by Isaacs and Lindenmann (1957) and independently by Nagano and Kojima (1958) as an entity with antiviral activity. In studies that followed, many properties of interferon were established (see De Clercq and Merigan 1970, Colby and Morgan 1971, Ng and Vilcek 1972, Finter 1973, Ho and Armstrong 1975, Baron and Dianzani 1977, Pestka 1978, 1981a, b, Stinebring and Chapple 1978, Stewart 1979, Vilcek et al. 1980, Yabrov 1980, Khan et al. 1980, Pestka et al. 1980, 1981a, b, Baron et al. 1981–1982, for reviews and general compendia). The interferons have been shown to have a large variety of actions. This brief report will summarize some of the past work on interferon and will discuss actions and properties of interferon to elucidate its mode of action and potential therapeutic value.

CLASSES OF INTERFERON

When interferon (IFN) was first discovered, its complexities were not anticipated. Several different classes have now been identified: leukocyte (α), fibroblast (β), and immune (γ) classes. The three classes represent protein molecules of different structure and antigenic properties.

PURIFICATION

Our purification of the interferons involved the introduction of new technology, that of high-performance liquid chromatography (HPLC) for the purification of proteins. Since these studies have been published in detail elsewhere, we would simply like to emphasize some major highlights.

Human Leukocyte Interferon

Leukocyte interferon was produced by incubating human white blood cells with Newcastle disease virus or Sendai virus for 6 to 24 hr (Familletti and Pestka 1981, Familletti et al. 1981, Waldman et al. 1981, Hershberg et al. 1981). The antiviral activity was found in the cell culture medium after overnight incubation of leukocytes. The use of casein (Cantell and Tovell 1971), a single protein, instead of serum, which contains many different and uncharacterized proteins, simplified the initial concentration and purification steps. We used leukocytes from normal donors as well as from patients with chronic myelogenous leukemia. After initial concentration steps, three HPLC steps were utilized to purify human leukocyte interferon (Rubinstein et al. 1978, 1979, 1981, Rubinstein and Pestka 1981). By applying normal-phase chromatography on LiChrosorb Diol (glycerol groups bonded to silica) between two reverse-phase chromatographic steps on LiChrosorb RP-8 (Figure 1), we were able to purify human leukocyte interferon to homogeneity in these three steps. Gradients of *n*-propanol were used for elution of interferon from these columns. The overall purification was about 80,000-fold, and the specific activity of pure interferon was $2–4 \times 10^8$ units/mg (Rubinstein et al. 1979).

During the purification of leukocyte interferon, it became evident that multiple species existed. Human leukocyte interferon is heterogeneous, and several bands containing antiviral activity ranging in molecular weight from 15,000 to 21,000 are observed on sodium dodecyl sulfate (SDS)-polyacrylamide gel electrophoresis (Stewart 1974). Heterogeneity of human leukocyte interferon has also been shown by isoelectric focusing (Stewart et al. 1977) and several types of chromatographic procedures (Rubinstein et al. 1981, Törma and Paucker 1976, Chen et al. 1976, Jankowski et al. 1976, Grob and Chadha 1979). As noted above, our initial work with HPLC revealed three major groups of interferon species that were separated on elution from LiChrosorb Diol (Rubinstein et al. 1979, 1981). These groups were further resolved into several homogeneous components. Although others had reported heterogeneity in crude human leukocyte interferon preparations (Törma and Paucker 1976, Chen et al. 1976, Jankowski et al. 1976, Grob and Chadha 1979), it was not thought to be due to amino acid sequence heterogeneity. In fact, a number of groups reported that leukocyte interferon contained carbohydrate and that heterogeneity was due to differences in carbohydrate content of the protein (Bose et al. 1976, Bridgen et al. 1977, Stewart et al. 1977, Bose and Hickman 1977). However, five purified species of leukocyte interferon examined contained no detectable carbohydrate (Rubinstein et al. 1981). Nevertheless, we have recently detected carbohydrate in several species of leukocyte interferon (Labdon et al. 1984).

By analogous procedures, additional leukocys2te interferon species were isolated from cultured myeloblasts (Hobbs et al. 1981, Hobbs and Pestka 1982). Since our initial purification, other reports (Allen and Fantes 1980, Zoon 1981a,b, Berg and Heron 1981) have also described multiple species of leukocyte interferon. Allen and Fantes (1980) also found no carbohydrate on the species of leukocyte interferon

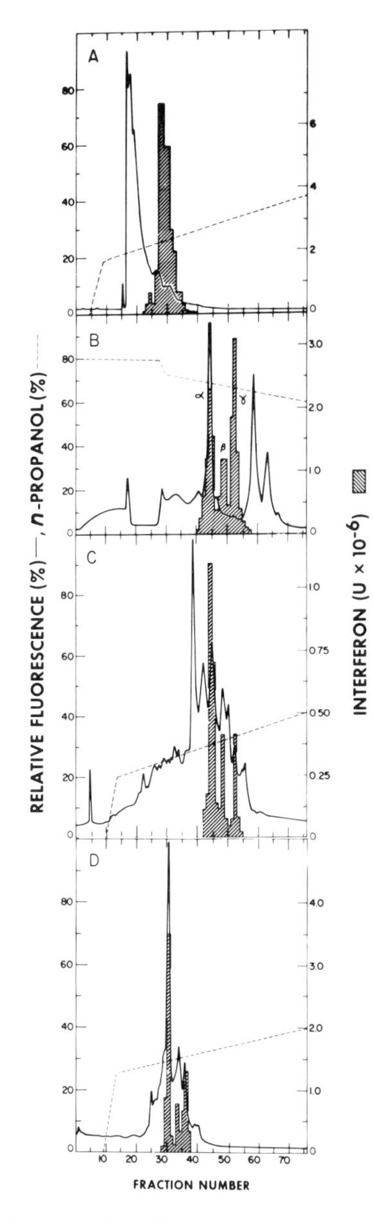

FIG. 1. High-performance liquid chromatography of interferon. **A:** Chromatography on LiChrosorb RP-8 at pH 7.5. **B:** Chromatography on LiChrosorb Diol at pH 7.5. **C:** Chromatography on LiChrosorb RP-8 at pH 4.0. **D:** Rechromatography on LiChrosorb RP-8. The conditions were similar to those of Step C. The gradations on the abscissa correspond to the end of the fractions. (Reproduced from Rubinstein et al. 1979.)

they purified. The dogma that interferons are glycoproteins has been so universal that leukocyte interferons are still considered glycoproteins despite the data to the contrary for many of the species. However, at least one minor species of leukocyte interferon appears to be glycosylated (Labdon et al. 1984), although we are only now beginning to determine its carbohydrate content. Levy et al. (1980, 1981) and Shively et al. (1982) reported amino acid sequences of three species of human leukocyte interferon. So far, the sequences of two have been determined almost completely. Additional sequences were reported by Zoon (1981b). Allen and Fantes (1980) reported the sequences of tryptic fragments obtained from a mixture of several leukocyte interferon species. All these sequences are sufficiently different to establish very clearly that they compose a family of closely related proteins. The total number of human leukocyte interferon proteins that exist in specific cells and the nature of their structural and functional differences will require the elucidation of their complete primary structures and the nucleotide sequences of their corresponding cDNAs and genes.

Human Fibroblast Interferon

Several laboratories reported the purification and partial structural analysis of human fibroblast interferon by the use of SDS-polyacrylamide gel electrophoresis as the last step in the purification (Knight 1976, Berthold et al. 1978). To obtain a product free of salt and solvent, we developed a simple two-step purification procedure (Stein et al. 1980, Kenny et al. 1981). The first step in the purification from the crude interferon-containing medium involved blue-Sepharose chromatography, a procedure described previously (Jankowski et al. 1976, Knight et al. 1980). The second step involved HPLC on octyl silica. The amino acid composition and 19 residues of the amino-terminal sequence of human fibroblast interferon were determined (Stein et al. 1980). The sequence was identical to the first 13 amino-terminal residues reported by Knight et al. (1980) and Okamura et al. (1980) and to the first 10 amino-terminal residues reported by Friesen et al. (1981). Unlike human leukocyte interferon, which was isolated as several different species (Rubinstein et al. 1979, 1981, Hobbs et al. 1981, Hobbs and Pestka 1982, Allen and Fantes 1980, Zoon 1981a, Berg and Heron 1981), only a single human fibroblast interferon species has so far been isolated.

CLONING AND EXPRESSION OF HUMAN INTERFERONS IN BACTERIA

The specific procedures for cloning leukocyte and fibroblast interferons have been described previously (Pestka 1983). Fibroblast interferon sequences were first cloned by Taniguchi et al. (1979, 1980) and subsequently by others (Derynck et al. 1980, Houghton et al. 1980, Maeda et al. 1980, Goeddel et al. 1980). Human leukocyte interferon recombinants were obtained by Nagata et al. (1980a) and then by Maeda et al. (1980). These initial recombinants were used to search through cDNA libraries and genomic libraries to isolate various human leukocyte interferons. A summary of the human leukocyte interferon sequences so far obtained is shown in Figure 2.

Analysis of the coding regions of the leukocyte interferon genes that were isolated in our laboratory and others (Maeda et al. 1980, 1981, Maeda S, McCandliss R, Chiang T-R, Costello L, Levy WP, Chang NT, Martin-Zanca D, Liu X-Y, and Pestka S, in preparation, Nagata et al. 1980a, b, Goeddel et al. 1980, 1981, Streuli

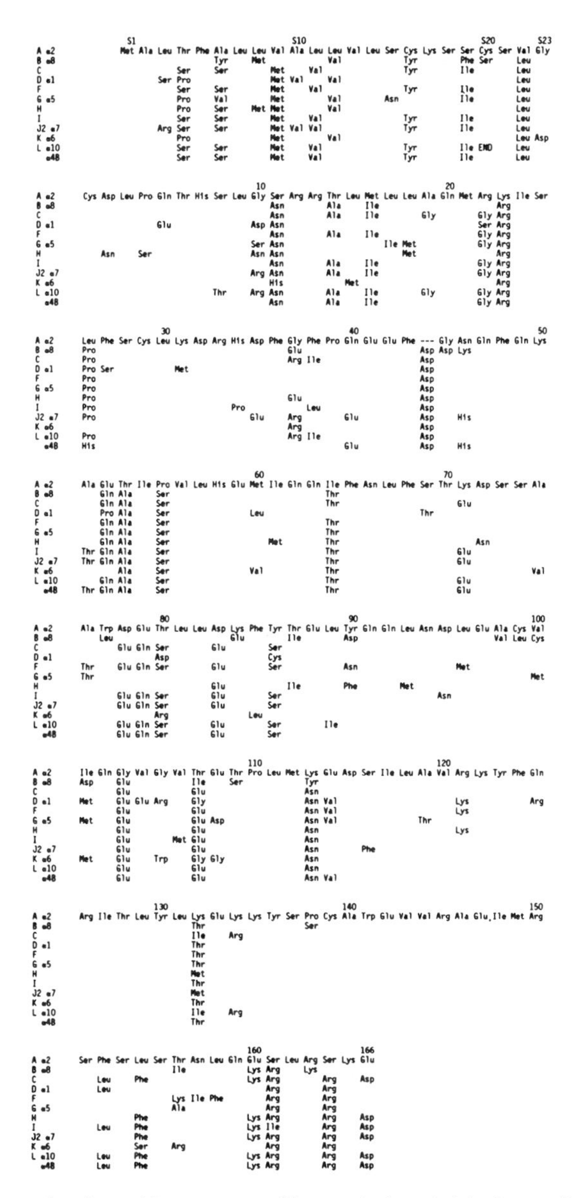

FIG. 2. Summary of amino acid sequences of human leukocyte interferons. Sequences were derived from the respective DNA sequences. The entire sequence of IFN-αA is given, including the sequence of the precursor signal peptide (S1-S23). Corresponding residues of the other species are shown only where they differ from those of IFN-αA. Since IFN-αA has only 165 amino acids, whereas the other species have 166 amino acids, a gap in the IFN-αA sequence has been introduced between residues 43 and 44, to provide maximum homology with the other species. Sequences A-L are from the laboratory of the authors (Maeda et al. 1980, 1981, Pestka 1983) and from that of Derynck et al. (1982). Sequences prefixed by "α" are from the laboratory of Weissmann et al. (1982). Sequences that differ by only a few amino acids are here listed together (e.g., there is a single amino acid difference between A and α2). The α5 sequence is complete, whereas the G sequence begins at residue 34, since only a partial-length clone of this recombinant was isolated; identity of G with α5 is proposed on the basis of the extant sequences.

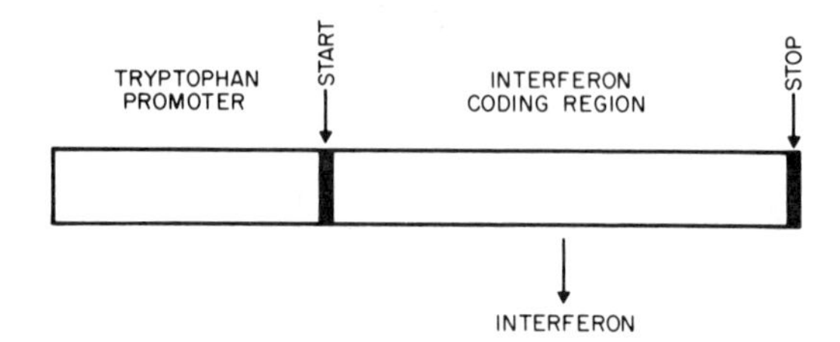

FIG. 3. Regulation of interferon expression in *E. coli.* The tryptophan promoter-operator containing the ribosome-binding site was ligated to the interferon-coding sequence that was constructed to express mature leukocyte interferon. Leukocyte interferon (IFN-αA) expression was thus regulated by the *trp* promoter-operator.

et al. 1980, Mantei et al. 1980, Lawn et al. 1981a, b, Brack et al. 1981, Ullrich et al. 1982) has shown that the interferon genes comprise a family of homologous proteins. As shown in Figure 2, these proteins are highly related and yet differ from each other in amino acid sequences.

Expression of Human Leukocyte Interferon in *Escherichia coli*

The first recombinant isolated in our laboratory corresponded to leukocyte A interferon (Maeda et al. 1980); the others were subsequently identified (Maeda et al. 1981, Goeddel et al. 1981). The precursor form of the recombinant (IFN-αpreA) was expressed in *Escherichia coli* under control of the *E. coli* tryptophan operon (Goeddel et al. 1980). The mature leukocyte interferon IFN-αA was expressed directly by reconstruction of the recombinant. The leader sequence of IFN-αA was removed, and an ATG-translational initiation codon was placed immediately preceding the codon for amino acid 1 (cysteine) of the mature leukocyte interferon IFN-αA. The intact coding sequence was then attached to the tryptophan promoter (Figure 3). Current expression levels under the control of the tryptophan promoter yield 50 mg per liter of IFN-αA. In an analogous manner, several of the α interferons, as well as β interferon, have been expressed in *E. coli* (Pestka 1983).

PURIFICATION OF RECOMBINANT HUMAN LEUKOCYTE INTERFERON

Monoclonal antibodies to the human leukocyte interferons were used to purify recombinant human leukocyte A interferon (IFN-αA) produced in bacteria (Staehelin et al. 1981a, b). In the procedure, *E. coli* containing IFN-αA are broken, and unbroken cells and cellular debris are removed by centrifugation. The IFN-αA and soluble bacterial proteins remain in the cell lysate. Nucleic acids (DNA and RNA) are precipitated by combination with polymin P. The soluble proteins remaining in the lysate can then be concentrated or, without concentration, passed directly through a column containing a monoclonal antibody to human leukocyte interferon. The antibodies bind only the interferon; all other components and proteins pass through the column. After the column is washed, the IFN-αA bound to the column is removed by elution with an acidic solution (Staehelin et al. 1981a, b). A virtually pure interferon solution is eluted from the monoclonal antibody column. The column is then washed and neutralized so that it can be used repeatedly. The interferon solution is concentrated by passage over a column of carboxymethylcellulose (Staehelin et al. 1981a, b). The activity of the purified interferon made in bacteria is similar to that of the same human leukocyte interferon species synthesized by human cells.

TARGET SIZE OF LEUKOCYTE, FIBROBLAST, AND IMMUNE INTERFERONS

Irradiation of macromolecules permits the estimation of their molecular weights (target size) independent of their isolation and purity. The target size of proteins has been estimated quite accurately by electron bombardment (Kempner and Schlegel 1979, Kempner et al. 1980, Nielsen et al. 1981). Thus, we determined the molecular weights of the functional units of the human interferons by determining their target size (Pestka et al. 1983) for antiviral activity. An example of the target size analysis is shown in Figure 4. The data appear to represent a single-hit inactivation and provide an estimate for the target molecular weight of natural fibroblast interferon as 42,000.

In the same manner, we have determined the target sizes of natural and recombinant human leukocyte interferon, natural and recombinant human fibroblast interferon, and natural and recombinant human immune interferon. A summary of the inactivation curves for the natural interferons is shown in Figure 5. It can be seen that the human leukocyte, fibroblast, and immune interferons appear to have target molecular weights of 20,000, 42,000, and 63,000, respectively. Analogous studies with the corresponding recombinant species are shown in Figure 6. In this case, the target molecular weights of the leukocyte, fibroblast, and immune inter-

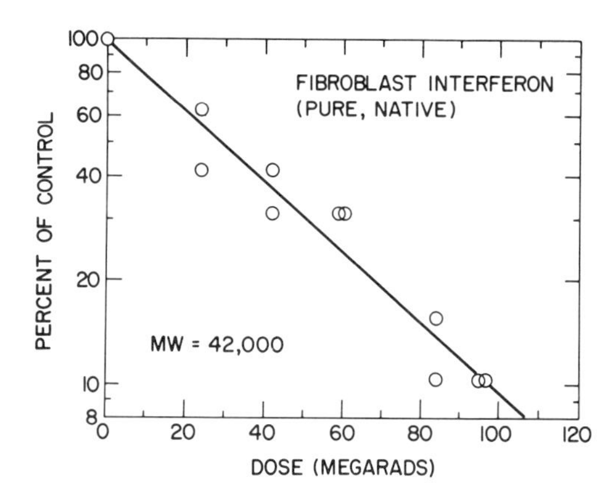

FIG. 4. Loss of antiviral activity of homogeneous human native fibroblast interferon as a function of ionizing radiation. The fibroblast interferon was purified as described (Stein et al. 1980, Kenny et al. 1981). The procedures are described in detail elsewhere (Pestka et al. 1983).

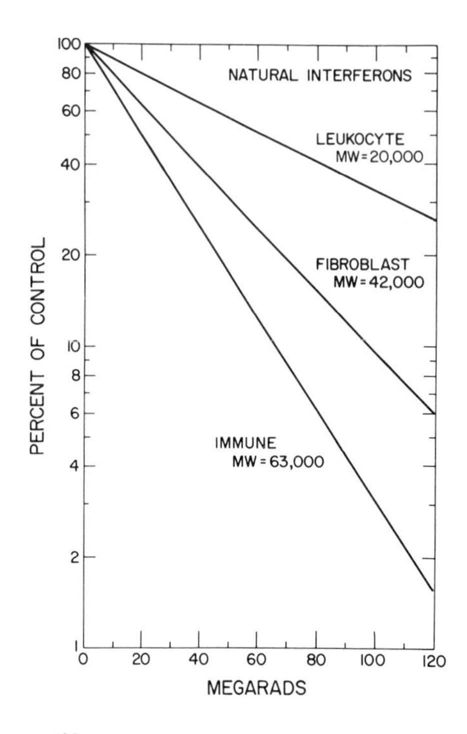

FIG. 5. Summary of the inactivation curves for the natural interferons. Loss of antiviral activity due to increasing doses of ionizing radiation is shown. Data from Pestka et al. (1983).

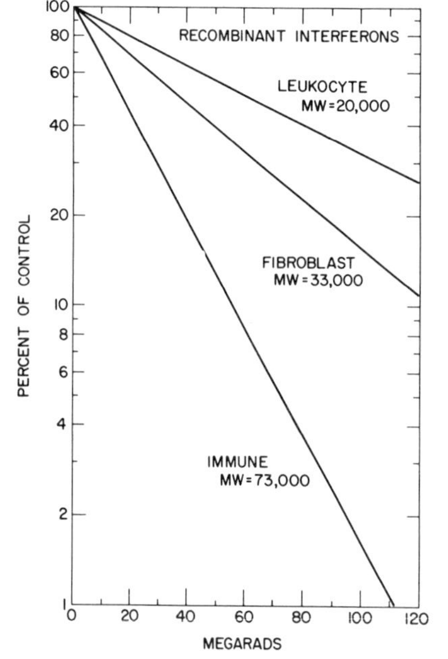

FIG. 6. Summary of the inactivation curves for the recombinant interferons. Loss of antiviral activity due to increasing doses of ionizing radiation is shown. Data from Pestka et al. (1983).

TABLE 1. *Target molecular weights of interferons*

Interferon	Experiment	Target M_r ± 2 S.E.	Monomer M_r
Leukocyte			
IFN-α (crude)	1	20,000 ± 2,200	17,000–24,000
IFN-αA (recombinant, pure)	1	23,000 ± 2,200	19,219
IFN-αA (recombinant, pure)	2	20,000 ± 2,300	19,219
Fibroblast			
IFN-β (crude)	1	31,000 ± 2,000	~24,000
IFN-β (crude)	3	42,000 ± 1,900	~24,000
IFN-β (pure, native)	3	42,000 ± 3,200	~24,000
IFN-β (recombinant, pure)	3	33,000 ± 2,300	20,004
Immune			
IFN-γ (crude)	1	63,000 ± 2,800	20,000;25,000
IFN-γ (crude)	3	66,000 ± 4,100	20,000;25,000
IFN-γ (recombinant)	3	73,000 ± 5,800	17,126

Summary of target molecular weights of interferon as determined by Pestka et al. 1983.

ferons are calculated to be 20,000, 33,000, and 73,000, respectively. A summary of the results is presented in Table 1.

There appears to be good agreement between the target molecular weight and the known molecular weight of the leukocyte interferons both in crude and pure forms. The leukocyte interferons are molecules that are generally not glycosylated (Pestka 1983, Rubinstein et al. 1981, Allen and Fantes 1980) and in solution are predominantly monomeric (Pestka et al. 1983).

In the case of fibroblast interferon, the target molecular weight of 31,000 to 42,000 is significantly larger than the experimentally determined molecular weight of 21,000 to 24,000 (Knight and Fahey 1981, Friesen et al. 1981, Pestka 1983). These results indicate that the antiviral activity of fibroblast interferon is the result of two monomers. Studies of fibroblast interferon have shown that oligomers of the molecule coexist with the monomers (Friesen et al. 1981, Knight and Fahey 1981). The target size of 31,000 to 42,000 daltons suggests that the dimer may be the predominant active molecular form even in solutions of the crude interferon. Although native fibroblast interferon is glycosylated (Knight 1976, Friesen et al. 1981) (whereas recombinant fibroblast interferon produced in *E. coli* is not), the glycosylation does not appear to be a significant factor in these determinations. This result is consistent with the observation that the radiation sensitivity of invertase is independent of the presence of attached oligosaccharide (Lowe and Kempner 1982).

In two different preparations of natural immune interferon, the target molecular weights ranged from 63,000 to 66,000 (Pestka et al. 1983). The molecular weight of these glycosylated molecules in nondenaturing solutions determined by independent means appears to be about 45,000 (Yip et al. 1981, Langford et al. 1979, de Ley et al. 1980). The recombinant immune interferon monomer exhibits a molecular weight of 17,000 (Pestka 1983, Gray et al. 1982, Devos et al. 1982).

The target molecular weight of recombinant immune interferon was determined to be 73,000 (Figure 6). These results indicated that the functional form of immune interferon in solution may be a tetramer.

In summary, the functional unit of leukocyte interferon is the monomer, that of fibroblast interferon is predominantly a dimer, and that of immune interferon a trimer or tetramer.

NEW RECEPTORS FOR INTERFERON APPEAR ON DIFFERENTIATED HL60 CELLS

The human leukemic cell line HL60 has proved to be a useful system for studying myeloid differentiation (Collins et al. 1978, Gallagher et al. 1979). These cells, which are predominantly a promyeloblast type, can be induced to mature to a neutrophil-like state by dimethylsulfoxide (Me_2SO) and retinoic acid (Honma et al. 1980, Breitman et al. 1980), or to a macrophagelike cell with 12-*O*-tetradecanoyl-phorbol-13-acetate (TPA) (Huberman and Callahan 1979, Rovera et al. 1979a, b, Lotem and Sachs 1979). In studies of the binding of interferon to cells differentiated in the presence of Me_2SO, it was found that more [^{125}I]IFN-αA bound to differentiated cells than to undifferentiated HL60 (Figures 7, 8). It is apparent that the increase in interferon binding is one of the earliest events that can be detected on differentiation. Whether this increased ability to bind interferon is an essential

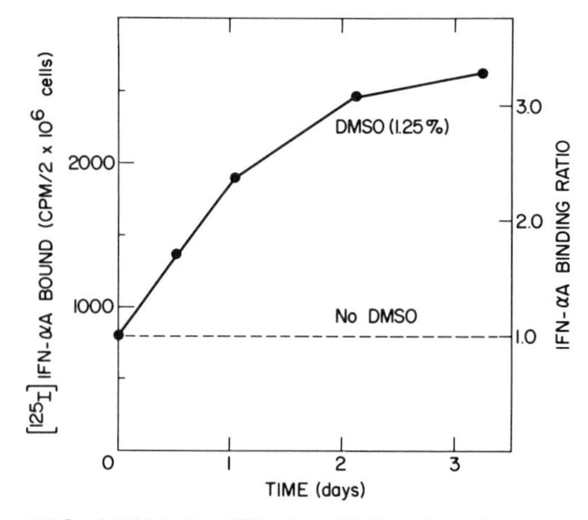

FIG. 7. Binding at 23°C of ^{125}I-labeled IFN-αA to HL60 cells in the presence and absence of Me_2SO for the indicated times. After growing in the absence or presence of Me_2SO for the indicated times, cells were washed and concentrated to 2×10^7 cells/ml. Approximately 2×10^6 cells in 100 μl of medium were added to 10 μl of ^{125}I-labeled IFN-αA (about 90,000 cpm), with or without competing nonradioactive IFN-αA (about 0.5 μg; approximately 500-fold excess). ^{125}I-labeled IFN-αA bound in the presence of nonradioactive IFN-αA ("nonspecific binding") was subtracted from counts bound in the absence of unlabeled IFN-αA to calculate specific binding. All points, including nonspecific binding, were measured in duplicate. Binding was for 1 hr at 23°C. (---), cells grown in the absence of Me_2SO. (●——●), cells grown in the presence of Me_2SO. The ratio of binding of ^{125}I-labeled IFN to HL60 cells grown in the presence of Me_2SO to those grown in its absence is shown on the right ordinate. Data from Langer and Pestka (1984).

FIG. 8. Schematic illustration of increase in interferon receptors on HL60 cells after differentiation.

event in differentiation is not known, but the role of interferons as growth modulators and immunomodulators suggests it may be (Tomida et al. 1982).

INTERFERON AFFECTS TERMINAL DIFFERENTIATION

In studies of human melanoma cells, it was found that interferon is a potent stimulator of differentiation. In four human melanoma cell lines examined, recombinant human fibroblast interferon (IFN-β) was more active than the recombinant human leukocyte interferons IFN-αA, IFN-αD, or IFN-αA/D *(Bgl)* (a hybrid protein made as described below) in inhibiting cellular proliferation. Mezerein, an analog of the tumor promoter TPA, can itself inhibit growth of cells, as well as stimulate differentiation of melanoma cells. The combination of interferon and mezerein resulted in a very potent synergistic inhibition of cellular proliferation in all four melanoma cell lines examined. The results indicated that the antiproliferative effect of interferon towards melanoma cells can be enhanced by treatment with mezerein and that this synergistic effect is associated with an enhancement of terminal differentiation. Since a major problem limiting the utility of interferon as an antitumor agent in vivo may be the development of resistant cell populations, the combination of interferon with agents capable of inducing tumor cell differentiation may help to circumvent this problem and thereby increase its efficacy as an antitumor agent (Fisher PB, Prignoli DR, Hermo H Jr, Weinstein IB, Pestka S, in preparation).

RELATIVE ACTIVITIES OF INTERFERON

Hybrid leukocyte interferons (Streuli et al. 1981, Weck et al. 1981, Rehberg et al. 1982) were prepared from recombinant human leukocyte interferons A and D (IFN-αA and IFN-αD). These hybrid interferons are schematically illustrated in Figure 9 (Rehberg et al. 1982).

The specific molecular activity of interferon is defined as the molecules/cell necessary to elicit a specific effect. This is an operational definition and is not dependent on knowledge of the specific amount of binding of interferons to cells. What is necessary is a supply of pure interferon to determine the concentration of that interferon at the time the experiment is initiated. Comparison of the specific molecular activities of the interferons provides a useful way to compare the relative effects of interferons on different cellular activities.

Comparison of the antiviral activities of the parental interferons IFN-αA and IFN-αD and the hybrid interferons on different species (Table 2) shows that new

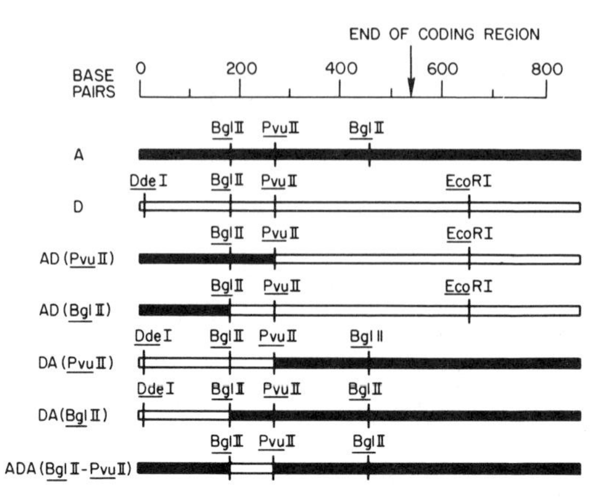

FIG. 9. Schematic illustration and restriction maps of the coding regions for the mature proteins of plasmids pIFN-αA, pIFN-αD, and hybrids constructed from these two molecules.

activities of the interferons were generated in the hybrids. Although the antiviral activities on bovine MDBK cells were essentially similar for all these molecules, their activities on human, mouse, feline, and rat cells differed markedly from the parental molecules. For example, the hybrids IFN-αA/D *(Bgl)* and IFN-αA/D *(Pvu)* were active not only on human and bovine cells, but, unlike the parental molecules, also on mouse and feline cells.

The antiproliferative activities of these interferons were determined on Daudi lymphoblastoid cells (Evinger et al. 1981). The quantity of each interferon that produced 50% inhibition of growth of the cells is shown in Table 3. It is evident that the number of antiviral units necessary to produce this inhibition provides a spurious picture of which interferons are most active on a molecular basis (Table

TABLE 2. *Molecules/cell for 50% inhibition of viral cytopathic effect for various cell lines*

Interferon	Cell line: AG-1732 (human)	MDBK (bovine)	L-Cells (murine)	Felung (feline)	Rat C6 (rat)
A	4,900	1,100	3.1×10^7	8,700	6.1×10^8
D	360,000	6,600	4.3×10^6	3,700	1.1×10^6
A/D *(Bgl)*	5,500	1,200	3,300	450	210,000
A/D *(Pvu)*	4,100	2,400	48,000	1,800	$>1.7 \times 10^6$
D/A *(Bgl)*	1.5×10^6	1,800	$>7.2 \times 10^7$	42,000	$>3.8 \times 10^7$
D/A *(Pvu)*	590,000	5,400	7.1×10^6	12,000	$>3.0 \times 10^7$
A/D/A	36,000	2,000	1.1×10^6	1,500	$>1.8 \times 10^7$

The molecular weights of the various species are as follows: IFLrA, 19,244; IFLrD, 19,417; IFLrA/D *(Bgl)*, 19,420; IFLrA/D *(Pvu)*, 19,452; IFLrD/A *(Bgl)*, 19,240; IFLrD/A *(Pvu)*, 19,200; IFLrA/D/A, 19,211. The number of cells in the microtiter well was determined at the time of addition of cells: AG-1732, 1.2×10^4; MDBK, 2.7×10^4; L-cells, 2.5×10^4; Felung, 1.8×10^4; Rat C6, 5.4×10^4. Data from Rehberg et al. 1982.

TABLE 3. *Antiproliferative activity: Concentrations of interferon at 50% inhibition of control cell growth*

Interferon	Molecules/cell	Units/ml
A	13,000	23
D	490,000	12
A/D *(Bgl)*	9,300	21
A/D *(Pvu)*	30,000	62
D/A *(Bgl)*	910,000	4
D/A *(Pvu)*	4,200,000	73
A/D/A	60,000	17

The values in the table represent the concentrations of the interferons that inhibited growth of Daudi cells 50%. Antiviral units were determined on AG-1732 cells. Molecules/cell were calculated from the cell number present at the time of addition of the interferons. Data taken from Rehberg et al. (1982).

3). Comparison of the molecules/cell necessary for 50% inhibition of growth, however, permits an accurate assessment of their relative activities. The ratio of the quantity required for 50% antigrowth activity to the quantity required for 50% inhibition of viral cytopathic effect (antiviral activity) (Table 4) varied over a 12-fold range. In the case of IFN-αA/D *(Pvu)* and IFN-αD/A *(Pvu)*, seven- to eightfold more interferon was required for 50% inhibition of growth than 50% inhibition of viral cytopathic effect.

TABLE 4. *Ratio of specific molecular activities of interferons for antiproliferative and antiviral activity on human cells*

Interferon	AP/AV
A	2.7
D	1.3
A/D *(Bgl)*	1.7
A/D *(Pvu)*	7.3
D/A *(Bgl)*	0.6
D/A *(Pvu)*	7.1
A/D/A	1.7

The specific molecular antiproliferative activity (molecules/cell) for inhibition of growth of human lymphoblastoid Daudi cells (Table 3), AP, was divided by the specific molecular antiviral activity on human AG-1732 fibroblasts, AV. The ratio AP/AV is given in the table. Data from Rehberg et al. (1982).

TABLE 5. *Molecules/cell for 50% effect*

Interferon	AV	AP	NK
A	4,900	13,000	120
D	360,000	450,000	2.8×10^5
A/D *(Bgl)*	5,500	9,300	58
A/D *(Pvu)*	4,100	30,000	1,300
D/A *(Bgl)*	1.5×10^6	910,000	2.5×10^5
D/A *(Pvu)*	590,000	4.2×10^6	4.6×10^4
A/D/A	36,000	60,000	2.0×10^4

AV, antiviral activity; AP, antiproliferative activity; NK, stimulation of natural killer cell activity of human cells. The data are taken from Rehberg et al. (1982) and Ortaldo et al. (1983a, b).

Similar studies were performed to determine the specific molecular activity of these interferons on stimulation of natural killer (NK) cell activity (Table 5). The results indicated that, as had been determined with the natural interferons (Evinger et al. 1981, Ortaldo et al. 1983c), these recombinant interferons also show remarkable differences in the number of molecules/cell necessary for the various effects. A more detailed discussion of the NK cell stimulatory activity of these interferons is presented by Ortaldo et al. (1983a, b). By comparing the ratio of the specific molecular activities (AV/AP and AV/NK in Table 6), remarkable differences in these ratios are evident. In fact, there is approximately a 100-fold difference in the AV/AP and AV/NK ratios for IFN-αA, IFN-αA/D *(Bgl)*, and IFN-αD/A *(Pvu)*.

These disparities between antiviral, antiproliferative, and NK cell stimulatory activities indicate that these activities are mediated by different mechanisms. Similar suggestions were made previously when it was observed that the individual purified human leukocyte interferons exhibited different ratios of antiviral to antiproliferative activity (Evinger et al. 1981). Since the effects of interferon are mediated by a number of different mechanisms (Lengyel 1981, Lengyel and Pestka 1981, Maheshwari and Friedman 1981, Sreevalsan et al. 1981, Revel et al. 1981, Kerr and

TABLE 6. *Ratio of specific molecular activities*

Interferon	AV/AP	AV/NK
A	0.38	41
D	0.80	1.3
A/D *(Bgl)*	0.59	95
A/D *(Pvu)*	0.14	3.2
D/A *(Bgl)*	1.7	6.0
D/A *(Pvu)*	0.14	13
A/D/A	0.60	1.8

The AV, AP, and NK ratios were calculated from the data of Table 5.

TABLE 7. *Effect of interferons on human NK activity*

IFN-α	Lytic units/ml
A	2
B	1
C	2
D	7
F	89
I	11
J	>10,000
K	29

Data from Ortaldo et al. (1984).

Brown 1978), it is not surprising that some of the effects can be dissociated. Thus, after comparison of the antiproliferative and antiviral activities of human leukocyte, fibroblast, and immune interferons, Eife et al. (1981) observed that the activity ratios differed significantly for the three interferons. Effects of interferon on lytic viruses such as vesicular stomatitis virus were dissociated from those on Moloney leukemia virus (Epstein et al. 1981, Sen and Herz 1983, Herz et al. 1983). These results indicate that many of the effects of the interferons can be dissociated and are due to different molecular mechanisms. The individual interferons can apparently activate several pathways to different degrees.

Because the biological activities of these interferons seem to vary greatly for NK activity compared to antiviral or antiproliferative activity, we examined the effect of several IFN-α species for their ability to stimulate NK activity (Table 7). It is evident that IFN-αJ has little or no ability to boost NK activity compared to the other α interferons; IFN-αJ nevertheless exhibits relatively strong antiviral and antiproliferative activities (Ortaldo et al. 1984). Not only is IFN-αJ unable to stimulate NK activity significantly during a 2-hr incubation with lymphocytes, but IFN-αJ can block the activity of IFN-αA (Table 8). For blocking to occur, it is necessary that IFN-αJ be active and not heat-denatured. Taken together, these

TABLE 8. *IFN-αJ blocks NK activity by IFN-αA*

Interferon	Lytic units above control
IFN-αA	98
IFN-αJ	15
IFN-αA + IFN-αJ	26
IFN-αA (65°C)	−8
IFN-αA + IFN-αJ (65°C)	108

Lytic units were determined as described by Ortaldo et al. (1984). Lytic units in the control sample in the absence of interferon was subtracted from each value to derive the values in the table. IFN-αA (line 4) and IFN-αJ (line 5) were heat-inactivated at 65°C.

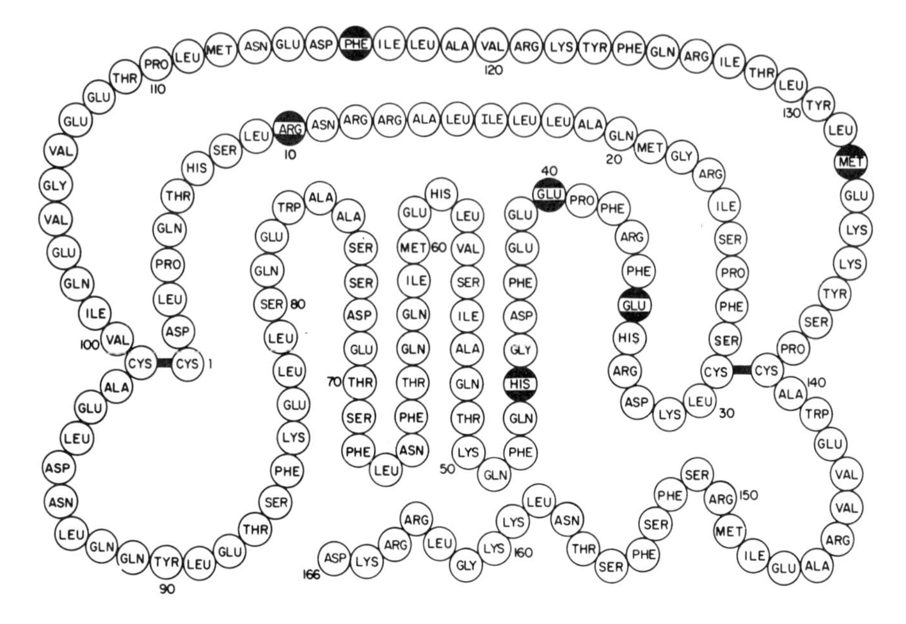

FIG. 10. Amino acid sequence of recombinant human leukocyte interferon J (IFN-αJ). Amino acids that differ from other leukocyte interferons are shadowed in black.

observations suggest that the failure of IFN-αJ to stimulate NK cells is not due to an inability to bind to its receptor (since it can block the action of IFN-αA). Comparison of IFN-αJ with known human leukocyte interferons shows that six positions of IFN-αJ are unique compared to the others (Fig. 10). Studies are under way to determine which, if any, of these amino acids alters the structure so that IFN-αJ is unable to stimulate NK activity effectively.

CONCLUDING COMMENTS

Although the interferon receptor has not been purified nor its structure determined, a number of observations reported here define some characteristics of the interferon receptor. Since the ratio of antiviral to antiproliferative and antiviral to NK cell activities of the various interferons varies over a wide range (Table 6), it is apparent that the interferon receptor cannot be a simple one that is turned on or off. Multiple receptors for interferon, each regulating a different biological activity, may exist. Alternatively, the interferon receptor may be a complex one able to transduce the differential signals controlling the individual biochemical events leading to antiviral activity, antiproliferative activity, stimulation of NK cells, and other activities to different degrees. Because IFN-αJ, only in the active form, blocks the activity of IFN-αA, we favor the hypothesis that the receptor is a complex one rather than that there are individual receptors for the leukocyte interferons. Purification and isolation of the receptor molecules will, however, be required before definitive resolution of these questions is obtained.

ACKNOWLEDGMENTS

We thank Bruce Kelder, Judith Altman, and Cynthia Rose for assistance in the interferon assays.

REFERENCES

Allen G, Fantes KH. 1980. A family of structural genes for human lymphoblastoid (leukocyte-type) interferon. Nature 287:408–411.

Baron S, Dianzani F (eds). 1977. The Interferon System: A Current Review to 1978. Tex Rep Biol Med 35:1–573.

Baron S, Dianzani F, Stanton J (eds). 1981-1982. The Interferon System: A Review to 1982. Tex Rep Biol Med 41:1–715.

Berg K, Heron I. 1981. Antibody affinity chromatography of human leukocyte interferon. Methods Enzymol 78:487–499.

Berthold W, Tan C, Tan YH. 1978. Purification and in vitro labeling of interferon from a human fibroblastoid cell line. J Biol Chem 253:5206–5212.

Bose S, Hickman J. 1977. Role of the carbohydrate moiety in determining the survival of interferon in the circulation. J Biol Chem 252:8336–8337.

Bose S, Gurari-Rotman D, Rüegg UT, Corley L, Anfinsen CB. 1976. Apparent dispensability of the carbohydrate moiety of human interferon for antiviral activity. J Biol Chem 251:1659–1662.

Brack C, Nagata S, Mantei N, Weissmann C. 1981. Molecular analysis of the human interferon-alpha gene family. Gene 15:379–394.

Breitman TR, Selonick SE, Collins SJ. 1980. Induction of differentiation of the human promyelocytic leukemia cell line (HL-60) by retinoic acid. Proc Natl Acad Sci USA 77:2936–2940.

Bridgen PJ, Anfinsen CB, Corley L, Bose S, Zoon KC, Rüegg UT, Buckler CE. 1977. Human lymphoblastoid interferon: Large scale production and partial purification. J Biol Chem 252:6585–6587.

Cantell K, Tovell DR. 1971. Substitution of milk for serum in the production of human leukocyte interferon. Appl Microbiol 22:625–628.

Chen JK, Jankowski WJ, O'Malley JA, Sulkowski E, Carter WA. 1976. Nature of the molecular heterogeneity of human leukocyte interferon. J Virol 19:425–434.

Colby C, Morgan MJ. 1971. Interferon induction and action. Annu Rev Microbiol 25:333–360.

Collins SJ, Ruscetti FW, Gallagher RE, Gallo RC. 1978. Terminal differentiation of human promyelocytic leukemia cells induced by dimethyl sulfoxide and other polar compounds. Proc Natl Acad Sci USA 75:2458–2462.

De Clercq E, Merigan TC. 1970. Current concepts of interferon and interferon induction. Annu Rev Med 21:17–46.

de Ley M, Van Damme J, Claeys H, Weening H, Heine JW, Billiau A, Vermylen C, De Somer P. 1980. Interferon produced in human leukocytes by mitogens: Production, partial purification and characterization. Eur J Immunol 10:877–883.

Derynck R, Content J, De Clercq E, Volckaert G, Tavernier J, Devos R, Fiers W. 1980. Isolation and structure of a human fibroblast interferon gene. Nature 285:542–547.

Derynck R, Gray PW, Yelverton E, Leung DW, Shepard HM, Lawn RM, Ullrich A, Najarian R, Pennica D, Hagie FE, Hitzeman RA, Sherwood PJ, Levinson AD, Goeddel DV. 1982. Synthesis of human interferons and analogs in heterologous cells. *In* Ahmad F, Schultz J, Smith EE, Whelan WJ (eds), From Gene to Protein: Translation into Biotechnology. Academic Press, New York, pp. 249–262.

Devos R, Cheroutre H, Taya Y, Degrave W, Van Heuverswyn H, Fiers W. 1982. Molecular cloning of human interferon cDNA and its expression in eukaryotic cells. Nucleic Acids Res 10:2487–2501.

Eife R, Hahn T, DeTavera M, Schertel F, Holtmann H, Eife G, Levin S. 1981. A comparison of the antiproliferative and antiviral activities of α-, β- and γ-interferons: Description of a unified assay for comparing both effects simultaneously. J Immunol Methods 47:339–347.

Epstein DA, Czarniecki W, Jacobsen H, Friedman RM, Panet A. 1981. A mouse cell line, which is unprotected by interferon against lytic virus infection, lacks ribonuclease F activity. Eur J Biochem 118:9–15.

Evinger M, Rubinstein M, Pestka S. 1981. Antiproliferative and antiviral activities of human leukocyte interferons. Arch Biochem Biophys 210:319–329.

Familletti PC, Pestka S. 1981. Cell cultures producing human interferon. Antimicrob Agents Chemother 20:1–4.

Familletti PC, McCandliss R, Pestka S. 1981. Production of high levels of human leukocyte interferon from a continuous human myeloblast cell culture. Antimicrob Agents Chemother 20:5–9.

Finter NB (ed). 1973. Interferons and Interferon Inducers. American Elsevier, New York.

Fisher PB, Prignoli DR, Hermo H Jr, Weinstein IB, Pestka S. 1984. Effects of combined treatment with interferon and mezerein on melanogenesis and growth in human melanoma cells. J Interferon Res (in press).

Friesen H-J, Stein S, Evinger M, Familletti PC, Moschera J, Meienhofer J, Shively J, Pestka S. 1981. Purification and molecular characterization of human fibroblast interferon. Arch Biochem Biophys 206:432–450.

Gallagher R, Collins S, Trujillo J, McCredie K, Ahearn M, Tsai S, Metzgar R, Aulakh G, Ting R, Ruscetti F, Gallo R. 1979. Characterization of the continuous, differentiating myeloid cell line (HL-60) from a patient with acute promyelocytic leukemia. Blood 54:713–733.

Goeddel DV, Leung DW, Dull TJ, Gross M, Lawn RM, McCandliss R, Seeburg PH, Ullrich A, Yelverton E, Gray PW. 1981. The structure of eight distinct cloned human leukocyte interferon cDNAs. Nature 290:20–26.

Goeddel DV, Yelverton E, Ullrich A, Heyneker HL, Miozzari G, Holmes W, Seeburg PH, Dull T, May L, Stebbing N, Crea R, Maeda S, McCandliss R, Sloma A, Tabor JM, Gross M, Familletti PC, Pestka S. 1980. Human leukocyte interferon produced by *E. coli* is biologically active. Nature 287:411–416.

Gray PW, Leung DW, Pennica D, Yelverton E, Najarian R, Simonsen CC, Derynck R, Sherwood PJ, Wallace DM, Berger SL, Levinson AD, Goeddel DV. 1982. Expression of human immune interferon cDNA in *E. coli* and monkey cells. Nature 295:503–508.

Grob PM, Chadha KC. 1979. Separation of human leukocyte interferon components by concanavalin A-agarose affinity chromatography and their characterization. Biochemistry 18:5782–5786.

Herz RE, Rubin BY, Sen GC. 1983. Human interferons-α and γ-mediated inhibition of retroviral production in the absence of an inhibitory effect on VSV and EMCV replication in RD-114 cells. Virology 125:246–250.

Hershberg RD, Gusciora EG, Familletti P, Rubinstein S, Rose CA, Pestka S. 1981. Induction and production of human interferon with human leukemic cells. Methods Enzymol 78:45–48.

Ho M, Armstrong JA. 1975. Interferon. Annu Rev Microbiol 29:131–161.

Hobbs DS, Pestka S. 1982. Purification and characterization of interferons from a continuous myeloblastic cell line. J Biol Chem 257:4071–4076.

Hobbs DS, Moschera JA, Levy WP, Pestka S. 1981. Purification of interferon produced in a culture of human granulocytes. Methods Enzymol 78:472–481.

Honma Y, Takenaga K, Kasukabe T, Hozumi M. 1980. Induction of differentiation of cultured human promyelocytic leukemia cells by retinoids. Biochem Biophys Res Commun 95:507–512.

Houghton M, Stewart AG, Doel SM, Emtage JS, Eaton MAW, Smith JC, Patel TP, Lewis HM, Porter AG, Birch JR, Cartwright T, Carey NH. 1980. The amino-terminal sequence of human fibroblast interferon as deduced from reverse transcripts obtained using synthetic oligonucleotide primers. Nucleic Acids Res 8:1913–1931.

Huberman E, Callahan MF. 1979. Induction of terminal differentiation in human promyelocytic leukemia cells by tumor-promoting agents. Proc Natl Acad Sci USA 76:1293–1297.

Isaacs A, Lindenmann J. 1957. Virus interference. I. The interferon. Proc R Soc London Ser B 147:258–267.

Jankowski WJ, von Muenchhausen W, Sulkowski E, Carter WA. 1976. Binding of human interferon to immobilized Cibacron Blue F3GA: The nature of molecular interaction. Biochemistry 15:5182–5187.

Kempner ES, Schlegel W. 1979. Size determination of enzymes by radiation inactivation. Anal Biochem 92:2–10.

Kempner ES, Miller JH, Schlegel W, Hearon JZ. 1980. The functional units of polyenzymes. J Biol Chem 255:6826–6831.

Kenny C, Moschera JA, Stein S. 1981. Purification of human fibroblast interferon produced in the absence of serum by Cibacron Blue F3GA-agarose and high performance liquid chromatography. Methods Enzymol 78:435–447.

Kerr IM, Brown RE. 1978. pppA2′p5′A2′p5′A: An inhibitor of protein synthesis synthesized with an enzyme fraction from interferon-treated cells. Proc Natl Acad Sci USA 75:256–260.

Khan A, Hill NO, Dorn GL (eds). 1980. Interferon: Properties and Clinical Uses. Leland Fikes Foundation Press, Dallas, Texas.
Knight E Jr. 1976. Interferon: Purification and initial characterization from human diploid cells. Proc Natl Acad Sci USA 73:520–523.
Knight E Jr, Fahey D. 1981. Human fibroblast interferon: An improved purification. J Biol Chem 256:3609–3611.
Knight E Jr, Hunkapiller MW, Korant BD, Hardy RWF, Hood LE. 1980. Human fibroblast interferon: Amino acid analysis and amino terminal amino acid sequence. Science 207:525–526.
Labdon JE, Gibson KD, Sun S, Pestka S. 1984. Some species of human leukocyte interferon are glycosylated. Arch Biochem Biophys 232:422–426.
Langer JA, Pestka S. 1984. Changes in binding of alpha interferon IFN-αA to HL60 cells during DMSO-induced differentiation. Arch Biochem Biophys, submitted.
Langford MP, Georgiades JA, Stanton GJ, Dianzani F, Johnson HM. 1979. Large scale production and physicochemical characterization of human immune interferon. Infect Immun 26:36–41.
Lawn RM, Adelman J, Dull TJ, Gross M, Goeddel DV, Ullrich A. 1981a. DNA sequence of two closely linked human leukocyte interferon genes. Science 212:1159–1162.
Lawn RM, Gross M, Houck CM, Franke AE, Gray PV, Goeddel DV. 1981b. DNA sequence of a major human leukocyte interferon gene. Proc Natl Acad Sci USA 78:5435–5439.
Lengyel P. 1981. Enzymology of interferon action—A short survey. Methods Enzymol 79:135–148.
Lengyel P, Pestka S. 1981. Interferon as an example of posttranslational modification and control. *In* Prockop DC, Champe PC (eds), Gene Families of Collagen and Other Proteins. Elsevier-North Holland, Amsterdam, pp. 121–126.
Levy WP, Shively J, Rubinstein M, Del Valle U, Pestka S. 1980. Amino-terminal amino acid sequence of human leukocyte interferon. Proc Natl Acad Sci USA 77:5102–5104.
Levy WP, Rubinstein M, Shively J, Del Valle U, Lai C-Y, Moschera J, Brink L, Gerber L, Stein S, Pestka S. 1981. Amino acid sequence of a human leukocyte interferon. Proc Natl Acad Sci USA 78:6186–6190.
Lotem J, Sachs L. 1979. Regulation of normal differentiation in mouse and human myeloid leukemic cells by phorbol esters and the mechanism of tumor promotion. Proc Natl Acad Sci USA 76:5158–5162.
Lowe ME, Kempner ES. 1982. Radiation inactivation of the glycoprotein, invertase. J Biol Chem 257:12478–12480.
Maeda S, McCandliss R, Gross M, Sloma A, Familletti PC, Tabor JM, Evinger M, Levy WP, Pestka S. 1980. Construction and identification of bacterial plasmids containing nucleotide sequence for human leukocyte interferon. Proc Natl Acad Sci USA 77:7010–7013; [correction 78:4648 (1981)].
Maeda S, McCandliss R, Chiang T-R, Costello L, Levy WP, Chang NT, Pestka S. 1981. Sequences of a human fibroblast interferon gene and two linked human leukocyte interferon genes. *In* Brown D, Fox CF (eds), Developmental Biology Using Purified Genes. Academic Press, New York, pp. 85–96.
Maheshwari RK, Friedman RM. 1981. Assay of effect of interferon on viruses that bud from plasma membrane. Methods Enzymol 79:451–458.
Mantei N, Schwarzstein M, Streuli M, Panem S, Nagata S, Weissmann C. 1980. The nucleotide sequence of a cloned human leukocyte interferon cDNA. Gene 10:1–10.
Nagano Y, Kojima Y. 1958. Inhibition de l'infection vaccinale par le virus homologue. C R Soc Biol 152:1627–1629.
Nagata S, Mantei N, Weissmann C. 1980a. The structure of one of the eight or more distinct chromosomal genes for human interferon-α. Nature 287:401–408.
Nagata S, Taira H, Hall A, Johnsrud L, Streuli M, Ecsödi J, Boll W, Cantell K, Weissmann C. 1980b. Synthesis in *E. coli* of a polypeptide with human leukocyte interferon activity. Nature 284:316–320.
Ng MH, Vilcek J. 1972. Interferons: Physicochemical properties and control of regular synthesis. Adv Protein Chem 26:173–241.
Nielsen TB, Lad PM, Preston MS, Kempner E, Schlegel W, Rodbell M. 1981. Structure of the turkey erythrocyte adenylate cyclase system. Proc Natl Acad Sci USA 78:722–726.
Okamura H, Berthold W, Hood L, Hunkapiller M, Inoue M, Smith-Johannsen H, Tan YH. 1980. Human fibroblastoid interferon: Immunosorbent column chromatography and N-terminal amino acid sequence. Biochemistry 19:3831–3835.
Ortaldo JR, Herberman RB, Harvey C, Oscheroff P, Pan Y-CE, Kelder B, Pestka S. 1984. A species

of human α-interferon which lacks the ability to boost human natural killer (NK) activity. Proc Natl Acad Sci USA 81:4926–4929.

Ortaldo JR, Mason A, Rehberg E, Kelder B, Harvey C, Oscheroff P, Pestka S, Herberman RB. 1983a. Augmentation of NK activity with recombinant and hybrid recombinant human leukocyte interferons. *In* DeMaeyer E, Schellekens H (eds), The Biology of the Interferon System. Elsevier Scientific Publishers, Amsterdam, pp. 353–358.

Ortaldo JR, Mason A, Rehberg E, Moschera J, Kelder B, Pestka S, Herberman RB. 1983b. Effects of recombinant and hybrid recombinant human leukocyte interferons on cytotoxic activity of natural killer cells. J Biol Chem 258:15011–15015.

Ortaldo JR, Mantovani A, Hobbs D, Rubinstein M, Pestka S, Herberman RB. 1983c. Effects of several species of human leukocyte interferon on cytotoxic activity of NK cells and monocytes. Int J Cancer 31:285–289.

Pestka S. 1978. Human interferon: The proteins, the mRNA, the genes, the future. *In* Weissbach H (ed), Dimensions in Health Research: Search for the Medicines of Tomorrow. Academic Press, New York, pp. 29–56.

Pestka S (ed). 1981a. Interferons Part A, Methods in Enzymology, Vol. 78. Academic Press, New York.

Pestka S (ed). 1981b. Interferons Part B, Methods in Enzymology, Vol. 79. Academic Press, New York.

Pestka S. 1983. The human interferons—From protein purification and sequence to cloning and expression in bacteria: Before, between and beyond. Arch Biochem Biophys 221:1–37.

Pestka S, Evinger M, McCandliss R, Sloma A, Rubinstein M. 1980. Human interferon: The messenger RNA and the proteins. *In* Beers RF Jr, Bassett EG (eds), Polypeptide Hormones. Raven Press, New York, pp. 33–48.

Pestka S, Kelder B, Familletti PC, Moschera JA, Crowl R, Kempner ES. 1983. Molecular weight of the functional unit of human leukocyte, fibroblast, and immune interferons. J Biol Chem 258:9706–9709.

Pestka S, Maeda S, Hobbs DS, Chiang T-RC, Costello LL, Rehberg E, Levy WP, Chang NT, Wainwright NR, Hiscott JB, McCandliss R, Stein S, Moschera JA, Staehelin T. 1981a. The human interferons: The proteins and their expression in bacteria. *In* Walton AG (ed), Recombinant DNA. Elsevier Scientific Publishing Co., Amsterdam, pp. 51–74.

Pestka S, Maeda S, Hobbs DS, Levy WP, McCandliss R, Stein S, Moschera JA, Staehelin T. 1981b. The human interferons. *In* Mozes LW, Scott WA, Werner R, Schultz J (eds), Cellular Responses to Molecular Modulators. Academic Press, New York, pp. 455–493.

Rehberg E, Kelder B, Hoal EG, Pestka S. 1982. Specific molecular activities of recombinant and hybrid leukocyte interferons. J Biol Chem 257:11497–11502.

Revel M, Wallach D, Merlin G, Schattner A, Schmidt A, Wolf D, Shulman L, Kimchi A. 1981. Interferon-induced enzymes: Microassays and their applications; Purification and assay of (2′-5′)-oligoadenylate synthetase and assay of 2′-phosphodiesterase. Methods Enzymol 79:149–161.

Rovera G, O'Brien TG, Diamond L. 1979a. Induction of differentiation in human promyelocytic leukemia cells by tumor promoters. Science 204:868–870.

Rovera G, Santoli D, Damsky C. 1979b. Human promyelocytic leukemia cells in culture differentiate into macrophage-like cells when treated with a phorbol diester. Proc Natl Acad Sci USA 76:2779–2783.

Rubinstein M, Pestka S. 1981. Purification and characterization of human leukocyte interferons by high performance liquid chromatography. Methods Enzymol 78:464–472.

Rubinstein M, Levy WP, Moschera JA, Lai C-Y, Hershberg RD, Bartlett RT, Pestka S. 1981. Human leukocyte interferon: Isolation and characterization of several molecular forms. Arch Biochem Biophys 210:307–318.

Rubinstein M, Rubinstein S, Familletti PC, Gross MS, Miller RS, Waldman AA, Pestka S. 1978. Human leukocyte interferon purified to homogeneity. Science 202:1289–1290.

Rubinstein M, Rubinstein S, Familletti PC, Miller RS, Waldman AA, Pestka S. 1979. Human leukocyte interferon: Production, purification to homogeneity, and initial characterization. Proc Natl Acad Sci USA 76:640–644.

Sen GC, Herz RE. 1983. Differential antiviral effects of interferon in three murine cell lines. J Virol 45:1017–1027.

Shively JE, Del Valle U, Blacher R, Hawke D, Levy WP, Rubinstein M, Stein S, McGregor WC,

Tarnowski J, Bartlett R, Lee D, Pestka S. 1982. Microsequence analysis of peptides and proteins. IV. Structural studies on human leukocyte interferon. Anal Biochem 126:318–326.
Sreevalsan T, Lee E, Friedman RM. 1981. Assay of effect of interferon on intracellular enzymes. Methods Enzymol 79:342–349.
Staehelin T, Hobbs DS, Kung H-F, Lai C-Y, Pestka S. 1981a. Purification and characterization of recombinant human leukocyte interferon (IFLrA) with monoclonal antibodies. J Biol Chem 256:9750–9754.
Staehelin T, Hobbs DS, Kung H-F, Pestka S. 1981b. Purification of recombinant human leukocyte interferon (IFLrA) with monoclonal antibodies. Methods Enzymol 78:505–512.
Stein S, Kenny C, Friesen H-J, Shively J, Del Valle U, Pestka S. 1980. NH_2-terminal amino acid sequence of human fibroblast interferon. Proc Natl Acad Sci USA 77:5716–5719.
Stewart WE II. 1974. Distant molecular species of interferons. Virology 61:80–86.
Stewart WE II (ed). 1979. The Interferon System. Springer-Verlag, New York.
Stewart WE II, Lin LS, Wiranowska-Stewart M, Cantell K. 1977. Elimination of size and charge heterogeneities of human leukocyte interferons by chemical cleavage. Proc Natl Acad Sci USA 74:4200–4204.
Stinebring WR, Chapple PJ (eds). 1978. Human Interferon: Production and Clinical Use. Plenum, New York.
Streuli M, Hall A, Boll W, Stewart WE II, Nagata S, Weissmann C. 1981. Target cell specificity of two species of human interferon-α produced in *Escherichia coli* and of hybrid molecules derived from them. Proc Natl Acad Sci USA 78:2848–2852.
Streuli M, Nagata S, Weissmann C. 1980. At least three human type α interferons: Structure of α2. Science 209:1343–1347.
Taniguchi T, Ohno S, Fujii-Kuriyama Y, Muramatsu M. 1980. The nucleotide sequence of human fibroblast interferon cDNA. Gene 10:11–15.
Taniguchi T, Sakai M, Fujii-Kuriyama Y, Muramatsu M, Kobayashi S, Sudo T. 1979. Construction and identification of a bacterial plasmid containing the human fibroblast interferon gene sequence. Proc Jpn Acad Ser B 55:464–469.
Tomida M, Yamamoto Y, Hozumi M. 1982. Stimulation by interferon of induction of differentiation of human promyelocytic leukemia cells. Biochem Biophys Res Commun 104:30–37.
Törma ET, Paucker K. 1976. Purification and characterization of human leukocyte interferon components. J Biol Chem 251:4810–4816.
Ullrich A, Gray A, Goeddel DV, Dull TJ. 1982. Nucleotide sequence of a portion of human chromosome 9 containing a leukocyte interferon gene cluster. J Mol Biol 156:467–486.
Vilcek J, Gresser I, Merigan TC (eds). 1980. Ann NY Acad Sci 350:1–641.
Waldman AA, Miller RS, Familletti PC, Rubinstein S, Pestka S. 1981. Induction and production of interferon with human leukocytes from normal donors with the use of Newcastle disease virus. Methods Enzymol 78:39–44.
Weck PK, Apperson S, Stebbing N, Gray PW, Leung D, Shepard HM, Goeddel DV. 1981. Antiviral activities of hybrids of two major human leukocyte interferons. Nucleic Acids Res 9:6153–6166.
Weissmann C, Nagata S, Boll W, Fountoulakis M, Fujisawa A, Fujisawa J-I, Haynes J, Henco K, Mantei N, Ragg H, Schein C, Schmid J, Shaw G, Streuli M, Taira H, Todokoro K, Weideli U. 1982. Structure and expression of human IFN-α genes. Phil Trans R Soc Lond B 299:7–28.
Yabrov AA (ed). 1980. Interferon and Nonspecific Resistance. Human Sciences Press, New York.
Yip YK, Pang RHL, Urban C, Vilcek J. 1981. Partial purification and characterization of human γ (immune) interferon. Proc Natl Acad Sci USA 78:1601–1605.
Zoon KC. 1981a. Purification and characterization of human interferon from lymphoblastoid (Namalva) cultures. Methods Enzymol 78:457–464.
Zoon KC. 1981b. Purification, sequencing, and properties of human lymphoblastoid and leukocyte interferon. *In* DeMaeyer E, Galasso G, Schellekens H (eds), The Biology of the Interferon System. Elsevier Scientific Publishers, Amsterdam, pp. 47–55.

Mediators in Cell Growth and Differentiation,
edited by Richard J. Ford and Abby L. Maizel.
Raven Press, New York © 1985.

Possible Mechanisms of Interferon-Induced Growth Inhibition

Joyce Taylor-Papadimitriou, Nicolette Ebsworth, and Enrique Rozengurt

Imperial Cancer Research Fund, Lincoln's Inn Fields, London WC2A 3PX England

The interferons are the only naturally produced inhibitors of cell growth that have been well characterized at the molecular level and that can inhibit the growth of a wide variety of cell types (for reviews, see Taylor-Papadimitriou 1980, 1983). Since they can also affect differentiated function, particularly in the effector cells of the immune system (Balkwill 1979, Moore 1983), they represent one of the most interesting groups of cell regulatory molecules known today.

Because of their regulatory functions, the interferons have attracted much interest as potential anticancer agents. They have as yet been less exploited in basic studies on growth regulation in in vitro model systems. To some extent this reflects the complexity of the systems and the lack of a detailed molecular picture of how growth factors stimulate cell proliferation. Considerable progress has been made in the study of the growth factors themselves, their interactions with receptors, and the initial signals generated by this interaction (Rozengurt 1983). However, for the period between these early events and the replication of DNA and cell division, there is much less information.

It has been hypothesized for a long time that labile proteins are required for cells to progress through G_1 (Schneiderman et al. 1971). Recent studies have extended this hypothesis and attempted to characterize the new gene expression messages (Cochran et al. 1983) and labile proteins (Croy and Pardee 1983) being made. However, since the modulation of macromolecular synthesis is the result of the binding of growth factors to surface receptors, it is likely that these processes are elicited by the early signals evoked by the occupancy of the growth factor receptors. Thus, a conceptual framework for thinking about growth regulation is slowly emerging within which experimental approaches can be found to the problem of interferon-induced inhibition of cell growth.

Determining how interferons interfere with the mitogenic signal, or the cells respond to it, is a logical approach to investigating possible mechanisms involved in interferon-induced growth inhibition. Another approach is to consider the effects interferons are known to have on cell structures (membrane, cytoskeleton components) and on the metabolic profile of the cell (induction or inhibition of the

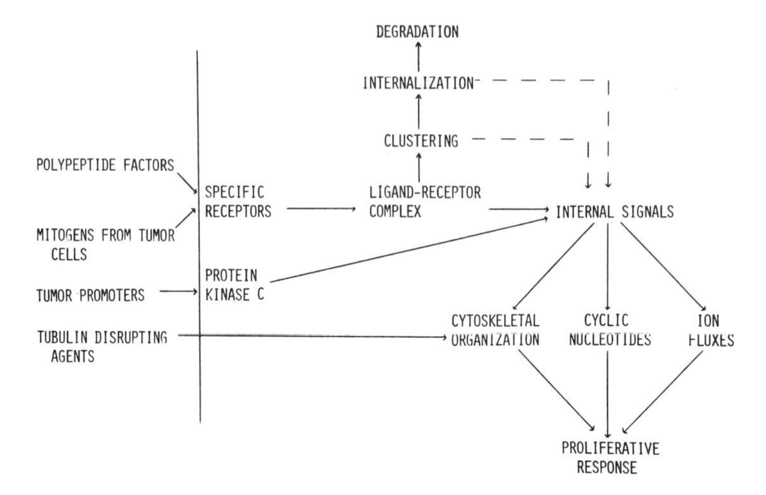

FIG. 1. A scheme for the generation of internal signals by different growth factors. The connections shown by dashed lines have not received experimental support.

synthesis of certain proteins) and to try to relate these to growth inhibition. In this chapter, we will consider some aspects of both of these approaches and the results they have yielded so far.

INHIBITION OF THE MITOGENIC SIGNAL

In investigations of the effects of interferons on the mitogenic signal, there are obvious advantages to working with synchronized systems in which molecular events can be examined. Quiescent 3T3 cells synchronized in G_1 can be stimulated to grow by the addition of serum or a combination of pure growth factors (Table 1), and a considerable amount of information is available about the initial signals that are generated by the different growth factors (Rozengurt 1983; Figure 1). Since interferon has been shown to inhibit entry into DNA synthesis in 3T3 cells stimulated by either serum (Balkwill and Taylor-Papadimitriou 1978, Sokawa et al. 1977) or growth factors (Sreevalsan et al. 1980), it has been possible to investigate in some detail this antagonistic effect of interferon to the growth-promoting signals of a range of growth factors.

Reversal of Interferon Action by Mitogens

One important finding of both practical and basic significance that has emerged from the work with 3T3 cells is that the degree of inhibition effected by interferon decreases as the number of growth factors used to stimulate the cells is increased. Those 3T3 cells stimulated from quiescence by only two mitogens are much more sensitive to the inhibitory effect of interferon on DNA synthesis than the same cells stimulated by five mitogens acting synergistically (Taylor-Papadimitriou et al. 1981). The practical implications of this observation are that to define the sensitivity

TABLE 1. *Chemically diverse factors stimulate DNA synthesis in Swiss 3T3 cells*

Polypeptide factors	EGF, PDGF, IGFs*
Neurohypophyseal hormones	Vasopressin, oxytocin, analogues
Tumor promoters	Phorbol esters (TPA), teleocidin
Regulatory peptides	Bombesin
Vitamin A derivatives	Retinoic acid
Permeability modulators	Melittin
Cyclic nucleotide–elevating agents	Cholera toxin, adenosine agonists, cAMP derivatives
Microtubule-disrupting agents	Colchicine, Colcemid, vinblastine, podophyllotoxin, nocodazole
Polypeptide released by transformed cells	Fibroblast-derived growth factor

*Epidermal growth factor, platelet-derived growth factor, insulin-like growth factors.

of a cell to interferon's growth inhibitory effect, it is necessary to define the growth conditions (Taylor-Papadimitriou and Rozengurt 1982). On a more basic level, it suggests that the signals induced by the various growth factors might converge at an interferon-sensitive step.

Since the stimulus for DNA synthesis by any combination of growth factors is inhibited by interferon, provided only two or three factors are used (Sreevalsan et al. 1980, Ebsworth et al., 1984), we must conclude that there is an interferon-sensitive step in the sequence of events leading from receptor interaction to DNA synthesis for each factor. Further, since any combination of five growth factors can reverse the growth inhibitory effect of interferon, it is plausible that the signals induced by the various factors converge before an interferon-sensitive event, such as, for example, induction of a specific protein whose level affects entry into DNA synthesis. According to this hypothesis, which has been previously discussed in detail (Taylor-Papadimitriou et al. 1981), as more factors are added, more of the protein would be induced, and with five factors, even though interferon treatment reduces the level of the induced protein, there would still be sufficient protein to stimulate entry into S.

Inhibition of Induction of Ornithine Decarboxylase

An increase in the level of the enzyme ornithine decarboxylase (ODC) is always found to accompany the proliferative response, and in stimulated 3T3 cells a peak of activity is seen 6 hr after addition of growth factors (Sreevalsan et al. 1979), the level of which increases dramatically as more growth factors are added (Sreevalsan et al. 1980). The production of putrescine, the rate-limiting step in the production of the polyamines, could therefore be a possible site of action for interferon. It has been possible to show that interferon can specifically inhibit the increase in ODC activity seen in 3T3 cells stimulated by growth factors (Sreevalsan et al. 1980), and, more recently, in human cells of fibroblastic (Sekar et al. 1983) and epithelial origin (Taylor-Papadimitriou et al. 1983). Moreover, whereas ODC induction by interferon is inhibited when several mitogens are used to stimulate 3T3 cells, the absolute ODC level in the highly stimulated, interferon-treated cell is much higher than the level in the comparable cell stimulated with fewer mitogens. It could therefore be argued that the levels of ODC regulate the rate of entry into S. Whether the inhibition of the increase in ODC activity is a primary or secondary event in interferon's action in inhibiting cell growth is, however, not clear. As will be seen below, interferon can be added quite late in G_1 (after the increase in ODC activity has occurred) and still inhibit entry into DNA synthesis. This fact would argue against inhibition of ODC's being the only site of interferon action in inhibiting the entry of quiescent 3T3 cells into DNA synthesis. ODC may, however, be representative of a group of proteins induced by mitogens and inhibited by interferons that are crucial to cell proliferation.

Growth Factors, Initial Signals, and Synergism

The use of pure growth factors to stimulate reinitiation of DNA synthesis in 3T3 cells maintained in serum-free medium revealed an important feature of their action: the existence of potent synergistic effects (see Rozengurt 1983; Table 1, Figure 1). The existence of such effects has several important implications. It suggests that extracellular factors bind to different receptors, generate multiple internal signals, and thereby activate various pathways leading to cell proliferation. There is now considerable evidence indicating that these pathways can involve (1) the opening of ion permeability pathways through the plasma membrane, (2) changes in the intracellular concentration of cyclic nucleotides, and (3) alterations in the organization of the cytoskeleton. It is reasonable, therefore, to consider the effect of interferon on these initial signals generated by different growth factors.

Ion Fluxes

An increase in the rate of Na^+ influx is one of the earliest events in the stimulation of a variety of quiescent cells into DNA synthesis and cell division by serum, PDGF, FDGF, vasopressin, phorbol esters, teleocidin, and melittin (Rozengurt 1981). Entry of Na^+ is coupled, at least in part, with the exit of H^+ via an amiloride-sensitive Na^+-H^+ exchange system. The stimulation of Na^+ influx is apparently sufficient to regulate intracellular pH and to enhance the activity of the Na-K pump, which increases K^+ in the cell (Rozengurt 1983). We have examined the effect of interferon on these early ion transport changes in 3T3 cells and found that interferon added for up to 24 hr prior to the growth factor has no effect on stimulation of the sodium pump. Figure 2 shows the level of ouabain-insensitive rubidium uptake in quiescent and stimulated 3T3 and the effect of interferon on these levels. Clearly, no inhibition of Rb^+ uptake occurred after interferon treatment; rather, a small stimulation was evident. We must conclude, therefore, that interferon does not inhibit the early changes in ion transport stimulated by some growth factors.

Cyclic AMP

Increased intracellular levels of cAMP have been shown to act as a mitogenic signal for quiescent Swiss 3T3 cells. The elevated cAMP levels stimulate DNA

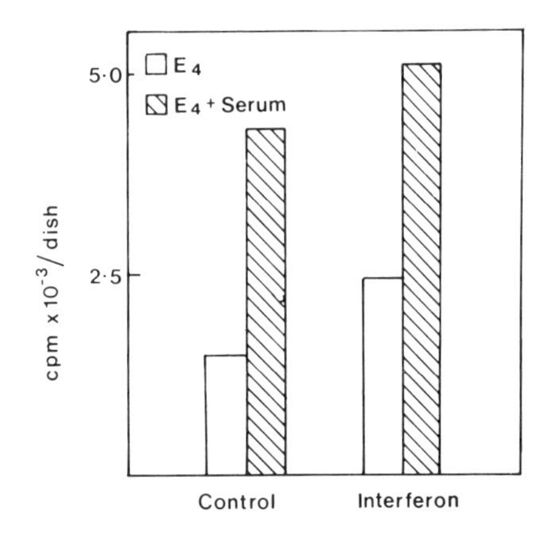

FIG. 2. Lack of inhibition by interferon of the sodium pump in quiescent and stimulated 3T3 cells. Quiescent cells were incubated with or without 2,000 units of L cell interferon for 24 hr, and 15 min after a change to fresh medium, or to medium containing serum, ouabain-sensitive uptake of Rb^+ was measured. Cells pretreated with interferon were also exposed to interferon after the medium change. The ordinate represents cpm over 15 min/3 cm dish; the protein content of each dish was 88 μg. E_4 = Dulbecco's modified Eagle's medium.

synthesis synergistically with other growth-promoting factors (Rozengurt 1982). The levels of the cyclic nucleotide were increased by a variety of ligands that stimulate adenylate cyclase, including cholera toxin, adenosine agonists, or cAMP derivatives. Since increased cAMP levels do not stimulate Na^+ entry into quiescent cells, it appears that cAMP delivers a mitogenic signal to Swiss 3T3 cells through a mechanism that does not involve a primary increase in Na^+-dependent ion fluxes. We have examined the effect of interferon on intracellular cAMP levels in quiescent 3T3 cells and in cells stimulated by cAMP-elevating agents and found the levels to be unaffected by interferon treatment (Ebsworth et al. 1984). Moreover, the DNA synthesis stimulated by elevating intracellular levels of cAMP in the presence of insulin or other growth factors affecting ion transport was sensitive to inhibition by interferon. These results indicate that, in 3T3 cells, cAMP plays neither a direct nor an indirect role in interferon's inhibition of cell growth. This is in contrast to the results obtained with macrophage cell lines, in which there is good evidence to suggest that cAMP is involved in mediating interferon-induced inhibition of cell growth (Schneck et al. 1982, and see below). This difference may be explained by the fact that cAMP normally inhibits cell growth in macrophages (whereas it is mitogenic in 3T3 cells) and that interferons recruit the normal growth regulation mechanisms used by the cell.

Since the early events (ion transport, cAMP elevation) occurring after stimulation of 3T3 cells are not affected by interferons, it is clear that binding of the relevant growth factors to the cell membrane is not inhibited by interferons, a fact that has been directly demonstrated for binding of ^{125}I-labeled EGF to 3T3 cells (Lin et al. 1980).

The Cytoskeleton

Since agents that disrupt the tubulin network in 3T3 cells can act synergistically with cAMP-elevating agents or growth factors affecting ion fluxes to stimulate DNA synthesis (Friedkin and Rozengurt 1981, Wang and Rozengurt 1983), the possibility exists that perturbation of the tubulin network is normally involved in growth regulation (see Table 1 and Figure 1). It has been known for some time that interferon action can be inhibited by agents that disrupt the tubulin network and can be enhanced by stabilizers of this network (Bourgeaude and Chany 1976, 1979). More recently, it has been shown that interferon treatment can result in an increase in the level of mRNA for tubulin (Fellous et al. 1982). In view of this, it is possible that the antagonistic effect between growth factors and interferon is operating, at least in part, at the level of the cytoskeleton. Our own experiments that show the reversal of interferon's growth inhibitory effect by increases in the number of mitogens were originally designed to test the effect of colchicine on interferon action. It was found that colchicine added with three or four other factors could indeed reverse the inhibitory effect of interferon, but replacing colchicine (as the fourth or fifth factor) with any other extra factor also resulted in reversal of its growth inhibition. Whether other growth factors have a destabilizing effect on the microtubules has not been investigated, but this remains a possibility.

One observation on the kinetics of inhibition by interferon and stimulation by colchicine of DNA synthesis is, however, potentially important. The stimulatory effect of colchicine is seen even when added 6 hr after the other mitogens have been added to 3T3 cells (Friedkin and Rozengurt 1981, Wang and Rozengurt 1983), and the inhibitory effect of interferon on DNA synthesis is also seen when added up to 8 hr after the growth factors (Figure 3). This observation tells us that some event is inhibited by interferon in late G_1, at a point where perturbation of the microtubules can affect the entry into DNA synthesis. Clearly, further investigation into the levels and state of aggregation of tubulin in interferon action is warranted now that the availability of cDNA probes and monoclonal antibodies makes monitoring of the mRNA and protein easier. In view of the effect of interferon added in late G_1, it is not surprising that no effects on the initial signals generated by the mitogens have been detected.

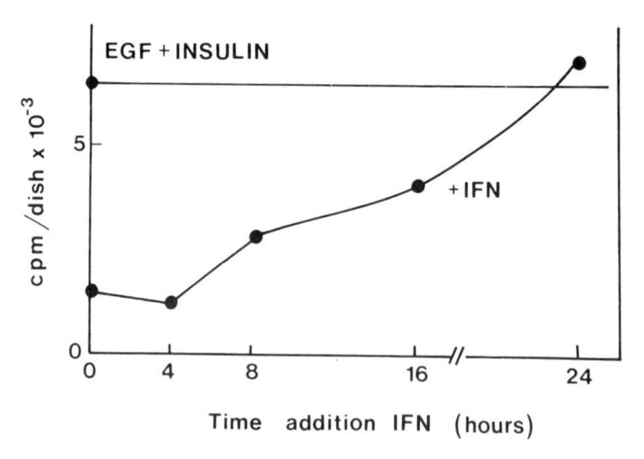

FIG. 3. Inhibition of DNA synthesis by interferon (1,000 units/ml) added at various times after stimulation of 3T3 cells by growth factors (EGF, 5 ng/ml, and insulin, 10 μg/ml). ^{3}HdT was added with the growth factors and total incorporation of radioactivity into TCA-insoluble material estimated 48 hr later. EGF, epidermal growth factor; IFN, interferon.

Implications of 3T3 Cell Results

From the data obtained following the inhibition by interferon of the mitogenic signal in synchronized 3T3 cells, it can be concluded that the early signals generated by a wide range of growth factors appear not to be inhibited by interferon. The inhibition of ODC seen in 3T3 cells treated with interferon is also unlikely to be solely responsible for the inhibition of entry into DNA synthesis, since addition of interferon after 6 hr (the peak of ODC activity) is effective in extending G_1. Furthermore, in contrast to polypeptide growth factors and tumor promoters, the stimulation of ODC induced by cAMP-elevating agents is not sensitive to inhibition by interferon (Lee et al. 1980), which, however, we have seen can inhibit the mitogenic signal produced by these agents (Ebsworth et al. 1984). An important clue appears to come from the observation that interferon added 6–8 hr after addition of growth factors can still inhibit entry into DNA synthesis. This clearly indicates that an event (or events) in late G_1 is a target for interferon action and suggests a possible involvement of the cytoskeleton, in particular the tubulin network. Further studies on the effect of interferon on the level and state of aggregation of the tubulin network are warranted. Interferon could also block the synthesis of some of the labile, G_1 proteins induced by growth factors and required for DNA synthesis.

CELLULAR CHANGES INDUCED BY INTERFERONS

General

Figure 4 gives a general outline of the major changes seen in cells treated with interferon. After interaction with the membrane receptor, signals are generated that produce (1) physical, chemical, and functional changes in the membrane; (2) changes in various components of the cytoskeleton; (3) the induction of a range of new proteins; and (4) inhibition of certain inducible proteins (for reviews, see Taylor-Papadimitriou 1980, Taylor-Papadimitriou et al. 1983). The question we must ask is: Which of these changes are related to the inhibition of cell growth induced by interferon treatment? One obvious approach to this question is to study the changes induced by interferon in cells that are sensitive to its growth inhibitory effect and compare them to those seen in resistant cells. In focusing on the cells, we need to remember that there are three classes of interferon, and common mechanisms may or may not be operative in the inhibition of cell proliferation by all three. In particular, since γ-interferons (class II) appear to interact with a different receptor from that recognized by the α and β type (class I) (Ankel et al. 1980, Aguet and Blanchard 1981, Branca and Baglioni 1981), induce a different group of proteins, and act synergistically with α and β (for review, see Epstein 1981), it is highly likely that different mechanisms are involved in the action of the class I and class II interferons.

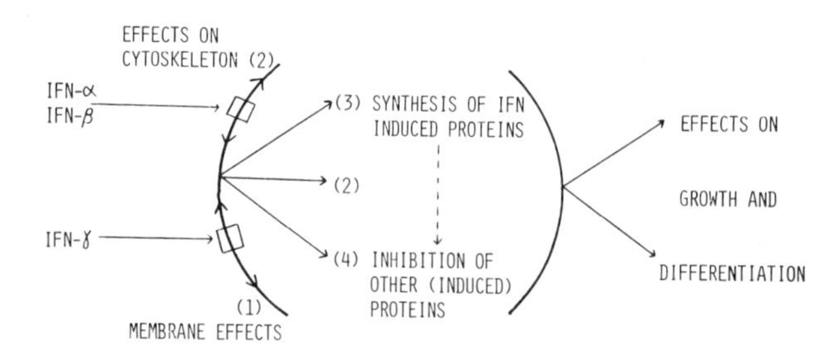

FIG. 4. Interferon effects that may be involved in inhibition of cell growth. IFN, interferon.

Studies with Cell Lines Resistant to Interferon's Growth Inhibitory Effect

Role of Interferon-Induced Proteins

The first interferon-resistant cell lines were derived from the mouse L1210 cell, but this cell has been shown to be deficient at the receptor level (Aguet 1980) and was therefore not useful for analyzing events subsequent to the receptor interaction. Other mutants that do have receptors have produced conflicting information regarding the function of the interferon-induced proteins. Several human cell lines (Verhaegen et al. 1980, Silverman et al. 1982, Vandenbussche et al. 1981, 1983, Tovey et al. 1983) and a clone of NIH 3T3 cells (Panet et al. 1981) have been isolated that are resistant to the growth inhibitory effect of interferon. The two most widely investigated proteins are the 2′5′ oligoadenylate synthetase (2′5′A synthetase) and the inducible protein kinase, both of which are activated by double-stranded RNA (for review, see McMahon and Kerr 1983).

In investigating the role of the 2′5′A system, most workers have not measured the level of nucleotide, but rather the synthetase itself or the nuclease activated by the nucleotide. These studies show that the synthetase and the kinase are inducible in the resistant human cell lines with the exception of HEC-1, which has high constitutive enzyme levels (Verhaegen et al. 1980). A complete analysis of the 2′5′A system has been done by Silverman and colleagues (1982) using sensitive and resistant Daudi cells. These workers also reported induction of the synthetase and kinase in resistant cells, but could not detect 2′5′A in either the sensitive or the resistant cells. The only indication that the 2′5′A system might be involved in sensitivity to interferon's growth inhibitory effect was the observation that the RNase activated by 2′5′A was not induced in the wild type cells. Another cell system in which this RNase was implicated is in a class of NIH 3T3 cells whose growth is not inhibited by interferon and that have low or undetectable levels of the nuclease.

In summary, the results from work on interferon-induced proteins in cell mutants resistant to interferon's growth inhibitory effect do not point to a consistent role for the 2′5′A synthetase or the protein kinase in interferon-induced growth inhibition, although some aspects of the 2′5′A system may be involved in some cell types. It has been suggested that the interferon-induced double-stranded RNA–

dependent kinase may phosphorylate ODC and thus inactivate it, in the same way that a polyamine-dependent kinase regulates the activity of ODC in the slime mold *Rhysarum polycephalum* (Sekar et al. 1982). This appealing idea has not yet received experimental confirmation.

Role of cAMP

As discussed above, the evidence obtained using 3T3 cells in which cAMP is a mitogenic signal indicates that this nucleotide is not involved in mediating the inhibition of growth induced by interferon in these cells. However, using a series of mutant macrophage cell lines, some of which have defective cAMP-dependent kinase activity, Schneck and colleagues (1982) obtained convincing evidence for a central role of cAMP in interferon-induced growth inhibition. In the parent cell line, cAMP has been shown to inhibit cell growth, and the mutant cell lines with a defective kinase show a much-reduced sensitivity to cAMP and to interferon, both as an inhibitor of cell growth and enhancer of cell function. Their results are in contrast to those reported by Bannerjee et al. (1983) using a series of mutant Chinese hamster ovary (CHO) cell lines. In this case, the mutant cells defective in the cAMP-dependent protein kinase showed no reduction in sensitivity to interferon as an inhibitor of cell growth. One must conclude from the studies relating to cAMP and interferon that the nucleotide may play a role in mediating the inhibition of cell growth induced by interferon in some cell types, but it cannot be considered to be a general second messenger in all systems (Tovey 1982).

Role of Thymidine Kinase

The uptake of thymidine is a controlling factor in the growth of cells in culture, providing the folic acid content of the medium is of the same order as that found in body fluids (Taylor-Papadimitriou and Rozengurt 1979). The inhibition of uptake of thymidine can thus inhibit cell growth. It is well documented that interferons can inhibit the uptake of thymidine in a variety of mouse and human systems (Brouty-Boye and Tovey 1978, Gewert et al. 1981), and it has been suggested that inhibition of cell growth by interferon may depend on a functioning thymidine kinase (TK). Again, results with mutant TK cell lines are conflicting. In mouse L929 cells, sensitivity to interferon's growth inhibitory effect has been correlated with a functioning TK gene (Mengheri et al. 1983). However, Gewert and colleagues (1984) working with S49T lymphoma mutants have found the TK^- cells to be equally sensitive to growth inhibition by interferon. It should be pointed out that for these studies the medium in which cells are grown is critical; it might be interesting to determine if interferon inhibits growth more profoundly in medium 199 (which contains low levels of folic acid) and if inhibition is reversed by exogenous thymidine. It is possible that modulation of thymidine incorporation is a result of interferon's inhibiting cell growth rather than a primary cause of this inhibition.

Studies with Other Cell Types Resistant to Interferon's Growth Inhibitory Effect

Embryonal Carcinoma Cells

Undifferentiated embryonal carcinoma cells are insensitive to both the antiviral and antigrowth effect of interferon (Burke et al. 1978). In this case, receptors are present and 2′5′A synthetase is induced, although the protein kinase appears not to be (Wood and Hovanessian 1979). Since the responsive differentiated cells do show inducible kinase activity, there is a possibility that the kinase is involved, although again it may just reflect a block in a controlling event in the interferon response.

Human α Interferons in Human and Bovine Cells

Until the human interferons were studied in detail, interferons were considered to be species specific. Human interferons of the α class (HuIFN-α), however, were found to induce an antiviral state in other species, bovine and feline cells being particularly sensitive. Bovine cells (usually the MDBK cell line) are widely used to assay HuIFN-α since all the different subclasses give a similar specific activity when titrated on these cells. In contrast, in human cells HuIFN-α1 is 100× less effective as an antiviral agent than HuIFN-α2 (Streuli et al. 1981, Rehberg et al. 1982). In spite of being highly sensitive to the antiviral effect of HuIFN-α, bovine cells are resistant to the antigrowth effect of this interferon (Taylor-Papadimitriou et al. 1982). Since all the bovine cells tested (some of which were very sensitive to the antiviral effect) were resistant to the antigrowth effect and most human cells (even those with low sensitivity to the antiviral effect) showed some degree of growth inhibition, it seems likely that it is not the response of the cell that is different in the two species but rather the signals generated by the membrane interaction of the HuIFN-α. We have based our experimental approach on this reasoning and have examined in detail the kinetics of binding of ^{125}I-labeled HuIFN-α2 to a range of human and bovine cells.

Previous data from studies of the antiviral action of HuIFN and their binding to bovine and human cells suggested that the interaction of HuIFN-α with the bovine and human membrane receptor is not identical. Thus, HuIFN-β has no antiviral action on bovine cells (Gresser et al. 1974) and cannot compete with HuIFN-α in binding to these cells (Yonehara et al. 1983a), although HuIFN-α and HuIFN-β appear to compete for the same receptor on human cells. Furthermore, the differences in activity of HuIFN-α1 (and various α hybrids) on human and bovine cells have been shown to be due to its affinity for the receptor (Yonehara et al. 1983b). Finally, a point not always considered is that binding data with bovine cells can be and have been obtained at 4°C (Arnheiter et al. 1983), whereas most of the work with human cells is done at 37°C or 21°C, and usually with the highly sensitive Daudi cell (Mogensen et al. 1981, Mogensen and Bandu 1983, Hannigen et al. 1983, Branca and Baglioni 1981). We have now compared the kinetics of binding of ^{125}I-labeled HuIFN-α2 to a range of bovine and human cell lines at 4°C in which the effector functions of the ligand-receptor complex cannot proceed (Taylor-Papadimitriou and Shearer 1984). Our results clearly show that the kinetics of binding of HuIFN-α2 to bovine cells follow the simple kinetics predicted from the binding of a single molecular species of ligand to a single high-affinity membrane receptor. The dissociation constant for HuIFN-α2 was found to be 3×10^{-11} for the three types of bovine cells listed in Table 2. This figure is similar to the K_D reported for MDBK cells (Yonehara et al. 1983b).

In sharp contrast to the results with bovine cells, the kinetics of binding of HuIFN-α2 to human cells are complex. At interferon concentrations of less than 100–200 units, receptor occupancy appears to increase the rate of binding of HuIFN-α2. This is seen clearly with two relatively insensitive cell lines (T47D and ICRF-23), and after disruption of the cytoskeleton (tubulin network) with a very sensitive cell line (BT20) (Shibata and Taylor-Papadimitriou 1981). Although it is not possible to draw definite conclusions from the kinetic data about the molecular basis for the complex kinetics seen with human cells, they suggest that formation of interferon dimers on the receptor could be involved. Stabilization of binding of one monomer of HuIFN-α2 by binding of a second could explain the positive cooperative binding we see with human cells (as is seen with the binding of the λ repressor to DNA, which shows similar kinetics (Johnson et al. 1981)). HuIFN-α molecules show a pH- and concentration-dependent reversible aggregation in forming dimers and even higher oligomers (Shire 1983), and this may reflect a functional role for dimer formation on the receptor.

Our results also show that in human cells, receptors bound to an organized tubulin network have a different affinity for HuIFN-α2 than those not bound. On the other hand, disruption of the tubulin network in bovine cells does not affect the dose-binding curve. That there are a relatively high number of receptors in BT20 cells and that at least some of these interact with the tubulin network may explain why the interferon-receptor interaction in these cells is effective in inhibiting cell growth even at low concentrations. Formation of interferon dimers from monomers bound to separate receptors might occur when receptors are more abundant and transposable through being bound to structural elements.

The possibility that more than one α monomer may bind to the receptor is interesting when we consider that the α and β interferons appear to share a common receptor on human cells and that HuIFN-β normally exists as a dimer (Pestka et al. 1983). If only monomers can bind to the bovine receptor, it is not surprising that HuIFN-β is not bound. The binding of two different α molecules to one receptor in human cells would also allow for a range of ligand interactions with the mixtures of HuIFN-α normally naturally produced by lymphocytes. Synergy has already been reported for α1 and α2, and this synergy may operate at the receptor level (Orchansky et al. 1983).

Binding studies done at 4°C should reflect the properties of the interferon-receptor interaction. At 37°C the effector functions of the complex can proceed, as can receptor movement in the cell membrane. It will be important to study the

TABLE 2. *Some characteristics of binding of HuIFN-α2 to human and bovine cells*

Cell line or strain	Ref.	Inhibition of cell growth by 100 units of HuIFN-α2	fmoles bound/10^7 cells after incubation with 100 units interferon	K_D HuIFN-α2
Bovine				
BEL (Diploid)	Obtained from Flow Labs	0	1–4	3×10^{-11}
BEK (Diploid)	Flow Labs	0	1–3	3×10^{-11}
SVLE1	Shearer and Taylor-Papadimitriou 1981	0	0.9	3×10^{-11}
Human				
BT20	Lasfargues and Ozzello 1958	60	2.6	
T47D	Keydar et al. 1979	6	0.6	
ICRF-23 (Diploid)	Taylor-Papadimitriou et al. 1982	15	0.45	

BEL, bovine embryo lung; BEK, bovine embryo kidney; SVLE1, SV40-transformed calf lens; BT20, primary breast carcinoma; T47D, metastatic breast carcinoma; ICRF-23, human embryo lung fibroblast.

reaction of interferon with the cell membrane at both temperatures to analyze further the differences in the interaction with the human and bovine receptor, and to see if these differences can be related to the resistance of the bovine cells to interferon-induced growth inhibition.

CONCLUSIONS

A consideration of the available data from studies investigating interferons as growth inhibitors indicates that definite conclusions regarding mechanisms cannot be made, although certain clues may be appearing that could point out directions for future work. In studies of the effect of interferon on the mitogenic signal, it would seem sensible to place some emphasis on studying the effect of interferons and growth factors on the cytoskeleton, particularly the tubulin network, which is known to have an important role in transport of messages through the cell. Also, although inhibition of induction of ODC may not result in an inhibition of cell proliferation, an understanding of the mechanism of inhibition of the induced increase in ODC activity could give important insights into how the interferons inhibit induced protein synthesis, a somewhat more defined phenomenon than cell growth.

Attempts to define cell responses crucial to interferon-induced growth inhibition using resistant and sensitive cells have not yet yielded consistent positive correlations, although it seems clear that the major interferon-induced proteins are not crucially involved. This approach, however, seems to us to be one that should still be pursued, although the investigation of the interferon-cell interaction should probably start at the receptor level, looking for initial signals that may be found only in the sensitive cells. It is a little naive to expect cells with functions as different as those of lymphocytes, keratinocytes, and muscle cells (for example) to use exactly the same mechanisms to control their growth and function. The molecular changes that characterize the response to interferons could therefore be different depending on the metabolic profile of the cell, even though the net result is an inhibition of cell growth. The receptors themselves may show subtle differences from one cell type to another, but it is likely that the signals generated by the receptor interactions are similar.

Perhaps the most important point to keep in mind when considering how to approach the study of interferons as inhibitors of cell growth is that progress is most likely to come in parallel with progress in our understanding of the mechanisms involved in growth regulation. It is therefore very suitable that this chapter should be read together with the other chapters on growth regulation that are also presented in this volume.

REFERENCES

Aguet M. 1980. High affinity binding of ^{125}I-labelled mouse interferon to a specific cell surface receptor. Nature 284:459–461.

Aguet M, Blanchard B. 1981. High affinity binding of ^{125}I-labelled mouse interferon to a specific cell surface receptor. Virology 115:249–261.

Ankel H, Krishnamurti C, Besancon F, Stefanos S, Falcoff E. 1980. Mouse fibroblast (type I) and immune (type II) interferons: Pronounced differences in affinity for gangliosides and in antiviral and antigrowth effects on mouse leukemia L-1210 cells. Proc Natl Acad Sci USA 77:2528–2532.

Arnheiter H, Ohno M, Smith ME, Gutte B, Zoon KC. 1983. Orientation of a human leukocyte interferon molecule on its cell surface receptor: Carboxy-terminus remains accessible to a monoclonal antibody made against a synthetic interferon peptide. Proc Natl Acad Sci USA 80:2539–2543.

Balkwill FR. 1979. Interferons as cell regulatory molecules. Cancer Immunol Immunother 7:7–14.

Balkwill FR, Taylor-Papadimitriou J. 1978. Interferon affects both G1 and S + G2 in cells stimulated from quiescence to growth. Nature 274:798–800.

Banerjee DK, Baksi K, Gottesman MM. 1983. Genetic evidence that action of cAMP-dependent protein kinase is not an obligatory step for antiviral and antiproliferative effects of human interferon in Chinese hamster cells. Virology 129:230–238.

Bourgeaude MF, Chany C. 1976. Inhibition of interferon action by cytochalasin B, colchicine and vinblastine. Proc Natl Acad Sci USA 153:501–504.

Bourgeaude MF, Chany C. 1979. Effect of sodium butyrate on the antiviral and anti-cellular action of interferon on normal and MSV-transformed cells. Int J Cancer 24:314–318.

Branca AA, Baglioni C. 1981. Evidence that types I and II interferons have different receptors. Nature 294:768–770.

Brouty-Boye D, Tovey MG. 1978. Inhibition by interferon of thymidine uptake in chemostat cultures of L1210 cells. Intervirology 9:243–252.

Burke DC, Graham CF, Lehman JM. 1978. Appearance of interferon inducibility and sensitivity during differentiation of murine teratocarcinoma cells in vitro. Cell 13:243–248.

Cochran BH, Reffel AC, Stiles CD. 1983. Molecular cloning of gene sequences regulated by platelet-derived growth factor. Cell 33:939–947.

Croy RG, Pardee AB. 1983. Enhanced synthesis and stabilization of Mr 68,000 protein in transformed BALB/c-3T3 cells: Candidate for restriction point control of cell growth. Proc Natl Acad Sci USA 80:4699–4703.

De Lean A, Rodbard D. 1979. Kinetics of cooperative binding. *In* O'Brien RD (ed), The Receptors. A Comprehensive Treatise, Plenum Press, New York, pp. 143–192.

Ebsworth N, Taylor-Papadimitriou J, Rozengurt E. 1984. Cyclic AMP does not mediate inhibition of DNA synthesis by interferon in mouse Swiss 3T3 cells. J Cell Physiol (in press).

Epstein L. 1981. Interferon gamma i. Is it really different? Interferon 3:13–44.

Fellous A, Ginzberg I, Littauer UZ. 1982. Modulation of tubulin mRNA levels by interferon in human lymphoblastoid cells. EMBO J 1:835–839.

Friedkin M, Rozengurt E. 1981. The role of cytoplasmic microtubules in the regulation of the activity of peptide growth factors. Adv Enzyme Regul 19:39–59.

Gewert DR, Shah S, Clemens MJ. 1981. Inhibition of cell division by interferons. Changes in the transport and intracellular metabolism of thymidine in human lymphoblastoid (Daudi) cells. Eur J Biochem 116:487–492.

Gewert DR, Cohen A, Williams BRG. 1984. The effect of interferon on cells deficient in nucleoside transport or lacking thymidine kinase activity. Biochem Biophys Res Commun 118:124–130.

Gresser I, Bandu M, Brouty-Boye D, Tovey M. 1974. Pronounced antiviral activity of human interferon on bovine and porcine cells. Nature 251:543–545.

Hannigan GE, Gewert DR, Fish EN, Read SE, Williams BRG. 1983. Differential binding of human interferon-α subtypes to receptors on lymphoblastoid cells. Biochem Biophys Res Commun 110:537–544.

Johnson AD, Poteete AR, Lauer G, Sauer RT, Ackers GK, Ptashne M. 1981. λ Repressor and cro-component of an efficient molecular switch. Nature 294:217–223.

Kerr IM, Brown RE. 1978. pppA2′p5′A2′p5′A: An inhibitor of protein synthesis synthesized with an enzyme fraction from interferon-treated cells. Proc Natl Acad Sci USA 75:256–260.

Keydar I, Chen L, Karby S, Weiss FR, Delarea J, Radu M, Chaitcik M, Brenner HJ. 1979. Establishment and characterization of a cell line of human breast carcinoma origin. Eur J Cancer 15:659–670.

Lasfargues EY, Ozzello L. 1958. Cultivation of human breast carcinomas. JNCI 21:1131–1147.

Lee EJ, Larkin PC, Sreevalsan T. 1980. Differential effect of interferon on ornithine decarboxylase activation in quiescent Swiss 3T3 cells. Biochem Biophys Res Commun 97:301–308.

Lin SL, Ts'o POP, Hollenberg MD. 1980. The effects of interferon on epidermal growth factor action. Biochem Biophys Res Commun 96:163–174.

McMahon M, Kerr IM. 1983. The biochemistry of the antiviral state. *In* Burke DC, Morris AG (eds), Interferons: From Molecular Biology to Clinical Application, 35th SGM Symposium. Cambridge University Press, Cambridge, pp. 89–108.

Mengheri E, Esteban M, Lewis JA. 1983. Thymidine kinase genes and the induction of anti-viral responses by interferon. FEBS Lett 157:301–307.

Mogensen KE, Bandu M-T, Vignaux F, Aguet M, Gresser I. 1981. Binding of ^{125}I-labelled human α-interferon to human lymphoid cells. Int J Cancer 28:575–582.

Mogensen KE, Bandu M-T. 1983. Kinetic evidence for an activation step following binding of human interferon-α2 to the membrane receptors of Daudi cells. Eur J Biochem 134:355–364.

Moore M. 1983. Interferon and the immune system. *In* Burke D, Morris AG (eds), Interferons: From Molecular Biology to Clinical Application, 35th SGM Symposium. Cambridge University Press, Cambridge, pp. 181–209.

Oleszak E, Inglot AD. 1980. Platelet derived growth factors (PDGF) inhibit antiviral and anticellular action of interferon in synchronised mouse or human cells. J Interferon Res 1:37–48.

Orchansky P, Goser T, Rubenstein M. 1983. Isolated subtypes of human interferon-α: synergistic action and affinity for the receptor. *In* DeMaeyer, Schellekens H (eds), The Biology of the Interferon System. Elsevier Scientific Publishers, Amsterdam, pp. 183–188.

Panet A, Czarniecki CW, Falk H, Friedman RM. 1981. Effect of 2′5 -oligoadenylic acid on a mouse cell line partially resistant to interferon. Virology 114:567–572.

Pestka S, Kelder B, Familletti PC, Moschera JA, Crowl R, Kempner ES. 1983. Molecular weight of the functional unit of human leucocyte, fibroblast, and immune interferons. J Biol Chem 258:9706–9709.

Rehberg E, Kelder B, Hoal EG, Pestka S. 1982. Specific molecular activities of recombinant and hybrid leukocyte interferons. J Biol Chem 257:11497–11502.

Rozengurt E. 1981. Stimulation of Na influx, Na-K pump activity and DNA synthesis in quiescent cultured cells. Adv Enzyme Regul 19:61–85.

Rozengurt E. 1982. Adenosine receptor activation in quiescent Swiss 3T3 cells. Exp Cell Res 139:71–78.

Rozengurt E. 1983. Growth factors, cell proliferation and cancer: an overview. Mol Biol Med 1:169–181.

Schneck J, Rager-Zisman B, Rosen OM, Blum BR. 1982. Genetic analysis of the role of cAMP in mediating effects of interferon. Proc Natl Acad Sci USA 79:1879–1883.

Schneiderman MH, Devvey WC, Highfield DP. 1971. Inhibition of DNA synthesis in synchronized Chinese hamster cells treated in G_1 with cycloheximide. Exp Cell Res 67:147–155.

Sekar V, Atmar VJ, Krim M, Kuehn GD. 1982. Interferon induction of polyamine dependent protein kinase activity in Ehrlich ascites tumour cells. Biochem Biophys Res Commun 106:305–311.

Sekar V, Atmar VJ, Joshi AR, Krim M, Kuehn GD. 1983. Inhibition of ornithine decarboxylase in human fibroblast cells by type I and type II interferons. Biochem Biophys Res Commun 114:950–954.

Shearer M, Taylor-Papadimitriou J. 1981. Anchorage-independent growth of normal calf lens epithelial cells of SV40 transformation on their growth properties. J Cell Physiol 108:385–391.

Shibata H, Taylor-Papadimitriou J. 1981. Effects of human lymphoblastoid interferon on cultured breast cancer cells. Int J Cancer 38:447–453.

Shire SJ. 1983. pH-dependent polymerization of a human leukocyte interferon produced by recombinant deoxyribonucleic acid technology. Biochemistry 22:2664–2671.

Silverman RH, Watling D, Balkwill FR, Trowsdale J, Kerr IM. 1982. The ppp(A2′p)nA and protein kinase systems in wild-type and interferon-resistant Daudi cells. Eur J Biochem 126:333–341.

Sokawa Y, Watanabe Y, Watanabe Y, Kawade Y. 1977. Interferon suppresses the transition of quiescent 3T3 cells to a growing state. Nature 268:236–238.

Sreevalsan T, Taylor-Papadimitriou J, Rozengurt E. 1979. Selective inhibition by interferon of serum-stimulated biochemical events in 3T3 cells. Biochem Biophys Res Commun 87:679–685.

Sreevalsan T, Rozengurt E, Taylor-Papadimitriou J, Burchell J. 1980. Differential effect of interferon on DNA synthesis, 2 deoxyglucose uptake and ornithine decarboxylase activity in 3T3 cells stimulated by polypeptide growth factors and tumor promoters. J Cell Physiol 104:1–9.

Streuli M, Hall A, Boll W, Stewart W II, Nagata S, Weissmann C. 1981. Target cell specificity of two species of human interferon-α produced in *E. coli* and hybrid molecules derived from them. Proc Natl Acad Sci USA 78:2848–2852.

Taylor-Papadimitriou J. 1980. Effects of interferon on cell growth and function. *In* Gresser I (ed), Interferon 2, 1980. Academic Press, London, pp. 13–46.
Taylor-Papadimitriou J. 1983. The effects of interferon on the growth and function of normal and malignant cells. *In* Burke D, Morris AG (eds), Interferons: From Molecular Biology to Clinical Application, 35th SGM Symposium. Cambridge University Press, Cambridge, pp. 109–147.
Taylor-Papadimitriou J, Rozengurt E. 1979. The role of thymidine uptake in the control of cell proliferation. Exp Cell Res 119:393–396.
Taylor-Papadimitriou J, Rozengurt E. 1982. Modulation of interferon's inhibitory effect on cell growth by the mitogenic stimulus. Tex Rep Biol Med 41:509–516.
Taylor-Papadimitriou J, Shearer M, Rozengurt E. 1981. Inhibitory effect of interferon on cellular DNA synthesis: Modulation by pure mitogenic factors. J Interferon Res 1:401–409.
Taylor-Papadimitriou J, Shearer M, Balkwill FR, Fantes KH. 1982. Effects of HuIFN-α2 and HuIFN-α (Namalwa) on breast cancer cells grown in culture and as xenografts in the nude mouse. J Interferon Res 2:479–491.
Taylor-Papadimitriou J, Kyriakides DA, Shearer M, Kortsaris A, Rozengurt E. 1983. Multiple biologic effects of interferon. *In* De Maeyer E, Schellekens H (eds), The Biology of the Interferon System, 1983. Elsevier Biomedical Press, Amsterdam, pp. 267–279.
Tovey MG. 1982. Interferon and cyclic nucleotides. Interferon 4:23–46.
Tovey MG, Dron M, Mogensen KE, Lebleu B, Mecht N, Begonlours-Guymarho J. 1983. Isolation of Daudi cells with reduced sensitivity to interferon. II. On the mechanisms of resistance. J Gen Virol 64:2649–2653.
Vandenbussche P, Divizia M, Verhaegen-Lewalle M, Fuse A, Kuwata T, De Clercq E, Content J. 1981. Enzymatic activities induced by interferon in human fibroblast cell lines differing in their sensitivity to the anticellular activity of interferon. Virology 111:11–22.
Vandenbussche P, Kuwata T, Verhaegen-Lewalle M, Content J. 1983. Effect of interferon on two human choriocarcinoma-derived cell lines. Virology 128:474–479.
Verhaegen M, Divizia M, Vandenbussche P, Kuwata T, Content J. 1980. Abnormal behavior of interferon-induced enzymatic activities in an interferon-resistant cell line. Proc Natl Acad Sci USA 77:4479–4483.
Wang Z-W, Rozengurt E. 1983. Interplay of cyclic AMP and microtubules in modulating the initiation of DNA synthesis in 3T3 cells. J Cell Biol 96:1743–1750.
Wood JN, Hovanessian AG. 1979. Interferon enhances 2-5A synthetase in embryonal carcinoma cells. Nature 282:74–76.
Yonehara S, Yonehara-Takahashi M, Ishii A. 1983a. Binding of human interferon-α to cells of different sensitivities: studies with internally radiolabeled interferon retaining full biological activity. J Virol 45:1168–1171.
Yonehara S, Yonehara-Takahashi M, Ishii A, Nagata S. 1983b. Different binding of human interferon α1 and α2 to common receptors on human and bovine cells. J Biol Chem 258:9046–9049.

Mediators in Cell Growth and Differentiation,
edited by Richard J. Ford and Abby L. Maizel.
Raven Press, New York © 1985.

Structure and Function of Human Interferon-Gamma

Jan Vilček, Hanna C. Kelker, Junming Le, and Y. K. Yip

Biological Response Modifiers Unit, Department of Microbiology and Kaplan Cancer Center, New York University Medical Center, New York, New York 10016

Interferons (IFNs) can be defined as a family of induced proteins sharing the capacity to exert pleiotropic effects on cell functions and to render cells resistant to virus infection. Although this definition encompasses all three major IFN species, IFN-alpha, -beta, and -gamma, many features distinguish IFN-gamma from the other two known IFN species. Unlike its two cousins, "immune IFN" (as IFN-gamma used to be called) is a typical lymphokine; it is produced by activated lymphocytes and it affects preferentially cellular functions related to immune reactions. IFN-gamma shares with the other IFNs the ability to induce cellular resistance to viruses, and there are other similarities in the biological activities of all three IFN species. However, IFN-gamma shows little, if any, structural relationship to IFN-alpha and -beta (Gray et al. 1982, DeGrado et al. 1982). IFN-gamma was also shown to bind to a different cell surface receptor (Branca and Baglioni 1981, Anderson et al. 1982). Moreover, the structural gene for IFN-gamma is located on chromosome 12, unlike the other known IFN genes, all of which are on chromosome 9 (Naylor et al. 1983). In addition, the gene for IFN-gamma contains three introns, making it distinct from the intron-less genes for IFN-alpha and -beta (Gray and Goeddel 1982).

Several years after the original description of IFN-gamma by Wheelock (1965), other investigators "rediscovered" IFN-gamma under the disguise of a protein that had come to be operationally known as macrophage activation factor, or MAF (Evans et al. 1972, Lohmann-Matthes et al. 1973). For years, MAF had been suspected to be a unique protein with an important role in the defense against malignant cells and intracellular pathogens. Only very recently has it become apparent that IFN-gamma and MAF are in fact one and the same molecule (Roberts and Vasil 1982, Nathan et al. 1983, Schultz and Kleinschmidt 1983, Pace et al. 1983, Le et al. 1983) and that most descriptions of MAF activities in the older literature can now be attributed to the presence of IFN-gamma.

It is also becoming apparent that IFN-gamma is responsible for a lymphokine activity earlier referred to by some investigators as "macrophage Ia^+ recruiting factor" (Scher et al. 1980, Steinman et al. 1980, Steeg et al. 1982). With the realization that IFN-gamma affects many functions of monocytes/macrophages and of other cells that are central to the regulation of the immune system, this protein suddenly has become an object of keen interest to a wide audience of immunologists.

PHYSICOCHEMICAL PROPERTIES OF THE HUMAN IFN-GAMMA MOLECULE

General Characteristics

Until recently, IFN-gamma was the least well characterized of the three IFN species. Before the successful purification of human IFN-gamma (Yip et al. 1982a), very little information was available on its physicochemical properties. It has been known that exposure of IFN-gamma to pH 2 leads to a substantial reduction or complete loss of biological activity (Wheelock 1965), and this property has been widely used to differentiate between IFN-gamma and the other, pH 2-stable IFN species. In native form IFN-gamma has a molecular weight of 40,000 to 60,000, i.e., one substantially higher than the other IFN species (reviewed by Yip et al. 1982b). Isoelectric focusing studies showed that, unlike the other IFNs, IFN-gamma is a highly basic protein (Yip et al. 1981b, Georgiades et al. 1980). Lectin-binding studies suggested that IFN-gamma is a glycoprotein (Yip et al. 1981b, Nathan et al. 1981).

Progress toward a better definition of the physicochemical properties of the molecule occurred when, in the process of the purification of human IFN-gamma, sodium dodecyl sulfate-polyacrylamide gel electrophoretic (SDS-PAGE) analysis unexpectedly revealed two protein bands with apparent molecular weights of 25,000 and 20,000 (Yip et al. 1982a, b). These findings suggested that the higher molecular weight form detected by gel permeation chromatography under nondenaturing conditions represented oligomers of the MW 25,000 and 20,000 molecules.

A major breakthrough in the understanding of the nature of IFN-gamma resulted from the successful cloning and expression of its cDNA by Gray et al. (1982). Sequence analysis of this cDNA revealed an open reading frame encoding 166 amino acids. Of these, 20 amino acids were postulated to form the signal peptide characteristic of secreted proteins, leaving 146 amino acids with a total molecular weight of 17,000 for the mature protein. (However, recent data on the amino acid sequence analysis of highly purified "natural" human IFN-gamma indicate N-terminus lacks the first three predicted amino acids (Rinderknecht et al. 1984).) The sequence shows a large excess of basic residues, in agreement with the highly basic nature observed by isoelectric focusing of the natural product. Sequence analysis also revealed two potential N-glycosylation sites, supporting earlier indirect observations on the glycoprotein nature of IFN-gamma made with the natural protein (Yip et al. 1981b).

Three Monomeric Forms of IFN-Gamma

Several laboratories recently succeeded in producing monoclonal antibodies (MoAb) to human IFN-gamma (Rubin et al. 1983, Novick et al. 1983, Le et al. 1984a).

Some of these MoAb have already proved helpful in the analysis of structural characteristics of IFN-gamma.

MoAb GIF-1, isolated by Rubin et al. (1983), is an antibody with neutralizing activity against natural IFN-gamma. However, the same antibody failed to either neutralize or bind *Escherichia coli*-derived recombinant human IFN-gamma (Nathan et al. 1983, Le et al. 1984b). Recombinant IFN-gamma from *E. coli* is known to differ from the natural molecule by the absence of carbohydrate. However, preliminary evidence suggests that the lack of reactivity of the recombinant molecule is not due to the absence of carbohydrate alone. More likely, the tertiary structure of the recombinant molecule is altered, perhaps because the folding of the molecule is different without the carbohydrate moiety or because the nascent polypeptide made in *E. coli* is not passed through the rough endoplasmic reticulum or Golgi apparatus.

MoAb GIF-1 and another MoAb specific for IFN-gamma, termed B3 (Le et al. 1984a,b), were also employed for immunoprecipitation of preparations containing IFN-gamma produced in cultures of human mononuclear cells endogenously labeled with [^{35}S]methionine (Kelker et al. 1984). Immunoprecipitated proteins were analyzed by SDS-PAGE. As expected, a 25,000 MW and a 20,000 MW band were revealed in autoradiograms of the immunoprecipitates separated by SDS-PAGE. However, somewhat unexpectedly, another band with the apparent MW of 15,500 was also found in the immunoprecipitates. The 15,500 MW band migrated faster than the 17,000 MW band of *E. coli*-derived human IFN-gamma. That the 15,500 MW band was indeed a form of IFN-gamma was confirmed when we demonstrated its biological activity in the IFN assay.

Thus, three monomeric forms of IFN-gamma were demonstrable in the immunoprecipitates of natural IFN-gamma (Table 1), and we proposed to term these IFN-gamma I, II, and III, corresponding to the 25,000, 20,000, and 15,500 MW bands of IFN-gamma (Kelker et al. 1984). IFN-gamma I is about three to four times more abundant than IFN-gamma II, based on determinations by biological activity (Yip et al. 1982b) or radioactivity (Kelker et al. 1984). IFN-gamma III is a minor species that accounts for only a small fraction of total biological activity ($<1\%$) recovered after SDS-PAGE. In contrast, about 5% of total radioactivity immunoprecipitated by MoAb was associated with the IFN-gamma III band (Table 1). The discrepancy between the amount of biological activity and radioactivity associated with this band could be due to less stability or lower intrinsic specific activity of the molecule.

How do these three forms arise? Many, if not all, of the differences can be explained by differential glycosylation. As illustrated in Table 1, treatment of natural IFN-gamma with a preparation of mixed glycosidases reduced the size of IFN-gamma I from 25,000 to 18,500 MW and that of IFN-gamma II from 20,000 to 15,500 MW. The size of IFN-gamma III (15,500 MW) was not altered by glycosidase treatment (Kelker et al. 1983, 1984). These results indicate that both IFN-gamma I and II are glycosylated, whereas IFN-gamma III apparently is not. However, it is not yet clear whether the molecular weight difference between the

TABLE 1. *Relative abundance of different monomeric subspecies in preparations of natural human IFN-gamma* and the effect of glycosidase treatment on their molecular sizes*

IFN-gamma subspecies	Apparent molecular size (× 10⁻³)		Percent total‡	
	Untreated	Glycosidase-treated†	Antiviral activity	Radioactivity
I	25	18.5	~75–80	~75
II	20	15.5	~20–25	~20
III	15.5	15.5	~0.2–0.5	~5

*Produced in cultures of mononuclear cells derived from plateletpheresis residues stimulated with a phorbol ester (TPA) and phytohemagglutinin as described (Vilček et al. 1980, Yip et al. 1981a).

†Treatment with a mixed glycosidase preparation (β-galactosidase, N-acetyl-β-galactosaminidase, and neuraminidase) did not decrease biological activity. The effect of this treatment on the molecular size of individual IFN-gamma subspecies was determined by SDS-PAGE (Kelker et al. 1983, 1984).

‡Approximate values based on the results of several experiments (Kelker et al. 1984).

products of glycosidase treatment of IFN-gamma I and II (18,500 vs. 15,500 MW) is due to the presence of glycosidase-resistant oligosaccharide in IFN-gamma I or to a difference in the protein moieties. Recent data obtained by Rinderknecht et al. (1984) make the latter possibility unlikely.

Unanswered Questions about Structural Properties of IFN-Gamma

As can be gleaned from the preceding discussion, the basis of the molecular heterogeneity of IFN-gamma has not been fully explained. A closely related question concerns the biological significance of the carbohydrate moiety in IFN-gamma. Since all indications are that *E. coli*–derived IFN-gamma exerts the same spectrum of biological activities as the natural molecule, it is clear that the carbohydrate is not a prerequisite for activity. Removal of carbohydrate by treatment with mixed glycosidases neither decreased the specific activity nor altered the strict species specificity of the IFN-gamma molecule (Kelker et al. 1983).

Whether the carbohydrate moiety plays an important role other than being responsible for much, if not all, of the molecular heterogeneity remains to be determined. No comparative data on the stability of the unglycosylated and fully glycosylated molecule have been published. The carbohydrate may yet prove to be important in determining the fate of the molecule in the intact organism, particularly after exogenous administration.

Another unanswered question is whether IFN-gamma undergoes proteolytic processing in addition to the cleavage of the signal peptide sequence. Detection of a naturally occurring 15,500 MW form (IFN-gamma III), with an apparent molecular weight lower than that of recombinant IFN-gamma (Kelker et al. 1984), suggests that this molecule may contain less than the full complement of 146 amino acids postulated on the basis of the DNA sequence (Gray et al. 1982). Naturally occurring IFN-alpha undergoes proteolytic processing, since molecules possessing full biological activity lack up to 10 COOH-terminal amino acids predicted from the cDNA sequence (Levy et al. 1981). Moreover, a form of recombinant human IFN-gamma lacking up to 11 COOH-terminal amino acids was found to be biologically active (Gray and Goeddel 1983).*

Finally, more information is needed about the quaternary structure of IFN-gamma. As mentioned earlier, in its native configuration IFN-gamma is likely to be an oligomer, most likely a dimer (Yip et al. 1982b, Devos et al. 1982). The nature of the interactions leading to oligomer formation has not been explained. Disulfide bonds are not likely to be involved in the formation of the oligomeric structure since treatment with the anionic detergent SDS alone, without the addition of a reducing agent, dissociates the molecule into the monomeric forms (Yip et al. 1982a,b). On the other hand, attempts to dissociate the oligomeric form by treatment with reducing agents or high salt in the absence of SDS have been unsuccessful (Yip, Y. K., unpublished data).

Available information suggests that human IFN-gamma may not be biologically active in monomeric form, but formal evidence confirming this notion has not been reported. Greater availability of purified IFN-gamma should facilitate future structural studies of this molecule.

SPECTRUM OF BIOLOGICAL ACTIVITIES OF IFN-GAMMA

Comparison with IFN-Alpha and -Beta

In view of the recent evidence that IFN-gamma is structurally very different from the other IFN species, one can justifiably ask whether it is at all appropriate to continue classifying IFN-gamma as an IFN. Should we perhaps reconsider the classification of this molecule?

There are many valid arguments for the continued use of the term IFN-gamma. Despite the structural differences mentioned in this article, all IFNs, including IFN-gamma, share a characteristic spectrum of biological activities (Table 2). Since a complete inventory of all common biological activities would be too long, we shall mention only a few examples. The most characteristic property of all IFNs is the induction of cellular resistance to virus infection. Although the exact mechanism of the antiviral action is not yet completely understood, it is known to depend on the induction of cellular protein synthesis. Among the IFN-induced proteins important for antiviral activity, the best-characterized one is the enzyme 2′5′oligoadenylate synthetase. IFN-gamma is active as an inducer of 2′5′oligoadenylate synthetase, although in most cell systems it may be less efficient in the induction

*Some of the questions asked in March 1984 (when this paper was written) have subsequently been answered by Rinderknecht et al. (1984). These authors have determined the entire amino acid sequences of the 25,000 and 20,000 MW bands of natural human IFN-gamma and shown that they are identical. The only demonstrable difference between IFN-gamma I and II was the amount of carbohydrate present. Rinderknecht et al. (1984) also showed that 9 to 16 of the amino acid residues predicted from the cDNA sequence were missing from the COOH-terminal end of both IFN-gamma I and II, suggesting that these molecules had been modified by proteolytic processing.

of this enzyme than IFN-alpha or IFN-beta (Baglioni and Maroney 1980, Wallach et al. 1982). The capacity to inhibit the growth of many cells in culture is also shared by all IFNs. Although IFN-gamma is more effective as a cell growth inhibitor in some cell lines (Rubin and Gupta 1980), in other types of cells it was found to be less active than IFN-alpha or -beta (De Ley et al. 1980, Czarniecki et al. 1984). Comparisons of IFN-gamma and of IFN-alpha or -beta in their ability to induce new proteins revealed that although all interferons induced a common set of new proteins, IFN-gamma also stimulated the synthesis of some proteins that are either not induced at all or stimulated to a much lesser extent by the other IFNs (Rubin and Gupta 1980, Hovanessian et al. 1980, Colonno and Pang 1982, Weil et al. 1982, Samuel and Knutson 1983). These findings might explain why IFN-gamma tends to have a broader range of biological activities than the other IFNs (Table 2).

It is apparent that despite the completely different primary structures, existence of different cellular receptors, and quantitative and qualitative differences in their biological activities, IFN-gamma and the other IFN species share a wide spectrum of activities. These similarities justify the continued classification of IFN-gamma as a member of the IFN family of proteins. (It remains unclear why, despite its structural uniqueness and different receptor specificity, IFN-gamma shares so many common activities with the other IFN species.)

However, it is also obvious that IFN-gamma is a *different* IFN. Investigators who early on strived to define the relationship of this protein to the other IFNs had intuitively used such terms as "interferon-like" (Wheelock 1965) or "type II interferon" (Youngner and Salvin 1973), which realistically convey the nature of this molecule. The IFN nomenclature adopted later (Committee on Interferon Nomenclature 1980) does not reflect the recently established fundamental differences between IFN-gamma and the other IFN species. Orchansky et al. (1984) have suggested the reintroduction of the terms Type I (for IFN-alpha and -beta) and Type II (for IFN-gamma). These terms would obviously not supplant the currently employed designations, but could be conveniently used for descriptive purposes.

TABLE 2. *Some important biological activities of IFN-gamma*

Activities shared by IFN-gamma and Type I IFNs:
Induction of antiviral state
Induction of 2′5′oligoadenylate synthetase
Induction of protein kinase
Inhibition of cell growth
Activation of natural killer (NK) cells
Activation of cytotoxic T-lymphocytes
Induction of class I histocompatibility antigens
Activities exhibited exclusively or preferentially by IFN-gamma:
Induction of Class II histocompatibility antigens (Ia or DR)
Induction of Fc immunoglobulin receptors on macrophages
Activation of macrophages for tumor cell cytotoxicity and antimicrobial activity
Stimulation of IL-1 and IL-2 synthesis
Inhibition of the growth of some nonviral intracellular pathogens in cells of nonmyelomonocytic origin

Immunoregulatory Activities of IFN-Gamma

The fact that IFNs can influence immune reactions has been recognized for a long time. Since IFN-gamma is typically produced by lymphocytes in the process of specific or nonspecific mitogenic activation, it has obviously been regarded as the IFN species most likely to play a physiological role in the regulation of immune responses.

Since highly purified preparations of IFN-gamma have become available for laboratory research—mainly as a result of the application of recombinant DNA technology—the amount of information about the various actions of IFN-gamma on immune reactions has been growing exponentially. A comprehensive review of these activities is beyond the scope of this article. A summary of the major types of demonstrated activities follows.

Activation of Cytotoxic T-Lymphocytes

It appears that IFNs (probably of all major types) can enhance cytotoxic T-lymphocyte (CTL) responses. Farrar et al. (1981) postulated that IFN-gamma has a mandatory role in IL-2-driven CTL responses, because antibody against IFN-gamma inhibited in vitro induction of CTL by interleukin-2 (IL-2). The basis of this synergism could be that IFN-gamma renders lymphocytes more responsive to IL-2, possibly as a result of its induction of IL-2 receptors (Johnson and Farrar 1983). It has been known for a long time that stimulation of the growth of resting T-lymphocytes requires not only IL-2 but also some other signals provided by the action of the mitogenic stimulus (antigen, lectin, etc.). IFN-gamma, a product of activated T cells, could represent the missing link. However, additional interactions may be involved because not only does IFN-gamma induce responsiveness to IL-2, but also IL-2 stimulates the synthesis of IFN-gamma by T cells (Farrar et al. 1981, Kasahara et al. 1983, Yamamoto et al. 1982, Pearlstein et al. 1983). Conversely, IFN-gamma was also shown to stimulate IL-2 production, possibly mediated by increased IL-1 synthesis by monocytes (Pearlstein, K.T., unpublished data).

Activation of NK Cells

All major IFN species were shown to stimulate natural killer (NK) cell cytotoxicity against tumor cells (Herberman 1984). Numerous reports described activation of NK cell populations in vitro by IFN-gamma preparations (reviewed by Epstein 1984). Some investigators found that IFN-gamma is not strikingly more potent as an NK cell activator than type I IFNs (De Ley et al. 1980), although one study reported IFN-gamma to be 50 times more active than type I IFN (Weigent et al. 1983a). On the other hand, IFNs are also known to protect some target cells from the cytotoxic action of NK cells (Trinchieri et al. 1981), and according to one report (Wallach et al. 1982) IFN-gamma was highly effective in mediating such protective activity.

As in the induction of CTL, an interaction of IFN and IL-2 may be involved in NK cell activation. IL-2 was shown to activate NK cells without the addition of IFN (Henney et al. 1981). However, in some instances the addition of IL-2 to cultured NK cells stimulated IFN-gamma synthesis (Handa et al. 1983). Thus, NK cell activation by IL-2 may be mediated by the induction of IFN-gamma (Weigent et al. 1983b). On the other hand, experiments in which antisera that neutralize IFN activity were added to mixed cultures of NK and target cells indicated that NK cell activation could occur in the absence of IFN, suggesting a separate role for IL-2 or some other regulatory factor (Blazar et al. 1984).

Modulation of Monocyte/Macrophage Functions

IFNs have been known to affect macrophage function since it was demonstrated that various IFN preparations stimulated enhanced phagocytic activity for colloidal carbon particles (Huang et al. 1971). Many other actions of IFN on macrophage functions associated with the process of "macrophage activation" have been reported (Stewart 1979, Vogel and Friedman 1984).

IFNs are known to increase phagocytosis of opsonized targets, probably as a result of an increased expression of IgG Fc receptors on macrophages (Vogel and Friedman 1984). IFN-gamma appears to be highly active as an inducer of Fc receptor expression (Guyre et al. 1983), and to promote the generation of macrophages from myelomonocytic precursor cells (Perussia et al. 1984).

Activation of macrophages is known to involve changes in the expression of class II histocompatibility antigens, and recently much attention has been focused on the role played by IFN-gamma in the control of Ia antigen expression on murine monocytes/macrophages (or of its human equivalent, HLA-DR expression). Ia antigens play a central role in the regulation of multiple accessory cell functions of macrophages that require the macrophage's direct interaction with T cells, e.g., macrophages' presentation of protein antigens to T lymphocytes or action as stimulator cells for allogeneic T cells in the mixed lymphocyte reaction (Unanue et al. 1984). It has been known for a long time that one or more lymphokines produced by T cells can induce Ia expression in mononuclear phagocytes and thereby transform them into immunologically functional accessory cells. The lymphokine responsible for this important function, formerly termed "macrophage Ia^+ recruiting factor" by some investigators (Scher et al. 1980), has recently been identified as IFN-gamma (Steeg et al. 1982, Kelley et al. 1984).

The ability of IFNs to induce cell surface changes is not restricted to monocytes and macrophages. Sonnenfeld et al. (1981) demonstrated the induction of H-2 antigens on murine lymphocytes by IFN-gamma, and Wallach et al. (1982) found that human lymphoid cells showed enhanced expression of HLA-A, -B, and -C and β_2-microglobulin after their exposure to recombinant or natural human IFN-gamma. These changes in major histocompatibility antigen expression were seen in both murine and human lymphocytes after treatment with low concentrations of IFN-gamma, ones not sufficient to produce antiviral activity in the common assays.

It is apparent that IFN-gamma is uniquely active as an inducer of class II histocompatibility antigens (Ia in the mouse and DR in man). On the other hand, according to a recent report (Kelley et al. 1984) type I and type II IFNs are similar in their activity as inducers of class I histocompatibility antigens (i.e., HLA-A, -B, and -C and β_2-microglobulin).

Effects on Immunoglobulin Synthesis

There have been many reports of the effects of various IFNs on antibody production (reviewed in Härfast et al. 1981). Both stimulatory and inhibitory effects were observed, but since experimental conditions varied, interpretation of the results is difficult. Administration of highly purified IFN-gamma to mice was recently shown to lead to enhanced antibody production, possibly as a result of increased antigen processing due to a stimulation of Ia expression on macrophages (Nakamura et al. 1984). IFN-gamma might also increase immunoglobulin synthesis by promoting B cell differentiation.

What Is the Primary Function of IFN-Gamma?

As stated in the Introduction, IFN-gamma is a typical lymphokine because it is produced by activated lymphocytes and it affects a variety of cellular functions related to immune reactions (Epstein 1981, 1984). Nevertheless, IFN-gamma shows some functional characteristics that make it distinct from other known lymphokines. The major function of most lymphokines is quite apparent from their known biological activities in vitro. However, IFN-gamma appears to have a broader spectrum of biological activities than other well-characterized lymphokines, and its major function in the intact organism has not yet been identified.

Most lymphokines can be classified as either growth factors or differentiation factors. Moreover, it appears that lymphokines usually exert their major activity on only one type of cell. A typical example of a lymphokine–growth factor is IL-2, or T-cell growth factor (TCGF): the major function associated with IL-2 is the stimulation of mitosis in T-cell subsets, usually in conjunction with an antigen or mitogen. Nonlymphoid cells may lack functional receptors for IL-2.

A typical example of a lymphokine–differentiation factor is monocyte-derived IL-1 or lymphocyte-activating factor (LAF): the major function of IL-1 is to stimulate IL-2 production by T cells. Other examples of lymphokines with well-defined functional activities and target cell specificities are the colony-stimulating factors.

In contrast to these lymphokines, IFN-gamma defies a simple functional classification. Depending on the target cell, IFN-gamma functions as a differentiation factor (e.g., in macrophages), as a modulator of gene expression in fully differentiated cells (Rubin and Gupta 1980, Hovanessian et al. 1980, Colonno and Pang 1982, Weil et al. 1982), as a potent cell growth inhibitor (Blalock et al. 1980, Rubin and Gupta 1980, De Ley et al. 1980) and, perhaps, as a growth factor (Guyre, T.M., personal communication). It appears that the repertoire of biological activities and the spectrum of target cell specificities of IFN-gamma are even broader than those of the other IFNs. (At the same time, the action of IFN-gamma appears to be quite strictly species-specific (Wiranowska-Stewart et al. 1980, De Ley et al. 1980, Kelker et al. 1983) and it is probably the most manifestly species-restricted type of IFN.)

Because of their broad range of biological activities (Table 2), IFNs might be likened to some polypeptide hormones, such as insulin. However, the primary function of insulin in the intact organism is the regulation of glucose transport and carbohydrate metabolism. Defects in insulin production or action are known to lead to well-defined pathological conditions. In contrast, there are no specific diseases known to result from a deficiency in IFN-gamma production or action. (The vital importance of type I IFNs also remains to be established. Individuals with spontaneously occurring (Mogensen et al. 1981, Panem et al. 1982) or iatrogenically induced antibodies to IFN-alpha (Trown et al. 1983) or IFN-beta (Vallbracht et al. 1981) have been identified. Despite the fact that such antibodies would effectively inactivate all endogenously produced homologous IFN species, these patients do not seem to suffer from virus infections more frequently or more severely than other individuals or show other abnormalities due to the lack of IFN.) Unlike insulin, IFN-gamma is not readily demonstrable in the intact organism. Although local production of IFN-gamma is likely to occur in the body during the activation of T cells by antigens, the small amounts that might be released into the circulation would probably elude detection.

A survey of the recent literature on its biological activities makes IFN-gamma stand out mainly in two respects: (1) As indicated earlier in this article, IFN-gamma is uniquely active as a modulator of monocyte/macrophage functions, and (2) it is uniquely active as an inhibitor of the growth of intracellular pathogens, e.g., *Toxoplasma gondii* (Nathan et al. 1983, Pfefferkorn and Guyre 1983), *Plasmodium berghei* (Ojo-Amaize, E., Nussenzweig, R.S., and Vilček, J., unpublished data) and *Chlamydia psittaci* (Rothermel et al. 1983).

It has become apparent that IFN-gamma is indistinguishable from a lymphokine activity operationally termed macrophage activation factor, or MAF (Mooney and Waksman 1970, Patterson and Youmans 1970, Evans et al. 1972, Lohmann-Matthes et al. 1973), and it appears that macrophage activation may be the major biological activity associated with IFN-gamma (Roberts and Vasil 1982, Nathan et al. 1983, Schultz and Kleinschmidt 1983, Pace et al. 1983, Le et al. 1983). This conclusion is based on the evidence that macrophage activation can generally be demonstrated at lower IFN-gamma concentrations than required for the expression of its antiviral activity (Roberts and Vasil 1982, Nathan et al. 1983, Le et al. 1983).

The functional expression of macrophage activation includes enhanced cytocidal and cytostatic activity against a variety of tumor cells. IFN-gamma, therefore, could play an important role in the defense against neoplasms. Another widely demonstrated function of activated macrophages is an enhanced capacity to kill intracellular pathogens, e.g., protozoa *(T. gondii)*, bacteria *(Mycobacterium, Salmonella, Brucella,* and *Listeria spp.)*, *Chlamydia*, and fungi. Is IFN-gamma important in the defense against intracellular pathogens? It is interesting that one report describes the presence of IFN-gamma in the serum of 70% of patients with *P. falciparum* malaria (Rhodes-Feuillette et al. 1981). Other observations point to a possible role of IFN-gamma in the defense against protozoan infections, including the ability of some protozoa, e.g., *P. berghei* sporozoites, to stimulate IFN-gamma production in T cells as a result of their mitogenic activity (Ojo-Amaize et al. 1984) and the ability of IFN-gamma to inhibit the intracellular multiplication of protozoa in cultured fibroblasts (Pfefferkorn and Guyre 1983, Ojo-Amaize, E., unpublished data).

In addition, as pointed out earlier, IFN-gamma is also a very potent inducer of Ia antigen (Steeg et al. 1982) or HLA-DR antigen (Basham and Merigan 1983, Kelley et al. 1984) and IgG Fc receptor (Guyre et al. 1983) expression on monocytes/macrophages. The recognition that IFN-gamma regulates Ia antigen expression is of major importance because the induction of Ia on antigen-presenting cells is a critical step in the control of immune responsiveness. A low level of Ia expression during neonatal development is thought to be an important factor diminishing immune responsiveness, just as increased Ia expression on macrophages may be a mechanism for the amplification of antigen presentation and subsequent immune responsiveness (Unanue et al. 1984). Similarly, the control of IgG Fc receptor expression is an important function affecting phagocytosis, clearance of immune complexes, and antibody-dependent cellular cytotoxicity (Vogel and Friedman 1984). Finally, recent data show that IFN-gamma might promote hematopoietic differentiation along the monocytic pathway (Perussia et al. 1984). Future studies will have to clarify how important IFN-gamma is as a natural regulator of Ia and Fc receptor expression and monocyte differentiation in the intact organism under "physiological" conditions.

ACKNOWLEDGMENTS

We thank Michele C. Maniscalco for editorial assistance and preparation of the manuscript. Work at the authors' laboratory was supported in part by U.S. Public Health Service Grants AI-12948 and AI-07057.

REFERENCES

Anderson P, Yip YK, Vilček J. 1982. Specific binding of ^{125}I-human interferon-gamma to high affinity receptors on human fibroblasts. J Biol Chem 257:11301–11304.

Baglioni C, Maroney PA. 1980. Mechanism of action of human interferons. Induction of 2′-5′-oligo(A) polymerase. J Biol Chem 255:8390–8393.

Basham TY, Merigan TC. 1983. Recombinant interferon-gamma increases HLA-DR synthesis and expression. J Immunol 130:1492–1494.
Blalock JE, Georgiades JA, Langford MP, Johnson HM. 1980. Purified human immune interferon has more potent anticellular activity than fibroblasts or leukocyte interferon. Cell Immunol 49:390–394.
Blazar BA, Strome M, Schooley R. 1984. Interferon and natural killing of human lymphoma cell lines after the induction of the Epstein Barr viral cycle by superinfection. J Immunol 132:816–820.
Branca AA, Baglioni C. 1981. Evidence that types I and II interferons have different receptors. Nature 294:768–770.
Colonno RJ, Pang RHL. 1982. Induction of unique mRNAs by human interferons. J Biol Chem 257:9234–9237.
Committee on Interferon Nomenclature. 1980. Interferon nomenclature. Nature 286:110.
Czarniecki CW, Fennie CW, Powers DB, Estell DA. 1984. Synergistic antiviral and antiproliferative activities of *Escherichia coli*- derived human alpha, beta, and gamma interferons. J Virol 49:490–496.
DeGrado WF, Wasserman ZR, Chowdry V. 1982. Sequence and structural homologies among type I and type II interferons. Nature 300:379–381.
De Ley M, Van Damme J, Claeys H, Weening H, Heine JW, Billiau A, Vermylen C, De Somer P. 1980. Interferon induced in human leukocytes by mitogens: production, partial purification and characterization. Eur J Immunol 10:877–883.
Devos R, Cheroutre H, Taya T, Degrave W, Van Heuverswyn H, Fiers W. 1982. Molecular cloning of human immune interferon cDNA and its expression in eukaryotic cells. Nucleic Acids Res 10:2487–2501.
Epstein LB. 1981. Interferon-gamma: Is it really different from the other interferons? *In* Gresser I (ed), Interferon 1981, Vol. 3. Academic Press, London, pp. 13–44.
Epstein LB. 1984. The special significance of interferon-gamma. *In* Vilček J, De Maeyer E (eds), Interferons, Vol. 2, Interferons and the Immune System. Elsevier, Amsterdam, pp. 185–219.
Evans R, Grant CK, Cox H, Steele K, Alexander P. 1972. Thymus-derived lymphocytes produce an immunologically specific macrophage-arming factor. J Exp Med 136:1318–1322.
Farrar WL, Johnson HM, Farrar JJ. 1981. Regulation of the production of immune interferon and cytotoxic T lymphocytes by interleukin 2. J Immunol 126:1120.
Georgiades JA, Langford MP, Goldstein JE, Blalock JE, Johnson HM. 1980. Human immune interferon: purification and activity against a transformed human cell. *In* Khan A, Hill NO, Dorn GL (eds), Interferons: Properties and Clinical Uses. Leland Fikes Foundation Press, Dallas, pp. 97–108.
Gray PW, Goeddel DV. 1982. Structure of the human immune interferon gene. Nature 298:859–863.
Gray PW, Goeddel DV. 1983. Cloning and expression of murine immune interferon cDNA. Proc Natl Acad Sci USA 80:5842–5846.
Gray PW, Leung DW, Pennica D, Yelverton E, Najarian R, Simonsen C, Derynck R, Sherwood PJ, Wallace DM, Berger SL, Levinson AD, Goeddel D. 1982. Expression of human immune interferon cDNA in *E. coli* and monkey cells. Nature 295:503–508.
Guyre PM, Morganelli PM, Miller R. 1983. Recombinant immune interferon increases immunoglobulin G Fc receptors on cultured human mononuclear phagocytes. J Clin Invest 72:393–397.
Handa K, Suzuki R, Matsui H, Shimizu Y, Kumagai K. 1983. Natural killer (NK) cells as a responder to interleukin 2 (IL-2). II. IL-2 induced interferon-gamma production. J Immunol 130:988–992.
Härfast B, Huddlestone JR, Casali P, Merigan TC, Oldstone MBA. 1981. Interferon acts directly on human B lymphocytes to modulate immunoglobulin synthesis. J Immunol 127:2146–2150.
Henney CS, Kuribayashi K, Kern DE, Gillis S. 1981. Interleukin-2 augments natural killer cell activity. Nature 291:335–338.
Herberman RB. 1984. Interferons and cytotoxic effector cells. *In* Vilček J, De Maeyer E (eds), Interferons, Vol. 2, Interferons and the Immune System. Elsevier, Amsterdam, pp. 61–80.
Hovanessian AG, Meurs E, Aujean O, Vaquero C, Stefanos S, Falcoff E. 1980. Antiviral response and induction of specific proteins in cells treated with immune T (type II) interferon analogous to that from viral interferon (type I)-treated cells. Virology 104:195–204.
Huang K-Y, Donahoe RM, Gordon FB, Dressler HR. 1971. Enhancement of phagocytosis by interferon-containing preparations. Infect Immun 4:581–588.
Johnson HM, Farrar WL. 1983. The role of a gamma interferon-like lymphokine in the activation of T cells for expression of interleukin 2 receptors. Cell Immunol 75:154–159.
Kasahara T, Hooks JJ, Dougherty SF, Oppenheim JJ. 1983. Interleukin 2-mediated immune interferon (IFN-gamma) production by human T cells and T cell subsets. J Immunol 130:1784–1789.

Kelker HC, Yip YK, Anderson P, Vilček J. 1983. Effect of glycosidase treatment on the physicochemical properties and biological activity of human interferon-gamma. J Biol Chem 258:8010–8013.
Kelker HC, Le J, Rubin BY, Yip YK, Nagler C, Vilček J. 1984. Three molecular weight forms of natural human interferon-gamma revealed by immunoprecipitation with monoclonal antibody. J Biol Chem 259:4301–4304.
Kelley VE, Fiers W, Strom TB. 1984. Cloned human interferon-gamma, but not interferon-beta or -alpha, induces expression of HLA-DR determinants by fetal monocytes and myeloid leukemic cell lines. J Immunol 132:240–245.
Le J, Prensky W, Yip YK, Chang Z, Hoffman T, Stevenson HC, Balazs I, Sadlik JR, Vilček J. 1983. Activation of human monocyte cytotoxicity by natural and recombinant immune interferon. J Immunol 131:2821–2826.
Le J, Barrowclough BS, Vilček, J. 1984a. Monoclonal antibodies to human immune interferon and their application for affinity chromatography. J Immunol Methods 69:61–70.
Le J, Rubin BY, Kelker HC, Feit C, Nagler C, Vilček J. 1984b. Natural and recombinant *Escherichia coli*-derived interferon-gamma differ in their reactivity with monoclonal antibody. J Immunol 132:1300–1304.
Levy WP, Rubinstein M, Shively J, Del Valle U, Lai C-Y, Moschera J, Brink L, Gerber L, Stein S, Pestka S. 1981. Amino acid sequence of a human leukocyte interferon. Proc Natl Acad Sci USA 78:6186–6190.
Lohmann-Matthes M-L, Ziegler FG, Fischer H. 1973. Macrophage cytotoxicity factor: a product of in vitro sensitized thymus-dependent cells. Eur J Immunol 3:56–58.
Mogensen KE, Daubas PH, Gresser I, Serendi D, Varet B. 1981. Patient with circulating antibodies to alpha-interferon. Lancet II:1227–1228.
Mooney JJ, Waksman BH. 1970. Activation of normal rabbit macrophage monolayers by supernatants of antigen-stimulated lymphocytes. J Immunol 105:1138–1145.
Nakamura M, Manser T, Pearson GDN, Daley MJ, Gefter ML. 1984. Effect of interferon-gamma on the immune responses in vitro and on gene expression in vitro. Nature 307:381–382.
Nathan I, Groopman JE, Quan SG, Bersch N, Golde DW. 1981. Immune (gamma) interferon produced by a human T-lymphoblastoid cell line. Nature 292:842–844.
Nathan CF, Murray HW, Wiebe ME, Rubin BY. 1983. Identification of interferon-gamma as the lymphokine which activates human macrophage oxidative metabolism and antimicrobial activity. J Exp Med 158:670–689.
Naylor SL, Sakaguchi AY, Shows TB, Law ML, Goeddel DV, Gray PW. 1983. Human immune interferon gene is located on chromosome 12. J Exp Med 157:1020–1027.
Novick D, Eshar Z, Fischer DG, Rubinstein M. 1983. Monoclonal antibodies to human interferon-gamma: production, affinity purification and radioimmunoassay. EMBO J. 2:1527–1530.
Ojo-Amaize E, Vilček J, Cochrane AH, Nussenzweig RS. 1984. *Plasmodium berghei* sporozoites are mitogenic for murine T cells, induce interferon and activate natural killer cells. J Immunol 133:1005–1009.
Orchansky P, Novick D, Fischer DG, Rubinstein M. 1984. Type I and type II interferon receptors. J Interferon Res 4:275–282.
Pace JL, Russell SW, Schreiber RD, Altman A, Katz DH. 1983. Macrophage activation: priming activity from a T-cell hybridoma is attributable to interferon-gamma. Proc Natl Acad Sci USA 80:3782–3786.
Panem S, Check IJ, Henriksen D, Vilček J. 1982. Antibodies to alpha interferon in a patient with systemic lupus erythematosus. J Immunol 129:1–3.
Patterson RJ, Youmans GP. 1970. Demonstration in tissue culture of lymphocyte-mediated immunity to tuberculosis. Infect Immun 1:600–603.
Pearlstein KT, Palladino MA, Welte K, Vilček J. 1983. Purified human interleukin-2 enhances induction of immune interferon. Cell Immunol 80:1–9.
Perussia, B, Dayton, ET, Fanning, V, Thiagarajan, P, Hoxie, J, Trinchieri, G. 1983. Immune interferon and leukocyte-conditioned medium induce normal and leukemic myeloid cells to differentiate along the monocytic pathway. J Exp Med 158:2058–2080.
Pfefferkorn ER, Guyre PM. 1983. Recombinant human gamma interferon blocks the growth of *Toxoplasma gondii* in cultured human fibroblasts. Fed Proc 42:964.
Rhodes-Feuillette A, Druilhe P, Canivet M, Gentilini M, Peries J. 1981. Présence d'Interferon circulant dans le sérum de malades infectés par *Plasmodium falciparum*. CR Acad Sci Paris 293:635–637.

Rinderknecht E, O'Connor BY, Rodriguez H. 1984. Natural human interferon-gamma. Complete amino acid sequence and determination of sites of glycosylation. J Biol Chem 259:6790–6797.

Roberts WK, Vasil A. 1982. Evidence for the identity of murine gamma interferon and macrophage activating factor. J Interferon Res 2:519–532.

Rothermel CD, Rubin BY, Murray HW. 1983. Gamma interferon is the factor in lymphokine preparations that activates human macrophages to inhibit intracellular *Chlamydia psittaci* replication. J Immunol 131:2542–2544.

Rubin, BY, Gupta SL. 1980. Differential efficacies of human type I and type II interferons as antiviral and antiproliferative agents. Proc Natl Acad Sci USA 77:5928–5932.

Rubin BY, Bartal AH, Anderson SL, Millet SK, Hirshaut Y, Feit C. 1983. The anticellular and protein-inducing activities of human gamma interferon preparations are mediated by the interferon. J Immunol 130:1019–1020.

Samuel CE, Knutson G. 1983. Mechanism of interferon action: human leukocyte and immune interferons regulate the expression of different genes and induce different antiviral states in human amnion U cells. Virology 130:474–484.

Scher MG, Beller DI, Unanue ER. 1980. Demonstration of a soluble mediator that induces exudates rich in Ia-positive macrophages. J Exp Med 152:1684–1698.

Schultz RM, Kleinschmidt WJ. 1983. Functional identity between murine gamma interferon and macrophage activating factor. Nature 305:239–240.

Sonnenfeld G, Meruelo D, McDevitt HO, Merigan TC. 1981. Effect of type I and type II interferons on murine thymocyte surface antigen expression: induction or selection? Cell Immunol 57:427–439.

Steeg PG, Moore RN, Johnson HM, Oppenheim JJ. 1982. Regulation of murine macrophage Ia antigen expression by a lymphokine with immune interferon activity. J Exp Med 156:1780–1793.

Steinman RM, Nogueira N, Witmer MD, Tydings JD, Mellman IS. 1980. Lymphokine enhances the expression and synthesis of Ia antigens on cultured mouse peritoneal macrophage. J Exp Med 152:1248–1261.

Stewart WE II. 1979. The Interferon System. Springer-Verlag, Vienna-New York.

Trinchieri G, Granato D, Perussia B. 1981. Interferon-induced resistance of fibroblasts to cytolysis mediated by natural killer cells: specificity and mechanism. J Immunol 126:335–340.

Trown PW, Kramer MJ, Dennin RA Jr, Connell EV, Palleroni AV, Quesada J, Gutterman JU. 1983. Antibodies to human leukocyte interferons in cancer patients. Lancet i:81–84.

Unanue ER, Beller DI, Lu C, Allen PM. 1984. Antigen presentation: comments on its regulation and mechanism. J Immunol 132:1–5.

Vallbracht A, Treuner J, Flehmig B, Joester K-E, Niethammer D. 1981. Interferon-neutralizing antibodies in a patient treated with human fibroblast interferon. Nature 289:496–497.

Vilček J, Sulea IT, Volvovitz F, Yip YK. 1980. Characteristics of interferons produced in cultures of human lymphocytes by stimulation with *Corynebacterium parvum* and phytohemagglutinin. *In* De Weck AL, Kristensen F, Landy M (eds), Biochemical Characterization of Lymphokines. Academic Press, New York, pp. 323–329.

Vogel SN, Friedman RM. 1984. Interferon and macrophages: Activation and cell surface changes. *In* Vilček J, De Maeyer E (eds), Interferons, Vol. 2, Interferons and the Immune System. Elsevier, Amsterdam, pp. 35–54.

Wallach D, Fellous M, Revel M. 1982. Preferential effect of gamma interferon on the synthesis of HLA antigens and their mRNAs in human cells. Nature 299:833–836.

Weigent DA, Langford MP, Fleischmann WR Jr, Stanton GJ. 1983a. Potentiation of lymphocyte natural killing by mixtures of alpha or beta interferon with recombinant gamma interferon. Infect Immun 40:35–38.

Weigent DA, Stanton GJ, Johnson HM. 1983b. Interleukin 2 enhances natural killer cell activity through induction of gamma interferon. Infect Immun 41:992–997.

Weil J, Epstein CJ, Epstein LB, Sedmak JJ, Sabran JL, Grossberg SE. 1982. A unique set of polypeptides is induced by gamma interferon in addition to those induced in common with alpha and beta interferon. Nature 301:437–439.

Wheelock EF. 1965. Interferon-like virus-inhibitor induced in human leukocytes by phytohemagglutinin. Science 149:310–311.

Wiranowska-Stewart M, Lin LS, Braude EA, Stewart WE II. 1980. Comparisons of the physicochemical and biological properties of human, murine and bovine interferons, types I and II. *In* De Weck AL, Kristensen F, Landy M (eds), Biochemical Characterization of Lymphokines. Academic Press, New York, pp. 331–338.

Yamamoto JY, Farrar WL, Johnson HM. 1982. Interleukin 2 regulation of mitogen induction of immune interferon (IFN-gamma) in spleen cells and thymocytes. Cell Immunol 66:333–341.
Yip YK, Pang RHL, Oppenheim JD, Nachbar MS, Henriksen D, Zerebeckyj-Eckhardt I, Vilček J. 1981a. Stimulation of human gamma (immune) interferon production by diterpene esters. Infect Immun 34:131–139.
Yip YK, Pang RHL, Urban C, Vilček J. 1981b. Partial purification and characterization of human gamma (immune) interferon. Proc Natl Acad Sci USA 78:1601–1605.
Yip YK, Barrowclough B, Urban C, Vilček J. 1982a. Purification of two subspecies of human gamma (immune) interferon. Proc Natl Acad Sci USA 79:1820–1824.
Yip YK, Barrowclough B, Urban C, Vilček J. 1982b. The molecular weight of human gamma interferon is similar to that of other human interferons. Science 215:411–413.
Youngner JS, Salvin SB. 1973. Production and properties of migration inhibitory factor and interferon in the circulation of mice with delayed hypersensitivity. J Immunol 111:1914–1922.

Differentiation in Normal and Neoplastic Cells

Mediators in Cell Growth and Differentiation,
edited by Richard J. Ford and Abby L. Maizel.
Raven Press, New York © 1985.

Differentiation Factors from Cell Lines

James D. Watson, Ross L. Prestidge, Roger J. Booth,
David L. Urdal,* Diane Y. Mochizuki,* Paul J. Conlon,*
and Steven Gillis*

*Department of Immunobiology, School of Medicine, Auckland University, Auckland, New Zealand and *Immunex Corporation, Seattle, Washington 98101*

The major classes of regulatory cells in the immune system are T lymphocytes, and the molecular basis of their effector function is currently viewed through the secretion of soluble mediators or lymphokines. At least four basic types of T cells have been postulated, namely helper (T_H), suppressor (T_S), cytotoxic (T_K), and delayed hypersensitivity (T_{DH}) cells. There exist specific assays in many antigen systems for the first three of these, but for T_{DH} cells there is no assay that can be confidently accepted as being selective for these cells. Delayed-type hypersensitivity (DTH) is an immunologically specific inflammatory reaction with a characteristic histologic appearance of infiltration with mononuclear cells. This reaction can be transferred in vivo by Lyt-1^+ T cells. However, DTH lacks a precise assay. Current in vivo footpad or ear tests or in vitro migration inhibition assays are difficult to correlate with the classic DTH reaction (Blanden 1974). This is compounded by the finding that cloned antigen-specific T_H (Bianchi et al. 1981) and T_K (Lin and Askonas 1981) cells transfer footpad swelling. These results imply DTH is mediated by T_H cells (Lyt-1^+2^-, H-2I restricted) as well as T_K cells (Lyt-1^-2^+, H-2K/D restricted).

Lymphokines are actively involved in DTH reactions. Elicitation of DTH requires the T cell–dependent release of serotonin by local mast cells (Askenase et al. 1980), which causes an increase in permeability of the local vasculature. The release of lymphokines by antigen-specific T cells recruits circulating monocytes and neutrophils, which are the majority of cells infiltrating the local DTH site.

Lymphokine-producing cells can be placed into two groups. Lyt-1^+ T cells have been shown to produce interleukin 2 (IL-2), colony-stimulating factors (CSF), interferon-γ (IFN-γ), macrophage-activating factor (MAF), migration inhibition factor (MIF), and a variety of B cell–stimulating factors (BSF). Lyt-1^-2^+ T cells appear to produce IFN-γ and MAF, as well as activities generally known as lymphotoxins and suppressor factors. The difficulty we face is that the phenotypes of T_H and T_{DH} cells are inseparable. Thus, the physiological function we attribute to some of the lymphokine activities defined by in vitro response effects depends very much on whether we view the Lyt-1^+ producer cell as belonging to a T_H or

T_{DH} class. We focus here on those murine T cells that produce IL-2 and CSF and the in vivo significance of these lymphokines.

REGULATION OF LYMPHOKINE SYNTHESIS

Coordinate Expression of Lymphokines

Since the identification of the murine T-cell lymphoma LBRM-33 as a cell line that can be induced to secrete IL-2 (Gillis et al. 1980), we have observed that LBRM-33 cells are capable of synthesizing at least two other classes of lymphokine activities, CSF and an MAF that enhances tumoricidal activity (Ralph et al. 1982). LBRM-33 cells have been phenotypically characterized as Lyt-1$^+$, Qa2.3$^+$, Qa3.2$^+$, Qa4$^+$, Lyt-5$^+$, and TdT$^+$ (Gillis et al. 1980), which are some of the features of less-mature T cells. This T-cell lymphoma has several properties of interest in studies of the regulation of lymphokine synthesis. First, the synthesis of at least two different lymphokines, IL-2 and CSF, is under coordinate control (Watson 1983). Second, the cells give rise to lymphokine nonproducer variants (Gillis and Mizel 1981). While LBRM-33 cells do not constitutively produce IL-2 or CSF, both lymphokines are secreted upon stimulation with phytohemagglutinin (PHA) or concanavalin A (ConA). When LBRM-33 cells are cloned, subclones can be isolated that do not respond to subsequent stimulation either with PHA or ConA by the production of IL-2 or CSF (Watson 1983). Some nonproducer subclones can be rescued to a producer phenotype by stimulation with both IL-1 and PHA (Watson 1983). In such cases, both IL-2 and CSF are produced. The high frequency of nonproducer variants of LBRM-33 cells (Watson 1983) may reflect that these tumor cells are poised between two differentiation states. The first is the more mature IL-2–producing state, and the second, a less mature, non-IL-2–producing state. The nonproducer state may be more stable, as we have not detected the reversion of nonproducer subclones of LBRM-33 cells to a stable producer state. The most significant observation remains that no LBRM-33 clones that produce IL-2 or CSF alone have been observed.

The stimulation of IL-2 and CSF production in LBRM-33 cells appears to be dependent upon the synthesis of new mRNA. This relation has been examined by isolating RNA from LBRM-33 cells and translating it in *Xenopus* oocytes. When poly A$^+$ RNA purified from nonstimulated LBRM-33 clone 5A4 cells was injected into oocytes, no IL-2 or CSF activity could be subsequently detected in the oocyte culture medium (Table 1). In contrast, when poly A$^+$ RNA purified from PHA-stimulated cells was injected into oocytes, high titers of IL-2 and CSF subsequently developed in the culture medium. These data indicate that LBRM-33 cells do not constitutively synthesize either IL-2- or CSF-specific mRNA. Thus, treatment of cells with PHA results in the induction of specific mRNA synthesis (Table 1).

Because the synthesis of IL-2 and CSF in the LBRM-33 lymphoma appears to be under coordinate control, the question of whether they are also coordinately synthesized in normal T lymphocytes arose. We approached this issue by examining

TABLE 1. *Coordinate induction of IL-2 and CSF-specific mRNA*

Source of poly (A)+ RNA*	CSF (units/ml)	IL-2 (units/ml)
LBRM-33	0	0
LBRM-33 + PHA (6 hr)	120	980

*20 ng RNA was injected into each oocyte. For each assay, 4–6 oocytes were incubated in 0–2 ml medium. The medium was assayed for CSF and IL-2 activity after 44 hr incubation.

lymphokine production in antigen-specific T-cell clones. To produce antigen-specific T-cell lines, thymocytes from BDF_1 (C57BL/6J × DBA/2J) mice were primed with heterologous sheep erythrocyte antigens (SRBC) in irradiated syngeneic recipients for 7 days, and spleen cells were cultured with SRBC according to procedures described elsewhere (Watson 1979). T cell lines were then established by passaging T cells with 0.01% SRBC and 3000 rad-irradiated BDF_1 spleen cells in culture with medium lacking IL-2. Cloning of SRBC-reactive T cells was accomplished by limiting dilution methods, plating cells at a concentration of 0.3 cells per well. As cell clones emerged, the supernatants of these lines were tested for the presence of IL-2 using a T-cell growth assay and for CSF using a CSF-dependent cell line, FDC-P2 (Dexter et al. 1980), in a microassay (Prestidge et al. 1984).

The most striking observation was the consistent finding of IL-2 and CSF activities together in supernatants from a high proportion of the clones. Of 75 clones tested, 52 secreted IL-2 and CSF into supernatants, and in only 3 clones assayed were these activities segregated—2 CSF and 1 IL-2. Some 20 clones produced neither IL-2 nor CSF. Thus, the findings are similar to those observed for the LBRM-33 tumor cell line. The coordinate expression of these two lymphokines in both normal and transformed T lymphocytes raises questions concerning regulation and structure. For example, do these lymphokines belong to a multigene family encoding growth factors that are controlled by a common regulatory gene? Are there similarities in the primary structure of IL-2 and CSF? These questions can only be answered by analysis of the amino acid sequences of IL-2 and CSF and by comparisons of the genes that encode these molecules.

Colony-Stimulating Factors and Interleukin 3

Although IL-2 has now been characterized as a single class of T-cell growth factors, the chemical nature of the activities that constitute CSF has long been an area of divergent views. The proliferation and differentiation of cells in various hematopoietic lineages can be induced in culture with factors from a variety of sources. A species known as CSF-1 acts only on macrophage precursors (Stanley and Heard 1977, Stanley et al. 1978), while other regulators act on erythroid (Iscove et al. 1982), granulocyte, megakaryocyte, eosinophil, and mast cell precursors, as well as on earlier stem cell progenitors (Burgess and Metcalf 1980, Burgess et al. 1977a,b, Metcalf et al. 1980). The principal T cell–derived factors with colony-stimulating activity that have been described are IL-3 and granulocyte/ macrophage CSF (GM-CSF). GM-CSF (Burgess and Metcalf 1980) is defined by its ability to produce mixed granulocyte and macrophage colonies from bone marrow in soft agar culture. The term IL-3 was introduced by Ihle and co-workers to describe a factor responsible for the induction or expression of an enzyme, 20α-steroid dehydrogenase (20α SDH) in cultures on nu/nu splenic lymphocytes (Ihle et al. 1981, Hapel et al. 1981, Lee et al. 1982, Ihle et al. 1982a, b). Initial studies of the tissue distribution of 20αSDH indicated a close association with T cells, and consequently IL-3 was thought to be a T cell–specific lymphokine. However, more recent data have indicated that cells from many different hematopoietic lineages can respond to IL-3.

To resolve the issue of the molecular nature of CSF species produced by T cells, we have now purified CSF species from LBRM-33 cells and compared T cell–derived CSF to that obtained from WEHI-3 cells (Prestidge et al. 1984). These CSF have been purified by sequential fractionation using salt precipitation, gel filtration, anion and cation exchange chromatography, and high-performance liquid chromatography (HPLC). Both LBRM-33 and WEHI-3 cells secrete a CSF species with similar chemical and biological properties. This CSF species exists in two forms, termed CSF-2α and CSF-2β, which appear to differ only in their glycosylation patterns. Both stimulate in vitro the growth of bone marrow cells in the granulocyte, macrophage, megakaryocyte, mast cell, and erythrocyte lineages, as well as the growth of a CSF-dependent cell line, FDC-P2 (Prestidge et al. 1984). These properties of CSF-2α and -2β are similar to those reported for IL-3, hematopoietic cell growth factor (Garland and Dexter 1983, Brazill et al. 1983), mast cell or persisting cell growth factor (Schrader 1981, Schrader et al. 1980, Schrader and Clark-Lewis 1982, Yung et al. 1981). However, LBRM-33 cells secrete another CSF species that is not produced by WEHI-3 cells. This CSF species is here termed CSF-2γ, and it stimulates the proliferation of granulocytes (G) and macrophages (M) from bone marrow, but does not support the growth of FDC-P2 cells (Prestidge et al. 1984).

To identify the bone marrow cells that mature in cultures containing these various CSF species, HPLC-purified species derived from LBRM-33 and WEHI-3 cells were separately tested in soft agar cultures and the composition of colonies analyzed. Selected colonies were applied to glass slides and incubated with Wright-Giemsa or nonspecific esterase stains. A total of 100–150 colonies were examined in each group. The majority of the colonies (around 90%) formed in the presence of CSF-2α or -2β were composed of granulocytes, macrophages, or a granulocyte/ macrophage mixture (Table 2). Two to four percent consisted of mast cells. The rest of the colonies were composed of mixed lineages other than macrophage, but were predominantly erythroid (hemoglobin-containing) or megakaryocytic. Essentially, the data derived from WEHI-3 and LBRM-33 CSF-2α and -2β were the same and are summarized in Table 2. In contrast, LBRM-33-derived CSF-2γ

TABLE 2. *Identification of bone marrow colonies*

CSF Origin	% Colonies					
	Granulocyte	Macrophage	Mixed G/M	Mast cell	Erythroid	Megakaryocytic
WEHI-3						
CSF-2α	30	30	28	4	1–2	1–2
CSF-2β	30	29	30	2	1–2	1–2
LBRM-33						
CSF-2α	30	34	34	4	1–2	1–2
CSF-2β	34	30	34	3	1–2	1–2
CSF-2γ	0	40	60	0	0	0

stimulated the growth of only two types of colonies, macrophage and mixed G/M types. No other colony types were observed (Table 2).

N-Terminal Amino Acid Sequence of CSF

The fractions containing CSF-2α activity after HPLC purification using 0.1% trifluoroacetic acid in acetonitrile (Prestidge et al. 1984) from either LBRM-33 or WEHI-3 sources were diluted 1:3 with 0.9 M acetic acid, 0.2 M pyridine and injected onto another C-18 column previously equilibrated with the pyridine-acetate buffer (Urdal et al. 1984a). A gradient of n-propanol was established to elute the protein from the column. CSF-2α eluted from the column 54 min into the gradient at 28% propanol. The recovery of CSF-2α following HPLC was essentially 100%, with 2 million-fold purification (Urdal et al., 1984b). Aliquots of the final HPLC run were analyzed by polyacrylamide gel electrophoresis. The bulk of the CSF-2α activity was found in a single major band of MW 24,500 (Urdal et al. 1984b). This fraction was dried under vacuum to a small volume and then subjected to automated amino terminal Edman degradation. A single major sequence was found for both LBRM-33 and WEHI-3 CSF-2α, suggesting that the proteins were homogeneous (Table 3). We estimated the specific activity of the final material as 1.1×10^7 U/μg.

Ihle and co-workers previously reported an N-terminal amino acid sequence for WEHI-3-derived IL-3 (Ihle et al. 1983). Of interest to us was the equivalence of residues 7–15 to the sequence we obtained, and it is also shown in Table 3.

TABLE 3. *N-terminal amino acid sequences of CSF-2α derived from LBRM-33 and WEHI-3 cells*

	1 5 10 15
LBRM-33	ala-ser-ile-ser-gly-arg-asp-thr-his-arg-lys/leu-thr-arg-thr-leu
WEHI-3	ala-ser-ile-ser-gly-arg-asp-thr-his-arg-lys/leu-thr-arg-thr-leu
Ihle-IL-3*	asp-thr-his-arg- leu -thr-arg-thr-leu

*Data taken from Ihle et al. (1983).

cDNA Sequence of IL-3

Mouse cDNA clones have now been reported for IL-3 (Fung et al. 1984, Yokota et al. 1984). One cDNA clone was derived from WEHI-3 cells (Fung et al. 1984), the other from a T cell clone Lyt-$1^{+}2^{-}$/9 (Yokota et al. 1984). Both cDNA clones contain a single open-reading frame consisting of 166 codons beginning with a methionine codon. Two additional methionine codons occur, 12 and 18 codons downstream from the first (Fung et al. 1984, Yokota et al. 1984). The amino acid sequences we have obtained for both LBRM-33- and WEHI-3-derived CSF-2α begin at residue 27 from the methionine codon, which appears to mark the start of the reading frame (Table 4).

Using this information from the published cDNA sequences, we can speculate about the nature and length of a signal peptide. If we use our sequence data, mature IL-3 would have a signal peptide of 26 amino acids, with its N-terminal amino acid being alanine at position 27. However, the Ihle sequence (Ihle et al. 1983) begins at position 33 with aspartic acid (Table 3). The possibility exists that either glutamine-alanine (26 - 27) or arginine-aspartic acid (32–33) can act as proteolytic cleavage sites. The former may mark the signal peptide, and additional proteolytic cleavage steps may be catalyzed by a number of proteases subsequently.

FUNCTION OF T-CELL-DERIVED LYMPHOKINES

IL-3 and GM-CSF

The functional and structural data discussed above indicate that CSF-2α and CSF-2β are identical to IL-3 and that CSF-2γ is similar to GM-CSF. It appears, therefore, that IL-2-producing T-cells produce two species of CSF. One species, most appropriately termed IL-3, exhibits multilineage colony-stimulating activity (Table 2) and stimulates the continuous growth of mastlike cell lines such as FDC-P2. The other, GM-CSF, is more restricted, yielding only mixed granulocyte/macrophage or pure macrophage colonies (Table 3), and does not support the continuous growth of cells in culture.

There is a striking difference between the in vitro activities of IL-3 and GM-CSF in bone marrow assays as demonstrated by the total number of colonies formed in soft-agar cultures (Figure 1). Saturating concentrations of IL-3 stimulate the growth of 150–200 colonies per 10^5 bone marrow cells, whereas GM-CSF stimulates the formation of 2- to 3-fold more colonies in identical culture conditions (Figure 1). Further, the colonies generated in the presence of GM-CSF are generally much larger than those observed in the IL-3-supplemented cultures. Thus, while both lymphokines give rise to macrophage and mixed granulocyte/macrophage colonies, one must ask whether these growth regulators act on common or different progenitor cells. Little is known of the nature of target cell populations for these two species of CSF.

TABLE 4.* *cDNA Sequence of IL-3*

ATG	GTT	CTT	GCC	AGC	TCT	ACC	ACC	AGC	ATC	CAC	ACC	ATG	CTG	CTC	CTG	CTC	CTG	ATG	CTC	TTC
MET	Val	Leu	Ala	Ser	Ser	Thr	Thr	Ser	Ile	His	Thr	MET	Leu	Leu	Leu	Leu	Leu	MET	Leu	Phe
1									10										20	
CAC	CTG	GGA	CTC	CAA	GCT	TCA	ATC	AGT	GGC	CGG	GAT	ACC	CAC	CGT	TTA	ACC	AGA	ACG	TTG	AAT
His	Leu	Gly	Leu	Gln	Ala	Ser	Ile	Ser	Gly	Arg	Asp	Thr	His	Arg	Leu	Thr	Arg	Thr	Leu	Asn
								30										40		
TGC	AGC	TCT	ATT	GTC	AAG	GAG	ATT	ATA	GGG	AAG	CTC	CCA	GAA	CCT	GAA	CTC	AAA	ACT	GAT	GAT
Cys	Ser	Ser	Ile	Val	Lys	Glu	Ile	Ile	Gly	Lys	Leu	Pro	Glu	Pro	Glu	Leu	Lys	Thr	Asp	Asp
							50										60			
GAA	GGA	CCC	TCT	CTG	AGG	AAT	AAG	AGC	TTT	CGG	AGA	GTA	AAC	CTG	TCC	AAA	TTC	GTC	GAA	AGC
Glu	Gly	Pro	Ser	Leu	Arg	Asn	Lys	Ser	Phe	Arg	Arg	Val	Asn	Leu	Ser	Lys	Phe	Val	Glu	Ser
						70										80				
CAA	GGA	GAA	GTG	CAT	CCT	GAG	GAC	AGA	TAC	GTT	ATC	AAG	TCC	AAT	CTT	CAG	AAA	CTT	AAC	TGT
Gln	Gly	Glu	Val	Asp	Pro	Glu	Asp	Arg	Tyr	Val	Ile	Lys	Ser	Asn	Leu	Gln	Lys	Leu	Asn	Cys
					90										100					
TGC	CTG	CCT	ACA	TCT	GCG	AAT	GAC	TCT	GCG	CTG	CCA	GGG	GTC	TTC	ATT	CGA	GAT	CTG	GAT	GAC
Cys	Leu	Pro	Thr	Ser	Ala	Asn	Asp	Ser	Ala	Leu	Pro	Gly	Val	Phe	Ile	Arg	Asp	Leu	Asp	Asp
				110										120						
TTT	CGG	AAG	AAA	CGT	AGA	TTC	TAC	ATG	GTC	CAC	CTT	AAC	GAT	CTG	GAG	ACA	GTG	CTA	GCC	TCT
Phe	Arg	Lys	Lys	Leu	Arg	Phe	Tyr	Met	Val	His	Leu	Asn	Asp	Leu	Glu	Thr	Val	Leu	Ala	Ser
			130										140							
AGA	CCA	CCT	CAG	CCC	GCA	TCT	GGC	TCC	GTC	TCT	CCT	AAC	CGT	GGA	ACC	CTG	GAA	TGT	TAA	
Arg	Pro	Pro	Gln	Pro	Ala	Ser	Gly	Ser	Val	Ser	Pro	Asn	Arg	Gly	Thr	Val	Glu	Cys		
		150										160								

*Data taken from Yokota et al. (1984).

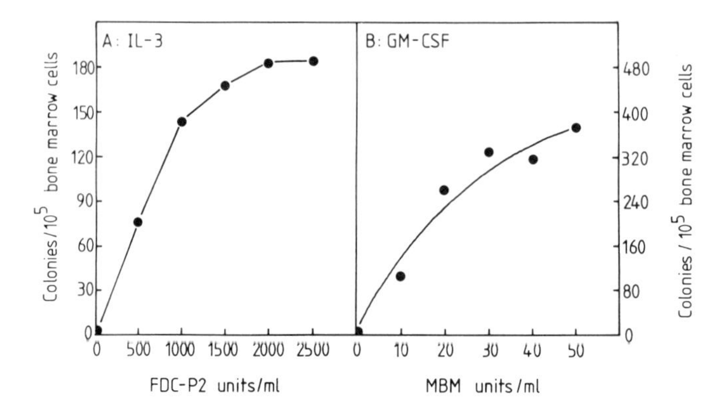

FIG. 1. Comparison of the colony-forming activities of interleukin 3 (IL-3) and granulocyte/macrophage colony-stimulating factor (GM-CSF). Both lymphokines were purified from LBRM-33 cells by high-performance liquid chromatography (Prestidge et al. 1984). Units of IL-3 were determined using FDC-P2 indicator cells, and units of GM-CSF were from the mouse bone marrow assay (Prestidge et al. 1984). BALB/c bone marrow cells were seeded in soft agar medium cultures with different concentrations of IL-3 or GM-CSF and colonies counted after 8 days of growth.

The Effector Function of Lymphokine-Secreting Cells

The recent isolation of cDNA that encodes three different lymphokines, IL-2 (Devos et al. 1983, Fujita et al. 1983), IL-3 (Fung et al. 1984, Yokota et al. 1984), and IFN-γ (Gray and Goeddel 1982) have resulted in a vision of new horizons in solving the problems that underlie the regulation of immune responses. However, we are still faced with a very basic question. What is the physiological significance of each of the lymphokine species identified in in vitro assays and now dissected further as a result of the advancing technology of molecular biology?

One of the major influences on current interpretation of the mode of action of a given lymphokine relates to the phenotype of the producer cell. The growth-promoting effects of IL-2, combined with the knowledge that IL-2-producer cells are Lyt-1^+, has led to the view that IL-2 secretion is a function of T_H cells. What then is the function of CSF also produced by these T_H cells?

As these putative T_H cells may also produce other lymphokines, such as MAF, IFNγ, and BSF (Prystowsky et al. 1982), it is difficult to accept that they are primary regulators of hematopoiesis, as well as of immune responsiveness. This is best illustrated in athymic (nude) mice, which lack these mature lymphokine-producing effector T cells and show no defects in the maturation of hematopoietic cells. One possibility to consider is whether T cells of the Lyt-1^+/Lyt-2^- phenotype, which can be induced to secrete IL-2 and CSF coordinately, are effector cells belonging to another T-cell lineage. Because of the difficulty in separating T_H and T_{DH} cells, the role of T_{DH} cells should be examined.

The function of IL-3 and GM-CSF may not be as primary regulators of the proliferation and differentiation of progenitors in the microenvironment of the bone marrow. Rather, these lymphokines may more appropriately perform a specialized role associated with the specific effector function of a T_{DH} cell, resulting in the rapid maturation of macrophages, granulocytes, mast cells, and other types at sites of inflammation.

If the class of T cell that secretes IL-2, CSF, MAF, and perhaps other growth regulators in a coordinate manner is in fact T_{DH}, then these molecules may be termed inflammatory lymphokines, functioning to stimulate the rapid proliferation and differentiation of cells at sites of inflammation, rather than "helper lymphokines." They may be viewed as forming a "backup" system to the helper cell, acting at a specialized site. Although T_{DH} cells may regulate the development of hematopoietic cells and lymphocytes in local lesions, other classes of hormones may be responsible for the regulation of cell differentiation at sites such as lymph nodes or bone marrow. Similarly, the role of IL-2 in the induction of T-cell responses may need reassessment. It has become dogma that IL-2 functions as a second signal in the induction of antigen-specific T-cell responses. What is important is that if IL-2 is a product of T_{DH} cells, there may exist T_H cells that produce other species of lymphokines that regulate the induction of T- and B-cell responses.

The problem then is the delineation of T_H and T_{DH} function. Do these classes of T cells share effector activities in common? Are there mechanisms of T-cell help that are of greater physiological significance than those helper mechanisms we deduce from the properties of IL-2 in cell culture bioassays? Are there classes of helper T cells that are responsible for regulating immunoglobulin isotype production by B cells? The resolution of the phenotypes of T_{DH} from T_H cells may lead to the search for other classes of mediators of lymphocyte as well as other blood cell development mechanisms.

ACKNOWLEDGMENTS

This work was supported by grants from the Medical Research Council of New Zealand and the Auckland Division of the Cancer Society of New Zealand, Inc.

REFERENCES

Askenase PM, Burdztajn S, Gershon MD, Gershon RK. 1980. T cell-dependent mast cell degranulation and release of serotonin in murine delayed-type hypersensitivity. J Exp Med 152:1358–1374.

Bianchi ATJ, Hooijkaas H, Brenner R, Tees R, Nordin AA, Schreier MH. 1981. In vivo properties of cloned antigen-specific helper cells. Nature 290:62–63.

Blanden RV. 1974. T cell response to viral and bacterial infection. Trans Rev 19:56–88.

Brazill CW, Haynes M, Garland J, Dexter TM. 1983. Procedure for the purification of interleukin 3 to homogeneity. J Biochem 210:747–751.

Burgess AW, Camakaris J, Metcalf D. 1977a. Purification and properties of colony-stimulating factors from mouse lung conditioned medium. J Biol Chem 252:1998–2003.

Burgess AW, Metcalf D, Russell SHM, Nicola NA. 1977b. Granulocyte/macrophage-, megakaryocyte-, eosinophil- and erythroid-colony stimulating factors produced by mouse spleen cells. Biochem J 185:301–314.

Burgess AW, Metcalf D. 1980. The nature and action of granulocyte-macrophage colony stimulating factor. Blood 56:947–958.

Devos R, Plaetinek G, Cheroutre H, Simons G, Degrave W, Tavernier J, Remault E, Fiers W. 1983. Molecular cloning of human interleukin 2 cDNA and its expression in *E. coli*. Nucl Acid Res 11:4307–4323.

Dexter TM, Garland J, Scott D, Scolnik E, Metcalf D. 1980. Growth of factor-dependent hematopoietic precursor cell lines. J Exp Med 152:1036–1047.
Fujita T, Chikako T, Matsui H, Taniguchi T. 1983. Structure of the human interleukin 2 gene. Proc Natl Acad Sci USA 80:7437–7441.
Fung MC, Hapel AJ, Ymer S, Cohen DR, Johnson RM, Campbell HD, Young IG. 1984. Molecular cloning of cDNA for murine interleukin 3. Nature 307:233–235.
Garland JM, Dexter TH. 1983. Relationship of hemopoietic growth factor to lymphocytes and interleukin 3: A short review. Lymphokine Res 2:13–22.
Gillis S, Mizel SB. 1981. T cell lymphoma model for the analysis of interleukin 1-mediated T cell activation. Proc Natl Acad Sci USA 78:1133–1137.
Gillis S, Scheid M, Watson J. 1980. Biochemical and biologic characterization of lymphocyte regulatory molecules. III. The isolation and phenotypic characterization of interleukin 2 producing T cell lymphomas. J Immunol 125:2570–2578.
Gray PW, Leung DW, Pennica D, Yelverton E, Najarian R, Simonsen CC, Derynck R, Sherwood PJ, Wallace DM, Berger SL, Levinson AD, Goeddel DV. 1982. Expression of human immune interferon cDNA in *E. coli* and monkey cells. Nature 295:503–508.
Hapel AJ, Lee JC, Farrar WL, Ihle JN. 1981. Establishment of continuous cultures of Thy1.2$^+$, Lyt1$^+$, 2$^-$ T cells with purified interleukin 3. Cell 25:179–184.
Ihle JN, Rebar L, Keller J, Lee JC, Hapel A. 1981. Interleukin 3: Possible roles in the regulation of lymphocyte differentiation and growth. Immunol Rev 63:5–32.
Ihle JN, Keller J, Henderson L, Klein F, Palaszynski E. 1982a. Procedures for the purification of interleukin 3 to homogeneity. J Immunol 129:2431–2436.
Ihle JN, Keller J, Greenberger JS, Henderson L, Yetter RA, Morse HC III. 1982b. Phenotypic characteristics of cell lines requiring interleukin 3 for growth. J Immunol 129:1377–1383.
Ihle JN, Keller J, Oroszlan S, Henderson LE, Copeland TD, Fitch F, Prystowsky MB, Goldwasser E, Schrader JW, Palaszynski E, Dy M, Lebel B. 1983. Biologic properties of homogeneous interleukin 3. J Immunol 131:282–287.
Iscove NN, Roitsch CA, Williams N, Guilbert LJ. 1982. Molecules stimulating early red cell, granulocyte, macrophage, and megakaryocyte precursors in culture: Similarity in size, hydrophobicity, and charge. J Cell Physiol (Suppl 1):65.
Lee JC, Hapel AJ, Ihle JN. 1982. Constitutive production of a unique lymphokine (IL-3) by the WEHI-3 cell line. J Immunol 128:2393–2398.
Lin YL, Askonas AB. 1981. Biological properties of an influenza A virus-specific killer T cell clone. J Exp Med 154:225–231.
Metcalf D, Johnson GR, Burgess AW. 1980. Direct stimulation by purified GM-CSF of the proliferation of multipotential and erythroid precursor cells. Blood 55:138–147.
Prestidge RL, Watson JD, Urdal DL, Mochizuki D, Conlon P, Gillis S. 1984. Biochemical comparison of murine colony stimulating factors secreted by a T cell lymphoma and a myelomonocytic leukemia. J Immunol (in press).
Prystowsky MB, Ely JM, Beller DI, Eisenberg L, Holdman J, Goldwasser E, Ihle J, Quintans J, Remold H, Vogel SN, Fitch FW. 1982. Alloreactive cloned T cell lines. IV. Multiple lymphokine activities secreted by helper and cytolytic cloned T lymphocytes. J Immunol 129:2337–2344.
Ralph P, Williams N, Nakoinz J, Jackson H, Watson J. 1982. Distinct signals for antibody-dependent and nonspecific killing of tumour targets mediated by macrophages. J Immunol 129:427–432.
Schrader JW. 1981. Stem cell differentiation in the bone marrow. Immunology Today 2:7–12.
Schrader JW, Arnold B, Clark-Lewis I. 1980. A Con A-stimulated T-cell hybridoma releases factors affecting haematopoietic colony-forming cells and B-cell antibody responses. Nature 283:197–199.
Schrader JW, Clark-Lewis I. 1982. A T cell-derived factor stimulating multipotential hemopoietic stem cells: Molecular weight and distinction from T cell growth factor and T cell derived granulocyte macrophage colony-stimulating factor. J Immunol 129:30–35.
Stanley ER, Heard PM. 1977. Factors regulating macrophage production and growth. Purification and some properties of the colony stimulating factor from medium conditioned by mouse L cells. J Biol Chem 252:4305–4312.
Stanley ER, Chen DM, Lin HS. 1978. Induction of macrophage production and proliferation by a purified colony stimulating factor. Nature 274:168–171.
Urdal DL, Mochizuki D, Conlon PJ, March CJ, Remerowski ML, Eisenman J, Ramthun C, Gillis S. 1984a. Lymphokine purification by reversed phase high performance liquid chromatography. J Chromatogr (in press).

Urdal DL, March CJ, Mochizuki D, Conlon PJ, Hopp TO, Watson JD, Gillis S. 1984b. Purification to homogeneity of a colony stimulating factor (CSF-2α) derived from a murine T cell lymphoma. Proc Natl Acad Sci USA (in press).
Watson J. 1979. Continuous cultures of helper T cells. J Exp Med 150:1510–1519.
Watson J. 1983. Biology and biochemistry of T cell-derived lymphokines. I. The coordinate synthesis of interleukin 2 and colony stimulating factors in a murine T cell lymphoma. J Immunol 131:293–297.
Watson JD, Prestidge RL. 1983. Interleukin 3 and colony stimulating factors. Immunology Today 4:278–280.
Yokota T, Lee F, Rennick D, Hall C, Arai N, Mosmann T, Nabel G, Cantor H, Arai K. 1984. Isolation and characterization of a mouse cDNA clone that expresses mast cell growth factor activity in monkey cells. Proc Natl Acad Sci USA (in press).
Yung YP, Eger R, Tertian G, Moore MAS. 1981. Long-term *in vitro* culture of murine mast cells. II. Purification of a mast cell growth factor and its dissociation from TCGF. J Immunol 127:794–799.

Mediators in Cell Growth and Differentiation,
edited by Richard J. Ford and Abby L. Maizel.
Raven Press, New York © 1985.

Modulation of Gene Expression During Terminal Cell Differentiation: Murine Erythroleukemia

Paul A. Marks, Takashi Murate, Tsuguhiro Kaneda,
Jeffrey Ravetch, and Richard A. Rifkind

DeWitt Wallace Research Laboratory and Sloan-Kettering Division, Graduate School of Medical Sciences, Memorial Sloan-Kettering Cancer Center, New York, New York 10021

The transition from precursor cells capable of proliferation to terminally differentiated cells involves the modulation of expression of a number of genes (Marks and Rifkind 1978, Pardee et al. 1978, Thomas et al. 1981, Linzer and Nathans 1983, Hofer et al. 1982, Shen et al. 1983). Among these are genes for "housekeeping" functions, e.g., ribosomal RNA and enzymes catalyzing reactions in various metabolic pathways, genes for proteins characteristic of the differentiated phenotype—e.g., α and β globin—and genes controlling factors determining progression through the cell division cycle—e.g., possibly p53 protein. In an effort to understand the regulation of gene expression involved in the transition of self-replicating precursor cells to cells expressing a terminal differentiated phenotype, including terminal cell division, we have been systematically studying the steps involved in inducer-mediated commitment of murine erythroleukemia cells (MELC) to terminal erythroid differentiation.

CHARACTERISTICS OF INDUCER-MEDIATED MELC ERYTHROID DIFFERENTIATION

Among the properties of MELC that make them suitable for studies of the molecular and cellular events involved in inducer-mediated transition to terminal differentiation is the fact that the cells can be maintained for essentially unlimited periods of time under appropriate in vitro conditions and as a relatively homogeneous population of cells (Marks and Rifkind 1978). MELC can be induced by a variety of agents (Table 1), including Me_2SO (Friend et al. 1971) and hexamethylene bisacetamide (HMBA) (Reuben et al. 1976), to express characteristics of terminal erythroid cell differentiation. Among these agents, HMBA

$$(CH_3\text{-}CH_2\text{-}\overset{\overset{\Large O}{\|}}{C}\text{-}NH\text{-}(CH_2)_6\text{-}NH\text{-}\overset{\overset{\Large O}{\|}}{C}\text{-}CH_2\text{-}CH_3)$$

has been found to be the most effective inducer of MELC.

TABLE 1. *Inducers of MELC erythroid differentiation**

Type	Inducer
Polar compounds	Dimethylsulfoxide Hexamethylene bisacetamide
Fatty acids	Butyric acid
DNA intercalators	Actinomycin
Modified bases	Azacytidine
Phosphodiesterase inhibitors	Methylisoxanthine
Ion-flux agents	Ouabain
Physical agents	uv, x-ray
Post-transcription-acting agent	Hemin

*References in review (Marks and Rifkind 1978).

Agents such as HMBA can induce a variety of transformed cell lines to express characteristics of terminal differentiation. Among these cell lines are transformed cells of various species, including mouse, rat, dog, and human, and ones derived from various cell types by chemical or viral transformation, as well as lines that appear to have developed spontaneously (Table 2). Such studies provide further evidence that the MELC system is suitable for analysis of the steps involved in the transition to terminal differentiation.

The virally transformed MELC appear to be blocked at a stage in erythroid lineage that is relatively late in erythropoiesis, e.g., erythroid colony-forming units (CFU-e) (Figure 1) (Marks and Rifkind 1978). Inducer-mediated differentiation of MELC is characterized by a number of early metabolic changes including alterations in membrane permeability to various ions (Bernstein et al. 1976, Mager and Bernstein 1978, Levinson et al. 1980), changes in cell volume, and transient increases in cAMP concentration (Figure 2) (Gazitt et al. 1978, Gazitt et al. 1978a). During these early metabolic changes, there is no detectable commitment to terminal cell division or increased expression of α and β globin genes. By approximately 12 to 15 hr in culture with an inducer such as HMBA, there begins the coordinated expression of a developmental program similar to that of normal terminal erythroid differentiation, including commitment to terminal cell division (Gusella et al. 1976, Fibach et al. 1977) and increased accumulation of α and β globin mRNA (Ostertag et al. 1972, Ross et al. 1972), α, β^{maj}, and β^{min} globins, hemoglobin major and hemoglobin minor (Ohta et al. 1976, Boyer et al. 1972), heme synthesizing enzymes (Sassa 1976), and membrane-associated erythroid-specific proteins, such as spectrin and glycophorin (Eisen et al. 1977). In addition, within 24 hr, the accumulation of a variety of other proteins is modulated: the accumulation of the chromatin-associated protein H1° increases (Chen et al. 1982) and that of the chromatin-associated protein p53 decreases (Shen et al. 1983). The function of H1° is not known, but p53 has been implicated as important in determining normal cell cycle progression from G_1 to S phase.

TABLE 2. *Cells inducible to terminal differentiation by HMBA and/or Me_2SO*

Cell type	Reference
Murine erythroleukemia	Marks and Rifkind 1978
HL60-human promyelocytic leukemia	Koeffler 1983, Lozzio and Lozzio 1975
	Collins et al. 1977
Rat mammary tumor	Warburton et al. 1981
Canine kidney epithelial carcinoma	Lever 1979
Mouse neuroblastoma	Palfrey et al. 1977
Human glioblastoma multiforma	Rabson et al. 1977
Human lung cancer	Tralka and Rabson 1976
Mouse teratocarcinoma	Paulin et al. 1979
Mouse embryonal carcinoma	McCue et al. 1983
Human lymphoma cells	Jaffee et al. 1981
Mouse liver tumor cells	Higgins 1982
Human colon cancer cells	Bennett and Hicks 1980
Human melanoma cells	Huberman et al. 1979

PROTEIN p53 AND INDUCER-MEDIATED MELC COMMITMENT TO TERMINAL CELL DIVISION

An early manifestation of termination of cell division in MELC appears to be a transient prolongation of the G_1 phase of the cell cycle, which can be detected after cell transit through one complete S phase in the presence of inducer (Terada et al. 1977). Induced MELC are arrested in a G_1-like phase of the cell cycle, after up to five terminal cell divisions. Evidence has accumulated to suggest that the synthesis of a labile protein during G_1 may control entry of cells into S phase (Rossow et al. 1979). The nuclear protein p53, initially recognized by its elevated

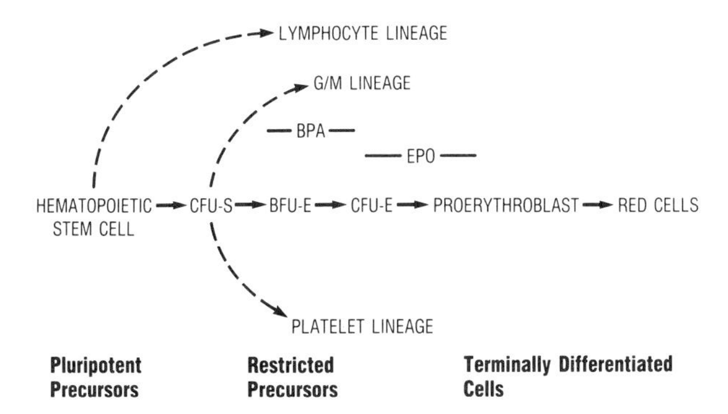

FIG. 1. Schematic representation of sequential stages of normal hematopoietic cell development. The pluripotent, hematopoietic stem cells produce the lymphocyte lineage and precursors (CFU-s) of granulocyte/macrophages, erythroid cells, and platelets. Erythroid precursors, BFU-e, differentiate to CFU-e: both are precursors of restricted potential and, in turn, generate proerythroblasts and terminally differentiated red cells.

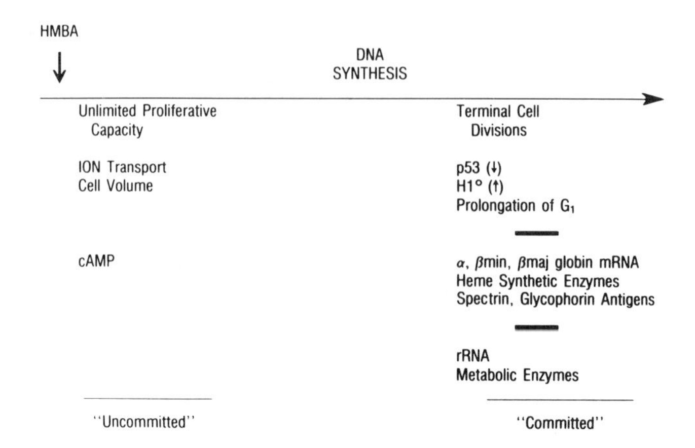

FIG. 2. Schema of inducer-mediated MELC erythroid differentiation. See text for details.

levels in transformed cells (DeLeo et al. 1979) and by its capacity to bind to simian virus 40 (SV40) tumor antigen (Lane and Crawford 1979, Linzer and Levine 1979), has been identified as possibly important in determining normal cell cycle progression from G_1 to S (Milner and Milner 1981, Mercer et al. 1982).

In our laboratory, we examined p53 protein in MELC DS19 and a MELC variant, R1, resistant to inducer-mediated commitment to terminal cell division. Levels of accumulation of newly synthesized p53 were determined by quantitative immunofluorescence employing monoclonal antibodies prepared to p53 and immunoprecipitation after [^{35}S]methionine incorporation (Shen et al. 1983).

A decrease in p53 synthesis and content was demonstrated during induced differentiation. The decrease in p53 was detected at all stages of the cell cycle, namely G_1, S, and G_2. Hemin induces MELC globin mRNA accumulation but no commitment to terminal cell division. In cells cultured with hemin, there is no detectable decrease in p53 content. A MELC variant cell line, R1 (Marks et al. 1983) resistant to HMBA-mediated commitment to terminal cell division, shows no detectable decrease in p53 synthesis or accumulation when placed in culture with the inducer (Shen et al. 1983).

These studies suggest a relationship between p53 and MELC cellular proliferative capacity. If, as suggested by several workers, a critical level of a labile protein, possibly p53, is required for the G_1-to-S transition, then inducer-mediated modulation of expression of the p53 gene could be part of the multigene modulation involved in the transition to terminally dividing cells.

EXPRESSION OF GLOBIN GENES DURING HMBA-INDUCED MELC DIFFERENTIATION

Different inducers cause different patterns of expression of α and β globin genes during inducer-mediated MELC differentiation (Nudel et al. 1977). For example, MELC cultured with HMBA, Me_2SO, or butyric acid accumulate α globin mRNA within 12 to 16 hr. Cytoplasmic accumulation of β globin mRNA sequences is not detected until about 8 hr later. By 48 hr there is approximately a 10-fold increase in globin mRNA content. By comparison, hemin rapidly (by 6 hr) and simultaneously initiates accumulation of both α and β globin mRNA. The principle β-like mRNA induced by polar planar agents is β^{maj} mRNA, whereas hemin causes the accumulation primarily of β^{min} globin mRNA. β^{min} mRNA is the principle β globin found in uninduced MELC.

The differences in the patterns of globin mRNA accumulation in MELC induced with different agents led us to study the mechanisms of action of the inducers at the molecular level. To investigate inducer-mediated effects at the transcriptional level, the technique of nuclear chain elongation was used to measure transcription activity (Profus-Juchelka et al. 1983). We compared the effects of two inducers, HMBA and hemin. The HMBA-mediated increase in accumulation of α and β globin mRNA was found to reflect primarily control of gene expression at the level of transcription. This was shown by the finding that in MELC cultured with HMBA, there was a marked increase in globin mRNA run-off transcription for both the α and β globin genes. By comparison, hemin had no detectable effect on the level of transcriptional activity of either the α or β globin genes. This later result suggests that hemin caused an increase in globin mRNA accumulation by acting primarily at a posttranscriptional level, influencing the processing, transport, or stability of the low, but constitutive, level of globin gene transcription.

INDUCER-MEDIATED CHANGES IN CHROMATIN STRUCTURE RELATED TO GLOBIN GENE EXPRESSION

Alterations in DNA and chromatin structure are associated with the transition from inactive to actively transcribed genes in differentiating cell systems (for references see review, Sheffery et al. 1983). These include changes in the pattern of DNA methylation (Felsenfeld and McGhee 1982, Doerfler 1981), binding of high mobility group (HMG) proteins 14 and 17 (Weisbrod and Weintraub 1979), sensitivity of chromatin to digestion by DNase I (Weintraub and Groudine 1976, Garel and Axel 1976), and the appearance of sites or regions that are hypersensitive to digestion by DNase I usually, but not invariably, upstream of the 5′ end of active or potentially active genes (Larsen and Weintraub 1982). Sites of S1 nuclease sensitivity have recently been demonstrated to be associated with, but not necessarily identical to, sites of DNase I hypersensitivity (Sheffery et al. 1984).

In our laboratory, several features of DNA and chromatin configuration were examined in detail with respect to inducer-mediated increases in globin gene expression (Sheffery et al. 1982, 1984). We evaluated alterations in chromatin structure in the domain of α_1 and β^{maj} globin structural genes by determination of the pattern of (1) DNA methylation, (2) sensitivity to DNase I, (3) sites hypersensitive to DNase I digestion, and (4) sensitivity to S1 nuclease digestion. Certain patterns of chromatin DNA associated with actively transcribed genes are found in globin domains of uninduced MELC and do not change upon induction. These include the pattern of DNA methylation about both the α_1 and β^{maj} genes and the general sensitivity of these genes to DNase I. Changes in chromatin structure in globin gene domains that are associated with inducer-mediated globin gene transcription include development of DNase I hypersensitive sites 5′ to both α_1 and β^{maj} genes and the development of S1 nuclease hypersensitive sites 5′ to both the α_1 and β^{maj} genes (Sheffery et al. 1984). Thus, certain changes in chromatin structure that appear to have occurred during the developmental history of the erythroid lineage are stably propagated in virus-transformed, developmentally arrested MELC. Additional changes in chromatin structure characteristic of the transition to actively transcribed genes occur only upon exposure to inducers. Temporal studies indicate that the inducer-mediated alterations in chromatin structure, specifically the appearance of the DNAse I hypersensitive sites and the S1 nuclease hypersensitive sites 5′ upstream from the 5′ Cap of both α_1 and β^{maj} globin genes, develop prior to a detectable increase in the rate of transcription of these genes (Sheffery et al. 1984). This suggests that the inducer-mediated alteration in chromatin structure is part and parcel of the changes necessary for the transition to actively transcribed globin genes.

INDUCER-MEDIATED COMMITMENT OF MELC TO TERMINAL CELL DIVISION

Inducer-mediated MELC commitment to terminal cell differentiation is a multistep process (Chen et al. 1982). The glucocorticoid dexamethasone and the tumor promotor 12-0-tetradecanoyl-phorbol 13-acetate (TPA) can suppress inducer-mediated MELC commitment to terminal cell division and hemoglobin accumulation (Rovera et al. 1977, Fibach et al. 1979, Chen et al. 1982). The inducer-mediated early changes in this multistep process of commitment are not suppressed by the steroid or TPA. MELC retain a "memory" for these early changes as evidenced by the fact that expression of commitment to terminal cell division occurs rapidly upon removal of the steroid or TPA (Chen et al. 1982). Expression of terminal cell division and hemoglobin accumulation are sensitive to suppression by the steroid or tumor promotor.

The following studies were designed to characterize those steps of HMBA-mediated MELC commitment to terminal cell division that are suppressed by dexamethasone and rapidly expressed upon removal of the steroid. In these studies, commitment of MELC to terminal cell division is operationally defined as the capacity of cells that have been exposed to inducer to express differentiated characteristics and loss of proliferative capacity in the absence of inducer (Fibach et al. 1977). Commitment at the single-cell level was assayed by culture of cells with inducer with or without steroid for various periods of time prior to transfer of cells to semisolid medium without inducer or steroid and then scoring the proportion of small (less than 64 cells), benzidine-reactive (i.e., hemoglobin-containing) colonies after 5 days of growth in semisolid medium.

We found that a substantial proportion of cells rapidly (within 2 hr) expressed commitment to terminal cell division on transfer of cells from medium with inducer

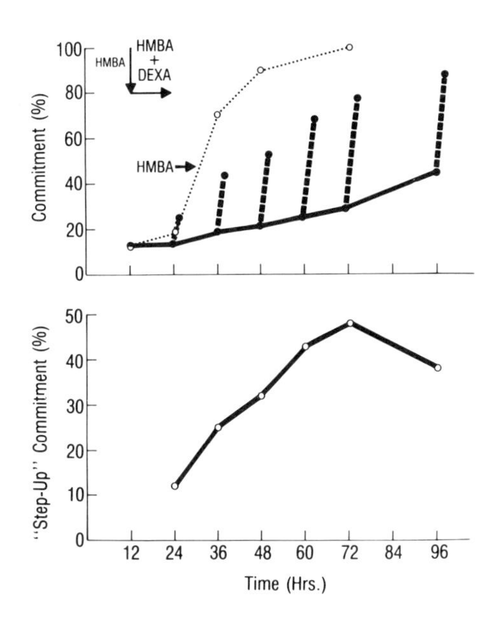

FIG. 3. Step-up expression of commitment in MELC in relation to duration of prior culture with HMBA and dexamethasone. Upper panel: MELC was cultured with 5×10^{-3} M HMBA for 12 hr (indicated by arrow), and 4×10^{-6} M dexamethasone was added and culture continued for up to 96 hr. At the times indicated (●–●), aliquots of the suspension culture were removed for commitment assay and for transfer to fresh media with 3×10^{-3} M HMBA alone for an additional 2 hr, at which time assay for commitment was again performed (●--●). MELC were also cultured with 5×10^{-3} M HMBA alone (o····o), and commitment assays performed at times indicated by open circles. Lower panel: Step-up expression of commitment plotted as the difference between commitment of MELC in culture with inducer plus steroid and after transfer to culture with 5×10^{-3} M HMBA for 2 hr (data derived from upper panel). (Reproduced from Murate et al. 1984.)

and steroid to fresh medium with inducer alone (Chen et al. 1982). The relationship between duration of culture of MELC with HMBA plus dexamethasone and "step-up" expression of commitment was examined by determining the magnitude of step-up expression after periods of prior culture with inducer plus steroid ranging from 12 to 96 hr (Murate et al. 1984). When cultured continuously with HMBA alone, the proportion of committed MELC exceeded 90% by 48 hr (Figure 3). Dexamethasone suppressed the expression of HMBA-mediated commitment throughout the period of observation, although the proportion of cells committed tended to rise gradually during this period of culture. The magnitude of step-up expression increased with increasing duration of time in culture with steroid and inducer up to 72 hr and then tended to plateau (Figure 3).

These findings indicate that step-up expression of commitment reflects a memory for inducer-mediated accumulation of a factor or factors required for commitment. Accumulation of these putative factors is not prevented by the steroid. These findings suggest that inducer-mediated changes essential for transition to terminal cell division occur in an increasing proportion of the cells with increasing time in culture with inducer plus steroid. Thus, an increasing proportion of the population of cells become synchronized at a step ready to express commitment. This step, the transition to terminal cell division, is blocked by the steroid.

EFFECT OF INDUCER ON STEP-UP EXPRESSION

The requirement for HMBA in step-up expression was examined by transfer of cells from culture with steroid plus inducer to medium without any addition, or one with HMBA, Me_2SO, or hemin (Figure 4). HMBA and Me_2SO are agents that

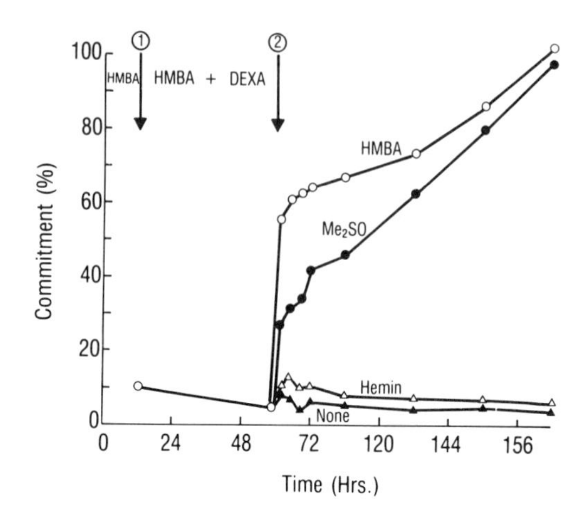

FIG. 4. Dependence of expression of step-up expression on inducer. MELC were placed in culture with 5×10^{-3} M HMBA; at 12 hr 4×10^{-6} M dexamethasone was added to the culture (arrow 1) and culture continued to 60 hr (arrow 2), at which time cells were recovered by centrifugation and resuspended in fresh medium with 5×10^{-3} M HMBA, 2.8×10^{-1} M Me_2SO, 0.1 M hemin, or fresh medium without addition. At the times indicated, aliquots of cells were removed for commitment assay. (Reproduced from Murate et al. 1984.)

induce MELC to both terminal cell division and hemoglobin accumulation. By comparison, hemin induces MELC to accumulate globin mRNAs, but not to commit to terminal cell division. Step-up expression was observed in cultures with HMBA or Me_2SO but not in those with hemin or those with no addition (Figure 4).

These findings indicate that step-up expression is dependent on the presence of inducers such as HMBA or Me_2SO that cause commitment to terminal cell division. The difference in the effect of HMBA and Me_2SO and that of hemin may reside in the difference in effect at the stage of step-up expression. Differences in effectiveness of various inducers of commitment to terminal cell division may also be related to differences in activity at this step. Me_2SO supports step-up expression but not as effectively as HMBA, and HMBA is known to be more effective than Me_2SO as an inducer of MELC (Reuben et al. 1976).

IS RNA SYNTHESIS REQUIRED DURING STEP-UP EXPRESSION OF COMMITMENT?

Actinomycin D at concentrations up to 100 ng/ml did not inhibit step-up expression (Murate et al. 1984). At this concentration of actinomycin D, total RNA synthesis is blocked more than 90%, while plating efficiency is decreased less than 25%.

Cordycepin at concentrations up to 20 μg/ml did not block step-up expression, although at these concentrations accumulation of newly synthesized poly(A) RNA was inhibited more than 85% (Figure 5). These findings suggest that step-up expression of commitment does not require new RNA synthesis.

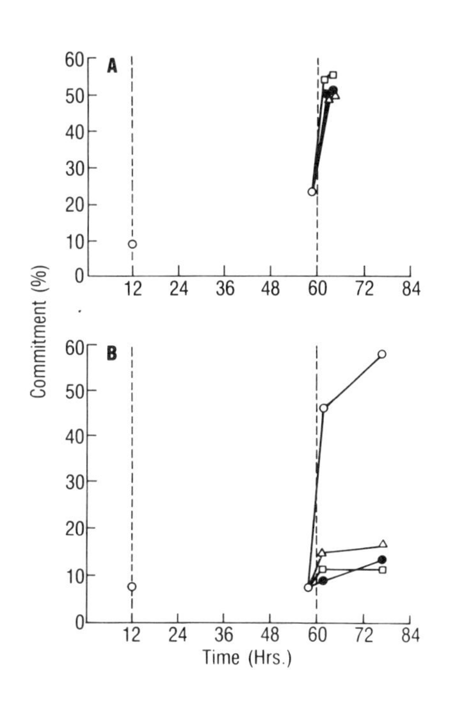

FIG. 5. Effect of cordecypin **(A)** and of cycloheximide **(B)** on step-up expression of commitment. **A:** After culture for 12 hr with 5×10^{-3} M HMBA, and 5×10^{-3} M HMBA plus 4×10^{-6} M dexamethasone for an additional 46 hr, the suspension culture was divided into three portions: to one there was no addition (□), to the second was added 10 μg/ml (●), and to the third 20 μg/ml cordecypin (Δ). At 60 hr, the cells were recovered from each culture by centrifugation and resuspended in fresh medium with HMBA alone (□) or HMBA plus 10 μg/ml cordecypin (●) or HMBA plus 20 μg/ml cordecypin (Δ). Assays for commitment were performed at the times indicated. **B:** Experimental design was similar to that described above except that at 59 hr, culture was divided into three portions: to one there was no addition, to another, 0.5 μg/ml, and to the third, 1.5 μg/ml cycloheximide were added. At 60 hr, cells in each of these cultures were recovered by centrifugation and resuspended as follows: culture without cycloheximide was divided into two portions, to one of which 5×10^{-3} M HMBA was added (○) and to the other, nothing (□). To cultures with cycloheximide were added 5×10^{-3} M HMBA plus 0.5 μg/ml cycloheximide (Δ) or HMBA plus 1.5 μg/ml cycloheximide (●), respectively. At each of the times indicated, assays for commitment were performed. (Reproduced from Murate et al. 1984).

IS PROTEIN SYNTHESIS REQUIRED DURING STEP-UP?

Cyclohexamide, at concentrations of 0.5 μg/ml and 1.5 μg/ml, essentially completely inhibited step-up expression of commitment (Murate et al. 1984) (Figure 5). These concentrations of cyclohexamide inhibit total protein synthesis 74% and 86%, respectively. These findings suggest that new protein synthesis is required for step-up expression.

Taken together, the studies with the inhibitors, actinomycin D, cordycepin, and cyclohexamide suggest that HMBA initiates a process leading to the accumulation of a factor or factors (which may be mRNAs) that are required for the transition to terminal cell division and whose expression is suppressed by dexamethasone.

STABILITY OF FACTORS REQUIRED FOR STEP-UP EXPRESSION

The stability of the cellular memory that is required for step-up expression was evaluated by transferring cells from cultures with inducer plus steroid to medium without any addition. At various intervals up to 132 hr, aliquots were removed and transferred to medium with HMBA for 2 hr (Figure 6). In these studies (Murate

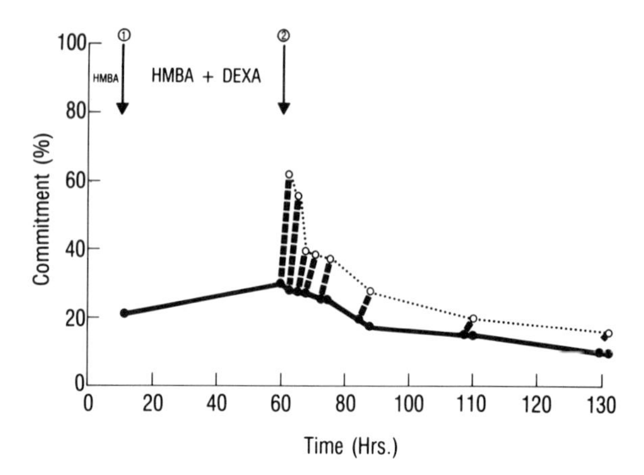

FIG. 6. Stability of HMBA-induced step-up expression. MELC were cultured with 5×10^{-3} M HMBA for 12 hr and then 4×10^{-6} M dexamethasone added (arrow 1) and culture continued for additional 48 hr. At 60 hr (arrow 2), the cells were transferred to fresh medium without addition. At the times indicated by the solid dots (●), aliquots of the cultures were removed and 5×10^{-3} HMBA added and the culture continued for 2 hr (○---○). At each of the times indicated by the solid and open dots, assays for commitment were performed. (Reproduced from Murate et al. 1984).

et al. 1984), at 60 hr, 30% of the cells were committed and step-up expression reached 62% of the cells. By 72 hr, the level of step-up expression was only 36% and decreased gradually thereafter up to 130 hr. These data indicate a relatively rapid decay in step-up expression, with an apparent half-life of the factors required for this phenomenon being approximately 12 hr, or essentially the duration of one cell cycle.

It appears that HMBA, in addition to initiating the sequence of events leading to the accumulation of commitment factors, directly or indirectly has an action required for step-up expression. Present data do not establish the site of such action of HMBA, but are consistent with its involving a stabilization of putative inducer-generated mRNAs.

INDUCER-MEDIATED GENE TRANSCRIPTION RELATED TO STEP-UP EXPRESSION

The studies just reviewed suggest that the dexamethasone-suppressed steps in HMBA-induced MELC commitment to terminal cell division involve a block in translation of inducer-mediated accumulation of mRNAs. It is likely that the transition from dividing to nondividing cells reflects modulation of expression of a number of genes. To examine this phenomenon further, the following strategy was employed to identify inducer-activated genes for factors critical to the transition to terminal cell division.

cDNA libraries were prepared from total poly(A) RNA purified from MELC (1) cultured for 60 hr without inducer, (2) cultured for 60 hr with HMBA, and (3) cultured for 60 hr with HMBA plus dexamethasone (Ravetch, Ramsey, Rifkind, and Marks, manuscript in preparation). These were then screened to detect sequences corresponding to genes expressed in cells cultured with inducer plus steroid and in cells cultured with inducer alone, but were not expressed in uninduced cells. These sequences should reflect genes whose transcription was increased by the inducer in the presence of the steroid but should exclude sequences that were (1) expressed in uninduced cells, (2) induced by the steroid alone, and (3) induced by HMBA in the absence of steroid. Approximately 6,000 clones were screened in this manner, and 20 clones were detected that met the above-outlined criteria. The 20 clones corresponded to two genes, and the transcriptional activity of each was enhanced by HMBA. One of the genes corresponded to a size of approximately 7.5 kb and was induced by HMBA to a level of transcriptional activity several times normal. The other gene corresponded in size to approximately 1.5 kb and showed an increase in rate of transcription of about 100-fold and corresponds by sequence analysis to the α globin gene.

These studies indicate that HMBA-mediated accumulation of mRNAs occurs in MELC cultured with inducer plus steroid. Relatively few genes are possible candidates for controlling products critical to step-up expression of commitment. In these studies, there may be inducible genes that are not detected in the screening process. Nevertheless, the results encourage us to continue our pursuit of the characterization of inducer-activated genes for factors that commit cells to terminal cell division.

SUMMARY

MELC are virus-transformed cells capable of indefinite proliferation that are blocked in differentiation at an early erythroid precursor stage, probably corresponding to CFU-e. A variety of agents, among them HMBA and Me_2SO, induce MELC to terminal differentiation and expression of characteristics similar to that associated with normal erythropoiesis. During inducer-mediated terminal differentiation, modulation of expression of a number of genes occurs. Studies to date have characterized inducer-mediated alterations in chromatin structure associated with activation of α and β^{maj} globin genes.

Inducer-mediated MELC terminal cell division is also associated with a decrease in the synthesis of the nuclear protein p53, a protein that has been implicated as a requirement for the progression from G_1 to S in the cell cycle. HMBA-mediated commitment to terminal cell division is suppressed by steroid. HMBA induces accumulation of mRNAs that may be required for commitment to terminal cell division and whose translation is suppressed by dexamethasone. At least two inducer-activated genes have been identified that may play a role in the transition to terminal cell division.

ACKNOWLEDGMENT

These studies were supported, in part, by the National Cancer Institute (PO1 CA31768-02 and CA08748) and the Bristol-Myers Cancer Grant Program. TK is a Brenner Fellow in cancer research.

REFERENCES

Bernstein A, Hunt DM, Crichley V, Mak TW. 1976. Induction of ouabain of hemoglobin synthesis in cultured Friend erythroleukemic cells. Cell 9:375–381.

Boyer SH, Wuu KD, Noyes AM et al. 1972. Hemoglobin biosynthesis in murine virus-induced leukemic cells in vitro: Structure and amounts of globin chains produced. Blood 40:823–835.

Chen ZX, Banks J, Rifkind RA, Marks PA. 1982. Inducer-mediated commitment of murine erythroleukemia cells to differentiation: A multistep process. Proc Natl Acad Sci USA 79:471–475.

Collins SJ, Gallo RC, Gallagher RE. 1977. Continuous growth and differentiation of human myeloid leukemic cells in suspension culture. Nature 270:347.

DeLeo AB, Jay G, Appella E, DuBois GC, Law LW, Old LJ. 1979. Detection of a transformation-related antigen in chemically induced sarcomas and other transformed cells of the mouse. Proc Natl Acad Sci USA 76:2420–2424.

Doerfler W. 1981. DNA methylation—a regulatory signal in eukaryotic gene expression. J Gen Virol 57:1–20.

Eisen H, Nasi S, Georopulos CP, Arndt-Jovin D, Ostertag W. 1977. Surface changes in differentiating Friend erythroleukemic cells in culture. Cell 10:689–695.

Felsenfeld G, McGhee J. 1982. Methylation and gene control. Nature 296:602–603.

Fibach E, Gambari R, Shaw PA, Maniatis G, Reuben RC, Sassa S, Rifkind RA, Marks PA. 1979. Tumor promoter-mediated inhibition of cell differentiation: Suppression of the expression of erythroid functions in murine erythroleukemia cells. Proc Natl Acad Sci USA 76:1906–1910.

Fibach E, Reuben RC, Rifkind RA, Marks PA. 1977. Effect of hexamethylene bisacetamide on the commitment to differentiation of murine erythroleukemia cells. Cancer Res 37:440–444.

Friend C, Scher W, Holland JG, Sato T. 1971. Hemoglobin synthesis in murine virus induced leukemia cells in vitro: Stimulation of erythroid differentiation by dimethyl-sulfoxide. Proc Natl Acad Sci USA 68:378–382.

Garel A, Axel R. 1976. Selective digestion of transcriptionally active ovalbumin genes from oviduct nuclei. Proc Natl Acad Sci USA 73:3966–3970.

Gazitt Y, Deitch AD, Marks PA, Rifkind RA. 1978. Cell volume changes in relation to the cell cycle of differentiating erythroleukemic cells. Exp Cell Res 117:413–420.

Gazitt Y, Reuben RC, Deitch AD, Marks PA, Rifkind RA. 1978. Changes in cyclic adenosine 3′:5′-monophosphate levels during induction of differentiation in murine erythroleukemic cells. Cancer Res 38:3779–3783.

Gusella J, Geller R, Clarke B, Weeks V, Housman D. 1976. Commitment to erythroid differentiation by Friend erythroleukemia cells: A stochastic process. Cell 9:221–229.

Higgins PJ. 1982. Response of mouse liver tumor cells to the differentiation-inducing agent dimethylsulfoxide. Pharmacol 25:170–176.

Hofer E, Hofer-Warbinek R, Darnell JE Jr. 1982. Globin RNA transcription: A possible termination site and demonstration of transcriptional control correlated with altered chromatin structure. Cell 29:887–893.

Huberman E, Heckman C, Langenbach R. 1979. Stimulation of differentiated functions in human melanoma cells by tumor-promoting agents and dimethylsulfoxide. Cancer Res 39:2618–2624.

Jaffee ES, Smith SA, Magrath IT, Freeman CB, Alabaster O, Sussman EH. 1981. Induction of complement receptors in human cell lines derived from undifferentiated lymphomas. Lab Invest 45:295–301.

Kim YS, Tsao D, Siddiqui B, Whitehead JS, Arnstein P, Bennett J, Hicks J. 1980. Effects of sodium butyrate and dimethylsulfoxide on biochemical properties of human colon cancer cells. Cancer 45:1185–1192.

Koeffler HP. 1983. Induction of differentiation of human acute myelogenous leukemia cells: Therapeutic implications. Blood 4:709–721.

Lane DP, Crawford LV. 1979. T antigen is bound to a host protein in SV40-transformed cells. Nature (London) 278:261–263.

Larsen A, Weintraub H. 1982. An altered DNA conformation detected by S1 nuclease occurs at specific regions in active chick globin chromatin. Cell 29:609–622.

Levinson R, Housman D, Cantley L. 1980. Amiloride inhibits murine erythroleukemia cell differentiation: Evidence for a Ca^{2+} requirement for commitment. Proc Natl Acad Sci USA 77:5948–5952.

Lever JE. 1979. Inducers of mammalian cell differentiation stimulate dome formation in a differentiated kidney epithelial cell line (MDCK). Proc Natl Acad Sci USA 76:301–316.

Linzer DI, Levine AJ. 1979. Characterization of a 54K dalton cellular SV40 tumor antigen present in SV40-transformed cells and uninfected embryonal carcinoma cells. Cell 17:43–52.
Linzer DI, Nathans D. 1983. Growth-related changes in specific mRNAs of cultured mouse cells. Proc Natl Acad Sci USA 80:4271–4275.
Lozzio CB, Lozzio BB. 1975. Human chronic myelogenous leukemia cell line with positive Philadelphia chromosome. Blood 45:321.
Mager D, Bernstein A. 1978. The program of Friend cell erythroid differentiation: Early changes in Na^+/K^+ ATPase function. J Supramol Struct 8:431–438.
Marks PA, Rifkind RA. 1978. Erythroleukemic differentiation. Ann Rev Biochem 24:419–448.
Marks PA, Chen ZX, Banks J, Rifkind RA. 1983. Erythroleukemia cells: Variants inducible for hemoglobin synthesis without commitment to terminal cell division. Proc Natl Acad Sci USA 80:2281–2284.
McCue PA, Matthaei KI, Taketo M, Sherman MI. 1983. Differentiation defective mutants of mouse embryonal carcinoma cells: Response to hexamethylene bisacetamide and retinoic acid. Dev Biol 96:416–426.
Mercer WE, Nelson D, DeLeo AB, Old LJ, Baserga R. 1982. Microinjection of monoclonal antibody to protein p53 inhibits serum-induced DNA synthesis in 3T3 cells. Proc Natl Acad Sci USA 79:6309–6312.
Milner J, Milner S. 1981. SV40-53K antigen: A possible role for 53K in normal cells. Virology 112:785–788.
Murate T, Kaneda T, Rifkind RA, Marks PA. 1984. Inducer-mediated commitment of murine erythroleukemia cells to terminal division: The expression of commitment. Proc Natl Acad Sci USA 81:3394–3398.
Nudel U, Salmon J, Fibach E, et al. 1977. Accumulation of alpha and beta globin messenger RNAs in mouse erythroleukemia cells. Cell 12:463–469.
Ohta Y, Tanaka M et al. 1976. Erythroid cell differentiation: Murine erythroleukemia cell variant with unique pattern of induction by polar compounds. Proc Natl Acad Sci USA 73:1232–1236.
Ostertag W, Melderis H, Steinheider G, Kluge N, Dube S. 1972. Synthesis of mouse hemoglobin and globin mRNA in leukemic cell cultures. Nature New Biol 239:231–234.
Palfrey C, Kimhi Y, Littauer UZ, Reuben RC, Marks PA. 1977. Induction of differentiation in mouse neuroblastoma cells by hexamethylene bisacetamide. Biochem Biophys Res Commun 76:937–942.
Pardee AB, Dubrow R, Hamlin JL, Kletzien RF. 1978. Animal cell cycle. Ann Rev Biochem 47:715–750.
Paulin D, Perreau J, Jakob H, Jacob F, Yaniv M. 1979. Tropomyosin synthesis accompanies formation of actin filaments in embryonal carcinoma cells induced to differentiate by hexamethylene bisacetamide. Proc Natl Acad Sci USA 76:1891–1895.
Profous-Juchelka HR, Reuben RC, Marks PA, Rifkind RA. 1983. Transcriptional and post-transcriptional regulation of globin gene accumulation in induced murine erythroleukemia cells. Mol Cell Biochem 3:229–232.
Rabson AS, Stern R, Tralka TS, Costa J, Wilczek J. 1977. Hexamethylene bisacetamide induces morphologic changes and increased synthesis of procollagen in cell line from glioblastoma multiforme. Proc Natl Acad Sci USA 74:5060–5064.
Reuben RC, Wife RL, Breslow R, Rifkind RA, Marks PA. 1976. A new group of potent inducers of differentiation in murine erythroleukemia cells. Proc Natl Acad Sci USA 73:862–866.
Ross J, Ikawa Y, Leder P. 1972. Globin messenger-RNA induction during erythroid differentiation of cultured leukemia cells. Proc Natl Acad Sci USA 69:3620–3623.
Rossow PW, Riddle VGH, Pardee AB. 1979. Synthesis of labile, serum-dependent protein in early G1 controls animal cell growth. Proc Natl Acad Sci USA 76:4446–4450.
Rovera G, O'Brien TG, Diamond L. 1977. Tumor promoters inhibit spontaneous differentiation of Friend erythroleukemia cells in culture. Proc Natl Acad Sci USA 74:2894–2898.
Sassa S. 1976. Sequential induction of heme pathway enzymes during erythroid differentiation of mouse Friend leukemia virus-infected cells. J Exp Med 143:305–315.
Sheffery M, Rifkind RA, Marks PA. 1982. Murine erythroleukemia cell differentiation: DNase I hypersensitivity and DNA methylation near the globin genes. Proc Natl Acad Sci USA 79:1180–1184.
Sheffery M, Rifkind RA, Marks PA. 1983. Chromatin structure, globin gene transcription and erythroid cell differentiation. *In* Ultman J, Rowley J (eds), Bristol-Myers Cancer Symposia, vol. 5. Academic Press, New York, pp. 291–306.

Sheffery M, Marks PA, Rifkind RA. 1984. Gene expression in murine erythroleukemia cells: Transcriptional control and chromatin structure of the alpha$_1$-globin gene. J Mol Biol 172:417–436.
Shen DW, Real FX, DeLeo AB, Old LJ, Marks PA, Rifkind RA. 1983. Protein p53 and inducer-mediated erythroleukemia cell commitment to terminal cell division. Proc Natl Acad Sci USA 80:5919–5922.
Terada M, Fried J, Nudel U, Rifkind RA, Marks PA. 1977. Transient inhibition of initiation of S phase associated with dimethylsulfoxide induction of murine erythroleukemia cells to erythroid differentiation. Proc Natl Acad Sci USA 74:248–252.
Thomas G, Thomas G, Luther H. 1981. Transcriptional and translational control of cytoplasmic proteins after serum stimulation of quiescent Swiss 3T3 cells. Proc Natl Acad Sci USA 78:5712–5716.
Tralka TS, Rabson AS. 1976. Brief communication: Cilia formation in cultures of human lung cancer cells treated with dimethylsulfoxide. J Natl Cancer Inst 57:1383–1388.
Warburton MJ, Head LP, Rudland PS. 1981. Redistribution of fibronectin and cytoskeletal proteins during the differentiation of rat mammary tumor cells in vitro. Exp Cell Res 132:57–66.
Weintraub H, Groudine M. 1976. Chromosomal subunits in active genes have an altered conformation. Science 193:848–856.
Weisbrod S, Weintraub H. 1979. Isolation of a subclass of nuclear proteins responsible for conferring a DNase I-sensitive structure on globin chromatin. Proc Natl Acad Sci USA 76:630–634.

Mediators in Cell Growth and Differentiation,
edited by Richard J. Ford and Abby L. Maizel.
Raven Press, New York © 1985.

Regulatory Proteins for Growth and Differentiation in Normal and Leukemic Hematopoietic Cells: Normal Differentiation and the Uncoupling of Controls in Myeloid Leukemia

Leo Sachs

Department of Genetics, Weizmann Institute of Science, Rehovot 76100, Israel

The cloning and clonal differentiation of normal hematopoietic cells in culture made it possible to study the controls that regulate growth (multiplication) and differentiation of different hematopoietic cell types (See Sachs 1974, 1978a, 1980, 1982a,b). We first showed (Ginsburg and Sachs 1963, Pluznik and Sachs 1965) and others then confirmed (Bradley and Metcalf 1966) that normal mouse myeloid precursor cells cultured with a feeder layer of other cell types can form clones of granulocytes and macrophages in culture. We also found that the formation of these clones is due to secretion, by cells of the feeder layer, of specific inducers that are required for the formation of clones and the differentiation of cells in these clones to macrophages or granulocytes for cells from mice (Pluznik and Sachs 1965, 1966, Ichikawa et al. 1966) and from humans (Paran et al. 1970). Since we first detected their presence in culture supernatants (Ichikawa et al. 1966, Pluznik and Sachs 1966), these protein inducers have been referred to by a number of names; I shall use the name macrophage and granulocyte inducers (MGI) (Table 1).

These proteins can be produced and secreted by various normal and malignant cells in culture and in vivo (Sachs 1974). Their production can be induced by a variety of compounds (Fibach and Sachs 1975, Weiss and Sachs 1978, Lotem and Sachs 1979, Falk and Sachs 1980), and some cells produce these proteins constitutively (Ichikawa et al. 1966, Landau and Sachs 1971, Austin et al. 1971, Stanley and Heard 1977, Lipton and Sachs 1981). The MGI are a family of proteins of a number of molecular forms with different biological activities. Our cell culture approach has led to the cloning and isolation of growth factors for all the hematopoietic cell types, including different types of lymphocytes (as this monograph shows).

NORMAL GROWTH- AND DIFFERENTIATION-INDUCING PROTEINS

The MGI proteins include some that induce cell growth (multiplication) and others that induce differentiation. Those that induce growth (they are also required for normal cell viability) we now call MGI-1. They include proteins that induce the formation of macrophage clones (MGI-1M) (Ichikawa et al. 1966, Stanley and Heard 1977, Lotem et al. 1980), granulocyte clones (MGI-1G) (Ichikawa et al. 1966, Lotem et al. 1980), or both types of clones (MGI-1GM) (Landau and Sachs 1971, Burgess et al. 1977, Lipton and Sachs 1981). MGI-1 has previously been referred to as mashran gm (Ichikawa et al. 1967), colony-stimulating factor (CSF) (Metcalf 1969), colony-stimulating activity (CSA) (Austin et al. 1971), and MGI (Landau and Sachs 1971) (Table 1). The existence of an antibody that does not react with all forms of MGI-1M shows that molecules of the same form of MGI-1 can have different antigenic sites (Lotem et al. 1980). The other main type of MGI, which we now call MGI-2 (Sachs 1980, Lotem et al. 1980, Liebermann et al. 1982), induces the differentiation of myeloid precursor cells, either leukemic (Fibach et al. 1972) or normal (Sachs 1980, Liebermann et al. 1982), without inducing colony formation. This differentiation-inducing protein (Fibach et al. 1972, Fibach and Sachs 1976) has also been referred to as MGI (Fibach et al. 1972), D factor (Maeda et al. 1977, Yamamoto et al. 1980), and GM-DF (Burgess and Metcalf 1980). It has been suggested that different forms of MGI-2 may induce differentiation to macrophages or differentiation to granulocytes (Liebermann et al. 1982). The regulation of MGI-1 and MGI-2 appears to be under the control of different genes (Falk and Sachs 1980). The differentiation-inducing protein MGI-2, but not the growth-inducing protein MGI-1, is a DNA-binding protein (Weisinger and Sachs 1983).

These macrophage and granulocyte inducers can be proteins or glycoproteins, depending on the cells in which they are produced, and the presence of carbohydrates does not appear to be necessary for their biological activity (Lipton and Sachs 1981). The molecular weights are generally around 23,000 or multiples of this number (Sachs 1978a, Nicola et al. 1979, Lotem et al. 1980). MGI-2 activity is more sensitive to proteolytic enzymes and high temperature than MGI-1 activity (Lipton and Sachs 1981), and MGI-2 has a shorter half-life in serum than MGI-1 (Lotem and Sachs 1981). The ready separability of the different forms of MGI seems to depend on the cells in which they are produced (Lotem et al. 1980). Further studies should determine whether different forms of MGI are derived from a common precursor and whether tumor cells with the appropriate gene rearrangements, and possibly even normal cells under certain conditions, may produce hybrid molecules of different forms of MGI, including hybrid molecules with MGI-1 and MGI-2 activity (Liebermann et al. 1982).

CONTROL OF GROWTH AND DIFFERENTIATION IN LEUKEMIA

Normal myeloid precursor cells isolated from bone marrow (Lotem and Sachs 1977a) require an external source of MGI-1 for their viability and growth. There

TABLE 1. *Terminology used for proteins that induce cloning and differentiation of normal macrophages and granulocytes*

Mashran gm (Ichikawa et al. 1967)
Colony stimulating factor (CSF) (Metcalf 1969)
Colony stimulating activity (CSA) (Austin et al. 1971)
Macrophage and granulocyte inducers (MGI) (Landau and Sachs 1971)
MGI-1 (=mashran gm, CSF, CSA) for cloning; MGI-2 for differentiation (Sachs 1980, Lotem et al. 1980, Liebermann et al. 1982, Lotem and Sachs 1982)

are, however, myeloid leukemia cells that no longer require MGI-1 for their viability and growth: these leukemic cells can multiply in the absence of MGI-1 (Sachs 1978a, 1980). When there is a limiting amount of MGI-1, these cells have a growth advantage over normal cells. Starting with a decreased requirement for MGI-1 eventually leads to a complete loss of this requirement. Other myeloid leukemia cells constitutively produce their own MGI-1 (Paran et al. 1968, Moore 1982), and these cells also have a growth advantage compared with normal cells that require an external source (Table 2). A change in the requirement of MGI-1 for growth, to either a partial or complete loss of this requirement, or the constitutive production of MGI-1 thus both give a growth advantage to leukemic cells.

The existence of myeloid leukemia cells that either no longer require MGI-1 for viability and growth or constitutively produce their own MGI-1 raises the question of whether these cells can still be induced to differentiate to mature cells by the normal differentiation-inducing protein MGI-2. We have shown that there are clones of myeloid leukemia cells that no longer require MGI-1 for growth, but can still be induced to differentiate normally to mature macrophages and granulocytes by MGI-2 (Figure 1) via the normal sequence of gene expression (Sachs 1978a, 1980, 1982a,b). These mature cells are then no longer malignant in vivo (Fibach and Sachs 1974, Lotem and Sachs 1981, 1984). Among the many differentiation-associated properties induced in these cells by MGI-2 is the ability to respond chemotactically (Figure 2) to a variety of chemoattractants (Symonds and Sachs 1979). After the myeloid leukemia cells are injected into embryos, these cells can participate in hematopoietic differentiation in apparently healthy adult animals (Gootwine et al. 1982, Webb et al. 1984).

TABLE 2. *Differences in growth requirement of normal and leukemic cells*

Type of myeloid cells	Requirement of MGI-1 for growth
Normal	External source
Leukemic	Decrease → no requirement or Constitutive production

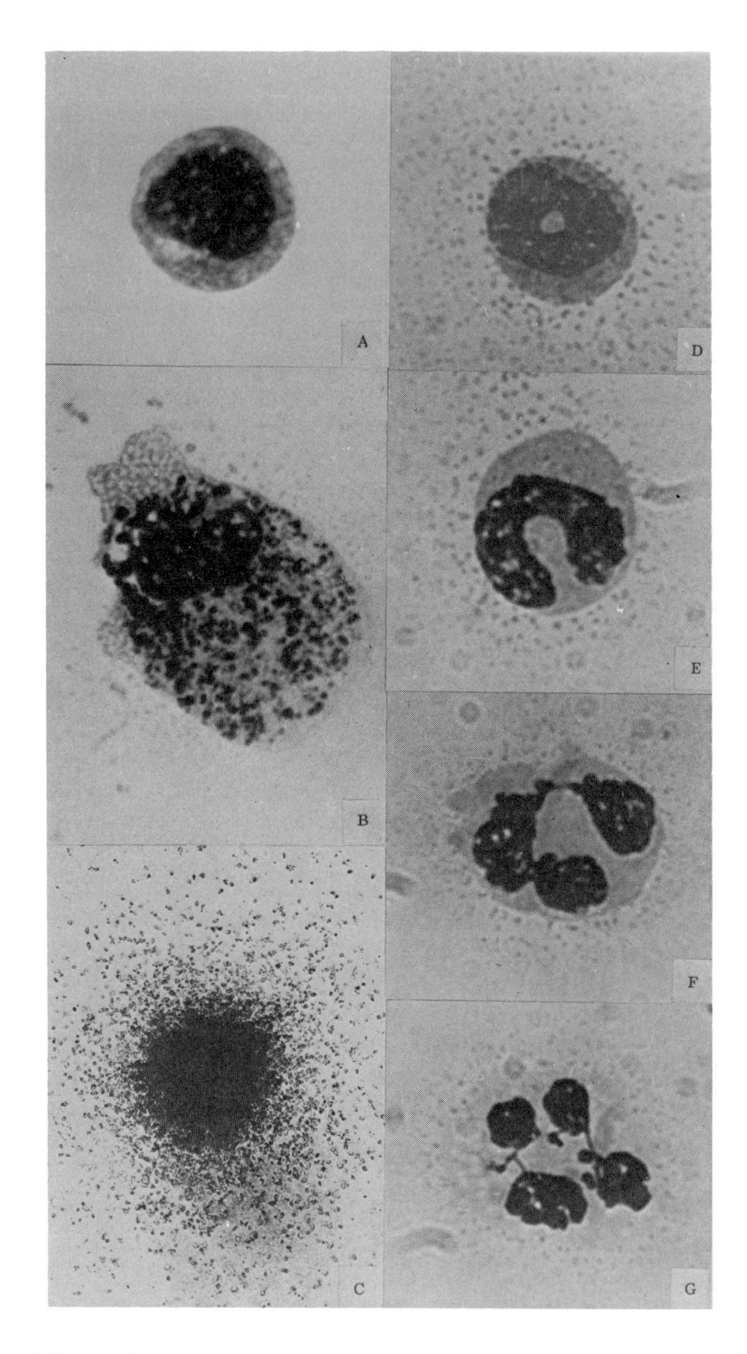

FIG. 1. Differentiation of MGI⁺D⁺ myeloid leukemia cells to mature macrophages and granulocytes by the normal myeloid differentiation-inducing protein MGI-2. Leukemic cell **(A)**, macrophage **(B)**, colony of macrophages **(C)**, stages in differentiation to granulocytes **(D-G)** (Reproduced from Nature 237;276–278, copyright 1972, Macmillan Journals Limited).

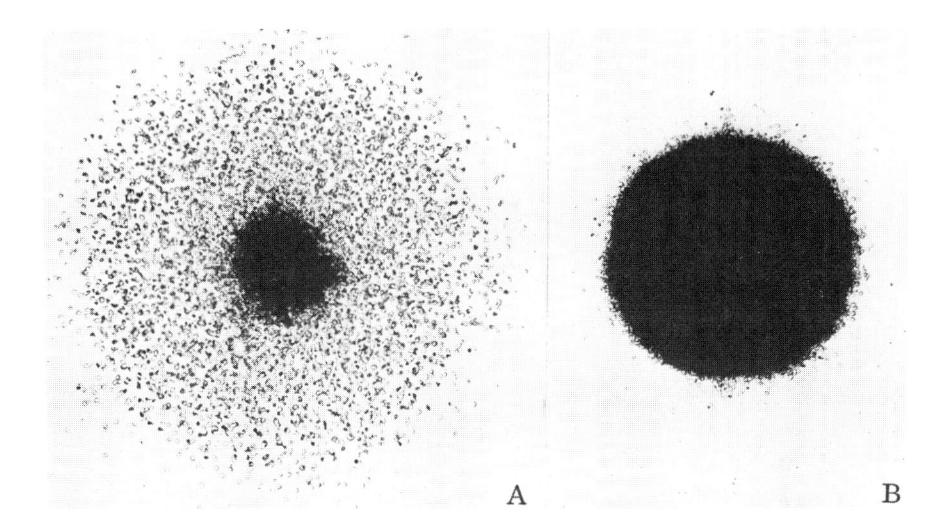

FIG. 2. Colony from an $MGI^{+}D^{+}$ clone **(A)** and from an $MGI^{-}D^{-}$ clone **(B)** cultured with MGI-2 (Hayashi et al. 1974). The differentiating cells in the $MGI^{+}D^{+}$ clone migrate from the colony when the cells are induced to respond to chemotactic stimuli (Symonds and Sachs 1979).

Injection of MGI-2 into animals or the in vivo induction of MGI-2 by compounds that induce the production of this differentiation-inducing protein results in an inhibition of leukemia development in animals with such leukemic cells (Lotem and Sachs 1981, 1984). There are also myeloid leukemia cells that constitutively produce their own MGI-1 and that can be induced to differentiate by MGI-2. Our results indicate that induction of normal differentiation in myeloid leukemia cells by MGI-2 is a potentially useful approach to therapy based on the induction of normal differentiation in malignant cells (Paran et al. 1970, Fibach et al. 1972, Sachs 1974, 1978a, b, Lotem and Sachs 1978b, 1981, 1983b, 1984).

Leukemic clones that can be induced to differentiate to mature cells by MGI-2 have been found in different strains of mice (Ichikawa 1969, Fibach et al. 1972, 1973, Ichikawa et al. 1976, Burgess and Metcalf 1980) and in humans (Paran et al. 1970, Lotem and Sachs 1979). They are referred to as $MGI^{+}D^{+}$ (MGI^{+} to indicate that they can be induced to differentiate by MGI-2; D^{+} for differentiation to mature cells). $MGI^{+}D^{+}$ leukemic cells have specific chromosome changes compared to normal cells (Hayashi et al. 1974, Azumi and Sachs 1977). These chromosome changes thus seem to involve changes in genes other than those involved in the induction of normal differentiation. Other clones of myeloid leukemia cells can also grow without added MGI-1, but they are either partly ($MGI^{+}D^{-}$) or almost completely ($MGI^{-}D^{-}$) blocked in their ability to be induced to differentiate by MGI-2 (Figure 3) (Fibach et al. 1973, Ichikawa et al. 1976, Hoffman-Liebermann and Sachs 1978, Simantov and Sachs 1978, Simantov et al. 1980, Lotem et al. 1980, Hoffman-Liebermann et al. 1981a). These differentiation-defective clones

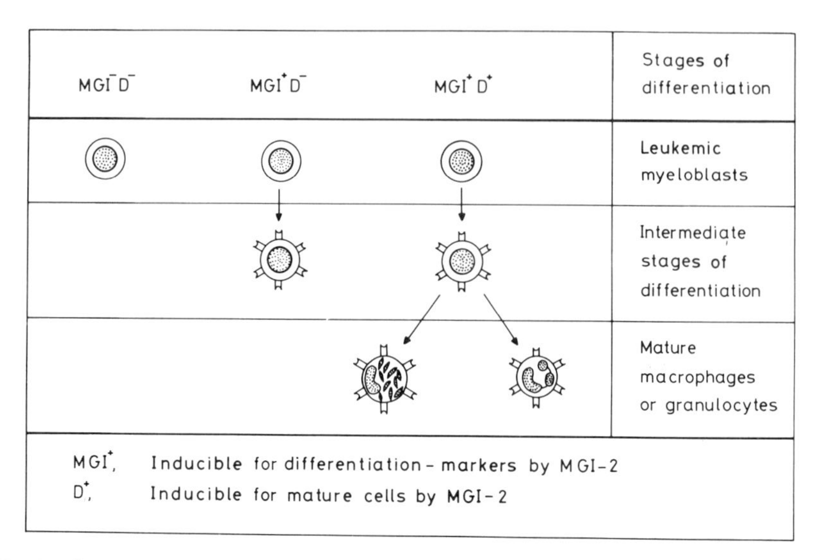

FIG. 3. Classification of different types of myeloid leukemic cell clones according to their capacity to be induced to differentiate by MGI-2 (Reproduced from Hayashi et al. 1974, with permission of UICC).

have specific chromosome changes compared with MGI^+D^+ cells (Hayashi et al. 1974, Azumi and Sachs 1977).

A variety of compounds other than MGI-2 can induce differentiation in MGI^+D^+ clones. Not all these compounds are active on every MGI^+D^+ clone, and they do not all induce the same differentiation-associated properties. The inducers include certain steroids, lectins, polycyclic hydrocarbons, tumor promoters, lipopolysaccharides, X rays, and cancer chemotherapy agents (Sachs 1978a, Lotem and Sachs 1982) (Table 3). The existence of clonal differences in differentiation response to X irradiation and cancer chemotherapeutic chemicals may help explain differences in response to therapy in different patients (Sachs 1978a). As a result of these experiments, we have suggested a form of therapy based on induction of differentiation (Paran et al. 1970, Fibach et al. 1972, Sachs 1974, 1978a, b, Lotem and Sachs 1978b, 1980, 1981). It would include prescreening in culture to select for the most effective compounds and using these compounds in a low-dose chemotherapy protocol aimed at inducing cell differentiation (Lotem and Sachs 1980). Since different myeloid leukemia clones respond differently to MGI-2 and other compounds, such differences will also occur in leukemic cells from different patients. Based on these suggestions (Sachs 1978a, b), some encouraging clinical results have been obtained by the use of low-dose cytosine arabinoside (Baccarani and Tura 1979, Housset et al. 1982, Michalewicz et al. 1984). Successful treatment of a patient with acute monoblastic leukemia by low-dose cytosine arabinoside is shown in Figure 4 (Michalewicz et al. 1984).

TABLE 3. *Compounds used in cancer therapy that at low doses can induce differentiation in clones of myeloid leukemic cells*

Adriamycin	Cytosine arabinoside
Daunomycin	Hydroxyurea
Methotrexate	Mitomycin C
Prednisolone	

ALTERNATIVE PATHWAYS FOR DIFFERENTIATION

Some of the compounds that induce differentiation in susceptible clones of MGI^+D^+ leukemic cells, including lipopolysaccharide, phorbol ester tumor promoters such as 12-0-tetradecanoyl-phorbol-13-acetate (TPA), and nitrosoguanide, can induce MGI-2 production in these clones. These compounds thus induce differentiation by inducing the endogenous production of the normal differentiation-inducing protein in the leukemia cells (Weiss and Sachs 1978, Lotem and Sachs 1979, Falk and Sachs 1980). Other compounds, such as the steroid dexamethasone, can induce differentiation in MGI^+D^+ clones without inducing MGI-2 (Falk and Sachs 1980). This steroid induces differentiation by another pathway of gene expression than MGI-2 (Lotem and Sachs 1977b, Cohen and Sachs 1981). The same applies to dimethylsulfoxide (DMSO).

In a line of human myeloid leukemia cells, DMSO induces the formation of granulocytes (Collins et al. 1978), whereas MGI-2 (Lotem and Sachs 1979, 1980) and the tumor-promoting phorbol ester TPA (Lotem and Sachs 1979, Rovera et al.

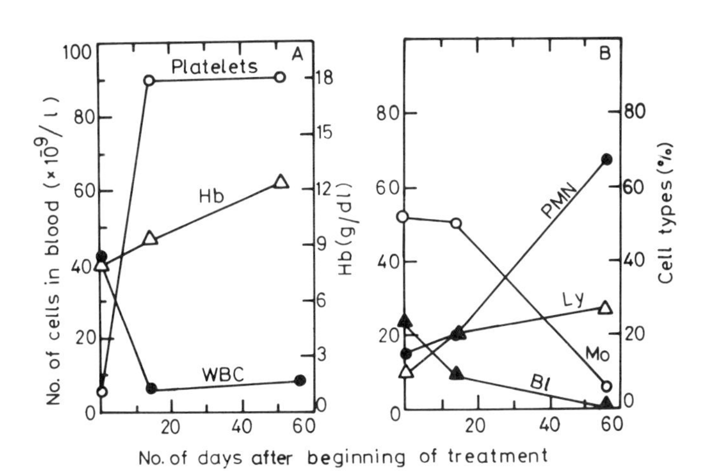

FIG. 4. Blood composition in a patient with acute monoblastic leukemia treated with a low dose of cytosine arabinoside. Hb, hemoglobin; WBC, white blood cells; PMN, polymorphonucleocytes; Ly, lymphocytes; Bl, blast cells; Mo, monocytes (Reproduced from Michalewicz et al. 1984, with permission of Pergamon Press).

1979), which induces the production of MGI-2 (Lotem and Sachs 1979), induce the formation of macrophages. Studies of the protein changes induced by these compounds, using two-dimensional gel electrophoresis, showed similar developmental programs for macrophage differentiation induced by MGI-2 and TPA, which differed from the beginning from the granulocyte program induced by DMSO (Liebermann et al. 1981). Unlike MGI-2 or DMSO, TPA induces rapid attachment of these myeloid leukemia cells to the Petri dish. Combined treatment with TPA and DMSO produced cell attachment, extensive spreading of the cells, the regulation of specific proteins, and expression of the macrophage program. The results indicate that cells in suspension can express either the macrophage or granulocyte program depending on the inducer, and that changes in cell shape associated with cell attachment can regulate specific proteins and restrict the developmental program to macrophages (Liebermann et al. 1981). The in vivo environment of cells in relation to the possibilities of cell adhesion may thus play a major role in determining the differentiation program of myeloid and other cell types.

DIFFERENTIATION BY COMBINED TREATMENT WITH DIFFERENT COMPOUNDS

Induction of differentiation in some myeloid leukemia clones requires combined treatment with different compounds (Krystosek and Sachs 1976, Lotem and Sachs 1978a, 1979, Symonds and Sachs 1982c, 1983). In these cases one compound induces changes not induced by another, so that the combined treatment results in new gene expression. This complementation of gene expression can occur at the level of both mRNA production and mRNA translation (Hoffman-Liebermann et al. 1981b). With the appropriate combination of compounds, we have been able to induce some differentiation-associated properties in all our MGI^-D^- leukemia clones (Symonds and Sachs 1982c, 1983). It will be interesting to determine whether the same applies to differentiation of erythroleukemic cells (Friend 1978, Marks and Rifkind 1978). It is possible that all myeloid leukemia cells no longer susceptible to the normal differentiation-inducing protein MGI-2 by itself can be induced to differentiate by choice of the combination of compounds appropriate for the required complementation. These can include hormones such as steroids (Lotem and Sachs 1974, 1975) or insulin (Symonds and Sachs 1982b, c) and different nonphysiological compounds (Sachs 1978a) with or without MGI-2.

COUPLING OF GROWTH AND DIFFERENTIATION IN NORMAL CELLS

We have developed a simple procedure for isolating normal myeloid precursor cells from the bone marrow (Figure 5) (Lotem and Sachs 1977a). Incubation of isolated normal myeloid precursors with either MGI-1M or MGI-1G (Lotem and Sachs 1980) induces the viability and growth of these normal precursors and results in differentiation to macrophages or granulocytes even without the addition of the differentiation-inducing protein MGI-2. The incubation of normal myeloid precur-

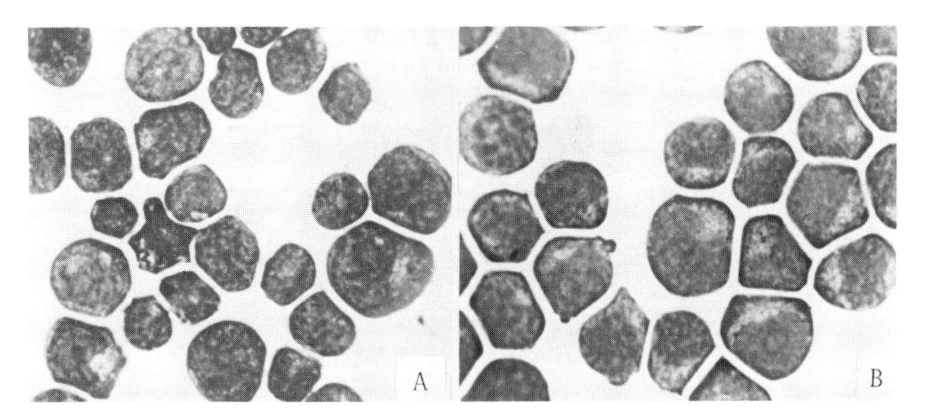

FIG. 5. Normal myeloid precursor cells isolated from normal bone marrow **(A)** and MGI$^+$D$^+$ myeloid leukemic cells **(B)** (Reproduced from Lotem and Sachs 1977a, with permission Alan R. Liss, Inc.).

sors with MGI-1 also results in the induction of MGI-2 (Sachs 1980, Liebermann et al. 1982, Lotem and Sachs 1982, 1983a). This induction of MGI-2 can be detected already at 6 hr after the addition of MGI-1 (Lotem and Sachs 1983a). MGI-2 induction can thus account for the induction of differentiation after the addition of MGI-1 to the normal cells. The induction of differentiation-inducing protein MGI-2 by growth-inducing protein MGI-1 thus appears to be an effective control mechanism for coupling growth and differentiation in normal cells.

Type of myeloid cells	Requirement of MGI-1 for growth	Induction of MGI-2 by MGI-1	Differentiation
Normal	+	Production of MGI-2 →	+
Leukemic	+ or –	No production of MGI-2	–
	Constitutive production MGI-1	Production of MGI-2 →	+ *

*Autoinduction of differentiation under specific conditions

FIG. 6. Differences in induction of MGI-2 by MGI-1 in normal and leukemic myeloid cells.

It has been shown that the receptor for epidermal growth factor has tyrosine-specific protein kinase activity (Ushiro and Cohen 1980). This activity has also been found for receptors for other growth factors such as insulin (Kasuga et al. 1983) and presumably also applies to the receptor for the myeloid cell growth-inducing protein MGI-1. The myeloid differentiation-inducing protein MGI-2, but not MGI-1, can bind to double-stranded cellular DNA (Weisinger and Sachs 1983). Therefore, growth and differentiation in normal myeloid cells are coupled by induction of a differentiation-inducing DNA-binding protein by a growth-inducing protein. This mechanism for coupling growth and differentiation may also apply to other cell types. Differences in the time of the switch-on of the differentiation inducer would produce differences in the extent of multiplication before differentiation. The platelet-derived growth factor is structurally related to the simian sarcoma virus oncogene *sis* (Doolittle et al. 1983, Waterfield et al. 1983). It will be interesting to determine whether MGI-1 and MGI-2 are structurally related to any of the known oncogenes.

The multiplication of normal cells is regulated at two control points. The first control requires MGI-1 to produce more cells that can then differentiate because of the MGI-2 induced by MGI-1. The second control is the stopping of cell division, which occurs as part of the program of terminal differentiation induced by MGI-2. Growth and differentiation in normal cells are thus coupled at both these points. Mature cells can also produce feedback inhibitors that interfere with MGI-1 induction of growth of the normal precursors (Ichikawa et al. 1967, Paran et al. 1969, Broxmeyer et al. 1978, 1980).

UNCOUPLING OF GROWTH AND DIFFERENTIATION IN LEUKEMIA

As pointed out earlier, there are $MGI^{+}D^{+}$ clones of myeloid leukemia cells that no longer require MGI-1 for growth but can still be induced to differentiate normally by MGI-2. These leukemic cells have thus uncoupled the normal requirement for growth from the normal requirement for differentiation. Experiments on the properties of these cells after induction of differentiation by MGI-2 have shown that the normal requirement for MGI-1 for cell viability and growth is restored in the differentiating leukemic cells (Fibach and Sachs 1976, Lotem and Sachs 1982, 1983a). MGI-1 added to normal myeloid precursors induces the production of MGI-2, so that cell differentiation is induced by the endogenously produced MGI-2. However, in these leukemic cells, MGI-1 did not induce the production of MGI-2 even though, like normal cells, they required MGI-1 for viability and growth. Therefore differentiation was not induced after MGI-1 was added (Lotem and Sachs 1982, 1983a). Another type of leukemic cell constitutively produces its own MGI-1 and also lacks induction of MGI-2 by MGI-1, so that the cells do not differentiate (Symonds and Sachs 1982a). The absence of MGI-2 induction by MGI-1 therefore uncouples growth and differentiation in these leukemic cells. The lack of an MGI-1 requirement for growth and the inability of the growth-inducing protein MGI-1 to induce the differentiation-inducing protein MGI-2 are thus mechanisms that uncouple growth and differentiation in $MGI^{+}D^{+}$ leukemic cells (Sachs 1980, Lotem and Sachs 1982, 1983a, Symonds and Sachs 1982a).

In leukemic cells that constitutively produce MGI-1, changes in specific components of the culture medium can result in an autoinduction of differentiation because of the restoration of MGI-2 inducibility by MGI-1, which in turn restores the normal coupling of growth and differentiation (Figure 6). These changes in the culture medium include the use of mouse or rat serum instead of horse or calf serum, serum-free medium, and removal of transferrin from serum-free medium (Symonds and Sachs 1982a). Autoinduction of differentiation in this type of leukemic cell may also occur under certain conditions in vivo.

The coupling of growth and differentiation in normal cells is regulated at two control points. The uncoupling of growth and differentiation in $MGI^{+}D^{+}$ leukemic cells occurs at the first control point, but in these leukemic cells the coupling at the second control found in normal cells, the stopping of multiplication in mature cells after the induction of differentiation by MGI-2, is maintained. There are differentiation-defective $MGI^{+}D^{-}$ leukemic cells that, like the $MGI^{+}D^{+}$ leukemic cells, no longer require the addition of MGI-1 for growth. However, MGI-2 induces only a partial differentiation in these cells: mature cells are not produced and the cells do not stop multiplying. In addition to an uncoupling of growth and differentiation at the first control point, $MGI^{+}D^{-}$ leukemic cells thus show a second uncoupling between the initiation of differentiation by MGI-2 and the stopping of cell multiplication that occurs as part of the normal program of terminal differentiation. It has been suggested that leukemia originates by uncoupling the first control and that uncoupling of the second control then results in a further evolution of leukemia (Sachs 1978a, 1980).

CHROMOSOME CHANGES IN THE LEUKEMIC CELLS

None of the clones of myeloid leukemic cells studied has a completely normal diploid chromosome banding pattern. There are also specific chromosome differences between the normal and leukemic cells and between $MGI^{+}D^{+}$, $MGI^{+}D^{-}$, and $MGI^{-}D^{-}$ clones (Hayashi et al. 1974, Azumi and Sachs 1977). Chromosome studies on normal fibroblasts, sarcomas, and revertants from sarcomas that have regained a nonmalignant phenotype have indicated that the difference between malignant and nonmalignant cells is controlled by the balance between genes for expression (E) and suppression (S) of malignancy. When there is enough S to neutralize E, malignancy is suppressed; when the amount of S is not sufficient to neutralize E, malignancy is expressed (Figure 7) (Rabinowitz and Sachs 1968, 1970, Hitosumachi et al. 1971, Yamamoto et al. 1973, Sachs 1974). Genes for expression of malignancy (E) are now called oncogenes (Bishop 1983, Land et al. 1983), and these experiments indicate there are other genes, S genes, that can suppress the activities of oncogenes.

Studies on the chromosomes of the myeloid leukemia clones suggest that this balance between different genes also applies to the origin of malignancy in these

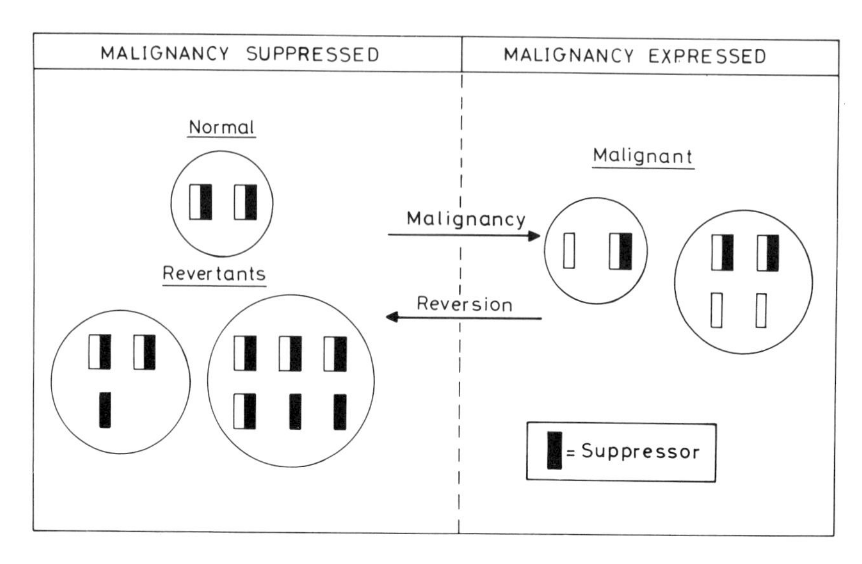

FIG. 7. Gene dosage and the expression and suppression of malignancy, from experiments with normal fibroblasts, sarcoma cells, and their nonmalignant revertants. Genes for □ expression and ■ suppression of malignancy (Reprinted by permission from Nature 225;136–139, copyright 1970, Macmillan Journals Limited).

leukemias and that the ability of the leukemic cells to be induced to differentiate by MGI-2 is also dependent on the balance between different genes. It has been shown that MGI^-D^- cells can give rise to MGI^+D^+ progeny by segregation of appropriate chromosomes and that these chromosome changes then restore the appropriate gene balance required for differentiation induction by MGI-2 (Table 4) (Azumi and Sachs 1977). Change in the balance of specific genes due to change in gene dosage is thus a mechanism that could uncouple growth and differentiation in myeloid leukemia. Specific changes in gene dosage have also been found in lymphoid leukemia (Dofuku et al. 1975, Klein 1979).

We have suggested from our chromosome data on these mouse myeloid leukemias that inducibility for differentiation by MGI-2 is controlled by the balance between genes on chromosome 2 and chromosome 12, and that these chromosomes also carry genes that control the malignancy of these cells (Azumi and Sachs 1977). In the leukemic cells we also found deletions of chromosome 2 and rearrangements with chromosome 12 (Azumi and Sachs 1977). It has since been found that the immunoglobulin heavy chain gene is on mouse chromosome 12 (D'Eustachio et al. 1980), that the c-*abl* gene is on mouse chromosome 2 (Goff et al. 1982), and that c-*abl* is involved (De Klein et al. 1982) in the translocation of the Philadelphia chromosome in human chronic myelocytic leukemia (Rowley 1977). The origin and further evolution of malignancy can involve, in different cases, changes in gene dosage, deletions, rearrangements, or mutation. Genetic changes in the regulation or structure of the normal genes that control growth and differentiation can produce the uncoupling required for the origin and further evolution of malignancy.

TABLE 4. *Chromosomes of $MGI^{+}D^{+}$ and $MGI^{-}D^{-}$ clones*

Cell type	Clone no.	No. of chromosomes in chromosome group*												Modal chromosome no.
		2	3	6	7	12	14	X	T(3;6)	T(3;12)	T(7;15)	T(12;15)	U	
$MGI^{+}D^{+}$	11	2	2	2	2	1 + B	2	2	0	0	0	0	0	40
$MGI^{+}D^{+}$	7-M9	1 + F	1	1	1	0	1	1	1	1	1	1	1	38
$MGI^{+}D^{+}$	7-M11	1 + F	1	1	1	0	1	1	1	1	1	1	1	38
$MGI^{+}D^{+}$	7-M16	1 + F	1	1	1	0	1	1	1	1	1	1	1	38
$MGI^{+}D^{+}$	7-M4	1 + F	1	1	1	1	1	1	1	0	1	1	1	38
$MGI^{+}D^{+}$	7-M5	1 + F	1	1	1	1	1	1	1	0	1	1	1	38
$MGI^{-}D^{-}$	7	1 + F	1	1	1	1	1	1	1	1	1	1	1	39

*All other chromosome groups had the normal diploid pattern. Translocation (T), deletion (F), insertion (B), unknown (U) (Azumi and Sachs 1977).

CONSTITUTIVE GENE EXPRESSION IN LEUKEMIA

Since there are leukemic cells that, unlike normal myeloblasts, no longer require MGI-1 for viability and growth, the molecular changes required for viability and growth that have to be induced in normal cells are constitutive in these leukemic cells. This also applies to leukemic cells that constitutively produce their own MGI-1. This suggests that myeloid leukemia can be the result of a change from an induced to a constitutive expression of genes that control cell viability and growth (Sachs 1978a, 1980).

Studies on changes in the synthesis of specific proteins in normal myeloblasts and MGI^+D^+, MGI^+D^-, and MGI^-D^- leukemic clones at different times after the addition of MGI-1 and MGI-2, using two-dimensional gel electrophoresis (Liebermann et al. 1980), have directly shown changes from inducible to constitutive gene expression in the leukemic cells. The results also indicated a relationship between constitutive gene expression and uncoupling of the initiation of differentiation by MGI-2 and the stopping of multiplication in the mature cells. The leukemic cells constitutively expressed changes in the synthesis of specific proteins that were induced in normal cells only by MGI-1 treatment. These protein changes, which included the appearance of some proteins and disappearance of others, were constitutive in all studied leukemic clones derived from different tumors. These have been called C_{leuk}, for constitutive for leukemia. Other protein changes were induced by MGI-2 in normal and MGI^+D^+ leukemic cells and were constitutive in MGI^+D^- and MGI^-D^- leukemic cells. More of these constitutive changes occurred in MGI^-D^- than in MGI^+D^- leukemic cells. These have been called C_{def}, for constitutive for differentiation defective (Figure 8) (Liebermann et al. 1980).

These results indicate that changes from an induced to a constitutive expression of certain genes is associated with the uncoupling of growth and differentiation,

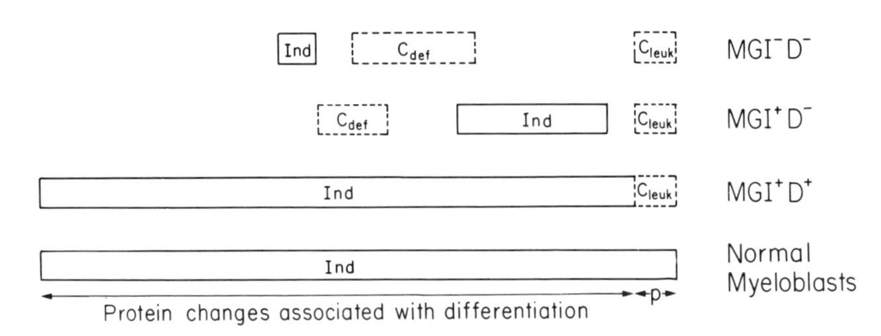

FIG. 8. Schematic summary of protein changes associated with growth (proliferation) and differentiation in normal myeloblasts and different types of myeloid leukemia cells. C_{leuk}, constitutive expression of changes in leukemic compared to normal myeloblasts; C_{def}, constitutive expression of changes in the differentiation defective MGI^+D^- and MGI^-D^- clones compared to MGI^+D^+ clones; Ind, MGI-induced changes; p, protein changes associated with proliferation to produce more cells before differentiation (Reproduced from Liebermann et al. 1980, with permission of Academic Press).

both at the control that requires MGI-1 to produce more cells and at the control that stops cell multiplication when mature cells are formed.

The protein changes during the growth and differentiation of normal myeloblasts seem to be induced by MGI-1 and MGI-2 as a series of parallel multiple pathways of gene expression (Liebermann et al. 1980). It can be assumed that the normal developmental program that couples growth and differentiation in normal cells requires synchronous initiation and progression of these parallel pathways. Constitutive gene expression for some pathways can be expected to produce asynchrony in the coordination required for the normal developmental program. Depending on the pathways involved, this asynchrony could then result in an uncoupling of the controls for growth and differentiation and produce different blocks in the induction and termination of the differentiation process.

We have been able to treat MGI^-D^- leukemic cells so as to induce the reversion of specific C_{def} proteins from the constitutive to the nonconstitutive state. This reversion was then associated with a gain of MGI-2 inducibility of various differentiation-associated properties. Reversion from the constitutive to the nonconstitutive state in these cells thus restored the synchrony required for differentiation induction (Symonds and Sachs, 1983).

These results led to the suggestion (Sachs 1978a, 1980, Liebermann et al. 1980) that myeloid leukemia originates by a change that produces certain constitutive pathways of gene expression, so that cells no longer require MGI-1 for growth or constitutively produce MGI-1 without inducing MGI-2. These leukemic cells can, however, still be induced to differentiate normally by MGI-2 added exogenously or induced in the cells in other ways. The differentiation program induced by MGI-2 can thus proceed normally when it is uncoupled from the growth program induced by MGI-1. This change in the leukemia cells can be followed by constitutive expression of other pathways, resulting in the uncoupling of other controls and an asynchrony that interferes with the normal program of terminal differentiation. These second changes then result in the further evolution of leukemia (Sachs 1980).

The MGI-1-independent growth of myeloid leukemic cells seems to proceed through stages (Collins et al. 1977), probably starting from a small decrease in the MGI-1 requirement and ending in complete independence from MGI-1. There presumably are also stages in the amount of constitutively produced MGI-1 and the degree of lack of inducibility of MGI-2 by MGI-1. These stages would produce different degrees of asynchrony, resulting in differences in the degree of hematological abnormality; the final stages of asynchrony would result in leukemia. There are probably a variety of tumors in which (A) the original malignancy has a normal differentiation program and the cells are malignant because the requirement for growth has been uncoupled from the requirement for differentiation by a change from inducible to constitutive expression of the genes required for growth and (B) the tumor evolves further because of changes from inducible to constitutive expression of other genes to produce asynchrony in the normal differentiation program, so that mature, nondividing cells are not formed by the physiological inducer of differentiation. Even these tumors, however, may still be induced to differentiate to form nonmalignant cells, by treatment with compounds that can reverse the constitutive to the nonconstitutive state or induce the differentiation program by other pathways.

NEW TYPE OF THERAPY

Our results suggest some novel possibilities for the therapy of myeloid leukemia. The finding of MGI^+D^+ myeloid leukemia cells that can be induced to differentiate normally by MGI-2 suggests the use of MGI-2 injection, stimulation of the production of MGI-2 in vivo, or engrafting of MGI-2-producing cells to induce the normal differentiation of these leukemic cells in patients with myeloid leukemia. This approach would be an alternative to cytotoxic agents, which kill normal cells and tumor cells. There are membrane differences between the cells that differ in their response to MGI-2 that may be useful markers for predicting leukemic cell response to MGI-2 in vivo. MGI^+D^+ leukemic cells can be induced by MGI-2 to again require MGI-1 for cell viability and growth. This suggests that induction of differentiation of the leukemic cells to this stage may also result in the loss of viability and growth of the induced MGI^+D^+ leukemic cells in vivo if there is not enough MGI-1.

The induction of growth of the normal myeloid precursors by MGI-1 and their differentiation to macrophages or granulocytes by MGI-2 also suggest that injected MGI or MGI induced in vivo could be used to restore the normal macrophage and granulocyte population after cytotoxic therapy. MGI therapy may also be useful for treatment of nonmalignant macrophage or granulocyte diseases.

Our results may also help explain why some, but not all, patients respond to cytotoxic chemical and irradiation therapy. We have shown that low doses of chemicals and irradiation can induce differentiation in myeloid leukemia clones with the appropriate genotype, and that clonal differences in inducibility for differentiation are not necessarily associated with differences in response to the cytotoxic effects. The growth in vivo of leukemic cells with the appropriate genotype may thus be controlled by the therapeutic agents not only because of their cytotoxic effect, but also because they induce differentiation. Differences in the competence of cells to be induced by these agents may thus explain differences in response to therapy by different patients. The possible induction of MGI-2 by these compounds may also play a role in the therapeutic effects obtained in vivo.

These studies therefore suggest new forms of leukemia therapy based on the use of a normal regulatory protein such as MGI-2 to induce differentiation in malignant cells and of MGI-1 and MGI-2 to induce a more rapid recovery of the normal cell population after the present forms of therapy. In addition, it is possible to use other compounds that induce MGI-2 in vivo or can affect mutant malignant cells at differentiation sites that are no longer susceptible to the normal regulator. The malignant cells of each patient should be prescreened for the compounds to which they are susceptible. These possibilities may also be applicable to diseases in other types of cells whose growth and differentiation are controlled by other normal regulators.

ACKNOWLEDGMENT

This research is now being supported by contact with the National Foundation of Cancer Research, Bethesda, and by grants from the Jerome and Estelle R. Newman Assistance Fund and the Julian Wallerstein Foundation.

REFERENCES

Austin PE, McCulloch EA, Till JE. 1971. Characterization of the factor in L cell conditioned medium capable of stimulating colony formation by mouse marrow cells in culture. J Cell Physiol 77:121–134.

Azumi J, Sachs L. 1977. Chromosome mapping of the genes that control differentiation and malignancy in myeloid leukemic cells. Proc Natl Acad Sci USA 74:253–257.

Baccarani M, Tura S. 1979. Correspondence. Differentiation of myeloid leukemic cells: New possibilities for therapy. Br J Haematol 42:485–487.

Bishop JM. 1983. Cellular oncogenes and retroviruses. Ann Rev Biochem 52:301–354.

Bradley TR, Metcalf D. 1966. The growth of mouse bone marrow in vitro. Aust J Exp Biol Med Sci 44:287–300.

Broxmeyer HE, Smithyman A, Eger RR, Myers PA, de Sousa M. 1978. Identification of lactoferrin as the granulocyte-derived inhibitor of colony stimulating activity production. J Exp Med 148:1052–1067.

Broxmeyer HE, de Sousa M, Smithyman A, Ralph P, Hamilton J, Kurland JI, Bognacki J. 1980. Specificity and modulation of the action of lactoferrin, a negative feedback regulator of myelopoiesis. Blood 55:324–333.

Burgess AW, Metcalf D. 1980. Characterisation of a serum factor stimulating the differentiation of myelomonocytic leukemic cells. Int J Cancer 26:647–654.

Burgess AW, Camakaris J, Metcalf D. 1977. Purification and properties of colony-stimulating factor from mouse lung conditioned medium. J Biol Chem 252:1998–2003.

Cohen L, Sachs L. 1981. Constitutive gene expression in myeloid leukemia and cell competence for induction of differentiation by the steroid dexamethasone. Proc Natl Acad Sci USA 78:353–357.

Collins SJ, Ruscetti FW, Gallagher RE, Gallo RC. 1978. Terminal differentiation of human promyelocytic leukemia induced by dimethylsulfoxide and other polar compounds. Proc Natl Acad Sci USA 75:2458–2462.

Collins SJ, Gallo RC, Gallagher RE. 1977. Continuous growth and differentiation of human myeloid leukemic cells in culture. Nature 270:347–349.

De Klein A, Van Kessel AD, Grosveld G, Bartman CR, Hagemeijer A, Bootsma D, Spurr NK, Heisterkamp N, Groffen J, Stephenson JR. 1982. A cellular oncogene is translocated to the Philadelphia chromosome in chronic myelocytic leukaemia. Nature 300:765–767.

D'Eustachio P, Pravtcheva D, Marcu K, Ruddle FH. 1980. Chromosomal location of the structural gene cluster encoding murine immunoglobin heavy chains. J Exp Med 151:1545–1550.

Dofuku R, Biedler JL, Sprengler BA, Old LJ. 1975. Trisomy of chromosome 15 in spontaneous leukemia of AKR mice. Proc Natl Acad Sci USA 72:1515–1517.

Doolittle RF, Hunkapiller MW, Hood LE, Devare SG, Robbins KC, Aaronson SA, Antoniades HN. 1983. Simian sarcoma virus onc gene, v-sis, is derived from the gene (or genes) encoding a platelet-derived growth factor. Science 221:275–277.

Falk A, Sachs L. 1980. Clonal regulation of the induction of macrophage and granulocyte inducing proteins for normal and leukemic myeloid cells. Int J Cancer 26:595–601.

Fibach E, Sachs L. 1974. Control of normal differentiation of myeloid leukemic cells. IV. Induction of differentiation by serum from endotoxin treated mice. J Cell Physol 83:177–185.

Fibach E, Sachs L. 1975. Control of normal differentiation of myeloid leukemic cells. VIII. Induction of differentiation to mature granulocytes in mass culture. J Cell Physiol 86:221–230.

Fibach E, Sachs L. 1976. Control of normal differentiation of myeloid leukemic cells. XI. Induction of a specific requirement for cell viability and growth during the differentiation of myeloid leukemic cells. J Cell Physiol 89:259–266.

Fibach E, Landau T, Sachs L. 1972. Normal differentiation of myeloid leukemic cells induced by a differentiation-inducing protein. Nature New Biol 237:276–278.

Fibach E, Hayashi M, Sachs L. 1973. Control of normal differentiation of myeloid leukemic cells to macrophages and granulocytes. Proc Natl Acad Sci USA 70:343–346.

Friend C. 1978. The phenomenon of differentiation in murine erythroleukemic cells. Harvey Lect 72:253–281.
Ginsburg H, Sachs L. 1963. Formation of pure suspension of mast cells in tissue culture by differentiation of lymphoid cells from the mouse thymus. JNCI 31:1–40.
Goff SP, D'Eustachio P, Ruddle FH, Baltimore D. 1982. Chromosomal assignment of the endogenous proto-oncogene c-abl. Science 218:1317–1319.
Gootwine E, Webb CG, Sachs L. 1982. Participation of myeloid leukaemic cells injected into embryos in haematopoietic differentiation in adult mice. Nature 299:63–65.
Hayashi M, Fibach E, Sachs L. 1974. Control of normal differentiation of myeloid leukemic cells. V. Normal differentiation in aneuploid leukemic cells and the chromosome banding pattern of D^+ and D^- clones. Int J Cancer 14:40–48.
Hitosumachi S, Rabinowitz Z, Sachs L. 1971. Chromosomal control of reversion in transformed cells. Nature 231:511–514.
Hoffman-Liebermann B, Sachs L. 1978. Regulation of actin and other proteins in the differentiation of myeloid leukemic cells. Cell 14:825–834.
Hoffman-Liebermann B, Liebermann D, Sachs L. 1981a. Control mechanism regulating gene expression during normal differentiation of myeloid leukemic cells. Differentiation defective mutants blocked in mRNA production and mRNA translation. Dev Biol 81:255–265.
Hoffman-Liebermann B, Liebermann D, Sachs L. 1981b. Regulation of gene expression by tumor promoters. III. Complementation of the developmental program in myeloid leukemic cells by regulating mRNA production and mRNA translation. Int J Cancer 28:615–620.
Housset M, Daniel MT, Degos L. 1982. Small doses of Ara-C in the treatment of acute myeloid leukemia: Differentiation of myeloid leukemia cells? Br J Haematol 51:125–129.
Ichikawa Y. 1969. Differentiation of a cell line of myeloid leukemia. J Cell Physiol 74:223–234.
Ichikawa Y, Pluznik DH, Sachs L. 1966. In vitro control of the development of macrophage and granulocyte colonies. Proc Natl Acad Sci USA 56:488–495.
Ichikawa Y, Pluznik DH, Sachs L. 1967. Feedback inhibition of the development of macrophage and granulocyte colonies. I. Inhibition by macrophages. Proc Natl Acad Sci USA 58:1480–1486.
Ichikawa Y, Maeda N, Horiuchi M. 1976. In vitro differentiation of Rauscher virus induced myeloid leukemic cells. Int J Cancer 17:789–797.
Kasuga M, Fujita-Yamaguchi Y, Blithe DL, Kahn CR. 1983. Tyrosine-specific protein kinase activity is associated with the purified insulin receptor. Proc Natl Acad Sci USA 80:2137–2141.
Klein G. 1979. Lymphoma development in mice and humans: Diversity of initiation is followed by convergent cytogenetic evolution. Proc Natl Acad Sci USA 76:2442–2446.
Krystosek A, Sachs L. 1976. Control of lysozyme induction in the differentiation of myeloid leukemic cells. Cell 9:675–684.
Land H, Parada LF, Weinberg RA. 1983. Cellular oncogenes and multistep carcinogenesis. Science 222:771–778.
Landau T, Sachs L. 1971. Characterization of the inducer required for the development of macrophage and granulocyte colonies. Proc Natl Acad Sci USA 68:2540–2544.
Liebermann D, Hoffman-Liebermann B, Sachs L. 1980. Molecular dissection of differentiation in normal and leukemic myeloblasts: Separately programmed pathways of gene expression. Dev Biol 79:46–63.
Liebermann D, Hoffman-Liebermann B, Sachs L. 1981. Regulation of gene expression by tumor promoters. II. Control of cell shape and developmental programs for macrophages and granulocytes in human myeloid leukemic cells. Int J Cancer 28:285–291.
Liebermann D, Hoffman-Liebermann B, Sachs L. 1982. Regulation and role of different macrophage and granulocyte proteins in normal and leukemic myeloid cells. Int J Cancer 29:159–161.
Lipton J, Sachs L. 1981. Characterization of macrophage and granulocyte inducing proteins for normal and leukemic myeloid cells produced by the Krebs ascites tumor. Biochim Biophys Acta 673:552–569.
Lotem J, Sachs L. 1974. Different blocks in the differentiation of myeloid leukemic cells. Proc Natl Acad Sci USA 71:3507–3511.
Lotem J, Sachs L. 1975. Induction of specific changes in the surface membrane of myeloid leukemic cells by steroid hormones. Int J Cancer 15:731–740.
Lotem J, Sachs L. 1977a. Control of normal differentiation of myeloid leukemic cells. XII. Isolation of normal myeloid colony-forming cells from bone marrow and the sequence of differentiation to mature granulocytes in normal and D^+ myeloid leukemic cells. J Cell Physiol 92:97–108.

Lotem J, Sachs L. 1977b. Genetic dissection of the control of normal differentiation in myeloid leukemic cells. Proc Natl Acad Sci USA 74:5554–5558.
Lotem J, Sachs L. 1978a. Genetic dissociation of different cellular effects of interferon on myeloid leukemic cells. Int J Cancer 22:214–220.
Lotem J, Sachs L. 1978b. In vivo induction of normal differentiation in myeloid leukemic cells. Proc Natl Acad Sci USA 75:3781–3785.
Lotem J, Sachs L. 1979. Regulation of normal differentiation in mouse and human myeloid leukemic cells by phorbol esters and the mechanism of tumor promotion. Proc Natl Acad Sci USA 76:5158–5162.
Lotem J, Sachs L. 1980. Potential pre-screening for therapeutic agents that induce differentiation in human myeloid leukemic cells. Int J Cancer 25:561–564.
Lotem J, Sachs L. 1981. In vivo inhibition of the development of myeloid leukemia by injection of macrophage and granulocyte inducing protein. Int J Cancer 28:375–386.
Lotem J, Sachs L. 1982. Mechanisms that uncouple growth and differentiation in myeloid leukemia: Restoration of requirement for normal growth-inducing protein without restoring induction of differentiation-inducing protein. Proc Natl Acad Sci USA 79:4347–4351.
Lotem J, Sachs L. 1983a. Coupling of growth and differentiation in normal myeloid precursors and the breakdown of this coupling in leukemia. Int J Cancer 32:127–134.
Lotem J, Sachs L. 1983b. Control of in vivo differentiation of myeloid leukemic cells. III. Regulation by T lymphocytes and inflammation. Int J Cancer 32:781–791.
Lotem J, Sachs L. 1984. Control of in vivo differentiation of myeloid leukemic cells. IV. Inhibition of leukemia development by myeloid differentiation-inducing protein. Int J Cancer 33:147–154.
Lotem J, Lipton J, Sachs L. 1980. Separation of different molecular forms of macrophage and granulocyte inducing proteins for normal and leukemic myeloid cells. Int J Cancer 25:763–771.
Maeda M, Horiuchi M, Numa S, Ichikawa Y. 1977. Characterization of a differentiation stimulating factor for mouse myeloid leukemic cells. Gann 68:435–447.
Marks P, Rifkind RA. 1978. Erythroleukemic differentiation. Ann Rev Biochem 47:419–448.
Metcalf D. 1969. Studies on colony formation in vitro by mouse bone marrow cells. I. Continuous cluster formation and relation of clusters to colonies. J Cell Physiol 74:323–332.
Michalewicz R, Lotem J, Sachs L. 1984. Cell differentiation and therapeutic effect of low doses of cytosine arabinoside in human myeloid leukemia. Leukemia Res (in press).
Moore MAS. 1982. G-SCF: Its relationship to leukemia differentiation-inducing activity and other hemopoietic regulators. J Cell Physiol (Suppl) 1:53–64.
Nicola NA, Burgess AW, Metcalf D. 1979. Similar molecular properties of granulocyte-macrophage colony stimulating factors produced by different organs. J Biol Chem 24:5290–5299.
Paran M, Ichikawa Y, Sachs L. 1968. Production of the inducer for macrophage and granulocyte colonies by leukemic cells. J Cell Physiol 72:251–254.
Paran M, Ichikawa Y, Sachs L. 1969. Feedback inhibition of the development of macrophage and granulocyte colonies. II. Inhibition by granulocytes. Proc Natl Acad Sci USA 62:81–87.
Paran M, Sachs L, Barak Y, Resnitzky P. 1970. In vitro induction of granulocyte differentiation in hematopoietic cells from leukemic and nonleukemic patients. Proc Natl Acad Sci USA 67:1542–1549.
Pluznik DH, Sachs L. 1965. The cloning of normal "mast" cells in tissue culture. J Cell Comp Physiol 66:319–324.
Pluznik DH, Sachs L. 1966. The induction of clones of normal "mast" cells by a substance from conditioned medium. Exp Cell Res 43:553–563.
Rabinowitz Z, Sachs L. 1968. Reversion of properties in cells transformed by polyoma virus. Nature 220:1203–1206.
Rabinowitz Z, Sachs L. 1970. Control of the reversion of properties in transformed cells. Nature 225:136–139.
Rovera G, Santoli D, Damsky C. 1979. Human promyelocytic leukemic cells in culture differentiate into macrophage-like cells when treated with a phorbol diester. Proc Natl Acad Sci USA 76:2779–2783.
Rowley JD. 1977. Mapping of human chromosomal regions related to neoplasia: Evidence from chromosomes 1 and 17. Proc Natl Acad Sci USA 74:5729–5733.
Sachs L. 1974. Regulation of membrane changes, differentiation, and malignancy in carcinogenesis. Harvey Lect 68:1–35.

Sachs L. 1978a. Control of normal cell differentiation and the phenotypic reversion of malignancy in myeloid leukemia. Nature 274:535–539.
Sachs L. 1978b. The differentiation of myeloid leukemia cells. New possibilities for therapy. Br J Haematol 40:509–517.
Sachs L. 1980. Constitutive uncoupling of pathways of gene expression that control growth differentiation in myeloid leukemia: A model for the origin and progression of malignancy. Proc Natl Acad Sci USA 77:6152–6156.
Sachs L. 1982a. Control of growth and differentiation in leukemic cells: Regulation of the developmental program and restoration of the normal phenotype in myeloid leukemia. J Cell Physiol (Suppl)1:151–164.
Sachs L. 1982b. Normal developmental programmes in myeloid leukemia: Regulatory proteins in the control of growth and differentiation. Cancer Surveys 1:321–342.
Simantov R, Sachs L. 1978. Differential desensitization of functional adrenergic receptors in normal and malignant myeloid cells. Relationship to receptor mediated hormone cytoxicity. Proc Natl Acad Sci USA 75:1805–1809.
Simantov R, Shkolnik T, Sachs L. 1980. Desensitization of enucleated cells to hormones and the role of cytoskeleton in control of a normal hormonal response. Proc Natl Acad Sci USA 77:4798–4802.
Stanley ER, Heard PM. 1977. Factors regulating macrophage production and growth. Purification and some properties of the colony stimulating factor from medium conditioned by mouse L cells. J Biol Chem 252:4305–4312.
Symonds G, Sachs L. 1979. Activation of normal genes in malignant cells. Activation of chemotaxis in relation to other stages of normal differentiation in myeloid leukemia. Somat Cell Genet 5:931–944.
Symonds G, Sachs L. 1982a. Autoinduction of differentiation in myeloid leukemic cells: Restoration of normal coupling between growth and differentiation in leukemic cells that constitutively produce their own growth-inducing protein. EMBO J 1:1343–1346.
Symonds G, Sachs L. 1982b. Cell competence for induction of differentiation by insulin and other compounds in myeloid leukemic clones continuously cultured in serum-free medium. Blood 60:208–212.
Symonds G, Sachs L. 1982c. Modulation of cell competence for induction of differentiation in myeloid leukemic cells. J Cell Physiol 111:9–14.
Symonds G, Sachs L. 1983. Synchrony of gene expression and the differentiation of myeloid leukemic cells: Reversion from constitutive to inducible protein synthesis. EMBO J 2:663–667.
Ushiro H, Cohen S. 1980. Identification of phosphotyrosine as a product of epidermal growth factor-activated protein kinase in A-431 cell membranes. J Biol Chem 255:8363–8365.
Waterfield MD, Scrace GT, Whittle N, Stroobant P, Johnsson A, Wasteson A, Westermark B, Heldin CH, Huang JS, Deuel TF. 1983. Platelet-derived growth factor is structurally related to the putative transforming protein $p28^{sis}$ of simian sarcoma virus. Nature 304:35–39.
Webb CG, Gootwine E, Sachs L. 1984. Developmental potential of myeloid leukemia cells injected into mid-gestation embryos. Dev Biol 101:221–224.
Weisinger G, Sachs L. 1983. DNA-binding protein that induces cell differentiation. EMBO J 2:2103–2107.
Weiss B, Sachs L. 1978. Indirect induction of differentiation in myeloid leukemic cells by lipid A. Proc Natl Acad Sci USA 75:1374–1378.
Yamamoto T, Rabinowitz Z, Sachs L. 1973. Identification of the chromosomes that control malignancy. Nature New Biol 243:247–250.
Yamamoto Y, Tomida M, Hozumi M. 1980. Production by spleen cells of factors stimulating differentiation of mouse myeloid leukemic cells that differ from colony stimulating factor. Cancer Res 40:4804–4809.

Subject Index